Computational Social Sciences

Computational Social Sciences

A series of authored and edited monographs that utilize quantitative and computational methods to model, analyze and interpret large-scale social phenomena. Titles within the series contain methods and practices that test and develop theories of complex social processes through bottom-up modeling of social interactions. Of particular interest is the study of the co-evolution of modern communication technology and social behavior and norms, in connection with emerging issues such as trust, risk, security and privacy in novel socio-technical environments. Computational Social Sciences is explicitly transdisciplinary: quantitative methods from fields such as dynamical systems, artificial intelligence, network theory, agent-based modeling, and statistical mechanics are invoked and combined with state-of-the-art mining and analysis of large data sets to help us understand social agents, their interactions on and offline, and the effect of these interactions at the macro level. Topics include, but are not limited to social networks and media, dynamics of opinions, cultures and conflicts, socio-technical co-evolution and social psychology. Computational Social Sciences will also publish monographs and selected edited contributions from specialized conferences and workshops specifically aimed at communicating new findings to a large transdisciplinary audience. A fundamental goal of the series is to provide a single forum within which commonalities and differences in the workings of this field may be discerned, hence leading to deeper insight and understanding.

More information about this series at http://www.springer.com/series/11784

Frank Dignum
Editor

Social Simulation for a Crisis

Results and Lessons from Simulating the COVID-19 Crisis

Editor
Frank Dignum
Computing Sciences
Umeå University
Umeå, Sweden

ISSN 2509-9574 ISSN 2509-9582 (electronic)
Computational Social Sciences
ISBN 978-3-030-76399-2 ISBN 978-3-030-76397-8 (eBook)
https://doi.org/10.1007/978-3-030-76397-8

This Springer imprint is published by the registered company Springer Nature Switzerland AG
The registered company address is: Gewerbestrasse 11, 6330 Cham, Switzerland

This book is dedicated to all the people that suffer from the Covid-19 crisis. Both the patients and people that passed away, but also the health care workers working endless shifts, the many SME's that saw their business become unviable, the people that had to work at home while home schooling their children, the youth that has been deprived for a very long time of social contacts in a period of their lives where this is of utmost importance, and all those other groups somehow affected by this crisis.

Preface

This book is the result from an effort of fourteen people that have joined together to work voluntarily (no funding!) on the Agent-based Social Simulation for the Covid-19 Crisis (ASSOCC) framework in order to make a positive difference in this crisis using social simulation. The primary goal of this project was to have real world impact and support decision makers during the crisis. However, doing this project has generated so much valuable experiences for the social simulation community at large and especially for using social simulations for crisis situations that we decided to write this book.

The book is, just like the ASSOCC project, an exceptional case. It seemed not to be possible to have a book with fourteen authors. So, in the end we opted for the solution of an edited book, where I, as initiator of this project, ended up as editor. The separate chapters do have different subsets as official authors, but the whole team has contributed in many ways to all the chapters. To emphasize that the book is actually a joint effort the whole team is co-author of the introduction and conclusion chapter of the book.

Just like the ASSOCC project results have been remarkable, so is this book. If one takes into consideration that half the team consists of (young) Ph.D. students, it is amazing how they have been able to accomplish so much in such a short time. This certainly would not have been possible without Loïs Vanhée who was the chief architect of the implementation and managed to keep a very big and diverse group of code contributors in line. We have all worked countless hours on this project, but he has been always there for anyone at any time of the day to support, encourage and help.

For me, as initiator of the ASSOCC project, it has also been a rewarding experience. All members have been very committed and supportive of each other. They were willing to put up with all my demands and directives. I have learned a lot from all of the team members. I feel guilty sometimes, because most of the media attention has come my way rather than the whole team. However, I have also learned that having regular contact with the media can really increase the impact of our work. With this book I hope that we can give a good foundation for having social simulations being accepted as a valuable and even necessary contribution for crisis management. Both before and during the crisis.

Umeå, Sweden
April 2021

Frank Dignum

Acknowledgements

The simulations were enabled by resources provided by the Swedish National Infrastructure for Computing (SNIC) at Umeå partially funded by the Swedish Research Council through grant agreement no. 2018-05973. This research was conducted using the resources of High Performance Computing Center North (HPC2N).

The research presented in this book was partially supported by the Wallenberg AI, Autonomous Systems and Software Program (WASP) and WASP—Humanities and Society (WASP-HS) funded by the Knut and Alice Wallenberg Foundation.

We also want to acknowledge here the contributions of Annet Onnes who worked on the transport model. Moreover we like to thank Virginia Dignum who set up and maintained the ASSOCC website, which was essential to distribute our results quick and to a large audience.

Acknowledgements

Contents

Part I ASSOCC Theory and Platform

1 **Introduction** ... 3
Frank Dignum, Loïs Vanhée, Maarten Jensen, Christian Kammler, René Mellema, Fabian Lorig, Cezara Păstrăv, Mijke van den Hurk, Alexander Melchior, Amineh Ghorbani, Bart de Bruin, Kurt Kreulen, Harko Verhagen, and Paul Davidsson

2 **Foundations of Social Simulations for Crisis Situations** ... 15
Frank Dignum

3 **Social Simulations for Crises: From Theories to Implementation** ... 39
Maarten Jensen, Loïs Vanhée, and Christian Kammler

4 **Social Simulations for Crises: From Models to Usable Implementations** ... 85
Cezara Păstrăv, Maarten Jensen, René Mellema, and Loïs Vanhée

Part II Scenario's

5 **The Effectiveness of Closing Schools and Working at Home During the COVID-19 Crisis** ... 121
Mijke van den Hurk

6 **Testing and Adaptive Testing During the COVID-19 Crisis** ... 139
Christian Kammler and René Mellema

7 **Deployment and Effects of an App for Tracking and Tracing Contacts During the COVID-19 Crisis** ... 167
Maarten Jensen, Fabian Lorig, Loïs Vanhée, and Frank Dignum

8 **Studying the Influence of Culture on the Effective Management of the COVID-19 Crisis** ... 189
Amineh Ghorbani, Bart de Bruin, and Kurt Kreulen

9 Economics During the COVID-19 Crisis: Consumer Economics and Basic Supply Chains 231
Alexander Melchior

10 Effects of Exit Strategies for the COVID-19 Crisis 269
René Mellema and Amineh Ghorbani

Part III Results and Lessons Learned

11 The Real Impact of Social Simulations During the COVID-19 Crisis 319
Frank Dignum

12 Comparative Validation of Simulation Models for the COVID-19 Crisis 331
Fabian Lorig, Maarten Jensen, Christian Kammler, Paul Davidsson, and Harko Verhagen

13 Engineering Social Simulations for Crises 353
Loïs Vanhée

14 Agile Social Simulations for Resilience 379
Maarten Jensen, Frank Dignum, Loïs Vanhée, Cezara Păstrăv, and Harko Verhagen

15 Challenges and Issues for Social Simulations for Crises 409
Frank Dignum, Maarten Jensen, Christian Kammler, Alexander Melchior, and Mijke van den Hurk

16 Conclusions 427
Frank Dignum, Loïs Vanhée, Maarten Jensen, Christian Kammler, René Mellema, Fabian Lorig, Cezara Păstrăv, Mijke van den Hurk, Alexander Melchior, Amineh Ghorbani, Bart de Bruin, Kurt Kreulen, Harko Verhagen, and Paul Davidsson

Appendix A: Culture 439

Appendix B: General Parameters 445

Appendix C: Full Need and Actions Model 451

Contributors

Paul Davidsson Department of Computer Science and Media Technology, Internet of Things and People Research Center, Malmö University, Malmö, Sweden

Bart de Bruin Faculty of Technology, Policy and Management, TU Delft, Delft, The Netherlands

Frank Dignum Department of Computing Science, Umeå University, Umeå, Sweden

Amineh Ghorbani Faculty of Technology, Policy and Management, TU Delft, Delft, The Netherlands

Maarten Jensen Department of Computing Science, Umeå University, Umeå, Sweden

Christian Kammler Department of Computing Science, Umeå University, Umeå, Sweden

Kurt Kreulen Faculty of Technology, Policy and Management, TU Delft, Delft, The Netherlands

Fabian Lorig Department of Computer Science and Media Technology, Internet of Things and People Research Center, Malmö University, Malmö, Sweden

Alexander Melchior Department of Information and Computer Science, Utrecht University, The Netherlands Ministry of Economic Affairs and Climate Policy and Ministry of Agriculture, Nature and Food Quality, Utrecht, The Netherlands

René Mellema Department of Computing Science, Umeå University, Umeå, Sweden

Cezara Păstrăv Department of Computing Science, Umeå University, Umeå, Sweden

Mijke van den Hurk Department of Information and Computing Sciences, Utrecht University, Utrecht, The Netherlands

Loïs Vanhée Umeå University, Umeå, Sweden

Harko Verhagen Department of Computer and Systems Sciences, Stockholm University, Kista, Sweden

Part I
ASSOCC Theory and Platform

In part I, we lay the foundations of the ASSOCC platform for the simulations that we have run with ASSOCC. We describe the theories that are used for the agent deliberation processes and we describe extensively how these theories are implemented in a practical and efficient way. We also show how we provide a proper user interface to the simulations that provide decision-makers with possibilities to follow the runs and also analyze them in several ways.

Chapter 1
Introduction

Frank Dignum, Loïs Vanhée, Maarten Jensen, Christian Kammler, René Mellema, Fabian Lorig, Cezara Păstrăv, Mijke van den Hurk, Alexander Melchior, Amineh Ghorbani, Bart de Bruin, Kurt Kreulen, Harko Verhagen, and Paul Davidsson

Abstract The introduction of this book sets the stage of performing social simulations in a crisis. The contents of the book are based on the experience of creating a large scale and complex social simulation for the COVID-19 crisis. However, the contents are reaching much further than just this experience. We will show the general contribution that social simulations based on fundamental social-psychological principles can have in times of crises. In times of big societal changes due to a pandemic or other disaster, these simulations can give handles to support decision makers in their difficult task to act in a very short time with many uncertainties. Besides giving our results, we also will indicate why the results are trustworthy and interesting. Finally we also look what challenges should be picked up to convert the successful project into a sustainable research area.

F. Dignum (✉) · L. Vanhée · M. Jensen · C. Kammler · R. Mellema · C. Păstrăv
Department of Computing Science, Umeå University, SE-901 87 Umeå, Sweden
e-mail: dignum@cs.umu.se

L. Vanhée
e-mail: lois.vanhee@umu.se

M. Jensen
e-mail: maartenj@cs.umu.se

C. Kammler
e-mail: ckammler@cs.umu.se

R. Mellema
e-mail: renem@cs.umu.se

A. Ghorbani · B. de Bruin · K. Kreulen
Faculty of Technology, Policy and Management, TU Delft, Jaffalaan 5,
2628 BX Delft, The Netherlands
e-mail: a.ghorbani@tudelft.nl

B. de Bruin
e-mail: bdb785@gmail.com

K. Kreulen
e-mail: kurtkreulen@gmail.com

F. Dignum (ed.), *Social Simulation for a Crisis*, Computational Social Sciences,
https://doi.org/10.1007/978-3-030-76397-8_1

1.1 Crisis

In March 2020, the gravity of the pandemic caused by the corona virus slowly became apparent. While most people (including us) thought that the consequences would be limited to Wuhan it became clear that the virus had already spread throughout Europe as well. The reason why COVID-19 could wreak such a havoc is not because it is very virulent and kills its host in a short time. The reason why it can spread so easily and is so persistent is exactly that not everyone is affected at the same level and that it might take quite some days before symptoms become clear, if at all. This means that people can carry the virus and spread it without being aware of their infection for a considerable amount of time. The COVID-19 virus is, thus, placed between Ebola in one side and flu at the other side. Ebola is very virulent and because of that isolation of people being infected can be done quite effective. Hence, although the virus kills most people that it infects, it can usually be contained pretty well. The flu viruses are usually not well contained because they often have an incubation time of several days and symptoms only appear after some time as well. Thus, the virus can be spread quite easily during the incubation time and by patients that have relatively mild symptoms and keep going to work and other places where they meet other people. However, the flu viruses are not very lethal and, thus, the disruption of society is relatively small.

The characteristics of the COVID-19 virus made it difficult to contain. The standard procedure of the health care authorities in cases of a pandemic outbreak is to try to track and trace all contacts of an infected person and isolate these persons as quickly

F. Lorig · P. Davidsson
Internet of Things and People Research Center, Department of Computer Science and Media Technology, Malmö University, 205 06 Malmö, Sweden
e-mail: fabian.lorig@mau.se

P. Davidsson
e-mail: paul.davidsson@mau.se

M. van den Hurk · A. Melchior
Department of Information and Computing Sciences, Utrecht University, Princetonplein 5, 3584 CC Utrecht, The Netherlands
e-mail: m.vandenhurk@uu.nl

A. Melchior
e-mail: a.t.melchior@uu.nl

A. Melchior
Ministry of Economic Affairs and Climate Policy and Ministry of Agriculture, Nature and Food Quality, The Netherlands, Bezuidenhoutseweg 73, 2594 AC Den Haag, The Netherlands

L. Vanhée
GREYC, Université de Caen, 14000 Caen, France

H. Verhagen
Department of Computer and Systems Sciences, Stockholm University, PO Box 7003, 16407 Kista, Sweden
e-mail: verhagen@dsv.su.se

as possible. However, someone could have infected other persons in the previous six days while having been to a pub at Friday night, having gone shopping in a shopping mall on Saturday and visited a soccer match with 40.000 other people it becomes very difficult to trace all possible contacts that might have been infected. Although the traditional track and tracing is still valuable it is not enough in this situation and other measures are needed.

The interventions and measures taken have differed widely between countries all over the world. There have been debates about what are the "best" measures and countries have been blamed and praised (and sometimes both at different times) for the measures they took. Unfortunately, this book will not give an answer on which is the best measure to take. Basically, because we do not believe there is one best measure. The measures that can and should be taken depend on the country/region, its infrastructure, its culture and many other aspects of its society.

However, the realisation that the spread of the COVID-19 virus and also the success of measures in a country or region depends crucially on human behaviour led us to the conviction that social simulations could have a huge added value in this type of pandemic. Thus, on March 16, 2020 Frank Dignum wrote e-mails to Ph.D. students and other colleagues to see who would want to collaborate to build a simulation for the COVID-19 crisis. So, that is how the Agent-based Social Simulation of the Coronavirus Crisis (ASSOCC) project started. In all respects, it is an extra-ordinary project.

First of all, it is not funded! All members of the team participate on voluntary basis and do a lot of work in their spare time. Fortunately, enough work could be combined with "normal" research work in order to keep the project moving. However, we have all the time been carefully balancing between being enthusiastic and spending many extra hours on the project and preventing people from burn-outs due to an unrelenting schedule driven by the events during the COVID-19 crisis.

Secondly, the project did not have a project plan, not even a start and end date or predefined milestones. However, we agreed on the cognitive models that would serve as foundation for the simulations and we knew that we wanted to make not just one simulation on one aspect but rather a sandbox in which many scenarios could be developed and run. We also knew that interfacing for non-specialists would be important and, thus, we set up a separate module to provide an adequate interface to the simulation and its results. The deadlines for the project were set by the events of the crisis. After a quick set up of the basic components of the system we wanted to be ready and show results in time to inform the national discussions on the major measurements that were considered. Should schools be closed, people work at home, etc.

Thirdly, the members of the team did not apply for a position, but were all asked if they were willing to spare some time for the project. Each member contributes as far as possible next to a normal job. It means the members of the team are highly motivated and all believe that the ASSOCC approach is not just of some academic interest, but can be of added value in the real world. This commitment, not to a job, but to a common goal and ideal has made a huge difference in the outcome of the project! Without the dedication and countless hours spend on the project we could not have achieved any of the results in such a short time. Thus we see that the disadvantage of working in times of a crisis can also be an advantage as it focuses efforts and also shows very concrete the impact one can make with one's research. It maybe should be mentioned at this place that we did not have an epidemiologist on the team. The social simulations of the ASSOCC framework were about the COVID-19 crisis as a whole and not specifically the epidemiological part of it. So, we have regularly consulted with epidemiologists and used their models as part of our framework rather than incorporating the discipline itself in the team.

Despite the unusual circumstances in which the ASSOCC project has been conducted it has been very successful in a number of respects. First of all, we have achieved a number of interesting results from our simulations that proved to be a real contribution to the debates on measurements in diverse countries. That in itself is a good result for any social simulation project.

However, a more interesting result is that all the scenarios that were run on very different aspects of the crisis have been using the same implementation model! Thus, we have shown that one can base a simulation framework like ASSOCC on a fundamental model that connects different aspects of life in a coherent way and allows to make all kinds of combinations of factors to create new scenarios. An ultimate example of this is the curfew scenario which is not a separate chapter in this book, because it was run at request of some party in The Netherlands during the debate leading up to the curfew in February 2021. We were able to set up, run and analyse this scenario within two days (and come up with believable results that seem to be corroborated since by the real world situation)! It provides a powerful argument for the use of abstract models based on sound social-psychological principles in this type of simulations.

Maybe more important than the specific results that we got from our simulations were the lessons that we learned from running these simulations during a crisis. These lessons were the direct reason for writing this book as they seem to be valuable for the whole social simulation community. It would have been very nice and helpful if there would have been tools and methodologies available at the start of the ASSOCC project specifically for social simulations for crises. So, a large part of this book is dedicated to lessons learned from the ASSOCC project and especially discussing the biggest challenges when trying to create social simulations for crisis situations. In the rest of this chapter, we will already briefly position this type of agile social simulations in the field of simulations in general and give an overview of the contents of the book, describing the role of each chapter in the main message of book:

Agent-based Social Simulation can make a valuable contribution, not only to science, but also to society in times of crisis!

1.2 Simulations for Crisis Situations

Before getting into the details of the ASSOCC project and the rest of the book it is important to first place simulations for crisis situations in the broad spectrum of simulations being performed. One of the main determinants of a simulation is the purpose for which it is built. Reference [1] describes seven core purposes for simulations:

- prediction: anticipate well-defined aspects of data that are not already known
- explanation: establishing a possible causal chain from a set-up to its consequences in terms of the mechanisms in a simulation.
- description: an attempt to partially represent what is important of a specific observed case (or small set of closely related cases)
- theoretical exploration: establishing and characterising (or assessing) hypotheses about the general behaviour of a set of mechanisms (using a simulation).
- illustration: communicate or make clear an idea, theory or explanation
- analogy: use a simulation to describe another process that is hard to access
- social learning: encapsulating a shared understanding (or set of understandings) of a group of people.

So, what is the main purpose of a simulation for a crisis situation? Right away it becomes clear that in a crisis several of the above purposes are important if the simulation is to support the decision makers during the crisis. Decision makers want to have at least some form of predictions in order to shape their preferences between different courses of action (restrictions or policies). But the simulation should also be able to explain what is happening. In a fast moving world during a crisis the decision makers need to have some sort of understanding how their decisions affect the world. To a lesser extent, one would like the simulation to highlight which are the determining factors that will define the effects of decisions. Due to the high interdependency of many factors in a crisis it is often difficult to distinguish determinant variables from confounding factors. Simulations can be used to get a grip on this. An example of this is the question whether closing basic schools will effectively help contain the spreading the virus? Which are the determining factors and how will they be affected by the closure of schools? Finally, we have also actively used ASSOCC for social learning. Using the simulation we could show people why track and tracing apps might be handy for the health care organisations, but will have a very limited effect on the spread of the COVID-19 virus.

Given the above, very brief, description showing that simulations for crisis situations have inherently multiple purposes it is easy to understand that these simulations are also inherently complex. One could argue that separate simulations should be built for each purpose. However, it is very difficult to keep these simulations consistent and also how to combine results of the different simulations. Indeed, we see that the ASSOCC framework and system is inherently quite complex, but can indeed be used for several purposes due to the principled architecture and wide coverage of the model.

So, how does the complex ASSOCC framework fit in the classical taxonomy of simulations described in [2] as a prototypical simulation for crisis situations? We will briefly describe each dimension.

Abstract versus Descriptive: The Abstract versus Descriptive axis from [2] denotes two modelling purposes: simulating for the sake of reproducing a general phenomena, generally using on abstracted mechanisms (Abstract) or for the sake of reproducing a very specific situation, often including a wide array of detailed elements that are specific to the situation (Descriptive).

ASSOCC is in the middle of these two extremes. It should be abstract to model many possible situations in a quickly changing world in crisis. E.g. people will violate lockdown rules due to unfulfilled needs. But it also contains enough details to make the results relevant for decision makers at the time of the crisis. E.g. will track and tracing apps be useful, as studied in Chap. 7. This bipolar orientation is a central aspect of the design methodology for building simulations for crisis, as described in Chap. 14.

Artificial versus Realistic: The Artificial versus Realistic axis from [2] denotes the goal for building simulations for either observing the behaviour of possible societies (Artificial) or for replicating the behaviour of existing societies (Realistic).

ASSOCC is again in the middle of these poles. It is meant to simulate potential effects of policies during the crisis. In such it is meant to simulate possible societies and alternatives. But these societies should be clearly anchored in the current society. However, we do not try to just explain phenomena of the current situation and are thus not completely realistic.

Positive versus Normative: The Positive versus Normative axis from [2] denotes the goal of building simulations for either studying a phenomenon, with a generative social-science mindset (Positive) or to be used for guiding decision-makers (Normative).

In this dimension, ASSOCC is purely based on the normative pole. It is clear that simulations for crisis situations are meant for supporting the decision makers during the crisis.

Spatial versus Network: The Spatial versus Network axis from [2] distinguishes two modelling method concerns: whether the simulation is laid in a space such as a 2D grid or a map (Spatial) or whether distances are abstracted away (Network).

The ASSOCC framework is strongly based on the Network pole. However, this was a choice purely based on pragmatic arguments. Although a spatial map would be good to have, it would also make the simulation far more complex and inefficient. Thus, we chose to leave the spatial component out only for efficiency reasons and not for any conceptual reason.

Complex versus Simple Agents: The last dimension distinguishes whether agents rely on advanced cognitive models (Complex) or simplified if-then kind of statements (Simple).

ASSOCC is squarely positioned on the complex agent pole. We will argue in the next chapter why this is necessary for any simulation for a crisis situation.

If we take the position of a framework like ASSOCC with respect to all dimensions and compare it with other simulations, we see that it has a quite a unique position.

It has complex agents, based on an abstract model. However, the complexity is not primarily caused by trying to fit as closely as possible to all the details of a specific situation, but rather by the combination of many aspects of reality. Thus, ASSOCC does not require loads of data. We have used data mainly to calibrate certain aspects of the simulation rather than the simulation as a whole. ASSOCC is, in it present form, also not meant to give very detailed predictions. The scale of the simulations is too small to be able to do that. However, ASSOCC simulations can indicate some timelines and general trends. E.g. a curfew will reduce the number of newly infected people, but not enough to prevent a new wave after the curfew is lifted. So, other measures are needed in combination with a curfew. The positive thing is that we can show with the ASSOCC simulations that having a principled, abstract agent decision making model facilitates creating reasonable realistic simulations in a crisis situation. This property is especially important in these situations where data about the situation is scarce and normal behaviour is no longer normal. In these situations, having a model that is not very dependent on lots of empirical data is very useful! Thus, it seems that with the ASSOCC framework we have shown that social simulations for crisis situations do take a unique place in the field of simulations. And, moreover, this place requires some type of characteristics of the simulation that are not well supported by the common simulation tools yet, while crucial for working in crisis situations. We will use the rest of this book to argue why this is the case.

1.3 Guide to the Book

The rest of this book is split up in three parts. In the first part, we describe the background and foundations of the ASSOCC framework. In Chap. 2, we give a detailed overview of the theories that we have used to base the agent decision models on and also the arguments why we used exactly these theories. The main claim that we make is that an abstract model is needed for the decision models of the agents and we give some arguments why the theories and models that we have chosen are particularly well suited for simulations for crisis situations.

In Chap. 3, we give an extensive overview of the way the foundations have been implemented. We more or less follow the ODD protocol in describing the elements of the implementation, but adjust this to better explain the very extensive submodels of the ASSOCC framework. People that are mostly interested in the actual results of the simulations might want to skip this chapter. However, this chapter shows the actual complexity of the simulation and especially the agent decision making model. Anyone who wants to use the ASSOCC framework for their own purposes can find all the details necessary of all parts of the model to adjust them, discard them or extend them. It also important to re-iterate that this implementation is used for all the results shown in Part II of the book. Thus, it can also be used to analyse all kinds of details of these results.

In the last chapter of part I (Chap. 4), we describe the user interface module of ASSOCC. A unique feature where we create an interface for stakeholders from which

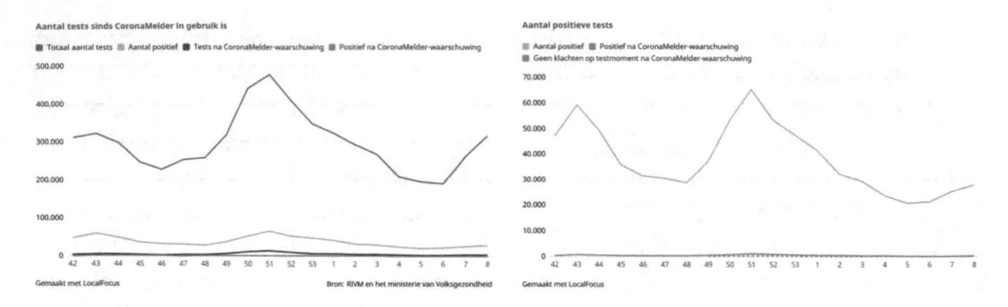

Fig. 1.1 Effect of track and tracing app in The Netherlands

they can see all scenarios, adjust some of the parameters in a controlled way and can explore the results of the simulations in various ways. A user interface like this is a necessity for any simulation as complex as the ones from the ASSOCC framework. We describe the architecture of the whole system in a way that others can use a similar set up if they want to provide a high level user interface for stakeholders of a simulation.

In Part II of the book, we collected six scenarios and their results that were run on the ASSOCC framework. Although many more scenarios could be run and included, we chose for these six scenarios as they are representative for certain types of applications and were in several cases directly used in national debates on the measures simulated in those scenarios.

Chapter 5 gives some insights in the effectiveness of closing (basic) schools. This was particularly relevant in the beginning of the COVID-19 crisis. Countries have chosen different strategies and the effectiveness of them is still not completely clear. In this chapter, we show which aspects play a role here and how their interdependence leads to some counter-intuitive results that still appear to be corroborated by reality.

The next chapter (Chap. 6), discusses some testing scenarios. This was done at the request of a regional government that wanted to know whether testing a large enough group randomly would work as well as giving priority to testing certain risk groups (like health care workers) regularly. Unfortunately, the results from the simulation were not very promising and this policy was never followed up.

The simulation results of the track and tracing apps have probably had the biggest societal impact of ASSOCC. We ran these simulations in April 2020 while the public debate in The Netherlands was questioning the benefits and fearing the consequences for privacy. Our results of the effectiveness of the apps deviated substantially from the most used epidemiological models. We predicted that the app would have a very limited effect on the spread of the virus. The following Fig. 1.1 which denotes the effect of the app in The Netherlands in the end of 2020 and beginning of 2021 shows that we were basically right with our prediction.

In the left figure the dark blue line indicates the number of tests being taken, the light blue line indicates the number of positive tests. At the bottom we see the red line denoting the number of tests taken after a warning from the app and the hardly visible pink line underneath indicates the number of those test that were positive. In the right figure the top blue line indicates the number of positive tests. The pink line shows the number of positive tests taken after a warning from the app and the purple line shows how many of those had no symptoms yet. Of course, one should also look at how many people actually used the app, which other measures were in place, etc. So, we will not make a scientific claim of having made the right prediction. But it gives a good indication and some of the analysis behind this figures is completely in line with the analysis that we will provide in Chap. 7.

All discussions about which country was taking the right measures at which time led us to investigate what could lead to the differences in effectiveness of measures between countries. Of course, countries differ in many aspects such as geography, population density, infrastructure, institutions and culture. Because taking up everything at the same time would take a multi-year project, we decided to pick one aspect that we already had previous experience in our simulations with: culture. So, in Chap. 8 we investigate the influence of culture on the effectiveness of the diverse measurements taken during the COVID-19 crisis.

At the beginning of the COVID-19 crisis, it was clear that this crisis was seen as a health crisis. Economic aspects were playing a minor role. Governments have given massive subsidies to industry to keep the economy from going bankrupt. However, after some time discussions started about which form of subsidy would be effective and how much and how long this financial support should be given. In Chap. 9, we investigate some economic aspects and effects of measurements of the government. Here, we also see that a macro perspective on the economy might lead to different measures than a social perspective that tries to support all individuals. More details on this are given in Chap. 9.

The last chapter of part II of the book appropriately investigates the consequences of different exit strategies. Which restrictions should be lifted first? In what order and when can restrictions be lifted without getting into a new wave of the pandemic. In Chap. 10, we look at several groups of exit strategies as they were applied around the world. Some exit strategies focus on getting the economic activity started again. Others mainly look at public life and how that can be restored safely. In this chapter, we see that some exit strategies have surprisingly similar consequences even though they are based on quite different principles. We analyse why this might be so and also give some heuristics that could be used to choose a good exit strategy.

After all the chapters of part II that gave an overview of the diverse set of scenarios that were run on the ASSOCC platform, in part III we turn to the analysis of the project as a whole. What did we learn from this experience and how can this help us for the future? In the first chapter of part III, Chap. 11, we discuss the actual impact we have had with the ASSOCC simulations. Not surprisingly, we were not part of governmental advisory committees. That could not be expected as newcomer in the field and in a time of crisis. However, our simulations have played a major role in the public debates in several countries in Europe and have indirectly also steered decisions that way. In this chapter, we discuss more in depth what we learned from the process, and what should be done for the future to get a place on the table for a next crisis.

When we started publicising our results and the media started picking up on that, (legitimate) questions were raised on the validity of our results. Especially the results of our simulations of the track and tracing app gave results that were at first sight counter intuitive. Because we did take these questions on validity serious we have done an extensive investigation into the validity of our simulations by comparing them to a state of the art epidemiological model [3]. In Chap. 12, we report on this comparison and how this can be used to show the validity of our simulations. It has been a long and difficult journey to get to all the details of both simulations and see exactly how they can be compared. But it is also worthwhile, because by itself also gave a better insight in the ASSOCC simulations. We actually would promote these kind of comparisons to be done more often.

Already quite early in the project we realised that scalability of the simulations would be an issue. Using NetLogo together with a complex cognitive agent model means that one can run maximally around 2000 agents in a run. But besides this, obvious limitation there were many issues to deal with while creating one of the most complex NetLogo simulations. In Chap. 13, we describe the software engineering aspects of running this big and complex project that also had to deal with external deadlines and an ever shifting focus on new aspects that became important during the crisis. The main reason we could manage this was that we had a very solid foundation to start with on which we could easily add and change all other components. Keeping very good software engineering principles in managing the code and coders was also of prime importance.

Many times, we have thought during the project how nice it would be if we already would have had some tools prepared beforehand. Although we did manage to build and adapt most support tools that we needed for the ASSOCC project it is clear that a better starting point would have helped in many ways to achieve even more, get quicker analysis, better communication, etc. In Chap. 14, we describe which are the main areas that have to be developed and what is needed for that in order to be ready for a next crisis. There are some fundamental conceptual and design aspects that can support a flexible and scalable simulation platform.

In Chap. 15, we recapitulate the challenges that were found during the project and indicate the most important research directions. These are not challenges for the ASSOCC project, but more fundamental issues for social simulations for crisis situations. They are about creating a flexible decision making mechanism for the agents that is also scalable. About which software engineering techniques can be used to support the scalability issues of these agile social simulations. In short, this chapter describes a first step towards a research agenda for the community that wants to give social simulations real impact on crisis situations.

The book is closed by Chap. 16, where we draw some general conclusions and give a vision of future work for social simulations for crisis situations based on the experiences of the ASSOCC project for the COVID-19 crisis.

References

1. B. Edmonds, Different modelling purposes. in *Simulating Social Complexity* (Springer, 2017), pp. 39–58
2. Nigel Gilbert, Agent-based social simulation: dealing with complexity. Complex Syst. Netw. Excell. **9**(25), 1–14 (2004)
3. R. Hinch et al., OpenABM-Covid19-an agent-based model for nonpharmaceutical interventions against COVID-19 including contact tracing. medRxiv (2020)

Chapter 2
Foundations of Social Simulations for Crisis Situations

Frank Dignum

Abstract Simulating human behaviour in times of crisis requires models of human decision that are include aspects beyond directly visible actions. In crisis times the behaviour of people will change based on the changing environment and needs. Without an underlying model that can represent how and when people will change their behaviour it becomes difficult to incorporate these behavioural changes in the simulation. In this chapter we will introduce the foundations of the model that we used to model the human behaviour for the COVID-19 crisis. We argue that these foundations are not only useful for this application but are broadly applicable for simulations that need to capture behavioural change due to crises or other external influences.

2.1 Introduction

During the COVID crisis it has become very apparent that the spread of the corona virus heavily depends on (changing) human behaviour. Where for other epidemics of less lethal viruses the human behaviour could be approximated using statistical models of normal behaviour, this was no longer sufficient for the corona virus. Due to a combination of a long incubation time where people are contagious but have no symptoms yet, the fact that many people do not show any easily recognisable symptoms at all, the fact that older people are much more likely to suffer severe consequences of being infected and the lethality of the virus meant that very strict restrictions were considered necessary to prevent the virus to spread to the most vulnerable groups and cause huge amount of deaths. Another important factor that made human behaviour and behaviour change important is that the pandemic and the various restrictions stretched over several months and thus impacted every aspect of life. Thus models of human behaviour during the crisis would also need to include different aspects of life, like social effects of long term isolation, economic consequences of closures of shops, public places, leisure places, etc.

F. Dignum (✉)
Department of Computing Science, Umeå University, 901 87 Umeå, Sweden
e-mail: dignum@cs.umu.se

F. Dignum (ed.), *Social Simulation for a Crisis*, Computational Social Sciences,
https://doi.org/10.1007/978-3-030-76397-8_2

Although the issues described above are special for the COVID crisis they are by no means exclusive for this crisis. In many crisis situations the above issues play a major role in the way a crisis evolves. There are several places in the world where regular natural disasters like war, draught, flooding or earth quakes create a crisis situation. In these situations the evacuation of people from the affected area, providing "temporary" shelters and recuperation of a "normal" life are important. Whereas modelling these crisis situations might at first focus on the evacuation process and the creation of the refugee camps, it should also include the social relations and status when allocating places in the camps. Moreover, the crisis is not finished with people having moved out of a disaster area. It is over when those people have some way of existence in another place or back in the original area after it has been restored. This longer term perspective might have a huge impact on how the short term aspects are or should be handled. E.g. where to place a refugee camp. In an example of flooding situations in Indonesia it is known that people are reluctant to evacuate from their homes out of fear of plundering and fear that they are not able to return to their often illegal dwellings. This hampers many long term solutions for this crisis. In all of these situations, social, economical and psychological aspects play a role and are not easily disentangled.

In this chapter we will first investigate what are the consequences of the above observations for the type of models that the social simulations for crisis situations should be based on. Next we will describe the foundations of a model that fulfils these requirements and is used in the ASSOCC project for the COVID-19 crisis. We will show later in this book that this model can be used to get insightful results in the COVID-19 crisis on many different aspects.

2.2 Crisis Situations Require Abstract Models

The points described in the previous section give strong arguments to create an agent model that is based on some fundamental abstract notions that can be used to link all of the different aspects mentioned above. This differs from social simulations that use (and often only require) statistics of real-world behaviours to model behaviours. E.g. if "23% of the people decide on A" this is modelled by having agents randomly 23% of the times decide on A. However, this behaviour cannot be explained afterwards, but more important we loose a possible consistency of behaviour. It might that in general 23% of the population has some property that makes it decide for A. (e.g. living in an area, having a certain profession, being of a certain age, etc.). This dependency of A on that property is now lost and the results of the simulation might differ substantially because of it. We will see some of this in the simulations about the effectiveness of the track and tracing apps in Chap. 7.

In Fig. 2.1 we very schematically compare the two approaches. We are aware that shows a very black and white picture and is grossly oversimplified. However, it shows the crux of the differences and the choice to be made.

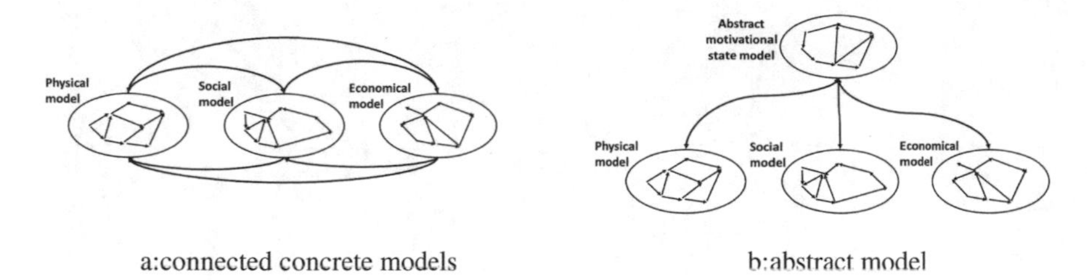

Fig. 2.1 Concrete models versus abstract model

In Fig. 2.1a one can see a few concrete models that each focus on a different aspect of reality. Each of these models can be validated on information from the real world. However, there also all kinds of dependencies between elements in the different models. These are denoted by the arrows connecting the models. Usually the existence of these dependencies is discovered through correlations in data and unexpected phenomena where a condition in one model will lead to a different action in another model than might be expected. E.g. when there is no money to buy petrol for transport (logistics model), evacuation by bus and car will not work even though it might be the most preferred option in terms of mobility and flexibility (preference model). The main disadvantage is that the dependencies between the models are generally not covered or through a collection of sometimes contradicting theories. This can lead to inconsistencies, incompleteness and ad-hoc solutions that are difficult to explain and justify.

The architecture where an abstract model is used that somehow governs the concrete models solves the ad-hoc representations of the inter-dependencies by using the abstract model that should have its own properties. This right away indicates the disadvantage of this approach. We need an abstract model that people can agree upon. Moreover, this abstract model can usually not directly be validated by information from the real world. It needs indirect validation through the other models. The advantage of using more concrete, simple models is that each of these models can be validated against (historical) data. But it should be noted though that in crisis situations the concrete models are often no longer correct.

Concrete models usually make implicit and (probably) unintended assumptions as they are usually based on stable situations where people will react in reasonably predictable ways to the situation. In these situations people will act according to standard social practices, norms, habits, etc. Thus simple models that connect the situation to an action are sufficient. This can be illustrated as in Fig. 2.2a where the red graph shows the actual behaviour of the people and the blue straight line shows the simple approximation. It works well in the left part of the graph, but gets worse results when the red graph changes direction (e.g. due to a crisis situation).

In a crisis the context of decision making changes drastically and thus the data used for previous situations does no longer predict the behaviour in the current context. Thus more abstract models might be necessary for these situations that contain several internal states (represented by additional parameters). It does not mean that these

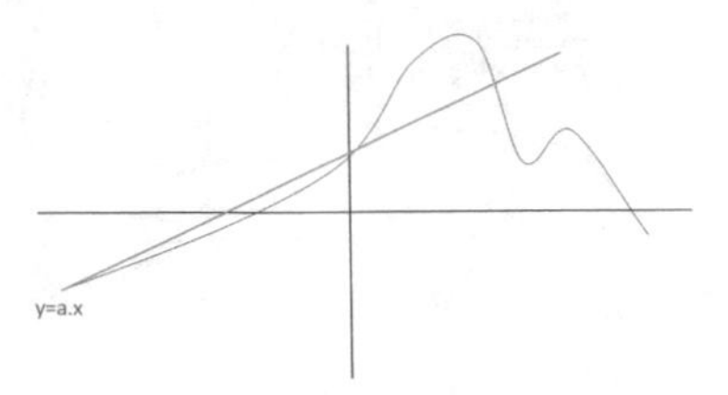

a:simple linear model

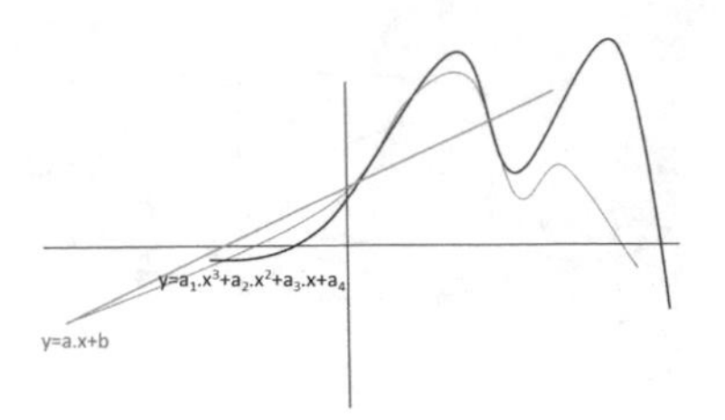

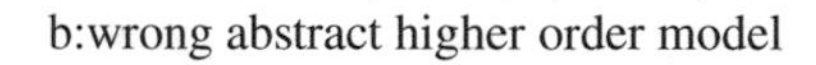
b:wrong abstract higher order model

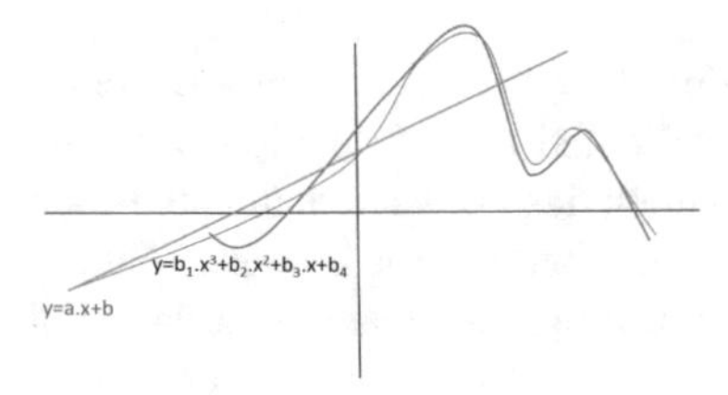

c:better abstract higher order model

Fig. 2.2 Simple model versus complex abstract model

abstract models will by definition give better results! If these abstract models are not well constructed or internal relations between concepts are not well defined they rather confuse than clarify behaviour. This is illustrated in Fig. 2.2b where we use a higher order function to approximate the real behaviour, but we apparently did not get all the parameters a_1 to a_4 right.

In Fig. 2.2c we have the parameters better calibrated and the resulting graph fits pretty well with reality. Of course, it is clear that to validate these more abstract models we need either a lot more data over many more situations or a good theory on how the parameters are related.

So, neither approach is per definition better than the other (unless we have a universally accepted abstract deliberation model). This is in line with [1] who argues that the type of model to be used depends on the situation that is to be simulated. Our argument is that the abstract model approach is more useful under the conditions of crisis situations that we want to model and simulate:

1. There are many dependencies that play a role in the situation and they clearly influence the outcome a lot.
2. changes in the environment (either natural or social) play a big role in the situation and the reaction to these changes depends on several dimensions.
3. the simulated situation spans over a time frame that is long enough to be more dependent on the interactions between the different dimensions.

With respect to 1. take as example the COVID crisis. The main perspective is that of health and thus an epidemiological model seems to be the most appropriate to base a simulation on. However, it soon became apparent that human behaviour is a determining factor in the spread of the virus and this behaviour could not easily be captured through the usual statistical estimates of behaviour. Thus at least a epidemiological model and behavioural model are needed and ways to connect these models.

With respect to 2. in any crisis situation the environment is changed in unexpected or new ways and society will react to those changes. However, the way people react to a crisis or a new policy/restriction in a crisis depends on many factors that are not always part of the same dimension. E.g. keeping isolation when having corona symptoms clearly is advantageous from the health perspective. However, old people that cannot see their (great)grandchildren due to this isolation might prefer to violate the isolation and take the risk of getting corona. So, the need for social contact overrides the health concerns.

Finally, in 3. we emphasise that dependencies between different social dimensions become more apparent and important over longer time periods. E.g. rescuing refugees from a war zone leads to a strong focus on safety and survival. However, in a long term perspective the refugees should also be given a perspective for the rest of their lives and their children's lives. This perspective leads to considerations of other aspects of life than just safety.

We argue that in the situation of the COVID crisis the above conditions are all present and play a big role. Therefore we took an abstract deliberation model as the basis for our agent models. We already stated that there is no universally accepted abstract model for agent deliberations. We also will not argue that the model that we will describe in this chapter and that forms the foundation of the rest of the project is the only possible model or the "best" model. However, we will argue that there are a number of characteristics that we would like to have from such an abstract model when we use it to model agents in crisis situations. We will discuss these characteristics in the next section. After that we describe the way we have filled in the foundations for our simulation model for the ASSOCC project. We do not claim to have the one and only foundation for abstract agent deliberation models. However, we claim that the considerations that we use to compose this foundation are important for any simulation for crisis situations. One might make different decisions on how to fill in the different components due to the importance of some aspects. However, some general properties of the model will be preserved if the considerations that we lay down are followed. That our foundations for the model do work can mainly be seen from the second part of this book in which several scenarios of the COVID crisis and their results are described. All of these scenarios have been made using the same conceptual model! This shows the power of the approach and also that the model gives at least interesting and explainable insights on several aspects of the crisis.

2.3 Foundational Concepts

Given that it makes sense to use a deliberation model for the agents in the simulation based on an abstract internal state, the question becomes which concepts should be included in this abstract internal state of the agent. Before discussing our choices for these concepts, we briefly discuss what kind of properties we would like to have for this abstract internal state of the agent.

In [2, (Chap. 2)] arguments were given for certain abstract agent architectures. They argued that, in certain environments, the agents should have goals or utility functions as internal state to guide them. The decisions on actions can then be related to the goals or utility function and the action that most contributes to a goal or utility is chosen. We will briefly discuss both and argue why they are not appropriate for agents in social simulations for crisis situations.

Goals: This internal state leads to a certain measure of consistent behaviour over time. Once a goal is chosen the actions can be chosen that lead to achieving that goal. However, this can lead to rigid behaviour when the goals are long term and gives little extra stability when the goals are achieved with a few actions each time. Compare a goal of "owning a free standing house with large garden" which can take many years to achieve and the goal to "get home before 17pm today", which is either achieved or unachievable after 17pm today. The first goal can easily be abandoned if e.g. I buy a very nice penthouse or if I need money to treat some health problems or for the study of my children.

Goals are particularly useful for establishing a kind of midway points in terms of the lifespan of the simulation. As a very vague heuristic one can take that goals should influence decisions over (courses of) actions that are of interest for the purpose of the simulation. E.g. for the COVID-19 crisis situation one could set as the goal of agents to have enough resources every week to survive while being healthy. To have enough resources the agent needs e.g. to buy food, for which it need money, for which it needs to work. To work the agent might have to go to the workplace and choose a mode of transport to get there. So, in this sense the goal determines a whole set of other choices.

In the same simulation the goal to get to work with a car is not very interesting. This only impacts the choices on which route to take with the car to get to work. These choices on the route have no influence on the spreading of the virus and do not impact the outcome of the simulation. Thus this goal is too detailed to be of any use.

As said before, if a goal is too general it also is not very useful as it will completely determine the whole behaviour of the agent during the simulation, making that behaviour very rigid. E.g. if the goal of an agent is to survive, and this is the only goal, then each agent will create a plan at the beginning of the simulation to survive and try to stick to that. This will lead to very rigid and unrealistic overall behaviour of the society.

When goals are used to determine a kind of midway points it follows that goals have to created and several goals can be active at any time. This means that goals have to be managed, prioritised, etc. This is at least a non-trivial issue and will heavily influence the outcome of the simulation. A related issue is where these goals are coming from. If a goal is achieved, it disappears, but which other goal comes in its place? How are goals generated? In practical uses of goals in software agents the agents get goals from another mechanism. However, if goals are to be generated by the agent itself it needs another underlying model that can be used for this. Until now there is no such model that we can use. So, it appears that using goals to model the internal state of an agent can be useful for some types of applications, but actually do not give the kind of stable and converging influence that we need for complex social simulations.

Utility functions: Debreu [3] gives a precise characterisation of utility functions for the first time. He does this in terms of complete orderings over sets of preferences. This implies that a utility function is also monotonic in the sense that adding more items to a set will increase the utility of that set. E.g. one has more utility when having €100 instead of €90. The utility function is used to optimise the choice at a certain moment over all alternatives. It is assumed these sets of preferences are given and static and the ordering does not change over time. Thus one makes the same choice every time one considers the same set of alternatives. Thus if I would choose to work at home and visit grandparents over working at the office and not visiting grandparents, this choice should be the same at all points in time. Utilities are often used in cases where concrete goals are not available and more general guidance of behaviour is needed. The main advantage is that a utility function is a numeric representation of the combined preferences of an agent, which makes it easy to use in algorithms and decision processes that try to optimise a choice between alternatives at one point in time. In economics (where utility functions are most often used).

However, this is also the disadvantage. All elements of the preferences over all aspects of the simulation should be combined in this utility function which is fixed for the whole period of the simulation. Thus, one needs to determine the criteria to be optimised and relate all behaviour to that criteria. Again this will work well for agents in a limited and static environment. However, if more aspects of life have to be combined in the same utility function it will quickly become a bottleneck with which it is difficult to balance all types of behaviour. Moreover, utility functions (based on preferences) are not static for people. They change in different circumstances. E.g. the need to socialise is not felt in normal life as we have ample opportunities to satisfy it. However, during a long lockdown it suddenly can become very salient as we miss seeing friends and family over a long period of time.

One of the most used properties of utilities in economics is that utility functions have a form of monotonicity with marginally decreasing benefit: more of something that is preferred implies a higher utility, but the added (marginal) utility becomes smaller the more one already has of a resource. E.g. one has more utility when having €100 instead of €90. But the increase in utility from +€10 is bigger when going from €90 to €100 than going from €999.990 to €1.000.000.

Although this works for many economic situations it is certainly not true for all our preferences. E.g. I might want to socialise and meet with a few people every day. However, meeting with more people does not by definition give more utility. Neither does meeting people more often. Thus, getting more of the same might have a negative effect on utility!

If we accept that utility functions can also decrease and maybe also are dependent of the time and context, the utility function becomes no more than the aggregated function that at any moment in time indicates our preferences. In this sense any decision making mechanism will be expressable as a utility function. But this utility function does not give us any structure or theory on its validity.

Finally, one can also have a more philosophical objection against using utilities. A utility function expresses the preferences of a person for making a choice. However, not all actions are deliberated upon and decisions are not always based on conscious deliberations over preferences. When a person has not eaten for a long time there is not just a preference to eat, but a necessity to eat –it seems odd to model that a starving agent will make a reasonable assessment of the utility of eating when offered a plate. So, persons have biological and psychological needs and drives that drive our behaviour and which are not always expressable as preferences over choices.

As we have shown in the examples above, in crisis situations the preferences involved in decision making are not stable over time (because the context can change drastically in unforeseen ways) and for several aspects of the decisions the property of monotonicity and marginal decreasing utility functions are not appropriate. Thus utility functions do not seem to be the ideal candidates for shaping the abstract internal state of the agents. Given that goals seem also too limited in scope to deal with quickly changing contexts and combinations of different aspects of life we now ask what properties we want to have for the abstract internal state of an agent that should inform the decision making of the agent. We claim that the following properties are important:

1. The concepts should direct behaviour over longer periods of time of the agents
2. Behaviour should not rigidly be tied to the concepts as to lead to unrealistic, deterministic behaviour
3. Context and environment can make certain behaviour more salient at certain moments thus priorities can shift for certain periods of time
4. The internal state should lead to a natural balance between different behaviours over the life span of the agent
5. The internal state should also lead to adaptive behaviour based on changes in the environment and lead to a (possibly new) balanced behaviour again.

We can summarise the above properties in the image of an internal state functioning as a kind of magnet for an ideal set of states that guides behaviour within certain boundaries and with respect for current environmental conditions. In the research that we have performed in social simulations we have used the idea of *value systems* as the basis for this abstract internal state. We will describe our use of the value concepts in Sect. 2.3.1. Note that we do not claim that everything can or should be translated to values, but just that a value system is a good foundation for such an internal state that can be complemented with other concepts. In particular we argue that the value system should be complemented with *motives* to model the internal drive for certain types of behaviour. The role of motives is further described in Sect. 2.3.2. Where values can be seen as a very abstract goal and also link to common value systems of communities, the motives are internal, personal drives that generate behaviour of a certain type. A third component that we use for the abstract model are the *affordances* which model the abstract relation of the environment to potential behaviour of the agent in that environment. Affordances are further discussed in Sect. 2.3.3. In Sect. 2.3.4 we discuss the final component of the foundations of the deliberation of the agents: *social structures*. They can be seen to mediate and direct all the other components in some standard behaviour. In some sense, constructs like norms, conventions and practices can be seen as shortcuts that direct behaviour in a practical way while keeping in line with all the more abstract influences.

2.3.1 Values

Values have been part of the psychological debate for a long time. However, their abstract nature causes people to use the term in different ways with different meanings. In [4] a good categorisation of the problems around this term is given. He indicates that firstly, people use *values* as a noun and as verb. When you use it as a noun it can be something as an abstract goal that one might want to achieve. But if you use *value* as a verb it implies that some evaluation is taking place. We will use values primarily in the second sense and show that if you use value systems as abstract evaluation of actions they also function as general, abstract goals, because they always generate a preference over possible actions.

The next category of problems stems from the fact that values are not independent of each other. They come in *value systems* where values have *priorities*. The third category of problems stem from the question "whose values are we talking about?". Are values something personal? Are they linked to a group or society? Or are there universal values? And how are all these values connected? Some of the issues stem from the fact that values are not directly observable variables that can be attributed (numerical) values through empirical investigation. Thus they remain abstract psychological concepts that some say have no ontological reason of existence, but only serve to facilitate some psychological models. However, over the years they have also proven their utility in explaining some human behaviour. According to [4, 6] there are at least five features that are generally ascribed to values:

1. Values are beliefs and thus it is possible to discuss what a value means and what importance is given to it.
2. Values are *trans-situational*. In other words, values transcend specific actions and situations and are therefore by nature abstract. The abstract nature of values distinguishes values from concepts like norms and attitudes, which usually refer to more concrete actions, objects, or situations.
3. Values guide selection and evaluation of behaviour and events. Values serve as standards or criteria.

4. Values are a motivational construct. Values are related to some fundamental needs that people want to satisfy.
5. Values are relatively ordered according to importance. The values people pursue are structured in a value system in which each value is given a relative importance to other values.

The first feature is mainly directed towards behavioural psychologists that argue that values themselves cannot be observed and thus have no place in the scientific theory of psychology. The fact that values are beliefs that can (and are) discussed and reasoned about means that they are part of our social reality.

The second feature of values is that they are (by nature) abstract, because they should transcend specific situations. This feature is one of the main obstacles to get a consensus on a definition of values and also leads to many arguments on the benefits of having the concept if it is so abstract that it never can be held directly responsible for any behaviour. However, we argue below (based on [9]) that it is possible to construct formal connections between abstract values and concrete behaviour in situations as well.

We take the feature of values as evaluation criteria as the central feature of values. Given a value it is possible to order states of the world based on their desirability according to that value. So, if we take safety as a value, than a world with less corona infections is preferred to one with more.

Taking this feature as central also shows right away the motivational feature of values. If, according to the value of safety we prefer states of the world that have less corona infections, than we can create goals to achieve these states. Note, however, that the value is a criteria to compare states and thus in principle is not a goal by itself. One might always find a state that is even better according to the value than the current state. And thus the value cannot be achieved as a goal can be achieved.

Finally, people have more than one value and those values are ordered in a value system. The value system with its priorities between the values determines overall which kind of states people prefer (over other states).

In principle there could be an infinite number of values, which would make it very hard to compare sets of values as they would possibly have completely disjunct values. However, Schwarz [7] has argued that in the end there are only a limited number of 10 abstract values that are universally recognised by people. These values are structured along two dimensions as shown in Fig. 2.3. This set of values has been empirically established through surveys in many countries all over the world and thus seems to have some validity. As can be seen from the figure the values are ordered along two dimensions: (1) openness to change versus conservatism and (2) self transcendence versus self enhancement. These two dimensions are actually closely related to the basic psychological needs or motives as distinguished by McCelland [5]: Achievement, Affiliation, Power and Avoidance. And can also be related to the basic needs as distinguished in Self Determination Theory [10]: Autonomy, Competence and Affection–discussion on the exact relations between all these theories is left for future work as it would be worth another book by itself. However, the fact that there are some close connections makes clear that the basic

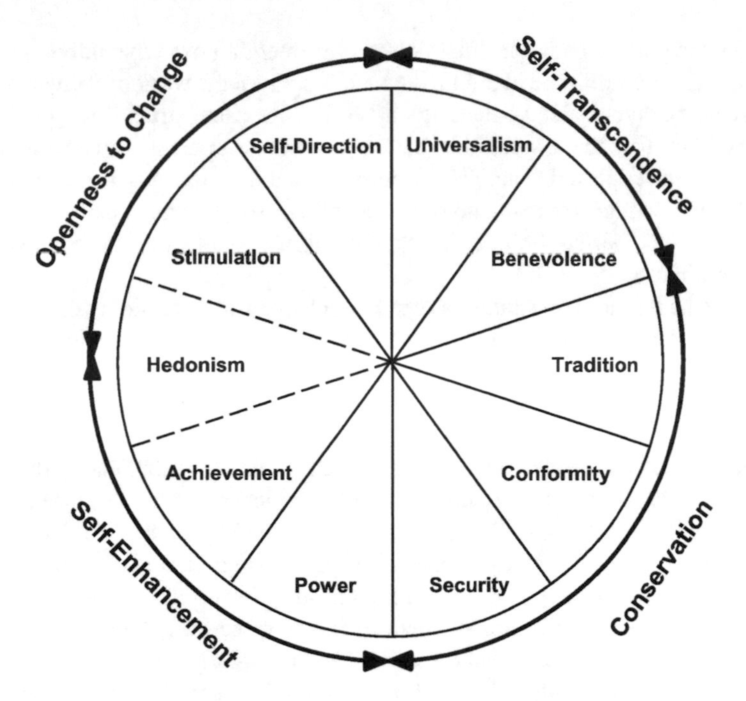

Fig. 2.3 Schwartz value circle

value system of Schwarz is closely tied to other psychological theories and gets indirect support through the validity of those theories as well. By aligning the 10 basic values of Schwarz with these psychological dimensions it becomes clear that there is to be an inherent balance in satisfying the values. When one prioritises the *Achievement* value it must be at the cost of *Benevolence*. Because we always want to keep some sort of balance in both dimensions it means that our behaviour cannot be exclusively directed towards one side of the dimension. I.e. even though we might prefer achievement over benevolence, we will not spent unlimited time and effort to reach some personal goals like e.g. a promotion at work. The effort will have to be balanced with effort spent on self-transcendent goals like being with the family. The exact balance between the two can differ for individuals, but the general structure of the values makes sure that resources are never exclusively directed into one direction and thus theoretically there is some optimum allocation of behaviour over the different abstract values for each individual. From the above discussion it seems values are only a cognitive concept and do not have a social or interaction perspective. However, values are also very often mentioned in discussions on societal level. E.g. talking about our "Western" values. The exact relation between these societal values and the private values has been the topic of much confusion [4]. While personal values can be based on psychological needs and motives, what would societal values be based on? Societal needs? The stance that we will very pragmatically take is that societal values emerge from the interactions between the individuals over many situations and times and places. Thus they are in some way emerging from the individual values and how these are exhibited in public interactions. From the work of Schwartz [8] we can assume that the societal values at least follow the same structure as the individual ones. Therefore the assumption that they somehow emerge from the individual values is not unsupported. There is currently no literature to support this view, but it would be interesting to investigate this interaction view of values using the tools of agent based social simulation platforms. We will use this feature to relate values to culture in Chap. 8.

Most of the research on values has been directed towards the nature of the concept and the foundations of it. It has been performed by psychologists interested in the role values have in influencing human behaviour rather than to formalise its meaning. Thus no formalisation is developed in the psychological literature. Within computer science, values have mainly be used in the area of automated argumentation. It was Perelman in [11] who stated that when arguments were opposing each other, it was usually not because one of the opponents made a logical error, but because they were supporting different values. Based on this statement, many years later Atkinson and Bench-Capon used values as means to prioritise arguments in an automated argumentation system [12]. However, although values are used in this work, they have not been given any formalisation by themselves nor is there any theory developed on how the values relate to each other formally. The first work in the actual formalisation of values is done by Weide [9]. This formalisation is based on two assumptions of values: Values should be used as criteria to guide behaviour and abstract values should thus somehow be related to concrete states of the world. We will not repeat the whole formal theory at this place but will illustrate how the theory can lead to a value structure that can be used to guide practical behaviour.

In Fig. 2.4 we show an example of a value structure. The top of the example value structure consists of three values of the Schwarz system. In principle all 10

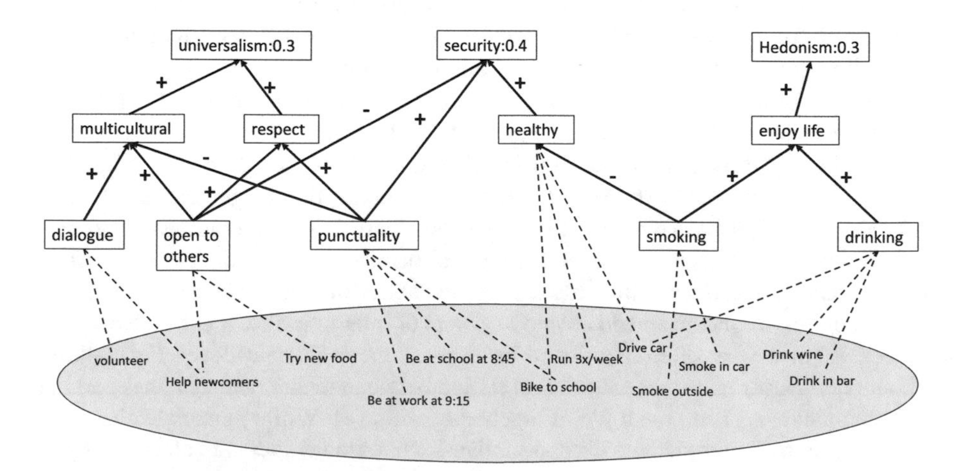

Fig. 2.4 Value system

values should be present in the value structure, but we wanted to keep the example simple enough to fit a page. The numbers with the top values indicate their relative normalised importance. Thus *security* is the most important value (with 0.4). Promote relations are depicted by arrows annotated with a "+". The demote relation is denoted by an arrow annotated with a "-". In the bottom of the figure the parameters that are considered in the world states are depicted. Dashed lines between the concrete values and the parameters indicate that they are part of their domain.

By making the set of parameters on the basis of which we measure the values explicit, it becomes now subject to discussion whether this is indeed the right set of parameters. E.g. should you check whether you eat new types of food to support your value of openness to others? And maybe punctuality should also check whether you are home in time for dinner. In a similar way one can now argue about a particular interpretation of an abstract value in concrete values. One can argue that other concrete values should be added, links should be added that either demote or promote another value, etc. E.g. One might argue that a concrete value "meticulousness" should be added that promotes security. It might demote punctuality as well. Or one can argue that "smoking" demotes respect for others. All these issues will determine the final valuation of the value structure. And it are often these aspects where people differ in their value structures and differences in preferences arise. We have used these ideas about values in several social simulation projects in which different perspectives of values and thus their interpretation into concrete behaviour is illustrated in simulations that support policy makers and NGO's (see [13, 14] for some examples).

2.3.2 Motives

The second type of drivers of behaviour in are the motives that all people have in common. This is based on the theory of McLelland [5]. The four basic motives that are distinguished are:

- achievement
- affiliation
- power
- avoidance

The achievement motive drives us to progress from the current situation to something better (whatever "better" might mean). The affiliation motive drives us to be together with other people and socialise. Thus we sometimes do things just to be doing it together with our friends and family. Note, this is unrelated with the fact that we might be dependent on other people for many things in life. The affiliation motive is purely driven by the need to be together and have positive interactions with other people. The power motive actually refers to having the ability (or skill) to control the environment. When this motive is very strong the agent wants to be as self-sufficient and autonomous as possible. The power motive thus does not mean we want power over others, but rather that we want to be autonomous. I.e. being able to do tasks without anyone's help. We can see the notion of social power as an extension of the power motive into having the ability to manipulate other people and in this way extend one's control over the environment.

The difference between the power and achievement motive are not always very clear and in many cases the combination of both can lead to specific actions. However, a good example to see the difference between the two motives is when a person practices a skill. The practice itself can give a person joy, because you feel that your skill improves and your abilities and control increase. Thus, playing a musical instrument can give joy without anyone else listening. In addition to this there might also be an achievement motive driving the practising in that it allows the person to be able to play well in a concert and achieve the goal of entertaining the audience.

Finally, the avoidance motive lets us avoid situations in which we do not know how to behave or what to expect from others. Thus we get anxious when we have to meet a group of new people or play a new piece of music. If we are not sure whether the outcome of the situation will be positive or negative there is a drive to avoid that situation. Thus we see that three of the four motives are drivers for new behaviour while the last motive balances these drivers and avoids us to get into too many new situations with uncertain outcomes and possibly get in dangerous or unpleasant states. Thus, like with the value system, there is an inherent balance between the motives that prevents people to consider only one motive. Just like with values each individual will have priorities that rank the motives. Thus one person will be more achievement driven, while another is more affiliation driven. However, all individuals will have all four motives and they all exert their influence on the behaviour of an individual to a certain measure related to their priority. Thus, each of these motives is active all the time and whenever possible it will drive a concrete behaviour.

Figure 2.5 shows how this influence of motives is supposed to drive behaviour according to McClelland [5]. In the figure it can be seen that the environment provides cues as to what type of behaviour is possible and opportune. This can be modelled through the affordances that the environment provides. The cues in the environment lead to some incentives becoming salient for these behaviours. E.g. seeing a picture of a grandparent or hearing them on the phone can lead to an incentive to visit them becoming salient. The motive of affiliation will then combine with this incentive to create a factual motivation to visit the grandparents. Of course, the environment can give many incentives and the motives can lead to several motivations for behaviour at any one time. The importance and the skills available will, in combination with the motivation, determine whether an impulse to act (an intention) arises. Thus, if the grandparents live in another country and travelling by plane is necessary to see them, it might be that in times of the Covid-19 crisis no impulse to act is formed on the motivation. The avoidance motive will possibly prevent an action to be taken as the possible danger for the grandparents to get infected when we visit is seen as too high. Or a combination of costs, time needed and health safety leads to inhibition of the action. Thus not every motivation leads to an intention to act. When an action is performed based on the motivation the person will monitor the results of the

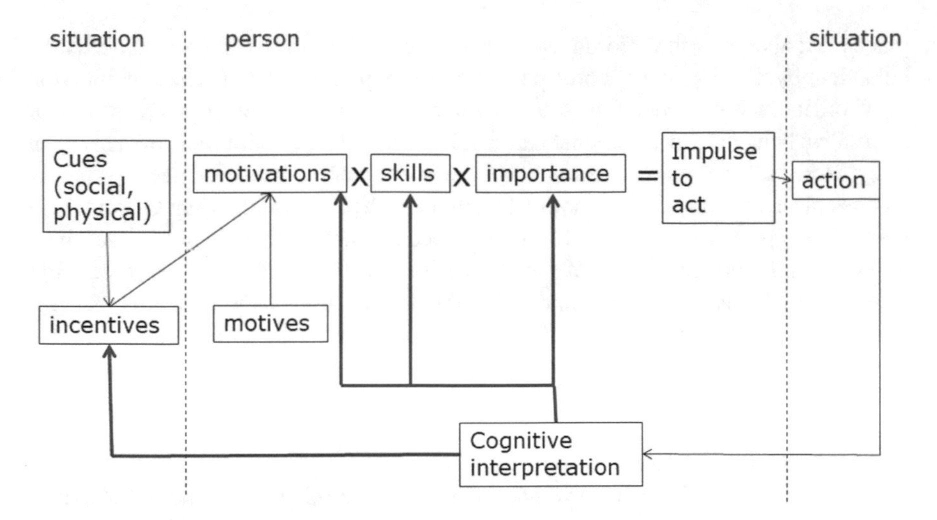

Fig. 2.5 Motives influencing behaviour

action and compare it with the expectations. Based on this comparison its perceived parameters can all be adjusted. Thus a successful visit to the grandparents can lead to the importance becoming bigger (or smaller if the need to see the grandparents is satisfied enough and only occasional visits are the norm), the skills needed are perceived to be more available, etc.

The way this figure combines with the previous section on values is through the incentives. The value system can provide the inherent incentives for certain types of behaviour. Thus in some way it are the values and affordances that determine together with the motives which concrete motivations are considered. In the next section we will discuss the influence of affordances on behaviour.

2.3.3 *Affordances*

The third type of elements that determine behaviour are the affordances [27] that a context provides. These affordances determine what kind of behaviour is available and also what type of behaviour is salient. E.g. in a bar one often drinks alcohol. Even though it is not obligatory it is salient and also afforded easily. The use of affordances is of great benefit when creating social simulations in which the persons are supposed to react to changes in the environment. These changes can be physical, but also social (such as new policies). E.g. suppose that young people have a big need for affiliation with other youth. Normally this motivation (the concrete motivation stemming from the affiliation motive and the incentive to meet potential partners) can be satisfied by going to a pub together with friends. However, if during the COVID-19 crisis the pubs get closed what will happen? Young people will look for other places that also have the affordance to meet peers. Thus, they can create alternative behaviour by meeting at e.g. home parties or go shopping together. The specification of affordances for elements in the environment allows for on the fly creation of alternative behaviour if the agents can combine affordances with their own skills in order to create (new) plans. Without the use of affordances one would need to specify all possible behaviours for each type of agent in each possible situation. Of course, in the end the type of affordances that are described for each element in the environment will determine the boundaries of the alternative behaviours that can be expected in a simulation. Thus we are not claiming that using affordances will create complete freedom of behaviour. We claim that specifying affordances and at the same time creating a planning mechanism in the agents that makes use of these affordances will create a more modular and flexible way of specifying potential behaviour of the agents which can determine new ways of behaving based on the current situation and available affordances.

A second aspect of affordances is that they also determine what kind of behaviour is salient. In a bathroom taking a shower is salient, at the office working is salient, sitting at a laid dining table makes eating salient, etc. Thus the affordances available in an environment also can give an easy focus and priority for some behaviour without having a fixed rule that deterministically decides for a behaviour in every situation. Thus an office place makes working salient, while a coffee corner in that space will make socialising salient in that place. However, in an office one can also socialise and work can be discussed in the coffee corner. We will see later that we assume that the people work in work places, but also use them to socialise to a certain extent. When people are forced to work home, they can still do the work, but miss the affordance to socialise. Thus they look for other actions to satisfy that need.

An aspect of affordances that will not really be discussed in this book in the context of the COVID-19 crisis, but is certainly of importance for natural disasters is that of engineering affordances. Just like policies can direct the behaviour of people through laws, subsidies, etc. government can also construct infrastructure in areas in a way that it naturally directs people in a certain way during special situations. E.g. in order to prepare for evacuations during a tsunami some streets can be made one way up the hill and wide enough to carry many vehicles at the same time. These streets will become the natural escape routes. Providing signs and logistic support along these streets will greatly improve the efficiency of an evacuation. So, experiments with different types of affordances in the infrastructure of an area will allow to prepare better for future natural disaster situations.

2.3.4 Social Structures

Individuals have to balance between their values, their motives and the affordances to determine what behaviour would be more appropriate in each situation. As one can imagine this is quite tricky and will take too much time and energy if done in every situation from scratch. Therefore, in human society social structures have emerged to standardise situations and behaviours in order to package certain combinations that will be acceptable and usually good (even if not optimal). These social constructs are things like: norms, conventions, social practices, organisations and institutions. Note that these constructs give general guidelines or defaults of behaviour, but are no physical restrictions on what is possible!

In normal circumstances the norms, conventions, practices, etc. are good guidelines of behaviour that will keep individuals aligned with the value priorities of society. Thus e.g. a norm to self isolate in case one has some corona symptoms is aligned with the value of safety which has a high priority during a pandemic. Thus following the norm is not just good for the sake of following a norm, but also assures that you align with the value priorities of society. In this way the social structures can be seen as a kind of short cuts for reasoning about values.

We have done a lot of research on the implementation of social structures in agents and agent based social simulations. See e.g. [15–20, 24–26]. The main issue with implementing social structures in artificial systems is to determine which properties of the real social structure one would really want to have in the artificial system. Thus, how realistic should these implementations be? E.g. if norms are only seen as constraints on behaviour one would miss the motivational component [24, 25]. In social simulations the social organisational property of the social structures becomes important. E.g. if the norm is to wear face masks it can lead to a salience of that action, but also to the fact that people start reacting negatively to persons not wearing a face mask. In most social simulations the implementation of norms has been quite minimalistic. However, in [15] we have shown how the existence of norms can crucially influence the effect of new policies. Thus they are of great importance for simulations of crisis situations where government measures are taken.

However, we should be aware of the differing perspectives on social structures when taking the human view on them and the computer science view on them. (see Fig. 2.6). As argued above, social structures in human societies have emerged in order to regulate and simplify very complex and dynamic environments. At any moment in time a person has a choice between dozens of behaviours to perform. The result of these behaviours can be influenced by the environment and other persons in many ways as well. Thus packaging behaviour in ways to create expectations of certain behaviour in certain situations will greatly support simplifying an otherwise daunting task of deliberation over all possibilities and choosing some kind of optimum.

If we take a computer science perspective on social structures it seems more or less opposite. These structures are not very precisely defined and can evolve over time and rather seem to complexify the deliberations of software agents rather than simplify them. Normally we only specify the actions available for an agent at a certain moment that we think are very important. Thus we tend to give software agents sparse possibilities for action. The actions and their results are usually very well defined as well. Thus the need for social structures to simplify the deliberation is absent. The social structures rather seem to be another layer that makes the deliberation more complex in ways that are not always very clear. In the following figure we see the approaches from the two perspectives on social structures. It is useful to keep this figure in mind when designing social simulations for crisis situations. We will give

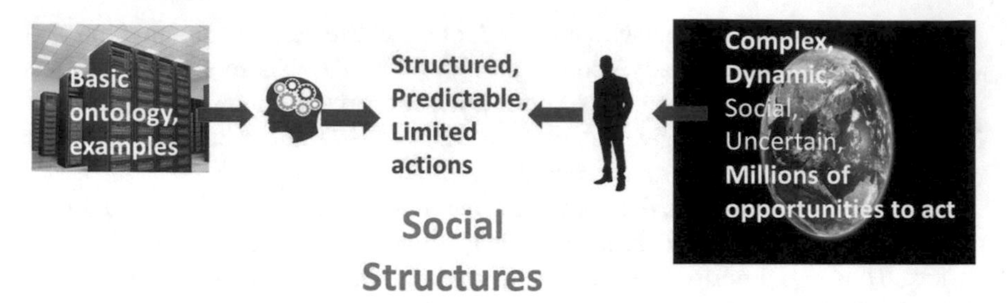

Fig. 2.6 Perspectives on social structures

one example on how this figure supports the design of a simulation. It relates to the use of norms in the simulation. Assume that we follow a data based approach to model the agents in the simulation. We see that people go to their workplace to work every day during the week (except Saturday and Sunday). They start work at 9 and go home at 17.

Now, the COVID-19 crisis forces a (partial) lockdown and people will work at home instead of at their workplace. Just taking the above data into account we would construct behaviour for the agents that makes them work 9–17 every day and stay at home during the lockdown. However, suppose we know that working from 9–17 every day is a norm rather than a given fact. During normal times everyone follows the norm and the norm seems of no consequence at all. However, during a lockdown it appears that people will consistently violate the norm. They miss their social contacts at the workplace and start finding excuses to violate the lockdown and socialise with other people during the day.

If we construct the agents based on statistical data only, this alternative behaviour is never modelled, because the norm is not violated during most of the times. However, if we take the norms seriously, we always have to give agents at least one alternative behaviour to violate a norm. Of course, this behaviour should not be a random action, but should be in line with the values and motives of the agents. This behaviour will actually not (or hardly) appear in the simulation in normal situations. However, it does provide suddenly a more realistic scenario when the circumstances change and norms get violated more often (as often happens in crisis situations!) (Fig. 2.7).

2.3.5 Combining the Elements

In the previous sections we have described the four components that we use as the foundations for the deliberations of the agents about their behaviour. In Fig. 2.7 we schematically show the relation between all the elements of the theoretical foundation of the deliberation model of the agents. From this figure we can see that we (indirectly) take three types of influences of behaviour into account. There is the influence on behaviour through the abstract values that are to some extend shared within a society.

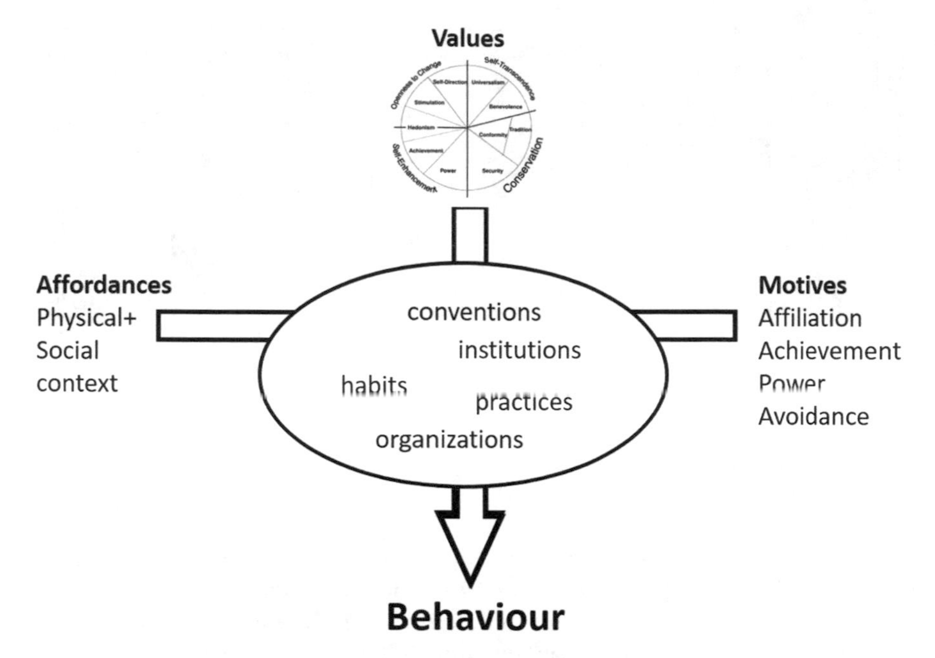

Fig. 2.7 Foundations of our model

They provide abstract goals that are acceptable from a societal perspective. A second influence are the motives, which are a very personal and individual drivers for the agent. These are mainly internally determined and updated based on the success rate of certain behaviour. The last influence is that of the environment of the agent, through the affordances that indicate what kind of behaviour should be considered in the first place in that situation. Finally, all these influences are regulated through the social structures that are in place and that filter and motivate certain behaviours to provide some consistency and create expectations that can be used to guide behaviours.

Although we have used parts of this whole architecture in several simulations over the past years [13–15, 21–23], implementing this whole architecture is usually too inefficient for any social simulation. Therefore we use this as theoretical starting point, but translate it into a simpler model that is more efficient and scalable. Thus for the ASSOCC simulation framework we fix the most important aspects of the values and motives described in the above architecture, into a set of *needs*, illustrated in the following Fig. 2.8.

From the figure it can be seen that we model the values and motives as needs that deplete over time if nothing is done to satisfy them. The model prevents that an agent will only look at the need with the highest priority and only at other ones when that need is completely satisfied. By calibrating the size and threshold and the depletion rate of each need we can calibrate and balance all the needs over a longer period, between different contexts and over several domains. E.g. using this model it becomes possible to decide for an individual whether it is more important to work

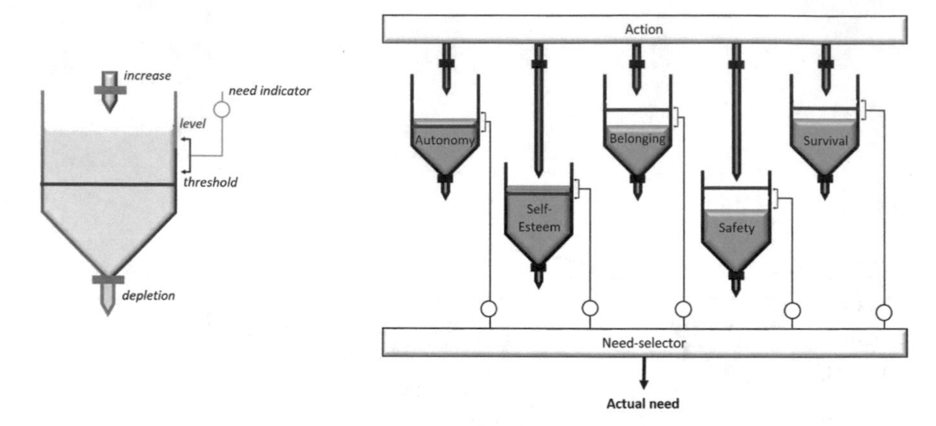

Fig. 2.8 Needs model

a bit more or go home and be with the family. This simple model is the crux behind combining health, wealth and social well-being in a simulation model.

The homeostatic nature of the needs model naturally leads to a kind of cyclic behaviour that is reflected in the patterns of life following habits, norms and practices as is common in normal life. Having a pattern of life in which all the needs are satisfied enough over time such that they never get completely depleted is something that is self reinforcing. As long as all the needs are satisfied the pattern of behaviour does not have to change to keep the system going. However, when something disrupts the normal pattern of behaviour the system is not right away broken. A disruption of the normal daily pattern of behaviour will lead to a change of how all needs are satisfied over time. If all needs are still satisfied within the time limits the disruption has no effect on other behaviour. However, if due to this disruption not all needs are satisfied in time anymore it will lead to some needs becoming salient at unusual times and this can lead to changing behavioural patterns. This change only happens if it is possible of course. If a direct change is not possible, it can lead to other changes for other needs and a ripple effect can lead to bigger behavioural changes. This will continue until some pattern is found that leads to a new cycle of behaviour in which all needs are satisfied in time again within certain boundaries.

The priorities between the needs is determined by the size of the containers of the need. The bigger the size, the more actions are needed to satisfy that need and thus how higher its priority. If the threshold of a need is higher it also will become salient quicker and thus the periods between paying attention to that need are smaller.

As can be seen from the above, this type of model is well suited to model the main characteristics of the value and motive systems. These systems do not directly steer the behaviour through fixed rules, but rather influence it indirectly by making some type of behaviour more desirable in some situations and times. Note that this homeostatic model also precludes the need of a planning system for the agents. This makes the system more flexible and simpler to maintain. Of course, it also comes with its own limitations. We will discuss some of those in Chap. 15.

The affordances in the ASSOCC framework are rather simple. Most locations in the environment of the agents allow for a limited number of actions. Those are the affordances of those locations. E.g. workplaces afford to work only. We do model the fact that people also socialise at work through the fact that the work action also contributes to the need of socialising when performed in a workplace. This aspect is the least well developed in the ASSOCC framework as it does not play a big role in the COVID-19 crisis.

The social structures have also not been represented in all details in the ASSOCC framework. However, we do represent that certain actions in certain situations will contribute to the conformity and belonging needs. Theses needs indirectly drive individuals to conform to norms. Thus normally the norm will be followed unless there is another need with higher salience that needs to be satisfied.

2.4 Conclusions

In this chapter we have described the foundations of the deliberation model that we have used for the ASSOCC framework. We claim that this foundation is able to provide a basic model for simulations of crisis situations in a more broad sense than just for the COVID-19 crisis. Depending on the type of crisis emphasis can be given to different aspects of these foundations. However, the foundations do fulfil the properties that we set out to satisfy for abstract models:

1. The concepts direct behaviour over longer periods of time of the agents
2. Behaviour is not rigidly tied to the concepts and does not lead to unrealistic, deterministic behaviour
3. Context and environment can make certain behaviour more salient at certain moments thus priorities can shift for certain periods of time
4. The internal state leads to a natural balance between different behaviours over the life span of the agent
5. The internal state leads to adaptive behaviour based on changes in the environment and leads to a (possibly new) balanced behaviour again.

We do not claim that our foundations are the only possible abstract framework to be used as a starting point for social simulations of crisis situations. However, we will show that it is a very workable model that does lead to simulations that are quite realistic and which at the same time can be explained through this framework in a very natural way. In the next chapter we will describe in more detail how the foundations are implemented in the ASSOCC framework.

Acknowledgements The authors would like to acknowledge the members of the ASSOCC team for their valuable contributions to this chapter. This research was partially supported by the Wallenberg AI, Autonomous Systems and Software Program (WASP) funded by the Knut and Alice Wallenberg Foundation.

References

1. N. Gilbert, Agent-based social simulation: dealing with complexity. Complex Syst. Netw. Excell. **9**(25), 1–14 (2004)
2. S. Russell, P. Norvig, Artificial intelligence: a modern approach (2002)
3. G. Debreu, Representation of a preference ordering by a numerical function, in Decision processes, ed. by R.M. Thrall, C.H. Coombs, H. Raiffa (Wiley, 1954), pp. 159–167
4. J. M. Rohan, A rose by any name? The values construct. Personal. Soc. Psychol. Rev. **4**(3), 255–277 (2000)
5. D.C. McClelland, *Human Motivation* (Cambridge University Press, 1987)
6. S. Schwartz, W. Bilsky, Toward a universal psychological structure of human values. J. Personal. Soc. Psychol. **53**(3), 550–562 (1987)
7. S.H. Schwartz, Universals in the content and structure of values: theoretical advances and empirical tests in 20 countries, in *Advances in Experimental Social Psychology*, ed. by M.P. Zanna, Vol. 25 (Academic Press, 1992), pp. 1–65
8. S. Schwartz, A theory of cultural values and some implications for work. Appl. Psychol. **48.1**, 23 (1999)
9. T.L. van der Weide, Arguing to motivate decisions. Ph.D. thesis. Utrecht University (2011)
10. R. Ryan, E. Deci, Self determination theory. *Basic Psychologicasl Needs in Motivation, Development, and Wellness* (Guilford Press, 2017)
11. C. Perelman, L. Olbrechts-Tyteca, *The New Rhetoric: A Treatise on Argumentation* (University of Notre Dame Press, 1969)
12. T. Bench-Capon, K. Atkinson, Action-state semantics for practical reasoning, in *The Uses of Computational Argumentation: Papers from the AAAI Fall Symposium (FS-09-06)* (AAAI Press, 2009), pp. 8–13
13. C. Boshuijzen-van Burken et al., Agent-based modelling of values: the case of value sensitive design for refugee logistics. J. Artif. Soci. Soc. Simul. **23.4**, 1–6 (2020)
14. S. Heidari, M. Jensen, F. Dignum, Simulations with Values, in *Advances in Social Simulation* (Springer, 2020), pp. 201–215
15. C. Pastrav, F. Dignum, Norms in social simulation: balancing between realism and scalability, in *Proceedings of the Fourteenth International Social Simulation Conference* (2018)
16. F. Dechesne et al., No smoking here: values, norms and culture in multi-agent systems. Artif. Intell. Law **21**(1), 79–107 (2013)
17. V. Dignum, F. Dignum, Contextualized planning using social practices, in *Coordination, Organizations, Institutions and Norms in Agent Systems X: COIN 2014*, ed. by A. Ghose et al. (Springer, 2015), pp. 36– 52
18. J. Vazquez-Salceda, V. Dignum, F. Dignum, Organizing multiagent systems. JAAMAS **11.3**, 307–360 (2005)
19. V. Dignum, F. Dignum, A logic of agent organizations. Log. J. IGPL **20**(1), 283–316 (2012)
20. L. Vanhee, H. Aldewereld, F. Dignum, Implementing norms? in *IEEE/WIC/ACM International Conferences on Web Intelligence and Intelligent Agent Technology*, vol. 3 (IEEE, 2011), pp. 13–16
21. G.J. Hofstede et al., Gender differences: the role of nature, nurture, social identity and self-organization, in *International Workshop on Multi-agent Systems and Agent-Based Simulation* (Springer, 2014), pp. 72–87
22. L. Vanhée, F. Dignum, J. Ferber, Modeling culturally influenced decisions, in *International Workshop on Multi-agent Systems and Agent-Based Simulation* (Springer, 2014), pp. 55–71
23. L. Vanhée, F. Dignum, Explaining the emerging influence of culture, from individual influences to collective phenomena. J. Artif. Soc. Soc. Simul. **21.4**, 11 (2018)
24. C. Castelfranchi et al., Deliberative Normative Agents: principles and architecture. Intell. Agents VI LNAI **1757**, 364–378 (2000)

25. S. Panagiotidi, S. Álvarez-Napagao, J. Vázquez-Salceda, Towards the norm-aware agent: bridging the gap between deontic specifications and practical mechanisms for norm monitoring and norm-aware planning, in *Coordination, Organizations, Institutions, andNorms in Agent Systems IX-COIN2013*, ed. by T. Balke et al., vol. 8386. Lecture Notes in Computer Science (Springer, 2013), pp. 346–363
26. F. Dignum, Autonomous agents with norms. Artif. Intell. Law **7.1**, 69–79 (1999). ISSN: 09248463. http://www.springerlink.com/index/N32XU121L58H417V.pdf
27. E.J. Gibson, A.S. Walker, Development of knowledge of visual-tactual affordances of substance, in *Child Development* (1984), pp. 453–460

Chapter 3
Social Simulations for Crises: From Theories to Implementation

Maarten Jensen, Loïs Vanhée, and Christian Kammler

Abstract This chapter describes how the general theories presented in the previous chapter have been used for the concrete ASSOCC software platform, which is used as the basis for all the scenarios described in Chaps. 5–10. We will describe the agent architecture and deliberation mechanism based on the needs. We also will introduce the environment which is modelled like a small town in which the agents live. The chapter also describes the epistemiological model that we use to represent the COVID-19 disease specific elements.

3.1 Introduction

As we already stated in Chap. 1, the ASSOCC framework, that is used to illustrate the issues of simulating for a crisis situation, is implemented in Netlogo. There were two reasons for this choice. First, the main software architect had a huge experience with Netlogo. Secondly, having the code in Netlogo could facilitate the uptake and reuse of the code by others in the social simulation community. However, implementing the framework in Netlogo also had as consequence that efficiency and scalability of the simulations became serious issues. Therefore the model as described in this chapter already shows many compromises where we chose to abstract away from some elements, simplify some others and also sometimes chose to refrain from including some aspects. Even given these restrictions the resulting framework is one of the most complex (if not the most complex) models ever built in Netlogo. This shows the difficult choices one has to make when simulating for crisis situations. In one hand one must incorporate as many aspects as possible, while at the other hand

M. Jensen (✉) · L. Vanhée · C. Kammler
Umeå University, Mit-huset, Campustorget 5, 901 87 Umeå, Sweden
e-mail: maartenj@cs.umu.se

L. Vanhée
e-mail: lois.vanhee@umu.se

C. Kammler
e-mail: ckammler@cs.umu.se

F. Dignum (ed.), *Social Simulation for a Crisis*, Computational Social Sciences,
https://doi.org/10.1007/978-3-030-76397-8_3

keep the model as simple as possible to ensure some level of efficiency. We will have a more in depth discussion about this aspect in Chap. 13, but will already point out some of the issues when relevant in this chapter.

The ASSOCC framework consists of multiple submodels, such as an agent model, disease model, economical model and transport model. The agent architecture includes the needs model as central deliberation mechanism. The disease model is based on the Oxford model [1], which is a very detailed instantiation of the standard epidemiological SEIR model [2] for the corona virus. In order to introduce all the parts of the ASSOCC framework we roughly follow the Overview, Design concepts, and Details (ODD) methodology [3]. The chapter starts with the purpose of the model, then we give an overview of all the elements of the framework. Each of these elements is described in more detail in the sections after. First we describe the environment such as the location, buildings and modes of transport. In section five we explain the disease model in detail and in section six we describe the representations of government guidelines and interventions. These sections are followed by a section that describes the agents and the behavioural model in detail. The last section describes how the model can be used for running experiments. Some more user specific parts of the ASSOCC framework such as the coupling and use of the Unity interface and the post simulation analysis tools are described separately in Chap. 4.

3.2 Purpose

The purpose of this framework lies in providing support for stakeholders for making informed decisions regarding the management of the COVID-19 disease. The framework is a sandbox that allows many different types of scenarios to be simulated and tested. In this regard, the core aspects that are part of the model are driven by both: (1) social aspects from psychology and sociology that are acknowledged to be of importance in a crisis (e.g. needs, social networks, norms, practices, habits, and values) and (2) the important features raised by decision makers and other stakeholders such as medical doctors (e.g. public measures, shortages, flattening the curve).

The ASSOCC framework aims to provide detailed models, capable of capturing the often-overlapping causes of the emerging phenomena, whereas most of available simulations focus on narrower aspects that may be grounded in statistics, but often fail to account for underlying causes. This way, ASSOCC offers a richer opportunity for studying the effects of decisions on many aspects (e.g. psychology, sociology, epidemiology, economy), as well as increasing the resilience of observations provided by the system by accounting for more sources of influence. This integrated approach allows us to see how health, the social system and the economic system influence each other when we adjust an aspect or introduce different types of government policies.

The R_0 factor is a good illustration of this complementary aim (i.e. the average total number of people infected by the first infected persons before anti epidemic measures against the specific disease have been taken). It is implicitly assumed that the behaviour of people in that condition did not change yet. Thus R_0 gives an indication of the inherent infectiousness of the virus. The same factor is now used as the R factor, denoting the average number of people that are currently infected by an infected person. Many models use this R factor as an input variable for describing disease dynamics. For the ASSOCC model (and in the real-world), R is an output or control variable, which depends on the interplay of other variables, such as the degree of infectiousness of people and their behaviour and amount of contact. Imagine that there is only one person left being infected and being contagious. If this person sits isolated at home R will be 0. However, if this person goes to a big party (because he has little symptoms and does not feel ill at all) he can infect a dozen other people during the party and R suddenly is 10 or 20! Thus, although the R factor gives some indication on how well the country is doing it can be very misleading as well. The ASSOCC model aims to play with more sensitive and advanced variables (e.g. the density of the population, the dynamics of contacts between people) and the subtle interplay of their dynamics (e.g. the influence of culture on R_0 correlated with how public measures impact people's psychological dynamics).

3.3 The Simulation Elements and Sub-models

This section introduces the elements and sub-models of the ASSOCC framework. The subsequent sections will describe each of the mentioned elements in more detail. The simulation represents a city with individuals, houses, schools, universities, work-places, two types of shopping places and two types of leisure places. The individuals will be called agents in this chapter as this chapter explains the implementation of the simulation. Figure 3.1 shows an overview of the city in our graphical interface made in Unity. The Unity interface loads and displays the data from the Netlogo simulation and is more extensively described in Chap. 4.

3.3.1 Simulation Elements

There are quite a number of aspects influencing the agents' behaviour. Figure 3.2 shows a general overview of all the elements incorporated in the simulation. Central are the agents that have a general and a social profile, epistemic model of the disease state, and a complex need based behaviour system. There are separate models for the epidemiological and economical aspects based on respective theories from those areas. We have a rather simple environment consisting of locations (that are called gathering points in the code) and transportation. Finally we also represent government interventions separately in order to easily add and retract them during the crisis.

In order to ensure a realistic simulation, it is necessary to incorporate an economic model in the simulation. What good is not getting sick if you cannot buy food? The

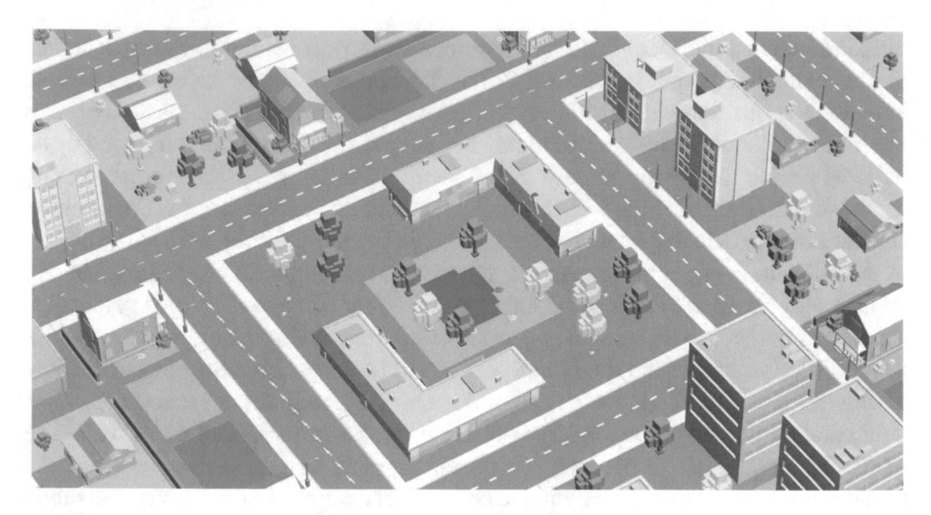

Fig. 3.1 Overview image of the simulated city in the Unity interface

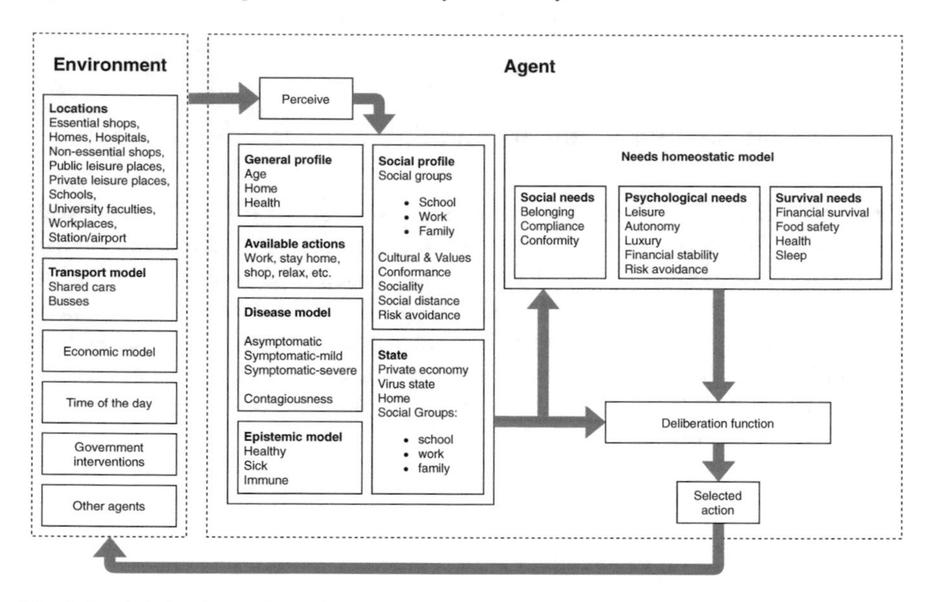

Fig. 3.2 High-level overview of the framework's components

economic model is a relatively simple micro-economic model. It is further explained in Chap. 9 Economics where we will explore the dependencies of the economy and health aspects of the COVID crisis. However it is also part of all the other scenarios. Here we just mention the main assumptions that are of importance for all scenarios. In the simulation the youth are not part of the economic process. It is simplified in a way that their parents are paying for them. Students and retired agents do not go to work, but they receive an allowance (scholarship or pension) from the government. People from working age all work at some location: essential-shops, non-essential-shops, workplaces (offices and factories), hospitals, schools and universities. Here it can be seen that we made a very simple categorisation of places of work. It is mainly based on categories that are important for the COVID-19 crisis. We did not include any pubs, restaurants, etc. as working place. Neither did we make a difference between contact and non-contact professions. These are distinctions that could be made in a larger scale simulation than what we could build in Netlogo. In a simulation with around 1000 agents these distinctions would not lead to more realistic simulations as there would be too little places in each category.

3.3.2 Temporal and Spacial Scale

The day is divided into four slices in the simulation: *morning*, *afternoon*, *evening* and *night*. Each of them have different implications for the agents. For example, in the night the agents sleep, while in the other parts of the day they go to their jobs or other places. The part of the day changes every tick, thus each day consists of four ticks. The days of the week are explicitly modelled and there is a difference between weekdays (when agents work) and weekends (no working).

The locations represent the spatial model. They do not have specific coordinates that influence the agents' decision making. Rather the agent knows at which building and which location type it is. The spatial scale could be compared to the size of a city. They are described in the next section.

3.4 The Simulated Environment

The environment the agents live in contains a variety of elements. The different locations provide the agents with a rich daily live of activities: going to work or school during the day, and going shopping or going for leisure in the evenings and weekends.

3.4.1 Types of Locations

Figure 3.1 details the different types of buildings and places, or locations (called gathering points in the implementation), based on stakeholder expectations. For example, both essential and non-essential shops as introduced as the effect of closing specific types of shops is important for stakeholders. Workplaces cover both, for example, offices and factories as the difference has low salience on disease spreading. The agents in a given location are assumed to come into contact with each other and could infect each other. As a simplification all agents engaged in a location are assumed to

Table 3.1 Description of the various location types: **N** Indicates the number of locations of that type, the station/airport does not have a number as it is just one

Location type	N	Description
Essential shops[a]	10	This is the place where agents buy food, they have to do this every once in a while
Homes[a]	391	The agents live in their homes, they are always at home during the night to sleep. The homes can be used as a workplace if the agent has to be in quarantine
Hospitals[a]	4	The hospital has some workers and treats agents who get severely infected, if they have enough capacity (number of beds)
Non-essential shops[a]	10	With non-essential shopping agents can satisfy the need for luxury
Private leisure places	60	These are places where agents can meet with their friends
Public leisure places	20	The public leisure places represent places such as parks where everyone can go to. The needs Self-Esteem, compliance are satisfied by going there. However risk-avoidance and safety depletes
Schools[a]	12	The youth go to school during working days. Needs such as compliance and autonomy are satisfied here
University faculties[a]	4	The students go to the university and can also satisfy autonomy and compliance here
Workplaces[a]	25	Where most of the worker agents work. Many needs can be satisfied through going to work
Station/airport (away)	–	This represents agents travelling outside of the city

[a]indicates a place where workers can work

be (equally) in contact with all others that are also engaged within these locations. As a side note, the agents can also infect each other when travelling between locations (detailed in the transport model section). Locations are at the core of the proximity model (a central factor for virus propagation in the disease model). Table 3.1 lists and describes all the location types. The number of the various location types is selected for creating a realistic average number of agents in a location at the same time. For example, twelve school locations represent that a school has twelve classes, but the city does not have twelve schools. With this number, around 25 children will be gathered in every school location during school time.

We simulate a city with different types of places where agents can have interactions. In real life there are a large variety of interactions, for example one could have a long meeting for a few hours when playing board games with friends but also just saying hello when meeting a neighbour at the staircase. While modelling these specific interactions can be interesting when studying corona virus spreading at the micro level, we study corona spread as a pandemic and not just in for example one building. To keep the implementation efficient and scalable we kept interactions more abstract. An agent can be interacting with another agent (and thus potentially infect another agent) when they are at the same location (building or vehicle) at the same period of the day. This has implications for the spread of the disease which are described in the disease model (Sect. 3.5).

3.4.2 Contacts and Expected Contacts

The *contacts of an agent* are the other agents that are at the same location for a tick. For example in a house with four agents each agent has three contacts. The *contacts at a location per tick* are thus calculated by multiplying the number of agents n with the number of contacts they each have $n - 1$, the formula is the following:

$$contacts = n \times (n - 1)$$

Many of the policies in real life are focused on decreasing the number of contacts agents have, since the COVID-19 virus is mainly spread through interactions between agents. To ensure a realistic amount of contacts we calculated the expected number of contacts per age group (Table 3.2) for each location type. We used these expected contacts to calibrate the number of location types and need model in the simulation. To give an example, there are twelve school locations and 312 children (with default settings of 1126 agents). This means about 26 children per class and there is one teacher (Worker) for each class. With $n = 27$ in a class we get $27 \times 26 = 702$ contacts per tick, since there is morning and afternoon school we get 1404 contacts per class on a regular school day. This leads to $12 \times 1404 = 16.848$ contacts in total for a school day. Per child this is 52 contacts per day. The twelve workers at school also have 52 contacts per school day. On average each child has 52 contacts at school, because all youth go to school. For workers the amount goes to $\frac{12 \times 52}{454} = 1.4$, because only twelve of the 454 workers work at school. Because the school is open five days per week we still have to multiply these amounts with $\frac{5}{7}$. Thus we get on average $\frac{5}{7} \times 52 = 37$ contacts per day for children at school and one contact per day for workers at school. On average the school contributes 10.65 contacts per agent day

Table 3.2 Expected contacts at locations per age group

Age	Home	School	Work	Uni	Hosp	Eshop	Nshop	Publs	Privls	Busses	Cars	Queue	**Total**
Youth	5.7	37	0	0	0.003	0	0	0.14	4.34	14.4	0	0	62
Student	1.8	0	0	49.5	0.018	2.71	5.82	0.17	4.63	19.2	3	0	87
Work	4.5	1	13	0.65	0.055	4.13	9.39	2.61	5.11	19.2	4.5	0	64
Retired	2.7	0	0	0	0.025	3.04	11.52	7.02	5.56	9.6	3	0	42
Average	4.2	10.65	5	4	0.03	2.6	7	2.81	4.97	15.6	2.77	0	**60**

(including weekend closing). In a similar way we get expected number of contacts for all other location types and age groups.

The expected contacts for homes are calculated based on the amount of agents per age group and the Great Britain household distribution, assuming that agents are at home 1.5 ticks per week day. One tick at night everyone is at home. The second tick in the evening depends on whether agents go to leisure places or shops instead. During the day youth, students and workers are at other places, but retired agents stay home unless they go to shops or leisure. In the weekend everyone is at home and only goes out for leisure or shopping. So, in the weekend we assume agents are home together 2.5 ticks per day. Averaging this means agents are at home about 1.8 ticks per day and multiplying this with the household distributions led to the values in the table.

The workplaces only have workers so the other age groups will have zero contacts here. There are 40 workplaces and about 390 workers working at workplaces (the rest works in shops, hospital, university). So we have roughly 10 workers per workplace. This gets to $10 \times 9 = 90$ contacts per workplace per tick. Since there are two ticks per day agents work, this gets to 180 contacts. In total this is $40 \times 180 = 7200$ contacts in workplaces. Thus on average this is around 18 contacts per worker and per agent this is on average 7 contacts per day. However taking again $\frac{5}{7}$ into account for weekends we get 13 contacts for workers and 5 contacts per day for agents at workplaces.

Contacts at other locations are calculated in similar ways, for example for universities we calculate students and a part of the workers that work at the universities. Youth and retired are not going to the universities so their contacts stay at zero. The youth also do not go to shops and use shared cars, while the retired are more frequently at the leisure places and non essential shops since they are not going to school, work or universities. To calculate the values for the shops and leisure places we ran the simulation and determined how often agents go to those places based on their needs. Then determined the contacts they would have and adjusted the number of location types where needed.

The number of private leisure locations was set to 60 to give a realistic amount of average contacts for all the agents. This led to a more or less realistic number of agents meeting about five friends a day at private leisure.

In the table it becomes clear that students have the most contacts per day on average with 87 (the universities contributing most to this) and retired the least with about 42, while youth and workers have about 62 and 64. These are only rough numbers but doing these calculations allowed us to re-calibrate the location numbers and needs model such that in a run without the COVID-19 virus the number of contacts become more realistic. We now have an average of about 55 contacts per day per agent during a simulation run, which seems very realistic.

3.4.3 Motivation to Go to Locations

The agents are motivated to go to certain locations based on their scheduling. In general workers go to the same work location, meeting the same agents each day in the morning. They spend the morning and afternoon periods at work. The night period is spent at home sleeping. The evening is used either at home, shopping, or at some leisure place. The standard schedules of agents will satisfy all of their needs through the activities they perform during the day. For example an agent going to its workplace in the morning of a working day will satisfy the needs: complying to rules, financial stability, belonging, autonomy and financial survival, while the risk avoidance need will be decreased. This is explained in more detail in 'The agents' section below.

3.4.4 Homes and Household Composition

The homes are the places the agents live. The agents sleep here during the night and are at home when they are not visiting a specific location. Table 3.3 shows the four different household compositions contained in the model.

Besides the relative group sizes, it is also the composition of the groups in households that has a large impact on the dynamics of the pandemic. This includes both the number of agents that are housing together but also the relative share of households in which different age groups live together. Multi-generational living, where children are sharing the household with elderly agents, create new paths of infection that seem to be important for the dynamics of the pandemic in a country. In their report on household size and composition around the world (2017), the united nations provided data on average household size, the relative number of households by age group, and the proportion of households with children as well as elderly persons. In Europe, the average household size varies from 2.1 up to 3.9 members, with a proportion of households with elderly agents varying from 30 to 46%.

Table 3.3 Household composition

Type	Implications
Adults rooming together	Two agents: either workers or students
Retired couple	Two agents: both retirees
Family	Four agents: two workers and two children
Multi-generation	Six agents: two workers, two children and two retirees

Table 3.4 Household distributions in the model per country. The data is set in the *country-specific-data.csv*

Country	Adults together	Retired couple	Family	Multi-generation
Belgium	0.278	0.315	0.371	0.036
Canada	0.44	0.23	0.31	0.02
Denmark	0.298	0.251	0.434	0.017
France	0.302	0.3	0.375	0.023
Germany	0.291	0.234	0.457	0.018
Great Britain	0.292	0.312	0.36	0.036
Italy	0.309	0.298	0.344	0.049
Korea South	0.352	0.431	0.163	0.054
Netherlands	0.272	0.276	0.432	0.02
Norway	0.253	0.256	0.473	0.018
Singapore	0.586	0.191	0.128	0.095
Spain	0.258	0.336	0.347	0.059
Sweden	0.295	0.27	0.419	0.016
U.S.A.	0.404	0.259	0.315	0.022

Statistical data that has been used for the definition of the scenarios comes from the UN report "Household Size and Composition Around the World 2017".[1] Based on that data the percentage of each household type is tied to the countries. Table 3.4 shows the countries and distributions used in the simulations. A country distribution was added when required.

3.4.5 Shopping Locations

We implemented essential shops and non-essential shops. The essential shops represent grocery store, butcher, baker and other food related shops. The agents can buy food here and thus satisfy the food safety need. Agents can order food online (regulated by the *food-delivered-to-isolators?* variable), however this is only possible when they are in isolation. The non-essential shops represent the other shops like clothing, electronic, jewellery, and other shops not related to food. The agents can satisfy the belonging and luxury needs by being at the non-essential shops. The number of shops for both types is 10 see Table 3.3. This does not mean we represent 10 different shops in real life, but rather we have so many shop locations such that the number of contacts becomes realistic.

[1]"Household Size and Composition Around the World 2017" https://www.un.org/en/development/desa/population/publications/pdf/ageing/household_size_and_composition_around_the_world_2017_data_booklet.pdf.

3.4.6 Leisure Places

The leisure places are either public or private leisure. The public leisure places represent parks and sports events. The number of public leisure places is 20 which is lower than the 60 private leisure places. This is done such that at public leisure places the groups that meet are bigger, as in a larger event for example watching a football match. In a football match one can expect to meet some other agents as well and not only the group of friends one is with. The private leisure places have a high number which leads to having only a couple of agents at each of the locations. The private leisure places represent activities with friends.

3.4.7 Hospital

For a small town or city we would expect to only have one hospital if it has a hospital at all. The reason we have four locations for the hospital is to make a split between the different sections in the hospital. Not everyone within a hospital gets into contact with everyone else. In our simulation some workers will work at the hospital and agents that become sick due to the COVID-19 virus (those who are in the hospitalised or hospitalised recovering state) will also stay in the hospital. The hospital does have a limited number of beds available (*#beds-in-hospital*) and when this number is reached very sick agents stay at home instead.

3.4.8 Transport Model

To take the probability into account of getting infected on the way between different locations (e.g., on the way to or from work), a transport module was added to the simulation model. When agents change location, there is a certain probability for taking a specific transport mode. The probability is different for different age groups. Currently, three different modes of travel exist: public transport (bus, train), ride sharing (taxi, uber, car-pooling), and solo transport (own car, bike, walk). Table 3.5 shows the different probabilities of using one of the three different modes of travel.

Table 3.5 Transport model probabilities dependent on age. The walking outside probably is dependent on the public transport and shared car probabilities

Transport type	Children	Students	Workers	Retired
Public transport	0.75	0.6	0.4	0.2
Ride sharing	0	0.1	0.15	0.5
Solo transport	0.25	0.3	0.45	0.3

The variables for the busses are listed below. When all the busses are full, the agents will start queuing.

`#bus-per-timeslot:`	30, this is the number of busses that are available each time the agents travel.
`#max-people-per-bus:`	20, the total number of agents allowed at a bus.

For the shared cars there is a maximum of five agents per car. There is no limit on shared cars.

The transport phase occurs between the activity-selection phase and the execution of this activity. At the end of this phase, we record the following variables, which will then be used by the contagion model:

1. Whether the agent went out with solo transport. If this is the case *travelling infection* will be calculated.
2. Whether the agent had to wait for transport, if this is the case *queuing infection* will be calculated.
3. Which mode of transport is used and who the agent travels with. These parameters are used to calculate *location infection*.

3.4.9 Density Factor

Table 3.6 shows the density factor which is an abstraction of micro level interactions, instead of explicitly modelling the interactions that happen within a location. For example parks have a low density for a couple of reasons: agents will be rather spread out, agents usually only have interaction with a few persons at the park (their friend/family group) and the park is outside which reduces the spread of corona as well. Homes have a high density, since: most agents will be rather close by at some points of the day (for example cooking or performing other activities), agents living in the same home usually have interaction with most other housemates and it is an inside environment where the corona virus can spread more easily than outside as it can linger in the air.

3.4.10 Migration

Most persons do not stay in the same city all of their lives. They could move outside of the city for various reasons: to visit a friend, to go working or for a holiday. The away location type represents this place outside of the city. When migration is enabled, dependent on the *probability-going-abroad* variable agents move to this location. The agents at this location cannot infect each other at this location, however they can get infected based on a probability (*probability-infection-when-abroad*).

Table 3.6 The density factor of the various location types and transportation options. These values are used in the contagiousness model

Location type	Density factor
Essential shops	0.30
Homes	1.00
Hospitals	0.80
Non-essential shops	0.60
Private leisure places	0.30
Public leisure places	0.10
Public transport	0.50
Queuing	0.60
Schools	1.00
Shared cars	0.80
Walking outside	0.05
Workplaces	0.20
University faculties	0.20

The agents also have a probability of returning to the city (*probability-getting-back-when-abroad*). By default migration is turned off, however there are some scenarios that use migration.

3.5 Disease Model

This section describes how the COVID-19 epidemiological model is implemented in the simulation. It consists of (1) contagiousness and spreading of the virus, how the virus transfers from agent to agent, (2) an agent disease transition model, showing the steps, time and probability of each of the disease stages in an agent, and (3) symptom recognition of the disease. The disease model follows the Susceptible, Exposed, Infectious, Resistant (SEIR) model [4]. Most agents in the simulation start out as susceptible while some agents get exposed. The exposed agents become infectious after a couple of days. In the infectious stage the agents may infect other agents when they come into contact with them, thus putting the other agents into the exposed stage. After enough time has passed an agent will either die from the disease or become resistant. A resistant agent in our model cannot be reinfected. The following section will describe the disease stage of the agents in more detail.

3.5.1 Contagiousness and Spreading of the Virus

The virus in our model can spread from agent to agent. In our implementation the virus mainly spreads through standard interactions at locations, in busses and in shared cars. For example in a shop where an infected agent enters, during that same tick the infected agent can infect other healthy agents (see Fig. 3.3). An immune agent cannot get reinfected similar to the SEIR model.

When transferring between locations and not using a bus or a shared car, the agents can also get infected by other agents. The *away* location is an exception compared to the other locations as here agents will get infected by a random probability and not through another infected agent. We will now explain in detail how agents can infect other agents.

3.5.1.1 Contagiousness from Agent to Agent

When an infectious agent is at the same location as a healthy, non immune, agent the infectious agent can infect the healthy agent. This is dependent on the following factors:

- The amount of ticks being in contact with an infected agent
- The location's density
- The number of agents at the location
- The state of the disease (infected, healthy, immune) and severity
- Whether the agent applies social distancing

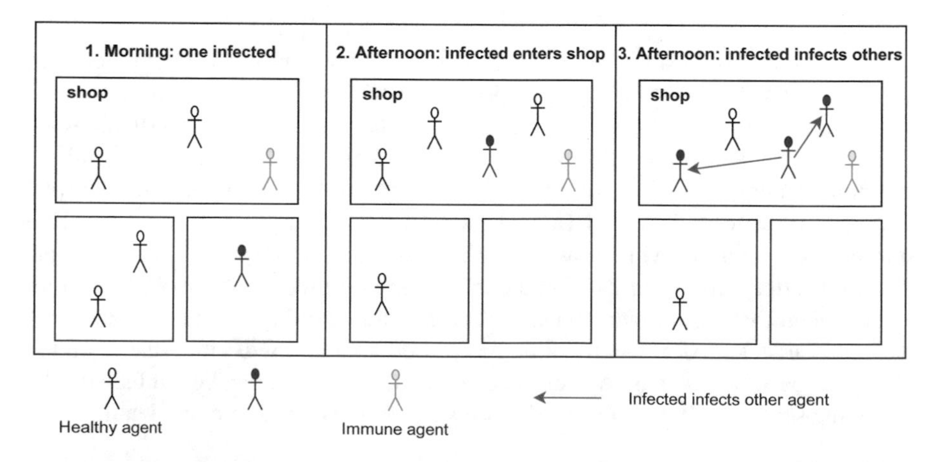

Fig. 3.3 The virus only transfers between agents at the same location. In the morning the infected agent is alone so he cannot infect other agents. In the afternoon the infected agent moves to the shop and during that tick infects two other agents

The spread of the disease is stochastically determined based on these factors. Agents at the same location have a chance of infecting each other based on the density of the locations and the infectiousness of the infected agent. The disease model has stochastic elements such as the severity of the disease, the survival chance, and the length of being in a specific disease state.

3.5.1.2 Disease Spread at a Location

The disease is spread at a location (including busses, shared cars and queues). To determine the spread for each tick at each location a list of the susceptible agents *susceptible-people* and contagious agents *contagious-people* is made. An agent is contagious when it is in any of the infected stages. For each contagious agent all the susceptible agents at that location are checked with a contamination function.

The contamination function (*risks-of-contamination*) determines whether the agent becomes infected. It requires the parameters the contagious agent, the susceptible agent and the location. It uses the *oxford-contagion-factor-between* function (explained in the next subsection) and multiplies this result with the *social-distancing-risk-mitigation-factor*. If the susceptible agent does not apply social distancing the probability will be multiplied by 1, otherwise it is multiplied by the *social-distancing-density-factor* parameter which is defined in the model setup. When this feature was setup, news were suggesting that social distancing will reduce the transmission to about 8%, therefore it is set to 0.08 in our model.

Whether social distancing is applied by the susceptible agent is dependent on the needs and current regulations. It will be explained in scenarios where it fulfils a role. The variable that indicates whether the agent performs social distancing or not is *is-I-apply-social-distancing?*.

3.5.1.3 Agent Contagiousness

For the actual contagiousness between a susceptible agent and an infected agent, we based our function on the infection dynamics in the Oxford model.[2] The function uses the following initial parameters: t is the time in days since the infected agent became infected, s_i indicates whether the infected is asymptomatic, mildy symptomatic or moderate to severely symptomatic, a_s is the age group of the susceptible agent and n is the type of location where the agents meet. The function we use is the following:

$$\lambda(t, s_i, a_s, n) = \frac{R \times S_{a_s} \times A_{s_i} \times B_n \times T_t}{I_{a_s}}$$

[2]OpenABM-Covid19: Agent-based model for modelling the COVID-19 and Contact-Tracing, https://github.com/BDI-pathogens/OpenABM-Covid19/blob/master/documentation/covid19.md.

Table 3.7 The relative susceptibility dependent on age of the agent, given by S_{a_s}

Age	Ages oxford	Values	Mean
Young	0–9, 10–19	0.71, 0.74	0.725
Student	20–29	0.79	0.79
Worker	30–39, 40–49, 50–59, 60–69	0.87, 0.98, 1.11, 1.26	1.055
Retired	70, 80+	1.45, 1.66	1.555

R scales the infection rate, in our model it is set to 11.5 which is higher than in the oxford model. We explain why it is set higher in the subsection below. $I_{a_s} = 3$ it represents the mean number of interactions at a location, this should not be confused with the contacts as contacts are an accumulation of every agent at a location. The S_{a_s} variable is the susceptibility according to age it is shown in Table 3.7[3] below. Since we have only four age categories in our model we group the Oxford model ages together and take the mean. A_{s_i} is the relative infection rate of the infected agent. This is 0.29 for asymptomatic infected, 0.48 for mildly symptomatic infected and 1 for moderate to severely symptomatic infected. B_n is the density factor of the location type. For the exact values see Table 3.6 in the previous section.

T_t is a gamma distribution of the contagiousness over time which has a *mean* of 6 and *standard deviation* of 2.5. An agent just infected will be minimally contagious, over time contagiousness is built up and after the peak contagiousness is gradually lost (see Fig. 3.4a[4]). The function is based on the function described in [5]. The actual implementation is rather than implementing a function, a set of discrete values is made that represent the upper cumulative distribution (see Fig. 3.4b). This was the most straight forward way of implementing and is possible since time in our simulation is represented by discrete values.

The final formula returning the probability of infection P that uses the calculated λ is:

$$P(t, s_i, a_s, n) = 1 - e^{-\lambda}$$

In practice the contagiousness is higher for elderly and lower for younger agents. Since young agents, as shown in the next section, have a lesser chance of getting severely sick. And the more sick an agent is the higher its contagiousness. Furthermore in Table 3.7 it is shown that older agents have higher susceptibility values.

[3]The parameters are inferred from the OpenABM-COVID-19 https://github.com/BDI-pathogens/OpenABM-Covid19/blob/master/documentation/parameters/infection_parameters.md.

[4]https://keisan.casio.com/exec/system/1180573216.

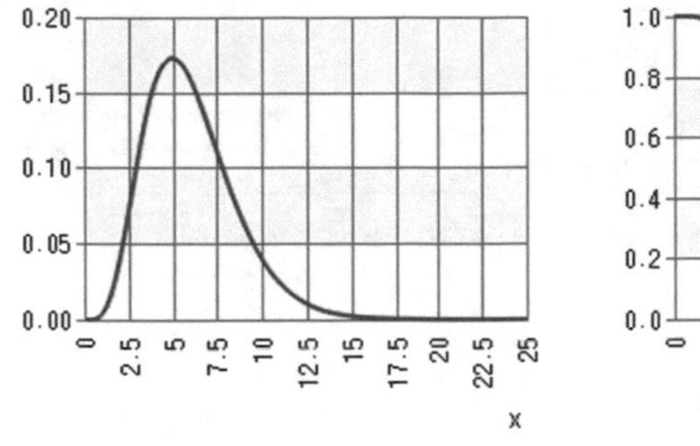

a: Gamma probability density

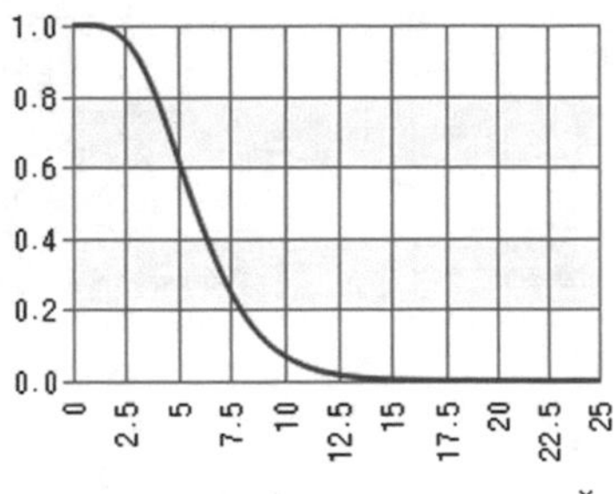

b: Gamma upper cumulative

Fig. 3.4 Left the relative contagiousness over time where x represents the day. Right the formula that represents the value of the actual implementation for a more straight forward implementation

3.5.1.4 The Contagiousness Value

Since we have a relatively small number of agents (around 1000) it is difficult to use the real statistical model of infections in our simulation. In reality in most days less than one in 1.000 persons gets corona every day. Therefore this can not (easily) be represented by a simulation with about 1000 agents. Small increases of going from 100 in 100.000 to 105 in 100.000 would just be abstracted to one agent out of 1000 in both cases. Using contagiousness levels based on the more realistic data will lead to the number of infected agents hardly ever increasing. Thus no pandemic will happen in the simulation. Therefore we increase the contagiousness to a higher level than it is in reality. In the contagiousness formula we use $R = 11.5$. This has some effect on the peak of the waves, as our simulation will have a much steeper peak and the absolute values are not comparable to real world data. We should instead evaluate the relative difference among runs within the simulation, for example with different tracking app users, where one setting could lead for example to a peak that is twice as high instead of saying for example at the peak 30% of the agents were infected (and comparing this to real data).

3.5.2 *Agent Disease Transition Model*

We replicated the disease model from an epidemiological COVID-19 model [1] to get a realistic representation of the disease and spread. This model takes into account that not everyone has the same disease severity, e.g. some agents stay asymptomatic while others get severely ill. Figure 3.5 shows the disease transitions which is divided into three disease processes. These are (1) stay asymptomatic through the whole infected period, (2) become mildly symptomatic and (3) become severely symptomatic (with possible hospitalisation and possible death). Agents start out healthy and will progress through these stages when getting infected. All agents who sur-

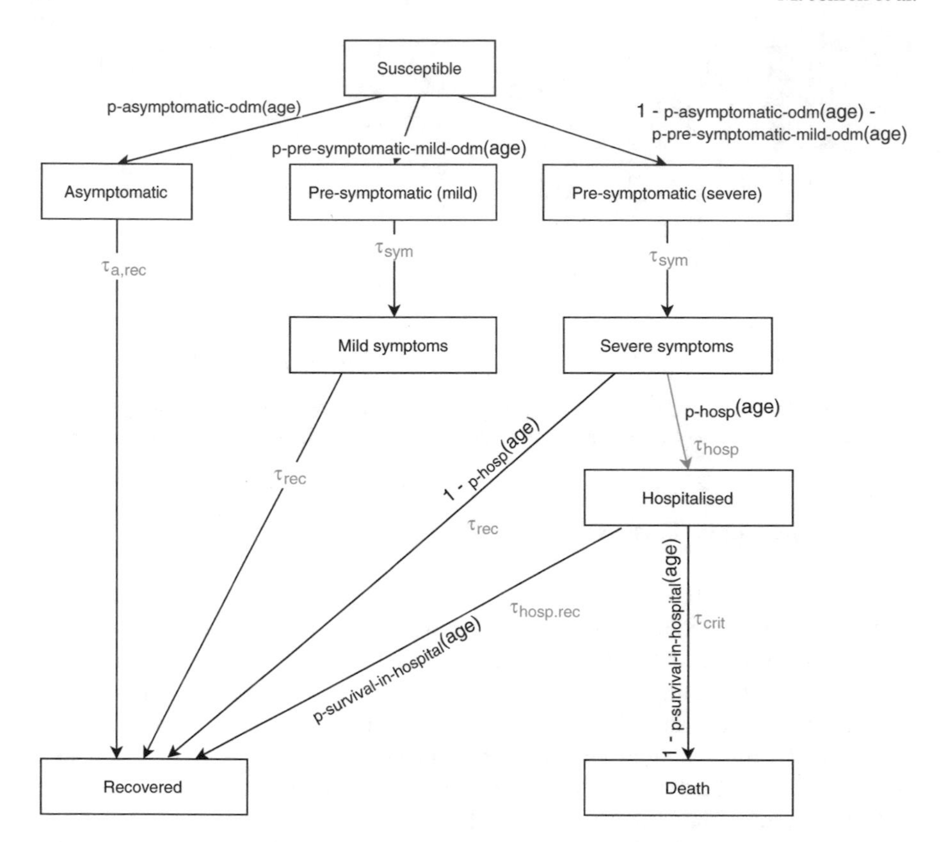

Fig. 3.5 The disease transition model used for our agents. Notice that some agents will stay asymptomatic and will only know they are infected through testing. The figure is adopted from the Oxford model [1] and adjusted to our model

vive the disease become immune after which they cannot be reinfected again for the remaining part of the simulation. Compared to the Oxford model we have removed the state of being 'Critical & ICU', because this does not really play a crucial role in our simulations and with about 1000 agents in the simulations there would usually be no agent in the ICU.

Each agent that has not been infected yet is susceptible at first. When the agent gets infected it will move to one of the three disease processes based on a probability that is dependent on the age. Figure 3.6 shows these probabilities.[5] We calculated the mean of the age categories from the Oxford model. For example the probability for a young agent to become asymptomatic, is the mean of the probabilities of becoming asymptomatic in the age group 0–9 and 10–19 from the Oxford model. The table shows that 18% of the infected agents stays asymptomatic regardless of age. While elderly agents have a much higher probability (0.48) of becoming severely

[5]The parameters for the disease transition model https://github.com/BDI-pathogens/OpenABM-Covid19/blob/master/documentation/parameters/disease_dynamics_parameters.md.

Disease probabilities for different symptom levels				
Type (netlogo name)	**Young (0-20)**	**Student (20-29)**	**Worker (30-69)**	**Elderly (70+)**
Asymp (**p-asymptomatic-odm**)	0.18, 0.18 **0.18**	**0.18**	0.18, 0.18, 0.18, 0.18 **0.18**	0.18 **0.18**
Mild (**p-pre-symptomatic-mild-odm**)	0.79, 0.79 **0.79**	**0.73**	0.68, 0.65, 0.59, 0.53 **0.6125**	0.41, 0.27 **0.34**
Severe (inferred from 1-Asymp-Mild)	**0.03**	**0.09**	**0.2075**	**0.48**
Severe that are hospitalized (**p-hosp**)	0.01, 0.03 **0.02**	**0.04**	0.06, 0.08, 0.12, 0.15 **0.1025**	0.16, 0.14 **0.15**
Hospitalised that become critical	0.05, 0.05 **0.05**	**0.05**	0.05, 0.063, 0.122, 0.274 **0.12725**	0.432, 0.709 **0.5705**
Hospitalized Critical that die	0.33, 0.25 **0.29**	**0.5**	0.5, 0.5, 0.69, 0.65 **0.585**	0.88, 1 **0.94**
Survival in hospital (**p-survival-in-hospital**)*	0.05 * 0.29 = **0.01425**	0.05 * 0.5 = **0.025**	0.12725 * 0.585 ≈ **0.074**	0.585 * 0.94 = **0.5499**

Fig. 3.6 The probabilities for transferring to a certain disease state. **p-survival-in-hospital* is determined by multiplying the 'hospitalised that become critical' and 'hospitalised critical that die' as we do not have the critical state in our model

symptomatic compared to young agents (0.48). The table also shows probabilities for agents that are severely symptomatic and will need to be hospitalised, if they are not hospitalised they recover. Further more there is a probability that the hospitalised agents will die.

Each of the disease stages has a transition time, which is shown in Table 3.8.[6] This is the time the agent stays in the disease stage before moving to the next one. We took into account the removal of ICU by combining the transfer time of the Hospitalised to 'Critical & ICU' with the 'Critical & ICU' to death transfer time (Fig. 3.5).

3.5.3 *Symptom Recognition of the Virus*

There is no testing when having symptoms because we already assume that agents who have symptoms believe (in most cases) that they have corona and stay in quar-

[6]The parameters for the disease transition model https://github.com/BDI-pathogens/OpenABM-Covid19/blob/master/documentation/parameters/disease_dynamics_parameters.md.

Table 3.8 Time span for different disease stages. $\tau_{hosp.rec}$ is calculated by taking the mean of τ_{crit} and $\tau_{crit.surv}$ from the Oxford model. τ_{crit} is calculated by taking the mean of τ_{crit} and τ_{death}

Name	Netlogo name	Mean	Standard deviation
$\tau_{a,rec}$	*tau-a-rec*	15	5
τ_{sym}	*tau-sym*	6	2.5
τ_{rec}	*tau-rec*	12	5
τ_{hosp}	*tau-hosp*	5	–
$\tau_{hosp.rec}$	*tau-hosp-rec*	$\frac{2.5+4}{2} = 3.25$	3.60555
τ_{crit}	*tau-hospital-to-death*	$\frac{2.5+6}{2} = 4.25$	2

antine. It saved us a separate testing action for most scenarios. However, this action is explicitly implemented for the track and tracing scenario.

3.5.3.1 Agent Epistemic Model

In addition to the actual disease model, an epistemic state is implemented in the agents. The agent can have the following beliefs about himself/herself: healthy, infected, or immune. However this does not necessarily have to entail the true state of the agent. An agent can believe it is healthy while actually being sick. However also the other way around. When an agent believes it has been sick and not shows any symptoms anymore, it will believe it is immune. The epistemic state is influenced by random events to occasionally create false beliefs in the agents.

3.6 Government Guidelines, Interventions and Scenarios

The simulation contains a number of government interventions. These can be used to fight the spread of the virus and will be explored in the various scenarios. Table 3.9 shows possible interventions present in our model.

Closing Specific Locations

Closing locations can be beneficial to reduce the spread of the virus, as agents get less contact with each other. This intervention is complex, because closing locations may have all kinds of consequences.

Universities and "workplaces" can be closed simply by flipping a switch (*close-universities?* and *close-workplaces?* respectively).

Schools can be closed based on several different criteria. It is possible to close schools as soon as one reported case of COVID-19 exists in a school. Furthermore,

Table 3.9 Policy description

Policy	Description
Closing specific locations	Closing locations, such as schools, universities or non-essential shops, can help to reduce the spread of the virus
Home-office	The agents should work from home. This is only possible for agents who work at some workplaces
Lockdown and global Quarantine	Every citizen is supposed to stay at home for a certain time period. This is to slow the spread of the virus, however agents can still break quarantine
Paying wages	Paying wages of workers of whom their workplaces do not have sufficient money can help reduce the financial stress on them and prevent the agents from going bankrupt
Phasing out strategies	Different phases with different conditions can be defined. It can then be specified which interventions are listed in which phase
Self-Isolation and Self-Quarantine	Agents that have COVID-19 stay at home (including home-office), but are still allowed to go shopping
Smartphone application	This is implemented as a tracking and tracing app that saves the contacts of an agent. When one of the contacts has the virus all the other contacts get a notification and may be tested
Social distancing	Agents are required to keep a certain distance to other agents, e.g. 2 m. If this is enabled there is a smaller risk of agents infecting each other
Testing for disease	This could be random testing but also trace testing. Agents tested positive are put into quarantine

it is possible to close schools based on the ratio of infected agents. This can be set between 0% of the population are infected and 100% of the population are infected.

Non-essential shops can be closed based on the ratio of infected agents using the slider *ratio-omniscious-infected-that-trigger-non-essential-closing-measure*, and also after a settable amount of days has past (*#days-trigger-non-essential-business-closing-measure*).

Home Office

People who work at the location "workplaces" are able to work from home. This policy is also called *working from home*. When working from home is the case, the productivity of the worker can be changed, as it can be assumed that a person is less productive when working from home. Nonetheless, this is a slider in our model (*productivity-at-home*) and users can also explore what would happen, if agents would be more productive at home than at work. Furthermore, it can be possible that agents break this recommendation and still go to work. However, adhering to government guidelines is closely coupled to the needs model and therefore, not too many agents will break these interventions. Working from home can be activated using the switch *recommend-working-from-home?*.

Lockdown and Quarantine

A global self-isolation (quarantine) can be enforced which means that everyone has to stay at home for 35 days. The threshold in our simulation is 2%. This measure gets activated once more than 2% of the population are infected or not, using the switch *all-self-isolate-for-35-days-when-first-hitting-2%-infected?*. Whether agents comply to the rule depends on their needs model. In general agents will comply to this measure but will deviate when some needs get very high. To further decrease the risk that agents break the self-isolation for shopping, it can be chosen to have online food delivery (*food-delivered-to-isolators?*). When activating this additional measure, agents do not have to go to essential shops anymore, as their necessities will be delivered to them. Another further option which exists is that retired agents can be forced to stay at home, even after the initial quarantine period of 35 days is over, using the switch (*keep-retired-quarantined-forever-if-global-quarantine-is-fired-global-measure?*). This can be beneficial, as retired agents are at higher risk of developing critical symptoms and die from COVID-19 compared to other age groups.

It is also possible to enact a hard lockdown. This hard lockdown can be active the whole time it is selected (total-lockdown option). It can also be active once more than 10% of the population are infected and then lockdown can end once less than 5% of the population are infected (lockdown-10-5 option). These two options, in addition to the option never, can be selected from the drop-down menu *global-confinement-measures*. During such a hard lockdown, only essential-shops and the hospital are open. All other locations are closed.

Paying Wages

The government can decide to pay the wages of workers when the workplace is out of money (*government-pays-wages?*). This can help to prevent workplaces from going bankrupt. The option can be activated or deactivated via a switch (*government-pays-wages?*) in the simulation interface. The agents get a percentage of their original wage (*ratio-of-wage-paid-by-the-government*).

Phasing out Strategies

It is not only necessary to investigate which strategies help to reduce the spread of the virus, but also how to exit from the strategies to enable the transition to a "normal" life again. To do so, different phases exist. The exit strategies are divided into three different phases with an additional ongoing crisis phase. It can then be selected which conditions are used for each phase. Examples of possible condition are: the number of infections decrease over a certain amount of days, a certain amount of days has passed since being in one phase (regardless of the amount of infections), or percentage of immune agents. All option can be found in the drop-down menu *condition-phasing-out*.

It can also be investigated what happens if the crisis is acknowledged or not. The options can be selected from the drop-down menu *condition-for-acknowledging-the-crisis* and the options are *never* or *"ratio infected>2%"*.

The different phasing out strategies are investigated in Chap. 10. Since this scenario has a lot of scenario specific parameters that are not shared by the other scenarios presented in this book, they will be explained in Chap. 10 as well.

Self-isolation and Self-quarantine

Agents stay at home when they are infected with COVID-19. A variety of different options to modify self-isolation exists. In addition to the agent, it is possible to force the household of an infected household member into isolation as well (*is-infected-and-their-families-requested-to-stay-at-home?*). In these cases we can avoid that agents break isolation for shopping, by using the option for online food delivery. This can be activated via the switch *food-delivered-to-isolators?*.

Not all people that have symptoms will go into self quarantine. To represent this phenomenon it is also possible to determine the percentage of the agents that goes into self-quarantine through (*ratio-self-quarantining-when-symptomatic*) or if one of their household members is symptomatic (*ratio-self-quarantining-when-a-family-member-is-symptomatic*.

It is also possible to observe what would happen if agents rigidly comply to the quarantine and do not break it. This can be done when using the switch *is-psychorigidly-staying-at-home-when-quarantining?*. Using this option, agents only have the option to stay at home and do not do anything else.

Social Distancing

Social distancing is used to reduce the density at the different locations. This means that the agents keep a greater distance to each other. This reduces the risk of getting infected. The density factor describes the local proximity at a specific location. The factor can be set manually, using the slider *social-distancing-density-factor*. Each location has a specific density factor which is set based on the size (compared to the real world). For example, universities are usually bigger places and have therefore a lower density factor, whereas shops are smaller places and have therefore a higher density factor. When using social distancing, the density factor for all places is set to the same value.

Social distancing can be automatically activated based on the proportion of infected agents. This can vary between 0% (i. e. social distancing is always active) and 100% (i. e. social distancing is only active if all of the population are infected). The slider to regulate this is called *ratio-omniscious-infected-that-trigger-social-distancing-measure*.

Scenarios

The policies are used in fixed combinations in the scenarios. Scenarios are simulation settings to analyse a specific country/strategy/aspect based on real world cases. The Netlogo code contains some scenario presets that load the relevant variables when selecting a scenario. A scenario can be selected with the *preset-scenario* list, then the button *load-scenario-specific-parameter-settings* loads the parameters and prepares the program for running the scenario. More specifically, first the baseline parameters are set. This will set the country (default = "Great Britain"), the household distribution, the disease model, migration model, contagion model, transport model, economic model, the agent values and needs, the social network model, distributes the workers and sets the interventions. Then the scenario specific parameters are set which will overwrite the baseline parameters where necessary. These scenarios typically set one or more of the following:

- The chosen country which will be used for the household distribution and the cultural model
- Measures or interventions that are active or inactive
- Worker distribution, how many workers work at specific locations
- Household number, which influences the number of agents
- Amount of locations per type and their density

When the scenario specific settings are set the household distributions and the dimension scores are loaded from a .csv file. This finalises the parameter loading process which allows the model to be run.

In the following chapters we will describe the results of several of these scenarios. Chapter 5 investigates the interventions of closing schools and the recommendation of working from home. The different testing possibilities and the difference between isolating only the infected person compared to isolating the household of an infected member are explored in Chap. 6. Chapter 8 looks into the cultural effects on the different interventions and Chap. 9 investigates the economical consequences. Finally, Chap. 10 explores different phasing out strategies.

General Remarks

In general, it is possible to combine the different governmental policies, as has been for example done in Chap. 10. Furthermore, it is possible to specify when agents become aware of existing measures. This can be done by a click on a button (*inform-people-of-measures*) to inform agents of the measures. Furthermore, the ratio of agents watching the news can be determined. New interventions are usually communicated via news and therefore agents become aware of them when watching the news. The ratio of agents watching the news can be determined using a slider and set between 0% and 100% (*percentage-news-watchers*). If not everyone is watching the news, the information of social distancing gets also distributed via the social network of an agent. An agent becomes aware of social distancing, if a member of the network is also doing social distancing. In addition, switches allow the user to determine, if agents are aware of working from home (*Aware-of-working-at-home-at-start-of-simulation?*) or social distancing at the start of the simulation (*Aware-of-social-distancing-at-start-of-simulation?*) respectively.

3.7 The Agents

In the ASSOCC model, the **agents** are represented by the set $\mathcal{AGT}=\{a_1, ...a_n\}$. We use a_i or just a to refer to an agent in $\mathcal{AGT}$. Each agent has an agent-related profile that contains the variables shown in Table 3.10.

We divided the ages of agents in four groups that loosely connect to the type of activity they perform normally, their susceptibility for getting the COVID-19 virus , conditions for spreading it, and how they are affected by potentially getting infected with the virus. We have that $age(a) \in \{youth, student, worker, retired\}$. The four groups represent children (0–19), students (20–29), workers (30–69), and retirees (70+). The names of the groups were chosen to stereotypically describe the agents in this group in relation to their behaviours and needs. Primarily, the age group determines where the agents spend their day. Whereas children go to school, students

Table 3.10 Important agent variables

Type	Implications
$age(a)$	The values are: youth (0–19), student (20–29), worker (30–69), retired (70+)
$home(a)$	Location of the home of the agent
$my\_hospital(a)$	Location of the hospital the agent will go to by default
$essential\_shops\_of(a)$	Denoting the essential shops the agent uses
$luxury\_shops\_of(a)$	Denoting the non-essential shops the agent uses
$public\_leisure\_places\_of(a)$	Indicating parks etc. the agent frequents
$private\_leisure\_places\_of(a)$	Indicating pubs, friends, etc. an agent visits
w_a^n for every need n	Indicating the importance of the different needs for an agent
$daily\_activity(a)$	Which denotes the main activity of an agent, like "work", "learn", etc.
$location\_daily\_activity(a)$	Which is the default location of the agent's daily activities
$can\_work\_from\_home(a)$	Indicating whether an agent can work from home or not

go to the university, workers go to the office, and retirees do not have any obligations. Instead, they might decide to start their day with a more relaxing activity such as taking a walk. Another important part of daily routines are shopping and leisure activities, where frequency and organisation of these activities also depends on the respective age group. The agents primarily have their decision making based on their schedule combined with a need based evaluation. The needs are adapted by the chosen culture, which has been explained in Sect. 3.7.3.

By fixing the different locations that an agent frequents we avoid having to decide and plan routes to a location for each activity an agent wants to do. Thus if the agent wants to get food it will go to one of the essential shops in its list and does not have to decide on which shop to go to and how to get there.

The interactions consist of agents being together at the same gathering point. This has two consequences for the agents (1) the needs are influenced and (2) the disease can be spread. Spreading the disease happens with a certain probability when one agent is infectious and the other agents do not have the disease and are not immune. When there are more infected agents the probability increases. The agents also have interaction with the places, at a shop the agent can buy food at a workplace the agent can work. There is no possibility for an agent to perform actions that directly change parameters of other agents.

3.7.1 Locations and Daily Schedule

The locations are the places where agents go to based on their daily schedule and needs (see also Table 3.1). The agents can go to a variety of locations for performing activities in it. The set of locations is represented by $\mathcal{L}$. Locations are partitioned in types where $\mathcal{L} = homes \cup workplaces \cup hospitals \cup schools \cup universities \cup essential\_shops \cup luxury\_shops \cup public\_leisure \cup private\_leisure$. The *away* location is not included here as agents will only go there based on a probability set by the user rather than based on their needs decision system.

A typical day is divided into four parts: morning, afternoon, evening and night. They all represent one tick formulated by **Tick (timestep):** $t \in \mathbb{N}$ represents the simulation tick (index of a simulation round). We write c_a^t to refer to the decision context of agent a at time t. We use c^t if the agent a is clear. An agent will often be at a location for only one tick at a time. One tick is also the least amount of ticks an agent spends at a gathering point where the agent has gone to. For example it is not possible for the agent to pass by the shop after work (afternoon) and then be home just before the evening. Instead if the agent goes to the shop after working (afternoon), it will spend the evening tick at the shop and afterwards go home. This is done for simplicity purposes.

Activities are represented by $\mathcal{A} = \{learn, rest, relaxing, work, get\_treated, shopping, get\_tested\}$. Each activity in this set can be performed with social distancing or not, which can be represented by the subscript $_{sd}$ and $_{nsd}$. E.g. $work_{sd}$ is working while keeping social distance. Each non-retired agent a has a usual daily activity

Table 3.11 The daily schedule of the agents dependent on the age of the agent

Functional group	Morning	Afternoon	Evening	Night
Youth (workdays)	School	School	Home/Leisure	Home
Youth (weekends)	Home/Leisure	Home/Leisure	Home/Leisure	Home
Student (workdays)	University	University	Free choice	Home
Student (weekends)	Free choice	Free choice	Free choice	Home
Worker (workdays)	Work	Work	Free choice	Home
Worker (weekends)	Free choice	Free choice	Free choice	Home
Retired (always)	Free choice	Free choice	Free choice	Home

represented by $daily\_activity(a)$. This activity depends on the age of the agent: $daily\_activity(a) = learn$ if $age(a) \in \{youth, student\}$, $daily\_activity(a) = work$ if $age(a) = worker$ and $daily\_activity(a)$ is undefined for $age(a) = retired$.

Because the results of many activities depend on the location where they are performed, we introduce *located activities*. Located activities are the set of all possible activities in all possible locations with or without social distancing, is represented by $\mathcal{LA} = \mathcal{L} \times \mathcal{A}$. The located activity currently performed in context c is $current\_LA(c)$.

The agents have an activity schedule dependent on their age group and type of day shown in Table 3.11. The schedules differentiate workdays and weekends for everyone except retired. On the weekends all agents have free choice, this means activities such as leisure at either private or public leisure, essential or non-essential shopping or being at home. The youth does not go shopping and can therefore only be at home or do leisure activities when they do not go to school. The schedule can of course change due to government policies, or when an agent or agents from the same household as the agent get infected.

3.7.2 Social Network

An agent is part of a social network. Two types of social networks are distinguished: the friends and family network. The generation of friend networks is based on the homophily principle, so similarities in values and age are taken into account. At the setup of the simulation the euclidean similarity with other agents at the same private leisure place, school, or university is calculated. Then, the seven agents who are most similar and have the same age are chosen to become friends. Other agents in such a network can be referred to as friends of each other. A friend network can also be setup at random. The variable *percentage-of-agents-with-random-link* determines how many agents have a random friend network instead of a similar value based network. An agent can be part of multiple social networks. Given the fact that agents can be part of multiple social networks, it can be possible that agents have more than seven friends. The family consists of all the agents that live in the same household.

Agents have a memory to keep track of what others in their network previously did. The memory consists of actions in combination with motives. These are taken into account when the expected satisfaction gains for conformity and belonging are computed. We distinguish two types of memories: one for the behaviour of the social network during weekdays and one for behaviour during the weekends, as action during weekdays differ from actions during the weekdays. Each memory consists of four elements, one for each slice of the day. Only the action that is chosen by the majority is saved.

3.7.3 Culture and Values

The ASSOCC model also contains a module that is built specifically to simulate the effects that culture may have on the spread and mortality of the coronavirus. This cultural submodel also enables users to explore the cross-country differences in the development and impact of the virus.

It is hypothesised that culture influences the values that agents hold. Specifically, culture informs the ranking or relative importance ascribed to a particular set of values by a given agent. Culture serves as a blueprint for agent's to base their value systems on. Culture is operationalised using the Hofstede Dimensions model and is country-based. Values are operationalised using the Basic Value Theory of Schwartz. Linking the country-level Hofstede Dimensions with agent-level Schwartz Values is done on the basis of theoretical work, and empirical data. The degree to which agents base their value systems on the country-based scores of the Hofstede Dimensions is moderated by the country's cultural tightness. Cultural tightness dictates the structural similarity in value systems of agents within a population and is calibrated on the basis of findings presented by [6] and [7]. In doing so, ASSOCC is able to represent both inter- and intra-cultural variation in agent value systems. The cultural profiles for the countries can be found in Chap. 8 in Table 8.7.

As noted, any given agent holds a set of values which dictates what its perceptions are of desirable states of reality (i.e. how reality ought to be). Values determine what an agent finds important, desirable or valuable. Values influence the agent's needs which in turn determine the agent's behaviour. Ultimately, the agent's behaviour determines how the virus is able to spread within an agent population over a certain span of time. Moreover, values may influence the social network topology via a process of preferential attachment (see section: Social Network Model), thereby influencing the way in which the virus percolates through the agent population. The cultural model is extensively described in Chap. 8.

3.7.4 Decision Making and Needs

The agents make their decisions based on their needs (Table 3.12). The abstract needs model introduced in Chap. 2 has been made more concrete into the following twelve needs in the implementation: $\mathcal{N}$={*food_safety*, *fin_survival*, *sleep*, *health*, *conformity*, *compliance*, *risk_avoid*, *fin_stability*, *belonging*, *autonomy*, *luxury*, *leisure*}. The needs that the agents have, are modelled using the homeostatic model described in Chap. 2. This means that every need is represented by a number between 0 and 1, which slowly decreases every tick using a decay function. The needs are inspired by Maslow's needs, however the strict hierarchical structure is not used. With the exception of physical needs (sleep, health, financial survival and food safety) which do get a higher weighting than the rest. The different parameters of the needs are stochastically fixed within some boundaries for each agent based on the cultural background of a scenario.

Table 3.12 The needs of the agents

Need	Description
autonomy	The need for autonomy and self-direction. This freedom of choice can be limited by government interventions
belonging	The belonging need, i.e. meeting friends and relatives
compliance	The need for complying with formal instructions, notably arising from the government and contractual obligations (e.g. work)
conformity	The need for conformity with regards to one's social network. Or in other words the social need of choosing an action that the majority of agents in its network has previously chosen
fin_stability	Represents the need for financial stability, with an income matching the expenses
fin_survival	The need of having enough money to buy food and other essentials to survive
food_safety	The need of having enough food at home, with decreasing reward as we narrow two weeks
health	The need for being in good health. When agents believe that they are infected, they want to either stay home or go to the hospital
leisure	The need of agents to have leisure time and relaxing. Resting at home is possible as well, but gives less satisfaction when compared to relaxing at a different location
luxury	The need for acquiring luxury goods
risk_avoid	The need for avoiding taking risks. The satisfaction of this need depends on the contagion status of the agent and the amount of agents at a gathering point. Furthermore, social-distancing is taken into account
sleep	The need of an agent to sleep, which is satisfied every night by sleeping. Furthermore, it is modelled that sleep when being sick is not as restful as when being healthy

The needs influence the decisions of the agents on which activity to perform next. The deliberation cycle starts with determining all possible actions at that moment. The action that has the highest expected combined(!) need satisfaction is executed. In general this might seem an inefficient way to determine the best action. However, not all actions are available at each moment and/or location. For example going to the shop when the shop is closed (at night or due to restrictions). As a consequence in the ASSOCC framework, built for the COVID crisis situation, the number of possible actions at each moment is extremely limited (around 2–5 maximum) and thus this set up is computationally more efficient than starting from e.g. the most pressing needs. Because agents look at the satisfaction of the combined needs the mechanism prevents agents from going to leisure places during work times (most of the times). The activity of working satisfies many needs such as financial needs, compliance, belonging (if friends are there and autonomy) and thus is often a good option if available. So only in exceptional situations will an agent not work.

Because one of the needs is the need for compliance, this system allows for both rule following, and rule breaking. If you adhere to a rule, the value of your need for compliance goes up, but if you break it, the value goes down. This means that agents can decide to break a rule if their highest desired need cannot be satisfied within the currently given rules. Thus often different needs interfere with each other and need to be balanced over time. Furthermore, current living situations, like living alone or working at home, will influence the needs of an agent. If an agent lives alone and works at home it will have very little contact with other agents and thus its need for belonging will grow over time, driving the agent to go out shopping or to a leisure activity.

The needs model is the main drive of the behaviour of the agents. This model is calibrated such that the agents mostly satisfy all their needs during their normal daily life patterns. However, when restrictions limit their behaviour or disease limits their capabilities, some needs may not be easily satisfiable. Depending on their needs, the agents will adapt their behaviour, some agents will stay in quarantine for the required amount of time, others will sneak out and go shopping (to satisfy the need for luxury goods) or meet friends (to satisfy belonging).

The inherent objective of an agent is to get a high quality of life. The quality of life is defined as the weighted average of the need satisfaction. Thus when the agent is able to satisfy all the needs all of the time it has a high quality of life. The agent does not have explicit goals but will get itself into situations where it can satisfy its needs, so for example the agent will work to get some money to be able to buy food and shop other goods (food safety or self-esteem needs). To assess the welfare of the whole population the averages of the needs can be seen in the interface.

Appendix C shows all connections between actions and the needs it fulfils. In the sections below we will describe in more detail how actions actually satisfy some needs and which conditions influence this satisfaction.

3.7.5 Agent Decision Model

The main decision consists in selecting which activity to perform in the next time step (which represents one fourth of a day, i.e. six simulated hours), where to perform it and whether to apply social distancing. As described in Sect. 2.3.1, needs are a central factor that drive this decision. As such, the ASSOCC agent decision model emphasises the representation of the dynamics of needs and their influence on decisions.

Contexts: A decision context of agent a is represented by a tuple c_a^t, which stores all information at the current time t in the context of the agent a. From a formal standpoint, the context is a large tuple (containing things like whether the agent is feeling sick, what is the time of the day). For each position in this tuple we have a function corresponding to the type of information of that position. Thus we use e.g. $food(c_a^t)$ (or $food(a)$ for short) to represent the amount of food agent a has at home. When the information of the context is agent independent such as e.g. the current time of the day we use $hour(c^t)$ and omit the subscript a.

3.7.5.1 Overview

As stated before, the ASSOCC decision model is built upon on the classic *generate-and-select* AI decision paradigm: the agent first generates all feasible options, then selects the one that best satisfies its needs. This decision process is coupled with the execution of the agents' actions and of world-dynamics, which influences the decision context of future decisions –here with the inclusion of the dynamics of need satisfaction resulting from the agents' actions. Whereas necessarily incomplete with regard to the details of the intrinsic complexity of real-world needs, the ASSOCC model still covers the core aspects of needs detailed in Sect. 2.3.5:

1. Decisions tend to satisfy needs
2. Need satisfaction fluctuates over time and tends to decrease when no specific action is taken
3. More deprived needs have greater influence on decisions
4. The added satisfaction decreases over time (diminishing return)
5. Needs have various degrees of importance and more important ones tend to be satisfied first
6. The possible distinctions that can occur between expected and actual need satisfaction

The rest of this section explains the decision making process: first the current satisfaction of needs is explained. The second part describes how the agents generate the alternatives to be considered. Finally, the selection is presented which introduces the process for evaluating and comparing the alternatives.

3.7.5.2 Current Satisfaction of Needs

As described in Chap. 2, the level of satisfaction of needs evolves over time, both due to internal psychological dynamics (e.g. lacking friends) and external contingencies (e.g. lacking food). The definition of need satisfaction is given in two parts. One for the intrinsic needs for which satisfaction decreases over time with a decay factor that depends on the specific need. For the extrinsic needs there is a separate formula per need as their decrease depends on explicit actions being taken or not (e.g. eating).

Intrinsic needs
The intrinsic needs are the following seven needs:
$\{sleep, conformity, compliance, risk\_avoid, belonging, autonomy, luxury\}$.

For example the need satisfaction of the sleep need will decrease over time during the day and evening when the agent is not sleeping. The more time passes without sleep the lower the need satisfaction. With the chosen decay rate it means agents will have a very low sleep satisfaction at the end of the night, making them choose sleep as preferred activity in the night. The satisfaction of the intrinsic needs of agent a at time t for need n ($CNS^n(c_a^t)$) is defined as:

$$CNS^n(c_a^t) = \gamma_n \times CNS^n(c_a^{t-1}) + SNSG^n(c_a^{t-1}, prev(c_a^{t-1})) \times \\ successful(c_a^{t-1}, prev(c_a^{t-1}))$$

This states that the current satisfaction of a need is the sum of the satisfaction of that need in the step before ($t-1$) multiplied by a decay factor γ_n and of the gains in satisfaction ($SNSG$) of the successful actions that contribute to that need in the last time step. Here $prev(c_a^t)$ is the located action performed by a at time t and $SNSG^n(c_a^t, la)$ is the satisfaction gain obtained for having performed la in context c_a^t. This satisfaction gain is only counted when the action was successful. I.e. $successful(c^t, la) = 1$ if la was successful in round t and $successful(c^t, la) = 0$ if the action failed (e.g. trying to reach a closed shop). If no actions contributed to the satisfaction of need n, then the satisfaction decreases over time.

The decay factors are different per need. Specifically:
$\gamma_{sleep} = \gamma_{conformity} = \gamma_{compliance} = 0.8$
$\gamma_{risk\_avoid} = 0.95$
$\gamma_{belonging} = \gamma_{luxury} = \gamma_{leisure} = \gamma_{autonomy} = 0.99$

The decay factor of *sleep* and *conformity* is stronger than the other needs. The satisfaction of the *sleep* need needs a stronger decay such that agents will be motivated to sleep every night. The *conformity* need's decay is stronger since it can apply to situations that are relevant almost every tick, e.g.: conforming to one's network and social distancing. Since conformity can be satisfied almost every tick the decay rate should also be strong such that in the next tick it also becomes a relevant need. The need *risk_avoid* is influenced by the disease model and social distancing, as it is also a more central need it has a stronger decay factor. The other needs are more occasional and do not have to be satisfied every tick, e.g. the luxury need can be satisfied a few times a week instead.

Extrinsic needs
The satisfaction of $\{food\_safety, fin\_survival, fin\_stability, health\}$ is tightly related to extrinsic, external, or bodily factors. For instance, the satisfaction of the need of feeling safe regarding food is directly related to the quantity of food in stock rather than a feeling of not having enjoyed a meal since a relatively long time: observation of the world (e.g. discovering reserves to be empty) can cause immediate rise and fall of the satisfaction of such needs. The extrinsic needs are defined using the following functions.

Food Safety:

$$CNS^{food\_safety}(c_a^t) = \frac{food(c_a^t)}{full\_reserve(c_a^t)}$$

where $food(c_a^t) \in \mathbb{N}$ represents the rations of food in the household of a at time t to be used for the whole household. $full\_reserve(c_a^t) \in \mathbb{N}$ is taken to be the amount of people in the household times 14 (having enough stock for 14 days for the whole household). This number is almost constant, but can change when persons in the household die or are in hospital for some weeks.

Financial Survival:

$$CNS^{fin\_survival}(c_a^t) = \frac{money(c_a^t)}{cost\_fully\_restocking\_food(c_a^t)}$$

where $money(c_a^t)$ represents the amount of available money of the agent a at time t to buy food (money in the bank account, minus fixed costs like rent). $cost\_fully\_restocking\_food(c_a^t)$ represents the cost for restocking food for the next two weeks, defined as $cost\_fully\_restocking\_food(c_a^t) = cost\_per\_item\_in\_es \times full\_reserve(c_a^t)$ where: $cost\_per\_item\_in\_es$ represents the cost per item in essential shops. Note that the variable $cost\_fully\_restocking\_food(c_a^t)$ can never be 0 as $full\_reserve(c_a^t) \geq 0$ due to having at least one person in the household of a and $cost\_per\_item\_in\_es > 0$.

Financial Stability:

$$CNS^{fin\_stability}(c_a^t) = \frac{money(c_a^t)}{wealth\_standard(c_a^t)}$$

where $wealth\_standard(c_a^t)$ represents how much money this agent possesses usually and wishes to possess for feeling stable. This parameter is stochastically fixed for each agent. More complex models could be used, but this simple solution appears to be enough for the simulation of the COVID crisis. The satisfaction fluctuates very slowly over time depending on the current monetary value of the agent. This need is only influenced by the amount of money and not by other aspects such as whether the agent safes money, owns a house or builds pension. Basically, these elements would only be needed to incorporated if the simulation would run over several years where these elements start making a difference between agents.

Health:

$$CNS^{health}(c_a^t) = health\_status\_factor(c_a^t) \times self\_care\_factor(c_a^t)$$

where $health\_status\_factor(c_a^t)$ represents the component of satisfaction due to the severity of the symptoms experienced by the agent and $self\_care\_factor(c_a^t)$ represents the component of satisfaction due to taking care of oneself when being sick. The $health\_status\_factor(c_a^t)$ has three levels: 1, 0.5, 0.2, corresponding to a healthy agent, a sick agent and a severely sick agent. These are the states connected to our disease model, as infected agents are either asymptomatic, mildly symptomatic or severely symptomatic.

- $health\_status\_factor(c_a^t) = 1$ if not $bel\_sick(c_a^t)$;
- $health\_status\_factor(c_a^t) = 0.5$ if $bel\_sick(c_a^t)$ but not $exp\_critical\_symptoms(c_a^t)$;
- $health\_status\_factor(c_a^t) = 0.2$ if $exp\_critical\_symptoms(c_a^t)$;

where $exp\_critical\_symptoms(c_a^t) \in \{\top, \bot\}$ represents whether critical symptoms are being observed and $bel\_sick(c_a^t)$ represents whether the agent believes to be sick. $bel\_sick(c_a^t)$ is $\top$ if for any t' with $t' > t - 14$, $experiencing\_symptoms(c_a^{t'}) \vee positive\_test\_received(c_a^{t'})$. $experiencing\_symptoms(c) = \top$ if the agent is in the state of the disease model where it is experiencing mild or critical symptoms. In specific scenarios this can also be caused by other diseases (flu or colds) with similar symptoms.

The $self\_care\_factor(c_a^t)$ represents whether the agent is resting at home or is in the hospital being treated while being ill. The self care factor is lower when the agent is sick and resting at home. If the agent is sick and not resting at home the self care factor is the lowest.

- $self\_care\_factor(c_a^t) = 1$ if the agent is not sick or if $current\_LA(c_a^t) = (my\_hospital(a), rest)$
- $self\_care\_factor(c_a^t) = 0.5$ if $current\_LA(c_a^t) = (home(a), rest)$
- $self\_care\_factor(c_a^t) = 0.2$ otherwise;
  where $my\_hospital(a) \in hospitals$ represents the preferred hospital location of the agent.

Having defined how the current satisfaction level of each need is calculated, we now proceed to describe how the agent generates the possible activities that it can perform on its current location. After that we can check which of these possible activities optimises the overall need satisfaction of the agent.

3.7.5.3 Generating Considered Located Activities

In general the next action of an agent is chosen as the action that will give most gain to the needs of the agent given the current context and weights of the needs. However, we do not have to consider all possible actions at every point in time. In many contexts the agent can only perform one or two actions out of all possible actions, e.g. at night time the agent by default will sleep. Thus we first create the set of possible actions that should be considered. $considered\_la(c_a^t)$ represents the set of activities being considered given a decision context c_a^t, where:

$$considered\_la(c_a^t) = available\_la(c_a^t) \setminus impossible\_la(c_a^t)$$

Impossible located actions are meant to filter out activities that are impossible because of locations that are closed due to some restriction or activities that cannot be performed while attending children.

$$impossible\_la(c_a^t) = \{(l, act) | l \notin open(c_a^t) \vee \neg performable\_with\_child(l, act)\}$$

$open(c_a^t)$ is the set of locations agent a expects to be open in the current context. Workplaces, schools, universities, luxury shops, and private leisure areas are assumed to be closed in case of lockdown. Some of these activities are also shut down by specific restrictions. Workplaces, schools, universities, essential shops, luxury shops, and private leisure areas are assumed to be closed during the night and workplaces, schools and universities are closed during the weekend.
$performable\_with\_child(l, a) = \perp$ for activities that cannot be performed when currently handling a dependable, e.g. attending children cannot be done in workplaces. Other activities such as for example shopping or going to leisure places can be done while attending children.

Available actions

The set $available\_la(c)$ is defined by the union of:

- Home activities:

$$\{(home(a), act) | act = rest \vee (act = work \wedge can\_work\_from\_home(a))\}$$

- Hospital activities:
  - Get a treatment: $\{(my\_hospital(a), get\_treated) | bel\_sick(c_a^t)\}$,
  - Get tested: $\{(my\_hospital(a), get\_tested)\}$

- Working:

$$(loc, work)|location\_daily\_activity(a) = loc$$

  Note that this also includes working at hospital, schools, universities and shops.
- School and university activities:

$$(loc, learn)|daily\_activity(a) = learn \wedge location\_daily\_activity(a) = loc$$

- Shop activities:

$$\{(sh, shopping)|sh \in essential\_shops\_of(a) \cup luxury\_shops\_of(a) \\ \wedge \neg is\_working\_time(c_a^t)\}$$

- Leisure areas, relaxing:

$$\{(l, relaxing)|\ l \in public\_leisure\_places\_of(a) \cup \\ private\_leisure\_places\_of(a) \wedge \neg is\_working\_time(c_a^t)\}$$

3.7.5.4 Selecting the Next Located Activity

Given the available actions as described in the previous section the agent selects the located action in $considered\_la(c_a^t)$ that is expected to maximise the need satisfaction given the context c_a^t. Formally:

$$la = argmax_{la \in considered\_la(c_a^t)} WSNSG(c_a^t, la)$$

where $WSNSG(c_a^t, la)$ is the total Weighted Subjective Need Satisfaction Gain expected to be acquired by performing la. $WSNSG$ combines all the needs using the following weighted sum:

$$WSNSG(c_a^t, la) = \sum_{n \in \mathcal{N}} w_a^n \times SNSG^n(c_a^t, la)$$

where w_a^n represents the relative importance of a need, defined based on the type of need (e.g. survival needs are more important than luxury needs), agent culture (average relative preference) and agent personality (moderate random deviations). $SNSG^n(c_a^t, la)$ is the subjective need satisfaction gain expected to be acquired by a for performing la in context c_a^t. $SNSG^n(c_a^t, la)$ combines an objective reward for performing la with a factor representing diminishing return due to the current satisfaction of n as:

$$SNSG^n(c_a^t, la) = ONSG^n(c_a^t, la) \times (1 - CNS^n(c_a^t))$$

where $ONSG^n(c_a^t, la) \in [0, 1]$ represents the objective satisfaction for need n expected to be acquired by a in the context c_a^t and $CNS^n(c_a^t) \in [0, 1]$ represents the current level of satisfaction for agent a in its context c_a^t. The rest of this section defines $ONSG^n(c, la)$.

3.7.5.5 Expected Reward per Activity

The description of the Objective Need Satisfaction Gain (ONSG) can be grouped around needs or activities. As the simulation has a smaller array of activities than needs and because the influence of activities over needs tends to be uniform across agents, we describe per activity how they satisfy the various needs.

The following definitions should not be seen as *the* way of implementing the satisfaction gains of needs in any simulation model. Rather they should be seen as a definition specifically suited for our current model. The ASSOCC model is a very broad model including many aspects from social contacts and epidemiology to basic needs and transport. This was only possible by keeping each aspect relatively simple. For example agents have money which is used to buy food or luxury goods, but the agents do not buy a house, car or invest their money into stocks. The most important aspects in the simulation such as the disease model and the overall agent behaviour model are developed in more depth.

Shopping activity Buying food (at the essential shop) causes the agent's reserves to be refilled thus satisfying food survival needs. While buying luxury goods (at the non-essential shop) satisfies luxury needs. Shopping does cost money though and causes the amount of money of the agent to be lowered, thus negatively affecting the financial safety and survival needs. The reward of shopping for the different needs is as follows:

- Food safety: The food safety need is directly related to the food (goods) bought at the essential shop.

$$ONSG^{food\_safety}(c_a^t, (l, shopping)) = \frac{nb\_of\_goods\_bought(c_a^t, l)}{full\_reserve(c_a^t)}$$

  where $l \in essential\_shops\_of(a)$ and $nb\_of\_goods\_bought(c_a^t, l)$ represents the number of goods bought when shopping in location l.
  $nb\_of\_goods\_bought(c_a^t, l) = full\_reserve(c_a^t) - food(c_a^t)$ (i.e. restock food for the next two weeks when going to essential shops).
- Financial survival: Agents can either buy food or luxury goods in this model. Since from these goods only food is relevant for survival, the financial survival is tied to being able to buy food for one self. The agents also have no medical expenses, which is realistic for most European countries as there is often free health care.

$$ONSG^{fin\_survival}(c_a^t, (l, shopping)) = \frac{money(c_a^t) - expected\_costs(c_a^t, l)}{expected\_costs(c_a^t, l)}$$

where $l \in essential\_shops\_of(a)$ and $money(c_a^t)$ is the amount of money agent a has and $expected\_costs(c_a^t, l)$ represents the expected money spent in l, defined as $expected\_costs(c_a^t, l) = nb\_of\_goods\_bought(c_a^t, l) \times cost\_per\_item(l)$. Thus, if the agent has more money than needed to buy all the food to replenish his stock of food then this action has a positive effect on financial survival! Basically, this is because there is money left for luxury goods. If the expected costs are high it means that the agent is in great need to get food. In this case having the expected costs also in the denominator means that the negative effect does not grow linearly. This means that the financial survival need will only in extreme cases prevent an agent to buy some food.

- Financial stability:

$$ONSG^{fin\_stability}(c_a^t, (l, shopping)) = -\frac{expected\_costs(c_a^t, l)}{wealth\_standard(c_a^t)}$$

where $wealth\_standard(c_a^t)$ represents the amount of money the agent wants to possess in order to feel stable.

- Luxury:

$$ONSG^{luxury}(c_a^t, (l, shopping)) = 0.06 \times nb\_of\_goods\_bought(c_a^t, l)$$

where $l \in luxury\_shops\_of(a)$ and $nb\_of\_goods\_bought(c_a^t, l) = 6$ by default. The *luxury* need can only be satisfied through buying goods at a non-essential shop. These values are very specific, but they are based on calibration of the model that takes care that agents do not safe up all excessive money, but also do not spend too much on luxury goods.

Work activity the agent obtains money in exchange for work during working hours, where $expected\_income(a)$ represents the expected income for a period of work and $is\_working\_time(c_a^t) \in \{1, 0\}$ represents the (contractual) obligation for performing the main activity (work or study) in the current period.
$is\_working\_time(c_a^t)$ is 1 if $day(c^t)$ is between Monday to Friday and $hour(c^t) \in \{morning, afternoon\}$ for agents a that are workers, teachers, students or children. Agents working in shops can work any day in the week and hospital personnel may work both during weekends and nights. All agents work five days per week for two periods per day.
$is\_working\_time(c_a^t)$ is 0 in all other cases.
Working gives the following rewards for the needs of the agent.

- Financial survival:

$$ONSG^{fin\_survival}(c_a^t, (l, work)) = \frac{expected\_income(c_a^t) \times is\_working\_time(c_a^t)}{cost\_fully\_restocking\_food(c_a^t)}$$

The satisfaction is increased proportionally to the amount of food that the work is expected to provide. Since the agents can only buy either food or luxury goods. Making this need related to food (which is required for the agent to live) makes it a realistic indication of financial survival.

- Financial stability:

$$ONSG^{fin\_stability}(c_a^t, (l, work)) = \frac{expected\_income(c_a^t) \times is\_working\_time(c_a^t)}{wealth\_standard(c_a^t)}$$

- Compliance:

$$ONSG^{compliance}(c, (location\_daily\_activity(c), work)) = 0.2 \times$$
$$is\_working\_time(c_a^t) \times loc\_complies\_quarantine(c_a^t, l)$$

where $loc\_complies\_quarantine(c_a^t, l) = 1$ when the agent is allowed to reach $l \in \mathcal{L}$ in context c_a^t without infringing public measures or $l = home(a)$ when the agent can work at home and is not allowed to go to its workplace.

- Autonomy:
There are multiple functions for the *autonomy* need, dependent of the location and whether the agent applies social distancing or not. The agent will gain more autonomy satisfaction from working at the workplace than at home, because the agent feels more restricted when having to work from home.
  - $ONSG^{autonomy}(c_a^t, (location\_daily\_activity(a), work_{sd})) =$
    $0.3 \times (1 - sd\_profile(a)) \times is\_working\_time(c_a^t)$.
    where $sd\_profile(a) \in [0, 1]$ represents how much a experiences the distancing from others as a limitation of its autonomy. The higher this value the more the agent feels limited and with a lower value the agents feels more free. This profile is only relevant when the agent itself applies social distancing. When the agent is not social distancing the following formula applics.
  - $ONSG^{autonomy}(c_a^t, (location\_daily\_activity(a), work_{nsd})) =$
    $0.3 \times is\_working\_time(c_a^t)$
  - $ONSG^{autonomy}(c_a^t, (home(a), work_{sd})) =$
    $0.05 \times sd\_profile(a) \times is\_working\_time(c_a^t)$
  - $ONSG^{autonomy}(c_a^t, (home(a), work_{nsd})) = 0.05 \times is\_working\_time(c_a^t)$
    When the agent cannot work from the workplace and has to work from home the autonomy satisfaction gained is very low.

Learn activity The learn activity is performed by youth and students at the schools and universities respectively. Because learning in our simulation is not related to getting a scholarship (agents do not have to pass exams to get scholarships) nor is there any other effect of learning on other actions, the effects are limited to the compliance and autonomy needs. Learning can only be done at the schools or universities.

- Compliance:

$$ONSG^{compliance}(c_a^t, (location\_daily\_activity(a), learn)) = \\ 0.1 \times is\_working\_time(c_a^t) \times \\ loc\_complies\_quarantine(c_a^t, location\_daily\_activity(a))$$

- Autonomy:
  - $ONSG^{autonomy}(c_a^t, (location\_daily\_activity(a), learn_{sd})) = 0.2 \times sd\_profile(a) \times is\_working\_time(c_a^t)$
  - $ONSG^{autonomy}(c_a^t, (location\_daily\_activity(a), learn_{nsd})) = 0.2 \times is\_working\_time(c_a^t)$

Relaxing and sleeping activity The satisfaction expected from resting depends on the hour of the day and the location where it is conducted.

- Sleep:

$$ONSG^{sleep}(c_a^t, (l, rest)) = effect\_time\_sleep(hour(c_a^t)) + \\ effect\_location\_sleep(l) + effect\_health\_sleep(experiencing\_symptoms(c_a^t))$$

  - where $effect\_time\_sleep(t)$ is the effect of the current time on the quality of sleep, modelled as: $effect\_time\_sleep(night) = 0.8$ and $effect\_time\_sleep(t) = 0.15$ if $t \neq night$. Having very high satisfaction during the night and much lower satisfaction at other times of the day, motivates the agents to rest during the night (without limiting it to only this period and creating flexibility when an agent is sick).
  - $effect\_location\_sleep(home(a)) = 0$ and $effect\_location\_sleep(my\_hospital(a)) = -0.1$
    This means that sleeping at home has a better effect than sleeping in the hospital. It is a (small) factor that motivates agents to go home at night whenever possible.
  - $effect\_health\_sleep(sick(a)) = -0.2$ and $effect\_health\_sleep(healthy(a)) = 0$, where
    $experiencing\_symptoms(c_a^t) \in \{sick(a), healthy(a)\}$ represents whether symptoms are being observed (true if the agent has mild or critical symptoms in the disease model and for specific scenarios that involving diseases with similar symptoms). Agents that are sleeping in a hospital bed or are sick receive less sleep satisfaction than other agents. This makes a sick agent require more frequent resting compared to a healthy or asymptomatic agent.

- Health:

$$ONSG^{health}(c_a^t, (l, rest)) = 0.1 \times exp\_critical\_symptoms(c_a^t)$$

  where $exp\_critical\_symptoms(c_a^t) = 1$ when critical symptoms are being observed and 0 otherwise. This motivates agents that are severely sick to rest, since it is better for their health. The lower satisfaction of 0.1 keeps agents from feeling completely healthy and rather continuously motivates them to rest.

- Autonomy:
  All actions can be done with or without social distancing. Therefore agents can also apply social distancing in for example the night, which represents for example sleeping in separate rooms. However this can limit the autonomy satisfaction gained dependent on the $sd\_profile(a)$.

  - $ONSG^{autonomy}(c, (home(a), rest_{sd})) =$
    $0.2 \times (1 - sd\_profile(a)) \times is\_night\_time(c^t)$.
    where $is\_night\_time(c^t) = 1$ if $hour(c) = night$ and 0 otherwise.
  - $ONSG^{autonomy}(c, (home(a), rest_{nsd})) = 0.2 \times is\_night\_time(c^t)$. The autonomy need is only satisfied when resting during the night.

- Leisure:
  Relaxing at home is less effective in satisfying the leisure need than being at a leisure place. Relaxing at leisure places gives a higher satisfaction to leisure, since the main role of leisure places for agents to have a place to relax and meet with friends. When relaxing at home in the night without social distancing it represents hanging out with friends, which also incurs an increase in leisure need satisfaction.

  - $ONSG^{leisure}(c, (home(a), rest_{sd})) = 0.1$.
  - $ONSG^{leisure}(c, (home(a), rest_{nsd})) = 0.2 \times is\_night\_time(c^t)$.
  - $ONSG^{leisure}(c, (l, rest_{sd})) = 0.6$ if
    $l \in public\_leisure\_places\_of(a) \cup private\_leisure\_places\_of(a)$.

Receive treatment activity:

- Sleep:

$$ONSG^{sleep}(c^t_a, (l, get\_treated) = effect\_time\_sleep(hour(c^t))$$

  Treating an agent has a positive effect on sleeping.

- Health:

$$ONSG^{health}(c^t_a, (l, get\_treated)) = gravity\_disease(c^t_a) \times hospitalization(is\_hospitalized(c^t_a))$$

  where $gravity\_disease(c)$ is 0.8 if $exp\_critical\_symptoms(c)$;
  0.3 if $bel\_sick(c)$ but not $exp\_critical\_symptoms(c)$;
  0 otherwise.
  $hospitalization(h) = 1$ if h is true, else $hospitalization(h) = 0$; In the formula $h = is\_hospitalized(c)$ and $is\_hospitalized(c) \in \{\top, \bot\}$ represents whether the agent is given a bed in a hospital. An agent attempting to get treated at the hospital may be denied a bed when the hospital is overrun.
  The *health* need is highly satisfied when receiving treatment as a severely sick agent. An agent believing to be sick but not showing critical symptoms will also

gain some *health* need satisfaction. $bel\_sick(c)$ represents the *belief* of being sick. $bel\_sick(c)$ is set to $\top$ when $experiencing\_symptoms(c)$ or upon receiving a positive test. The agent believes to be sick for 14 days from the last positive test or occurrence of $experiencing\_symptoms(c)$.

- Risk avoidance:
  $ONSG^{risk\_avoid}(c_a^t, (l, get\_treated)) = 0.4$ if $bel\_sick(c_a^t)$ and $is\_hospitalized(c_a^t)$
  The agents feel more safe and thus more risk avoiding when getting treated. However this satisfaction is only given to sick agents.

Consequences of deviations from health or daily schedule: With the influence of activities on the needs described, we will now describe the effect of deviations. For example a sick agent will decrease its health need satisfaction when doing activities instead of resting or getting treatment.

- $ONSG^{health}(c_a^t, (l, act)) = -0.1$ if $bel\_sick(c_a^t)$ but act is neither a resting nor a treatment activity. Performing other actions when feeling sick makes the agent feel less healthy.
- $ONSG^{compliance}(c_a^t, (l, act)) = -0.2$ if $age(a) = worker, is\_working\_time(c_a^t)$, and $act \neq work$. Agents that are not working during working time will get a decrease on compliance satisfaction. This is set to -0.2 so it is discouraging agents to not be working.
- $ONSG^{compliance}(c_a^t, (l, act)) = -0.1$ if $age(a) \in \{youth, student\}$, $is\_working\_time(c_a^t)$ and $act \neq learn$.
  While workers should work at working time, students and youth should learn at working time otherwise they loose compliance. Not learning gives a lower decrease of *compliance* satisfaction than not working. As usually in societies it is more frowned upon when not showing up at work than at a school or university. Although this is of course dependent on the culture and type of job. Time affects the available locations and activities (e.g. schools are closed during the night) as well as the effect of certain actions (e.g. sleeping is more effective during the night).
- $ONSG^{risk\_avoid}(c_a^t, (l, get\_treated)) = -0.1$ if $bel\_sick(c_a^t)$ and not $is\_hospitalized(c_a^t)$ When an agent is receiving treatment but not at the hospital, when there is not enough place for the agent, this will be seen as sub optimal. Therefore the *risk_avoid* need satisfaction is decreased.
- $ONSG^{autonomy}(c_a^t, (location\_daily\_activity(a), act)) = -0.1$ if $is\_working\_time(c)$ and $act \notin \{work, learn\}$. When the agent is not allowed to go to work, school or university during working time. The agents sees this as a constraint on its *autonomy* and therefore this need's satisfaction is decreased.

Effects of social distancing: Deciding to apply social distancing or not directly relates to the following three needs:

- Conformity:
  Conformity depends on whether the agent applies social distancing the same way other agents do in its $social\_network(c_a^t)$.
  $ONSG^{conformity}(c_a^t, (l, a_s)) = 0.1 \times conformity\_sd\_factor(c_a^t, s)$

where $conformity\_sd\_factor(c_a^t, s) = 1$ if
$s = normal\_sd\_social\_network(c_a^t)$ and 0 otherwise.
Here $normal\_sd\_social\_network(c) \in \{sd, nsd\}$ represents the most frequent attitude regarding social distancing usually adopted in the current time of the day by other agents in $social\_network(c_a^t)$.

- Compliance:
  If social distancing is required agents who apply social distancing get compliance satisfaction and agents who do not apply social distancing get a compliance penalty.

  – $ONSG^{compliance}(c_a^t, (l, ac_{sd})) = 0.1 \times social\_distancing\_conformity(c^t)$ where $social\_distancing\_conformity(c^t)$ is 1 if individuals are requested to apply social distancing behaviours and 0 otherwise.
  – $ONSG^{compliance}(c_a^t, (l, ac_{nsd})) = -0.1 \times social\_distancing\_conformity(c_a^t)$.

- Risk avoidance:
  An agent gets increased risk avoidance satisfaction when applying social distancing, otherwise it is decreased.

  – $ONSG^{risk\_avoid}(c_a^t, (l, ac_{sd})) = 0.1$.
  – $ONSG^{risk\_avoid}(c_a^t, (l, ac_{nsd})) = -0.1$.

Other rewards for social needs. Certain social needs are satisfied by the activity to be performed through indirect effects (e.g. is the agent's decision matching the decision of its friends in its social network):

- Conformity:
  This needs depends on two things: whether the agent quarantines the same way other agents do in $social\_network(c_a^t)$ and whether the agent performs the usual action in the location (e.g. work from home if most people in the network do that).

$$ONSG^{conformity}(c_a^t, (l, act)) = 0.2\times$$
$$conformity\_quarantining\_factor(c_a^t, location\_respects\_quarantine(c_a^t, l))+$$
$$conformity\_la\_factor(c, (l, act))$$

  where $location\_respects\_quarantine(c_a^t, l) \in \{com, viol\}$ represents that the agent in location l either complies (com) or violates ($viol$) the quarantine rule.
  $conformity\_quarantining\_factor(c_a^t, v) = 1$ if $v = sn\_complies\_quarantine(c_a^t)$ and 0 otherwise. Where $sn\_complies\_quarantine(c_a^t)$ represents whether a majority of the actions of agents in $social\_network(c_a^t)$ complies to the quarantine restrictions.
  $conformity\_la\_factor(c_a^t, (l, act))$ represents whether the located activity of agent a matches the decision of other agents in $social\_network(c_a^t)$. This factor is composed of three parts: both the location and action conform to the usual, the location conforms to the usual and the action conforms to the usual. The agent gets a higher conformity boost when at the same location performing the same activities as the social network. There is a slight conformity boost when the location or the activity are the same.

$$conformity\_la\_factor(c_a^t, (l, act)) = conforms\_to\_la(c_a^t, (l, act)) + conforms\_to\_loc(c_a^t, (l, act)) + conforms\_to\_act(c_a^t, l, act))$$

where

- $conforms\_to\_la(c_a^t, la) = 0.1$ if $la = normal\_LA\_for\_social\_network(c_a^t)$ and 0 otherwise
- $conforms\_to\_loc(c_a^t, (l, act)) = 0.15$ if $l = normal\_loc\_for\_social\_network(c_a^t)$ and 0 otherwise
- $conforms\_to\_act(c_a^t, la) = 0.15$ if $act = normal\_act\_for\_social\_network(c)$ and 0 otherwise

- Compliance:
When an agent is supposed to be in quarantine, the agent will get increased compliance satisfaction when complying, e.g. staying at home. If not complying the agent gets reduced compliance.
  - $ONSG^{compliance}(c_a^t, (l, ac)) = 0.2$ if $loc\_complies\_quarantine(c_a^t, l)$.
  - $ONSG^{compliance}(c_a^t, (l, ac)) = -0.2$ if not $loc\_complies\_quarantine(c_a^t, l)$.
- Risk avoidance:

$$ONSG^{risk\_avoid}(c_a^t, (l, ac)) = -(nb\_exp\_contacts(c_a^t, l) - 20) \times 0.01$$

where $nb\_exp\_contacts(c_a^t, l)$ represents the number of contacts agent a expects at location l. More than 20 contacts decreases the $risk\_avoid$ satisfaction, less than 20 increase the $risk\_avoid$ satisfaction. The number is set to 20 contacts as this was the median of contacts. At locations with a low amount of contacts (e.g. a home) one can expect 1-4 encounters and thus they are relatively risk free. At other places like work, schools and universities the amount of encounters could easily be 26 and thus these locations are seen as risk increasing.
- Belonging:
The belonging need is influenced by a combination of the number of contacts, the type of contacts and whether social distancing is observed.
$ONSG^{bel}(c_a^t, (l, act)) = 0.02 \times nb\_exp\_contacts(c_a^t, l) \times sat\_encounter\_type(c_a^t, l) \times sat\_social\_distancing(c_a^t, (l, act))$ where part of the belonging satisfaction comes from meeting family and friends. This satisfaction component is higher when the last meeting with them was longer ago.
  - $sat\_encounter\_type(c_a^t, l) = \frac{nb\_days\_last\_contact\_with\_family(c_a^t)}{28} \times expected\_family\_encounter(c_a^t, l) + \frac{nb\_days\_last\_contact\_with\_friends(c_a^t)}{28} \times expected\_friends\_encounter(c_a^t, l)$
  $expected\_family\_encounter(c_a^t, l) = 1$ if the agent expects to meet family in l and 0 otherwise.
  and $expected\_friends\_encounter(c_a^t, l) = 1$ if the agent expects to meet friends in l and 0 otherwise.

The factor of 28 indicates that we expect someone to see friends or family at least within four weeks.

- The last component of belonging satisfaction is $sat\_social\_distancing(c_a^t, (l, act))$ which represents the loss of proximity satisfaction caused by social distancing.
  $sat\_social\_distancing(c_a^t, act_{nsd}) = 1$ and
  $sat\_social\_distancing(c_a^t, act_{sd}) = 1 - sd\_profile(a)$

- Autonomy:
  When performing actions the agent gains some autonomy satisfaction. However the gain is limited when applying social distancing. The amount of reduction of autonomy gain by social distancing is given by the *sd_profile*() (the higher this number the more an agent experiences loss of autonomy through social distancing).
  - $ONSG^{autonomy}(c_a^t, (l, a_{sd})) = 0.1 \times (1 - sd\_profile(a))$
  - $ONSG^{autonomy}(c_a^t, (l, a_{nsd})) = 0.1$ if $l \neq home(a)$ and $lockdown(c^t)$
    This means that when there is a lockdown, the agent gains autonomy satisfaction when not social distancing at a different place than home. The autonomy increase comes from the agent doing what it wants to do rather than following the rules. Of course, this behaviour will have a negative impact on some of its other needs.

3.8 Conclusions

In this chapter we have given a quite detailed description of the implementation of the ASSOCC framework. This description serves a number of purposes. First of all it shows the internal mechanisms of the simulation framework that is use to create all the results of the different scenarios in Chaps. 5–10. Thus it provides the common ground for all these results and shows where they actually come from.

By giving this detailed descriptions we also show the complexity of the complete ASSOCC framework. This is the consequence of trying to combine many different aspects of life in the framework. Although we have chosen relatively simple solutions for most aspects of the framework it is the combination of all these aspects that makes the framework inherently complex.

The complexity is managed through the central needs model that combines the different aspects of life that an agent tries to balance. Having this central component that consists of a homeostatic model makes it possible to balance aspects without being to rigid. However, this model also had to be calibrated in order to create a kind of standard balance between the needs in common everyday life. Whenever a new aspect is added this system has to be recalibrated. Thus also the current system has its limitations and is pushed to the limit of its usability.

Due to the complexity of the model and its implementation we also had to explicitly do something about the usability of the framework. This goes beyond a kind of good engineering principles and creating nice graphs. We decided to use a combination of different platforms to take care of different aspects of the usability of the system. This will be described in detail in the next chapter.

Acknowledgements We would like to acknowledge the members of the ASSOCC team for their valuable contributions to this chapter. This research was partially supported by the Wallenberg AI, Autonomous Systems and Software Program (WASP) and WASP - Humanities and Society (WASP-HS) funded by the Knut and Alice Wallenberg Foundation.

References

1. R. Hinch et al., Effective configurations of a digital contact tracing app: a report to NHSX. In: en. In:(Apr. 2020). https://github.com/BDI-pathogens/covid-19_instant_tracing/blob/master/Report (2020)
2. R.C. Cope et al., Characterising seasonal influenza epidemiology using primary care surveillance data. PLoS. Comput. Biol. **14**(8) (2018)
3. Birgit Müller et al., Describing human decisions in agent-based models- ODD+ D, an extension of the ODD protocol. Environ. Modell. Softw. **48**, 37–48 (2013)
4. W.O. Kermack, A.G. McKendrick, A contribution to the mathematical theory of epidemics. in *Proceedings of the royal society of london. Series A, Containing papers of a mathematical and physical character*115.772 (1927), pp. 700-721
5. R. Hinch et al., OpenABM-Covid19-an agent-based model for nonpharmaceutical interventions against COVID-19 including contact tracing. medRxiv (2020)
6. M.J. Gelfand et al., Differences between tight and loose cultures: a 33-nation study. Science 332.6033 (2011), pp. 1100–1104
7. I. Uz, The index of cultural tightness and looseness among 68 countries. J. Cross-Cultural Psychol. **46**(3), 319–335 (2015)

Chapter 4
Social Simulations for Crises: From Models to Usable Implementations

Cezara Păstrăv, Maarten Jensen, René Mellema, and Loïs Vanhée

Abstract Simulations created for crises naturally have two important goals: the simulation must both be sound and solid from a scientific standpoint, but also should be exploitable at very short notice by stakeholders in the decision-support of the crisis. A central activity of building simulations during crises is conducting an advanced software project, for which implementing the central simulation model is only one of the many tasks. Taking a systems-design perspective, this chapter describes the needs, concerns, and solutions for achieving the goals raised by simulations during crises by illustrating how they were addressed by the ASSOCC software platform within the project. In particular, ASSOCC goes beyond classic social simulation standards by incorporating dedicated visualisation aspects, leading to an architecture that combines a simulation module (in NetLogo), a visualisation module (in Unity) and an analysis module (in R). This chapter explains what modules were required and for which purpose, what outcome to expect from developing such modules, and how to design and implement such a module and overarching architecture to interact with one another.

C. Păstrăv (✉) · M. Jensen · R. Mellema · L. Vanhée
Umeå University, Mit-huset, Campustorget 5, 90187 Umeå, Sweden
e-mail: cezarap@cs.umu.se

M. Jensen
e-mail: maartenj@cs.umu.se

R. Mellema
e-mail: rene.mellema@cs.umu.se

L. Vanhée
e-mail: lois.vanhee@umu.se

F. Dignum (ed.), *Social Simulation for a Crisis*, Computational Social Sciences,
https://doi.org/10.1007/978-3-030-76397-8_4

4.1 Introduction

The prime objective of the ASSOCC platform is to offer stakeholders a decision-support tool for better managing the COVID-19 crisis. While this goal is exhilarating as it gives scientists a chance to make a real positive impact on the world it also comes with a great responsibility to give the right support![1] We cannot just give some plausible results that academics can dissect, comment on and improve. The simulation results should be usable in practice and hopefully lead to positive impacts. Thus, in addition to offering the means for classic model-building, the ASSOCC platform needs to also[2]:

- be easy to adapt and expand
- deliver fast results while adhering to the highest scientific standards
- facilitate communication with stakeholders, both inside and outside academia

These requirements involve three very distinct objectives that can each be related to specific tools and approaches. Therefore, a natural architecture for covering these requirements, illustrated in Fig. 4.1 consists in combining three complementary modules, each designed to serve one goal: the *simulation module* (in NetLogo), described in Sect. 4.2, the *output analysis* (in R), described in Sect. 4.3, and the *visualisation module* (in Unity) described in Sect. 4.4.

In keeping with the fast development speed characteristic of crisis situations, the three modules are designed to evolve relatively independently, with the simulation module as the core that the other two modules draw data from for their own functionalities. Thus, the simulation module can operate independently of the other modules, ensuring that model development is not slowed down by the additional tasks of managing visualisation or data graphing, allowing us to bring the model up-to-date with the latest information and theories in the shortest possible time.

The visualisation and analysis modules are dependent on the simulation module, but are independent from each other, and thus the scarce development time and effort available in a crisis can be allocated to one or the other, as the circumstances dictate. Their dependence on the simulation module is fairly minimal. They use their access to the simulation module to draw data, such as simulation state data (e.g. number of infected people) and internally tracked statistical variables (e.g. number of contacts, infection tree). The simulation module is also accessible by other modules to alter the simulation setup before a run, to run the simulation step by step, and to trigger predefined commands (e.g. turning on/off the closing of schools). Therefore, the visualisation and graphic modules need to only keep track of the types and formats of the data provided by the simulation modules, as well as the kind of simulation commands it understands.

This separation allows our team to pursue the modelling, scientific, and stakeholder involvement goals of the ASSOCC project with high effectiveness—by using

[1] With possible severe legal consequences should the predictions reveal to be worse than predicted https://www.scientificamerican.com/article/italian-scientists-get/.

[2] A more exhaustive list of core criteria for developing simulations in crisis is provided in Chap. 13.

Fig. 4.1 High-level system diagram for all ASSOC modules and their connections. Unity and NetLogo communicate through a Java relay app. NetLogo writes simulation output to a csv file, which is the input for R

specialised tools (NetLogo, R, and Unity) and with minimal complexity and human coordination costs resulting from module interdependence.

This chapter shows how such a platform can be built and what it provides in terms of handling the critical task of collaborating with stakeholders without sacrificing the advantages of established simulation platforms with regards to performing effective high-quality scientific work. To our knowledge, no similar platforms were (and are) readily accessible for handling the requirements that we set out. Therefore, most of this platform had to be built from scratch during the crisis, towards achieving our immediate goals. This chapter provides the general background, purpose and technical solutions for building a crisis simulation platform. We use the ASSOCC platform as an example, but also indicate where lessons can be learned and issues are of general interest.

In the next three sections we describe the three main components of the ASSOCC platform. Then, Sect. 4.6 focuses on the management concerns that arise from simulating using multi-platform architectures during a crisis. This chapter is intended to be of specific interest for everyone aiming to build a simulation platform for crisis situations. In order to give maximal support to the general development of social simulation platforms for crisis situations the repository holding the code is available at https://github.com/lvanhee/COVID-sim.

4.2 The Simulation Module

Building a simulation model during a crisis involves to handle at the same time a number of non-trivial and sometimes self-opposing design challenges, as detailed in Chap. 13. Thus, deciding upon the right simulation platform is both critical and challenging, as the time pressure and the inherent complexity of simulations involves a very heavy path-dependency. The tool needs to correspond to the problem at hand, the team, the facilities and costs inherent to the tool, the constraints raised by the crisis, etc. As a matter of making such a decision in the context of crisis, we provide in Sect. 4.2.1 the rationale we followed for selecting the simulation platform (for ASSOCC it was NetLogo, which was not a trivial choice), then we describe in

Sect. 4.2.2 how this platform has been used and useful (and sometimes limiting) for our objectives before studying more analytically the relation between NetLogo, Crisis and Large-Scale Simulations in Sect. 4.2.3 challenges and limitations that we observed from using this platform in the context of a crisis.

4.2.1 Selecting the Simulation Platform

Selecting the right simulation platform is a critical decision to be made when initiating any large-scale simulation project.

NetLogo is used for implementing the simulations of the ASSOCC project. Whereas we do not claim that NetLogo is necessarily the best choice for the long term (many of its limitations and subsequent limitations for the ASSOCC project are described below), NetLogo was the best available platform in our case. This choice is based on the following arguments.

Pros: As a central argument, we preferred to have a limited working model that is easily kept up to date with emerging theories rather than approaches relying on slower cycles of development. As such, the NetLogo platform is very fit for building fast prototypes with short development cycles. The key benefits of this fast prototyping approaches in the situation of crisis are twofold. First, it offers high *reactivity*: we could quickly provide first results that could be checked for validity and engagement with stakeholders. Second, it offers means for creating *theoretical stability*. Crisis situations often involve unknown environments, with new data, models, and weakly-validated theories that emerge on a daily basis. Fast-prototyping allowed us to build and adapt relatively coherent theories and models relatively early on during the process. This activity is often referred to as bootstrapping in the social simulation community, i.e. using the model for helping the modeller to develop new theories and models). Thus, the output of the NetLogo model could be used on a daily basis to test newly emerging theories and to analyse unexpected emerging phenomena. Finally, we also based the decision on using NetLogo on which skills were available in our team and how much we wanted the results of this project to be available for the wider community. Both of these points favoured the decision to use NetLogo despite some of its shortcomings.

Cons: The disadvantages of NetLogo are mostly in terms of engineering effort, as the ASSOCC project ended up stretching far beyond what NetLogo is built for (hundreds of parameters, thousands of lines of code, millions of agent decisions). As concrete examples, NetLogo misses many productivity functions of other tools (e.g. step-by-step debugging, autocompletion, jump-to-definition); and involves compilation time that becomes crippling as the project grows (10+ seconds of compilation time every few minutes of programming time). This issue becomes salient relatively fast when applying the engineering discipline required for building a simulation such as ASSOCC that has a large code-base. In terms of limitations over what model can be produced, the raw user interface is insufficient for a project of this size. The graphical interface, very fit for small-scale projects, started to be clunky as we grew towards hundreds of parameters and dozens of output graphs. This is daunting even for the engineers and is completely inappropriate for engaging with stakeholders. Moreover, surprising computational restrictions made very difficult to scale up the number of agents and connect to external modules with satisfactory speed. The facilities for automating experiments are good but insufficient for our scale and thus also required development of external tools.

Although this decision appears to be the best option at the time of the crisis, NetLogo is not necessarily the best for everyone, in particular if some preparatory work can be produced ahead of the crisis. Therefore, as we can now further prepare, the NetLogo ASSOCC platform is now discontinued, serving as a necessary foundational step for understanding what is important for simulating in crisis. The next implementation we are working with is built in REPAST [1] and the main modules one can prepare before a crisis are described in Chap. 14.

Since our model implementation methodology corresponds to modelling methodologies already extensively described in the literature (e.g. short iterative approach), the core engineering insights that we developed during the ASSOCC project lies in how to push the inherent limits of NetLogo. In many regards, modelling in the context of a fast-evolving crisis is similar to building a car from spare parts without full knowledge of what the end-result needs to look like: we start with a bare-bones skeleton that has a few key functionalities, then we constantly add more and more parts and functions as new information comes in. In addition, we are doing all this while having limited intuition of what each new piece will add. We know that every added piece inevitably creates more grinding in the wheels, raises the probability that components will interact in unexpected ways, and brings the car closer to crumbling under its own weight. The engineering challenge is simple to articulate, but difficult to implement: keep the car working for as long as possible and drive it as far as possible.

The rest of this section describes how we coped with these issues in order to support the development of the code in the face of mounting size, complexity and computational cost. As a matter of appreciating the scale of the system, the size of the simulation is to date beyond 8000 lines of NetLogo code (excluding NetLogo GUI description), which makes the ASSOCC model one of the largest NetLogo models ever created.

4.2.2 *Use of NetLogo*

NetLogo contains many features that we extensively relied on for building the ASSOCC platform. Here is a description of the tools we used, following the listing from the Netlogo user manual.[3]

[3]NetLogo user manual version 6.2.0 https://ccl.northwestern.edu/netlogo/docs/.

4.2.2.1 General Considerations

NetLogo is free and available across platforms, which is a particularly beneficial feature as it dramatically cuts down the entry costs for using or developing the platform. Anyone can always immediately run the last version of the model without issues. In a crisis context, in which everyone is racing against time, cutting down such entry costs is decidedly important since adding extra overhead from dealing with sysadmins in charge of issuing software licences can be a bottleneck for the project development. Furthermore, our efforts regarding extensive collaboration and communication would have been rendered far less effective by us using proprietary or little-known platforms. By using a very popular tool like NetLogo, we made the software immediately available and understandable to a very wide range of prospective users and contributors, who might have otherwise been discouraged from participating.

4.2.2.2 Programming

The programming language is very suitable for the agent-based design we worked with: turtles, links, lists, turtle-sets, *ask* and *of* constructs are at the core of the model. Lambda expressions (constructs allowing the reuse of some code) were occasionally used, notably for reusing procedures while changing a small part of their internal behaviour (e.g. how to record or log certain operations performed within the procedure). Using this programming language is double edged: using it well requires some experience to use well, even for programmers used to classic programming languages, as NetLogo doesn't use the common object-oriented paradigm of languages like Java or C#. This becomes especially noticeable when the size of the model increases requiring more and more skill to keep the software well structured. One of the difficult issues in this respect is that when not using object-oriented programming paradigms it is easy to have duplicate code popping up while not being aware of it (due to different people working on several parts of the code at the same time). To keep the code well structured requires a high level of discipline in the team, good agreements on what are the practices to use in programming constructs and meticulously monitoring the code at all time to maintain its integrity.

4.2.2.3 Environment

The command center and the agent inspector features provided by Netlogo proved indispensable for improving debugging efforts. The import and export functions also contributed to this. They allowed us to save runs in which we observed unusual or unexpected behaviour, and then to pre-load these runs before a point of interest, thus allowing a more effective process of bug replication.

The Behaviourspace feature was extensively used for running batches of experiments. Using preset scenario settings, the model has been setup so it can easily simulate different scenarios. These preset scenarios load a default variable setup which sets the parameters for the household dependent on selected country, location numbers, location density, disease model, economic model, etc. There is a country preset which determines the culture variables and the household distribution. The agents will be spawned in the simulation, this is usually around a 1000 or for test runs around 300. They get a social network and assigned gathering places after spawning. The locations are spawned according to their preset number. After loading the default variable setup, the scenario specific presets are loaded which are described in more detail in their respective scenario chapter. The behaviour space will automatically load all the data in a single '.csv' file after the experiments are completed.

Though useful, it lacks certain functionalities which would have made the process more effective, such as a stop-and-resume option, or the option to do partial runs of all experiments, with the possibility of managing more detailed runs. We attempted to fill in these gaps, but had limited success due to technical limitations of the NetLogo API. The API is really geared towards batch runs and is not meant a more interactive connection between the behaviour space and the runs.

4.2.2.4 Display and Visualisation

The display options of the NetLogo platform were very important in our decision of using NetLogo. We want to offer users as much control over the dynamics of the simulation and have as few hidden assumptions as possible. The flexibility of the Netlogo GUI allowed us to easily include most of the many parameters of the model in the interface, allowing the user a very detailed control over the simulation. At the same time, having so many of the simulation aspects easily observable in the interface made the development and debugging process much more efficient, an issue which gained prominence as the model grew.

The main drawback of using Netlogo for visual display purposes is that the interface cannot easily be split into separate areas of interest. Rather, every button, slider, value field and chart are displayed together in one large, barely organised group. Given the size of our model and the number of displayed parameters, the effort of reading the interface grew to the point where a first time user would find the effort of navigating it rather off-putting. A secondary drawback is that the quality of the graphics is fairly low, doing a suitable, but limited job (Fig. 4.2).

4.2.2.5 Java API

We used the Java API to connect the NetLogo platform to the Unity GUI (see Sect. 4.5) and to apply external custom-made tools for running experiments. Unfortunately, while this API allows us to run NetLogo commands from Java, the interface suffers from unexpected performance issues. In our experiments, the same command ran ten times slower when using the API compared to when using the command line. This

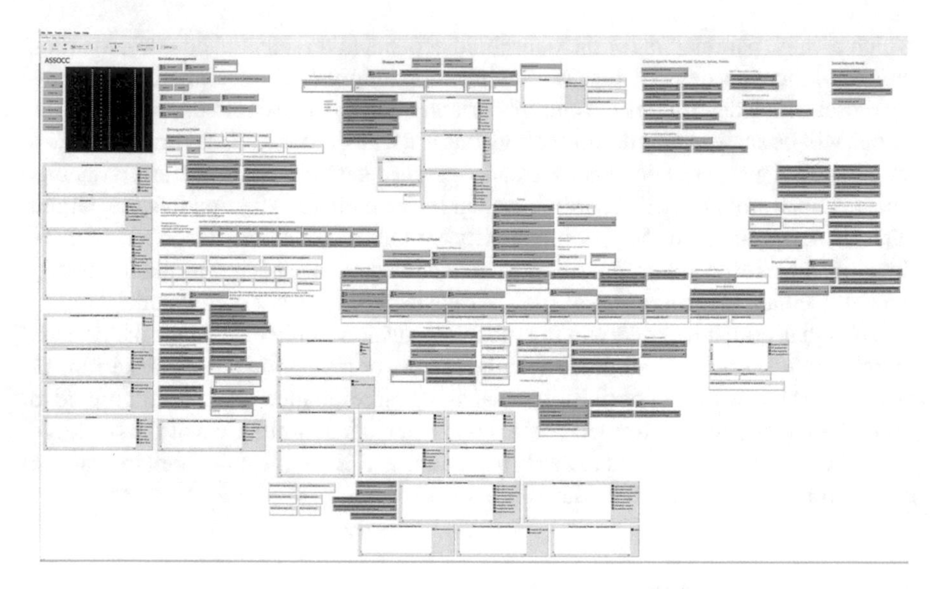

Fig. 4.2 Full NetLogo interface. This image is a composite of four different screenshots as it takes 4 screens to display the whole interface

limitation curtailed the relevance of using the API for running custom fine-grained experiments and required additional engineering efforts in building an efficient data pipeline to the Unity GUI.

4.2.2.6 High Performance Computing

In order to analyse some of the scenarios, many configurations had to be run many times. Given the time constraints of working under crisis conditions, speeding up this process as much as possible is very important. Therefore, we took advantage of the resources granted to us by the Swedish National Infrastructure for Computing (SNIC). In particular, the Abikso and Kebnekaise supercomputers at HPC2N in Umeå.

Running our experiments on the HPC2N saved us significant time, allowing us to quickly get results that would have taken days to produce otherwise. However, the process requires careful preparation due to certain Netlogo limitations and we run into three main issues that slowed us down and kept us from exploiting the full capabilities of the supercomputers.

The first, and most easily solved, problem that we ran into is that the script for running NetLogo headless (without GUI), which is needed on these machines, does not allow changing the JVM parameters by passing in arguments. We also could not modify the script to accept arguments because we were not granted this privilege for the supercomputer Netlogo installation. Thus we ended up starting Netlogo manually, which is a fairly easy fix, but not very efficient.

A second issue we ran into is that Netlogo only uses a single node for each simulation run, even when running on a machine that has multiple nodes available, as is the case for the supercomputers. This had implications for how we designed the experiments. We had to manually schedule experiments such that the runs would be divided over the nodes. Each node on the HPC could execute 28 runs in parallel. Thus, if we have behaviour spaces that would generate more runs, we had to make sure that the behaviour space was partitioned in a way that would create batches of multiples of 28 (or fewer) runs.

A third problem that we ran into is that the BehaviourSpace does not allow for conditional parameter values. This means that if we have a set of parameter values that only makes sense together with a set of values of another parameter, each value in this second set needs to go in its own BehaviorSpace experiment. To do otherwise, would result in Netlogo running every combination of parameter values, regardless of desirability, resulting in a large number of meaningless experiments taking up our already limited HPC resources. To overcome this issue, we had to split up the BehaviourSpace experiments for one scenario into multiple groups, which took time and effort we would have preferred to allocate elsewhere, given the crisis working conditions.

4.2.2.7 Testing

As simulation models grow in size and complexity, there is an increasing chance that any new added feature causes another feature to behave in a manner that is unacceptably unrealistic. E.g. results might show be a very realistic spread of infection among people in bars and restaurants due to a coincidental balance of other parameters rather than as a real consequence of the model. Such a faulty dynamic can be difficult to spot, until the results are completely changed by introducing another leisure place, like parks. Suddenly what seemed a very natural balance is disrupted and, given the complexity of the model, it can be hard to even realise that an obviously unrealistic behaviour is occurring at another side of the simulation, without significant impact on the aspects under scrutiny by the developer. However, the greatest care to such unrealistic behaviour is to be given as it can severely damage the relation with stakeholders, who are overly sensitive to such aspects.

To remedy this issue, some testing functions were added for automating sanity-checks, i.e. that the system behaved as expected, based on unit testing methods from software development. More precisely, every test consists of a NetLogo function that loads a certain set of parameters (e.g. the standard setup without active policy restrictions), runs the simulation and then performs some simplified checking over the output of the model (e.g. check whether there is a large wave of infection). In more advanced cases, a score in [0, 1] is recorded.

The results of this testing approach were limited, as the tests mostly revolved around basic safety scenarios, due to lack of engineering time for making it systematic for every scenario. It did achieve its goal in that at least the basic components are very stable and their stability can easily be verified after adding extra aspects.

The question of the relevance of testing in crisis scenarios remains open: on the one hand, tests do not contribute directly to the modelling effort and they tend to deprecate very fast, on the other hand, tests increase the coherence of the whole model, which is important for avoiding the accumulation of programming, modelling and theoretical errors and for maintaining high trust on the part of the stakeholders. In the case of relatively simple models, this accumulation of errors cannot go very far without being spotted and corrected, so its possible tests are a waste of precious time and effort. However, in the case of complex models that keep growing and changing according to the latest information available, errors can accumulate fast, remain hidden for much longer, and take considerable time and effort to fix. Under such circumstances, ensuring the core of model, at least, is free from any unwanted behaviours can prevent a model from collapsing too soon under the weight of its own complexity.

4.2.3 NetLogo, Crisis and Large-Scale Simulations

Under crisis conditions, the ability to quickly build and run models outweighs most other considerations, including limitations of scale, ease of interaction or a lack of convenient features for testing or defining experiments. A crisis is, by definition, short-lived since otherwise it would simply become the new status quo. Being able to respond and adapt quickly, if inelegantly, is far more important than any more comprehensive solutions after the crisis has ended. Furthermore, the ability to effectively share these models, not just the results, with the wider modelling community is of equal importance. Fast exchange of information ensures as many people as possible have access to the latest developments and no one wastes precious time reinventing the wheel or invests efforts attempting solutions that have already been shown not to work elsewhere.

Give these two main considerations, we chose to use Netlogo as a modelling platform. Despite its many limitations—such as failure to take full advantage of high performance computing facilities, limited API, limited inbuilt testing, and a GUI that is severely limited when it comes to effective stakeholder communication—it is very well suited for fast prototyping of small scale models, allowing us to quickly implement our model and then quickly adapt it as new information came in. At the same time, it is by far the most popular platform in the ABM academic community, which means it is the best way to effectively share our model and get useful feedback from others working on the same problems.

4.3 Output Analysis (with R)

Due to the large number of aspects and scenarios that can be run with the ASSOCC framework there is a clear need to systematically analyse and present the results of the simulations. In one scenario we have produced around 60 relevant graphs depicting different aspects of the results of the simulation for different settings. It is clear that one needs a separate analysis phase and tools to handle all this output. We chose R to do this mainly due its widespread use for statistical analysis and its easy availability. Note that using a tool like R is fine to create final results for scientific purposes, but cannot really be used to generate visualisations of the simulation at run time to stakeholders! In the next sections we describe briefly which were the main properties that we have used R for and how they were useful or necessary for simulations in crisis situations.

The standard NetLogo plots have to be predefined and are limited in their ability to display data. This works fine for smaller scale, focused simulations, but is not suitable for simulation scenarios as we performed within the ASSOCC framework. In this framework the specific variables of interest are very dependent on what scenario is analysed. Due to the large number of output parameters and the large number of possible results that can be generated from this output data we use a separate statistical package (R) to analyse and display the results. It also facilitates creating new analyses based on existing results. E.g. checking whether some needs are overall less satisfied when the virus is spreading fast. This kind of analysis can be initiated based on some additional request to check the overall mental state of the population or of some parts of it as a result of the crisis. Instead of creating new simulations we actually already have all this data available and can correlate it directly with other parameters of interest. Given the time constraints of working in crisis conditions, all this data processing also needs to be automated. Not only is R excellent at automating data processing, it is also a popular choice in the academic community for its ease of use and high quality plots.

4.3.1 Output Parameters

Since the ASSOCC framework contains many aspects, from health to economics and social well being, we also have a large set (hundreds) of output values. The most prominent are the number of infected agents, the amount of agents believing they are infected, fatal cases, number of tests performed, R0, the number of agents infected by others per age, and the number of contacts agents had. There are many more parameters of potential interest such as data pertaining to the needs model and the economical model, which have been explored in a subset of the scenarios.

4.3.2 Data Processing

4.3.2.1 Data Source

NetLogo's BehaviorSpace[4] can easily run experiments and generate standardised output data. The BehaviorSpace outputs results in .csv format with a choice of either a spreadsheet format or a table format. The spreadsheet output is more human readable as it calculates the min, mean, max, and final values for each of the output variables. The table output however is mainly used for further processing by data analysis software. Since our goal is to automate the process as much as possible, we chose the table preset which generates data as exemplified in Table 4.1. The format of the table is as follows: the first column represents the *run number* and the settings of the run, followed by the *simulation step* column, followed by a set of columns representing all the output variables. Since each row contains both run settings and the variable states for a step, it is easily processed by R, especially when using libraries such as *dplyr*.

4.3.2.2 Preparing and Cleaning the Data

Even though NetLogo's behaviour space gives well-formatted results as output, we still need to clean the data before use.

The data is cleaned by removing runs with too few infected (except for baseline runs, which contain no infected agents by design) and runs that have not been completed. Runs with too few infected agents occur when the infection fails to propagate through the agent population, which happens under certain starting conditions (such as all initial infected agents sharing a household and isolating early). Since we consider these runs to not be representative of the scenarios we want to simulate, we exclude them from the final results. Runs that are not complete happen when the simulation fails to reach the final tick or is stopped by the user because of observed bugs. These are removed because they are faulty and affect the reliability of the results.

The number of runs removed for any of the failure conditions is saved and plotted in a bar chart, which shows the total number of runs per setting and the numbers of runs that are approved. This allows us to check how many runs are missing and, therefore, whether we need to perform more runs or check the code for problems.

4.3.2.3 Data Processing

For each of the plots, the relevant parts of the full data are selected and manipulated. Since we have multiple runs of the same setting (for example 50) we take the mean per tick of the required data, resulting in one row of averaged data per tick per different

[4]https://ccl.northwestern.edu/NetLogo/docs/behaviorspace.html.

Table 4.1 NetLogo BehaviorSpace table output .csv example

[Run number]	Ratio-of-tracking-app-users	#random-seed	[Step]	#infected	#youngs-at-start	...
1	0	0	0	0	312	...
2	0.6	0	0	0	312	...
1	0	0	1	0	312	...
2	0.6	0	1	0	312	...
1	0	0	2	0	312	...
2	0.6	0	2	0	312	...
...	...	...	...	...	...	...

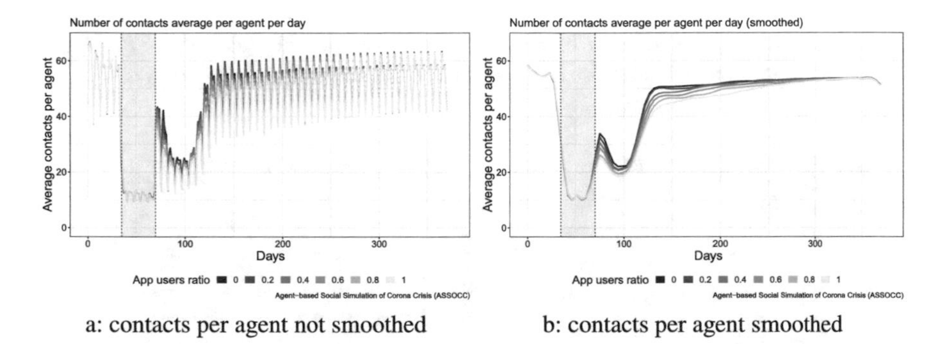

a: contacts per agent not smoothed   b: contacts per agent smoothed

Fig. 4.3 The average number of contacts per day per agent non-smoothed (left) and smoothed (right)

experiment setting. Despite this averaging, the resulting charts can remain fairly jagged, as can be seen in Fig. 4.3a. The graph is an average number of contacts per agent per day and the jagged pattern is caused by variations in weekend vs. weekday contacts. In this form, it is difficult to estimate at a glance whether the red line or the blue line represent more contacts, in addition the red line is partially hidden behind the blue, further complicating the task. To be able to see the general trend and compare the different lines we apply a smoothing function that is automatically given by the *ggplot* function. Figure 4.3b shows the smoothed result of the same plot, in which the trends of the lines is much more obvious. Now we can see that the red line actually stays higher than the blue line after the lockdown period (the red vertical block).

4.3.3 Presenting Results

This section presents the types of charts we generate, together with a brief explanation of specific data processing steps characteristic for each of them. For more advanced data visualisations, see the next section on our use of the Unity game engine. We will now describe the four general graph types that are generated with our R code.

4.3.3.1 Line Graph Per Day

Figure 4.4a shows a line graph which is the most common graph we are using. The data is either directly plotted as is done in the figure or it is manipulated to account for the difference between ticks and days, i.e. summing or averaging per four ticks. Each day contains four ticks so when plotting the data per day (as is done in Fig. 4.3 for contacts per day) we sum the data per four ticks to get the daily contacts.

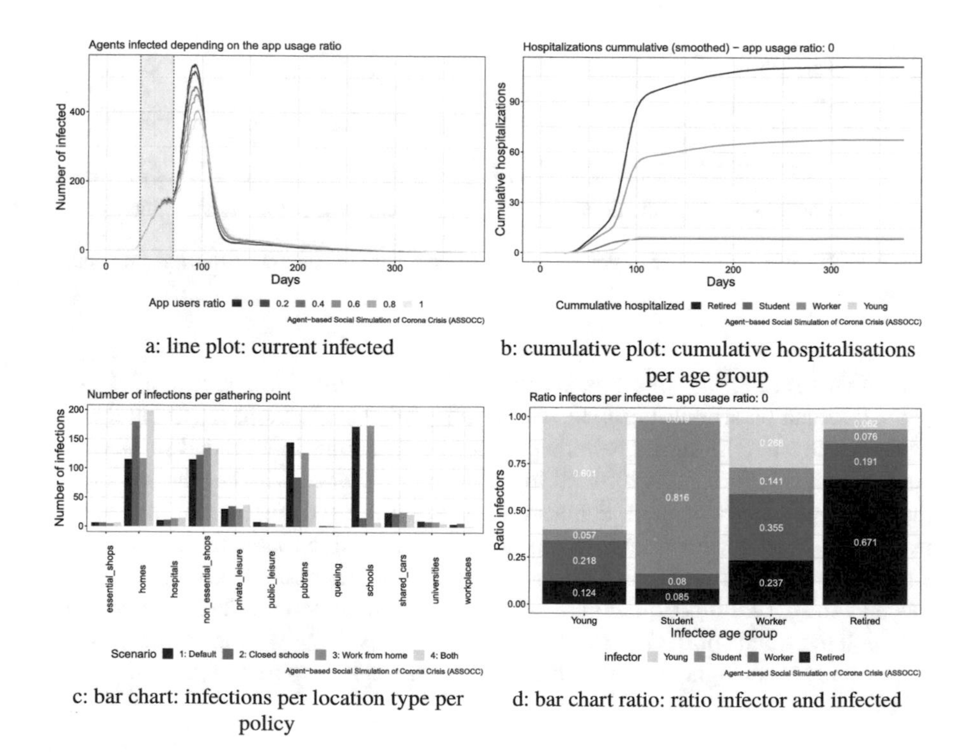

a: line plot: current infected

b: cumulative plot: cumulative hospitalisations per age group

c: bar chart: infections per location type per policy

d: bar chart ratio: ratio infector and infected

Fig. 4.4 The types of plots that the standardised R code generates

4.3.3.2 Line Graph Cumulative

Figure 4.4b is a line graph showing accumulated numbers. Dependent on the variable it is either directly plotted from the data or it is inferred when the data only has per tick values by accumulating the values.

4.3.3.3 Bar Chart Per Location Type

Figure 4.4c shows a bar chart of the number of infections per gathering point. This type of figure makes it easily possible to see at which location the most infections took place and how they differ among the different run settings.

4.3.3.4 Bar Chart Ratios

Figure 4.4d shows a bar chart with easily visible ratios. It shows per age group for the infected individuals the proportion of age groups they got infected by. For example the young are in 60.1% of the cases infected by other young (the light blue bar). Since there are different total amount of young, students, workers and retired every column is normalised. This makes the comparison between ages much easier as the bars have the same height.

4.3.4 Automated Generation of Plots and Data Files

The code can automatically generate the graphs in either one big .pdf or in separate .pdf's such that each plot has its own .pdf file. The former makes it is easy to check the simulation outcomes and share the results to colleagues, since it is only one .pdf file, while the latter is more convenient for use in (scientific) articles as each plot is saved in its own appropriately named file.

It can be tough to derive the exact values in a graph just from the graph alone. Therefore the code automatically saves the data used to generate the plots into separate .csv files. This makes it easy to write about the graphs, as specific data points (e.g. the data and the number of infections at the peak of an infection curve) in the graph can be mentioned based on the data in the accompanying .csv.

4.3.5 R, Crisis, Large-Scale Simulations

Given the high pace set by the crisis, there is a high pressure for analysing these models very fast, both for presenting results to stakeholders and grounding the behavior of the model in the highest (feasible) scientific standards. R is well suited for generating one specific plot, but the costs become prohibitive as many plots are to be produced due to everchanging input and output variables. The syntax and facilities around R just male difficult to automate this process (a large amount of effort has been put in this direction, but it would require one person working full-time on it). These efforts decreased towards the end of the project, as the core model became more stable and some data standardisation could finally be counted on. However, because R requires specialised knowledge to use effectively, the communication overhead between R developers and NetLogo developers remained rather higher than desirable, as the requests for new plots or updates of old plots kept coming in until the very end.

An ideal future development would be the access of interactive analysis platform, which would allow modellers and analysts to swiftly build on their own the plots they need with minimal requirement of understanding the technical intricacies of the model. This tool would relieve the pressure on the specialised programmers while at the same time cutting short the analysis cycles and overall (re)analysis efforts.

4.4 Visualisation Module

Stakeholders require different communication tools than the one used for scientific communication. NetLogo, for instance, offer a great compromise between the user-friendliness and expressivity for scientists to play with a simulation. However, this visual can be too cryptic for the stakeholder, who wants to be empowered by tools rather than submitted to technical (unintended) obfuscation. A significant communication effort is required for engaging stakeholders.

One of the main challenges of communicating to someone outside of one's field is that they lack the plethora of background information an expert takes for granted and unconsciously uses every time they assess or interpret a problem or result. It's one of the reasons academic communication is so impenetrable: much of the information goes unwritten or unsaid because it is assumed that everyone involved is already familiar with it. This means that effective communication to someone outside academia must determine and then explicitly present this missing information. The more engaging this presentation, the more effective the communication is likely to be, which only adds to the difficulty of the task.

We decided to use the Unity game engine to build a suitable interface that could display the simulations in real time as they were being run in Netlogo. We also wanted the interface to have some degree of interactivity so that users can freely navigate the simulation world, requesting more detailed information about aspects of the simulation that catch their attention, since this is a better way to increase engagement compared to passive observation (Fig. 4.5).

As a game engine, Unity can be used to build simulation worlds in vivid detail and display data in ways that agent-based simulation software are not designed to. That is not to say that Unity doesn't have its own drawbacks that would pose challenges should it be used as a simulation engine, but these are not the subject of this book.

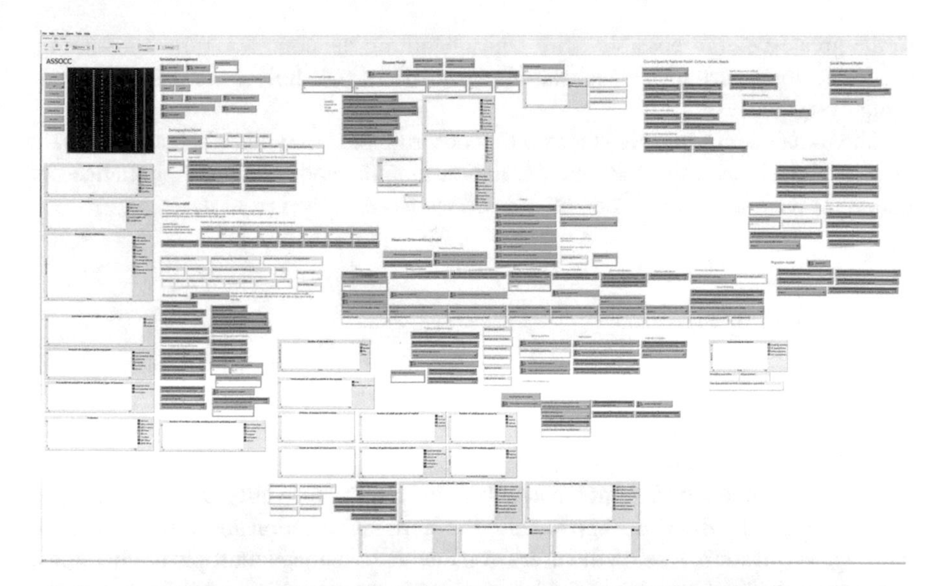

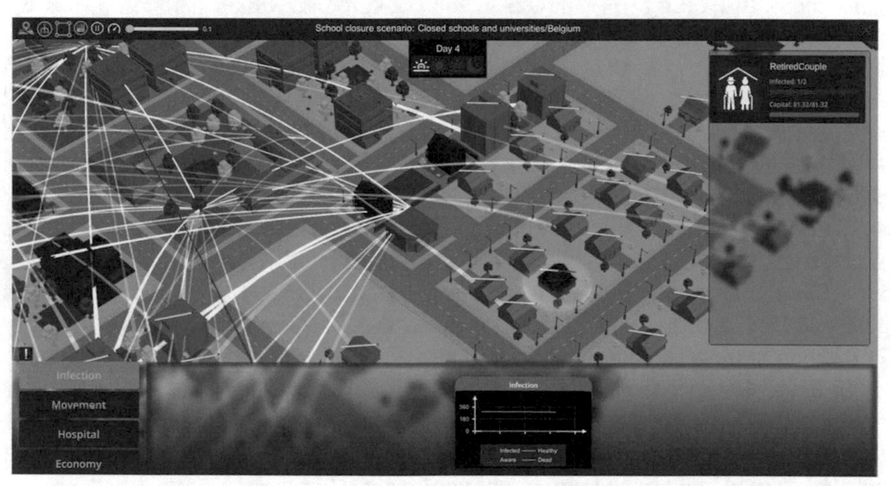

Fig. 4.5 NetLogo (up) versus Unity (down) interfaces

Rather, in our case, the main issue with using Unity as a simulations engine comes from our conflicting communication goals with academia and stakeholders. To be effective for stakeholders, the interface must go beyond the requirements of academic communication by including substantial additional information for the users. To be effective for academics, the model must be implemented in a platform that is widely used by academics. At the same time, if the user interface and the model development tasks are too strongly interconnected, the task becomes a nearly insurmountable time-sink, which is far from desirable when racing against the clock in a crisis. As such, the visualisation module is separated from the simulation module to the greatest extent possible, with Unity handling the complex, time-consuming, academically-superfluous visualisations designed for stakeholders, and Netlogo handling everything else.

The remainder of this chapter will first describe the Unity interface, then explain how it interacts with the Netlogo simulation module, and finally the challenges of making it work under crisis conditions and what we learned from it.

4.4.1 Ease of Use

Since the Unity app was designed with the needs of non-academic users in mind, the application is meant to be easy to use and intuitively organised, not just visually appealing, at every step.

Running simulations does not require keeping track of the Unity and NetLogo apps separately because the Unity app handles the start, initialisation and shutdown of the NetLogo application, freeing the user from having to manage multiple applications. Connecting to NetLogo is handled in the background, which also simplifies the user-side process to clicking one button. Despite this simplification of the process, NetLogo doesn't run in the background, so the user has full access to the original interface and code, as well as to the connection logs keeping track of the commands and data being passed between Unity and NetLogo. Thus if a user is proficient in NetLogo she can in principle access NetLogo during the run and check all separate parameters, adjust the once that are run-time adjustable, etc.

The Unity builds (as the version of the implemented ASSOCC framework to be accessed through Unity are called) we provide come bundled with the NetLogo files, as well as the Java relay app, which connects the two. As such, provided the user has Java and NetLogo installed on their machine, they can simply download the Unity build and run the executable, reducing the number of steps required to get started.

In the Unity app, the clutter of buttons, sliders and input fields present in the NetLogo GUI have been neatly organised by simulation scenario, which allows users to quickly select and parameterize the type of simulation they want to run without getting overwhelmed by the number of parameters (Fig. 4.6).

For the simulation world, we built a small city with houses, schools, shops, a hospital, university and park. The location is a floating island in space, contrasting the otherwise "realistic" setup, to emphasise the fact that we're not simulating a real place and the results are not supposed to be predictive of the course of the pandemic in any one real world region. The map layout is fixed and does not vary between scenarios, with the exception of some of the residential buildings, which change from houses to blocks of flats to accommodate the larger agent populations in some scenarios. The agents themselves are only represented by their trails as they move about the map. These trails are drawn as arcs between the agent's starting point and destination, white by default and red if the agent is infected.

Other data visualisations during the simulation reflect the global state such as charts, the infection tree, the social network, as well as of specific objects in the sim-

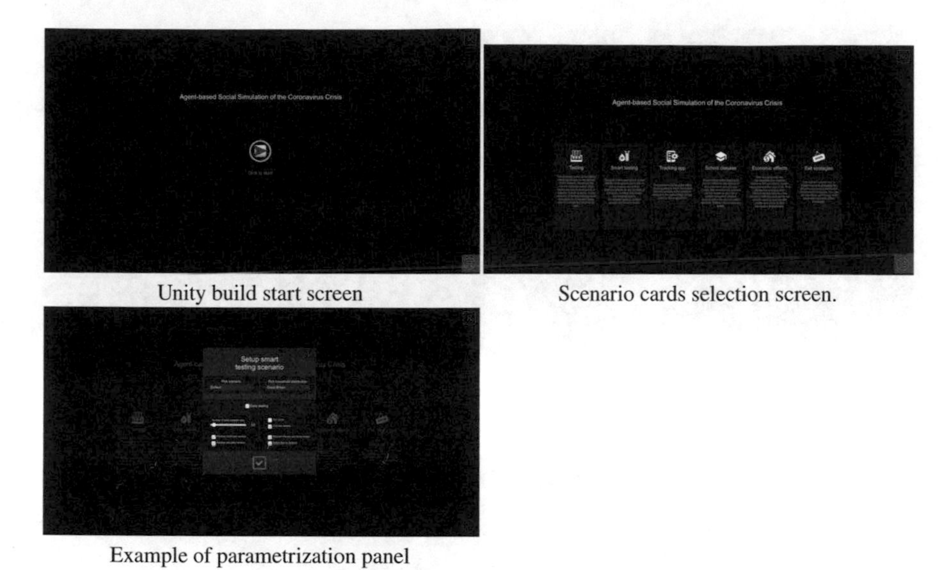

Unity build start screen

Scenario cards selection screen.

Example of parametrization panel

Fig. 4.6 Starting the unity interface

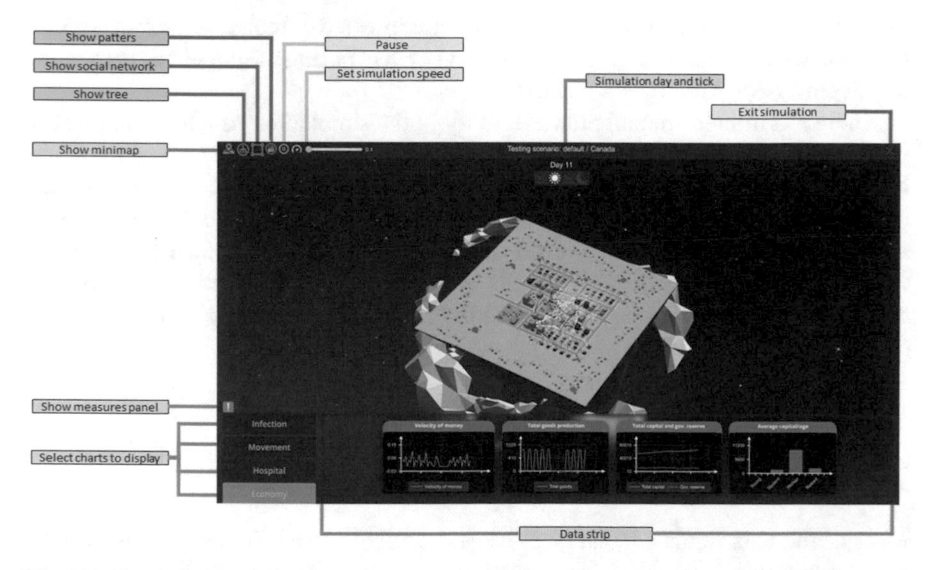

Fig. 4.7 Simulation map with legend

Fig. 4.8 Building data showing living arrangement and infection status for people living in the selected building (with the blue highlight)

ulation, such as how many people in a house are infected or their living arrangements based on what we found to be of interest to visualise during the simulation run while developing ASSOCC (Figs. 4.7 and 4.8).

There is a limited amount of interactivity in the simulation visualisation, most of it related to displaying/hiding information, either global or specific, about the simulation. The only exceptions are the simulation-speed slider and the world navigation controls. The slider allows users to slow down the simulation and then speed it up again. The world navigation controls allow the user to move around the map and zoom in and out as desired. While this is a step above what NetLogo offers, this setup doesn't allow users to intervene in the simulation at runtime, which NetLogo does allow.

4.4.2 *Visualisations*

Most of the data in the simulation is displayed using charts because charts are the obvious solution for most of the simulation's output. There is no need to complicate something that works perfectly fine (Fig. 4.9).

Since there are many charts in the app, we organised them by subject and used a tabbed structure to display them. Only one data tab is visible at any one time, but the user is free to click through them at their leisure.

In addition to classic charts, we built data visualisations that take full advantage of Unity's graphical versatility. These are the kinds of visualisations that would

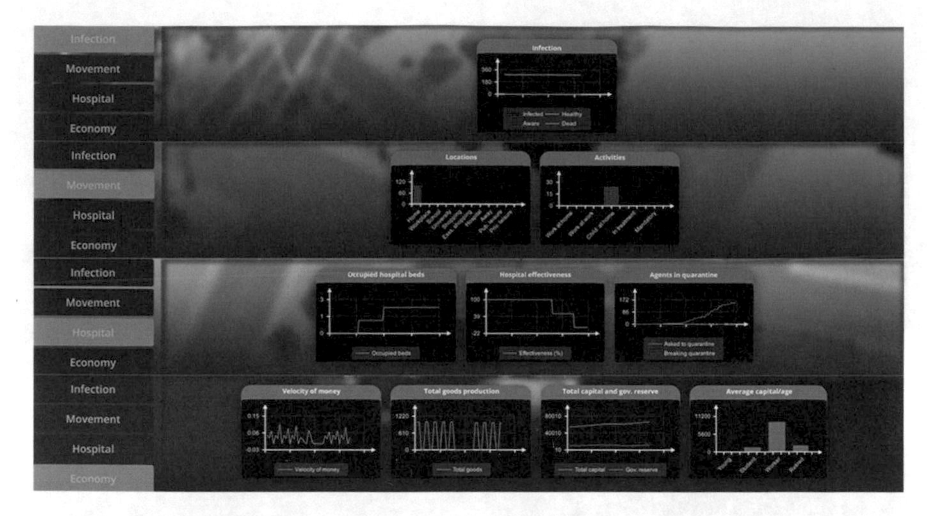

Fig. 4.9 Charts in the data strip

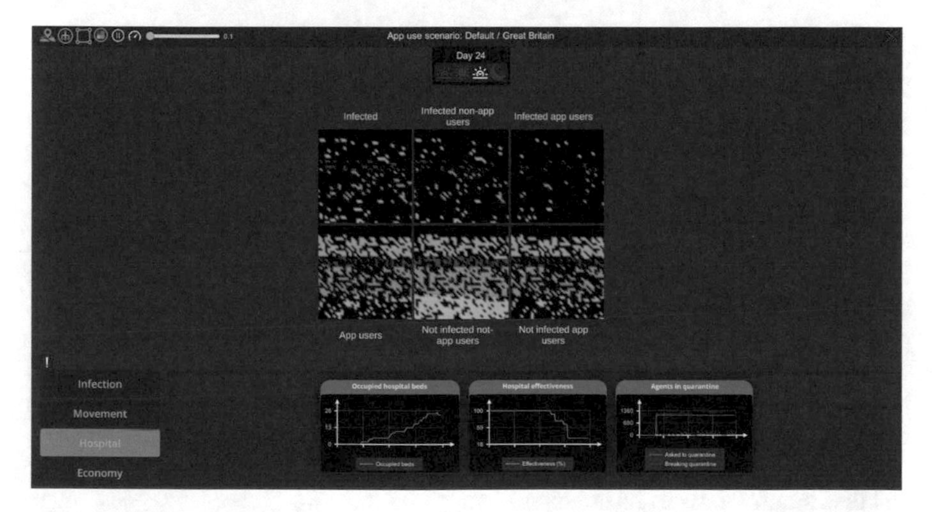

Fig. 4.10 Patterns visualisation

be nearly impossible to build with software dedicated to agent simulations such as NetLogo or Repast, or be made to work in real time in more graphically capable software.

E.g. the visualisation in Fig. 4.10 takes advantage of the human pattern matching and recognition ability. Each cell in each of the six matrices corresponds to an agent, and one cell corresponds to the same agent in every matrix. Each matrix is used to visualise a binary property of the agents. If a cell is black, the property is false, and if the cell is coloured, the property is true. The different colours of the cells represent the ages of the agents. In our case, the properties we are interested in are whether

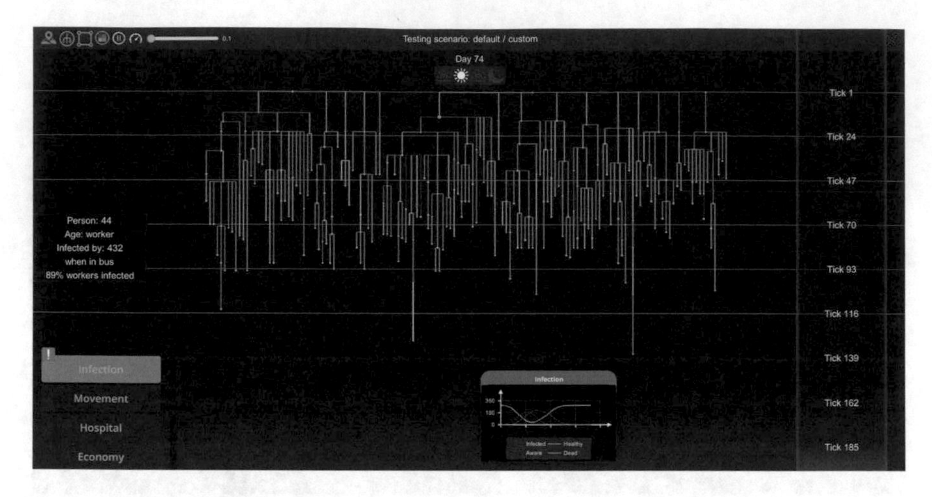

Fig. 4.11 Infection tree visualisation

an agent is a user of a track-and-trace app and whether it is infected. The colours green, pink, orange and blue correspond to the age categories of young, student, worker and retired, respectively. When using this visualisation, it is easy to estimate at a glance things like which age categories are using the app more, whether more infected agents are using the app than healthy agents, or whether there are more infected agents using the app than not using the app.

The infection tree in Fig. 4.11 is used to represent the path of the infection through the population of agents, showing whether some agents spread the infection more than others, which age categories are more likely to spread the infection, and which places see more infection spread. Each node in the tree is an agent and nodes are connected if one agent infected another, with the parent nodes infecting their child nodes. The colours of the nodes correspond to agent ages and link colours correspond to the location where the infection occurred.

The tree is interactive: clicking on a node will display information about the agent and the circumstances of its infection.

The social network visualisation is used to show the spread of infection in the social network of the agents. This network is defined in NetLogo and doesn't change throughout an agent's life. Each node is an agent and edges represent a social link between agents. The edges are blue by default, and turn red if one of the agents infected the other. In our model, social networks are built around workplaces and schools/universities. This results in small, fairly isolated groups of workers and retirees, and large, interconnected groups of young and student agents (Figs. 4.12 and 4.13).

Like the tree, the social network is interactive: clicking on a node will display information about the agent.

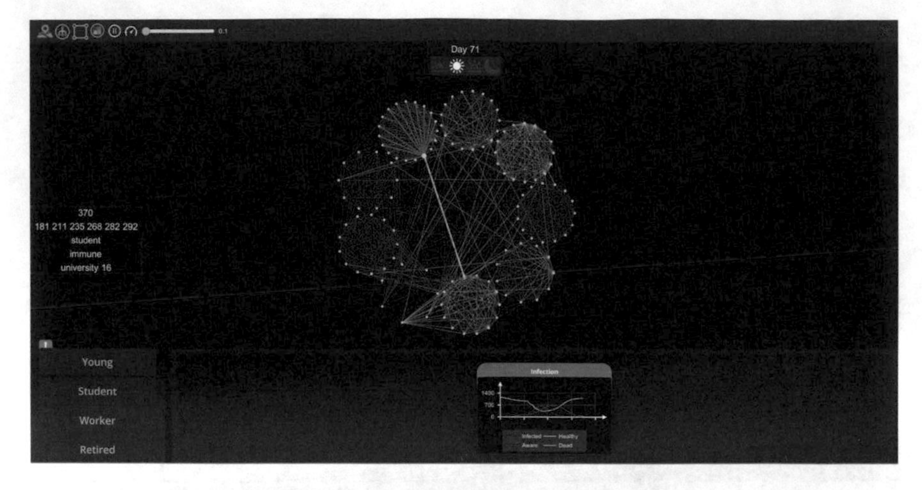

Fig. 4.12 Social network visualisation—student network

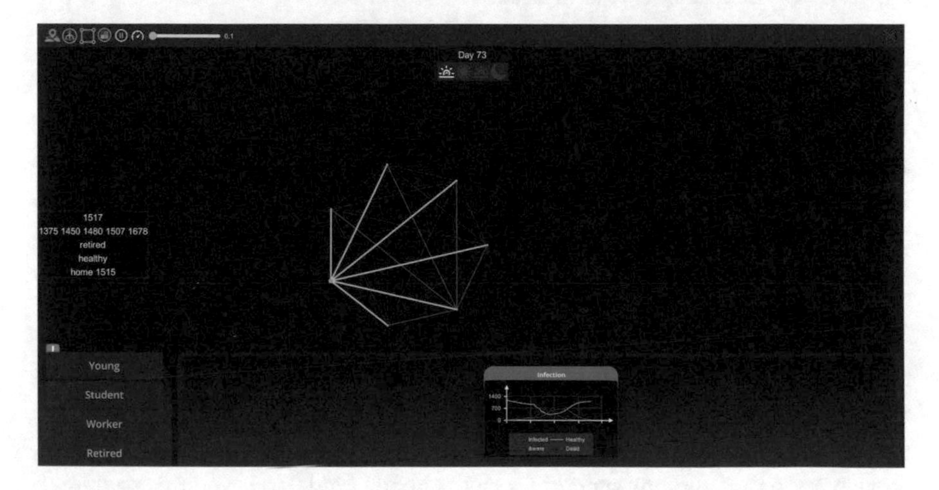

Fig. 4.13 Social network visualisation—retired group

4.4.3 Results Summary

The summary of the results of the simulation is a collection of charts showing the data for the simulation run that just ended. Some of these charts were present during the simulation run, but others are only generated once the simulation ends. We use a tabbed structure to organise the charts, similar to the one used for the data strip that is available throughout the simulation run (Fig. 4.14).

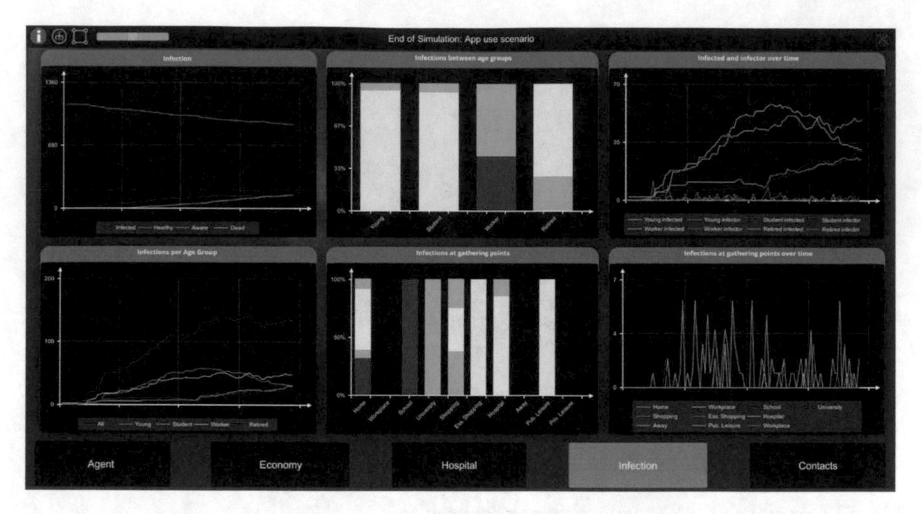

Fig. 4.14 Aftermath summary of the simulation

4.4.4 Implementation

This section describes some of the more salient implementation details regarding the visualisation module. The first part briefly touches on how the Unity visualisations were achieved, while the second part deals with how we connected Unity to NetLogo.

4.4.4.1 Visualisations

Charts

Building charts in Unity is a time consuming undertaking because Unity doesn't provide any out-of-the-box support for charts of any kind. However, there are many readily available assets, some of them free, in the Unity asset store that can be used to speed up the procedure considerably and we strongly recommend using them. We used GraphMaker[5] for all the charts in this application.

The Pattern Visualisation

This visualisation is not difficult to build, but requires some in-depth knowledge about how Unity represents and displays objects. The agents are mapped to vertices in the meshes that make up the mSectionatrix objects, and their properties are used to set the vertex colours. The six matrices exist on their own dedicated layer and are rendered with their own dedicated camera to ensure they do not overlap any other visual elements.

[5]Asset Store link: https://assetstore.unity.com/packages/tools/gui/graph-maker-11782.

The Contagion Tree Visualisation

This visualisation is a fairly challenging one to build because Unity doesn't provide any out-of-the-box support for tree or graph visualisations. We made our own tree data structure and our own heuristic layout algorithm for displaying the tree. The tree, like the pattern matrices, exists on its own layer, rendered by its own dedicated camera. The nodes and links are simple 3D geometric objects.

The Social Network Visualisation

This is another challenging visualisation because Unity doesn't provide support for graphs visualisations. Similarly to the tree, we built our own graph data structure and our own heuristic layout algorithm. The layout groups agents by age, and then groups them again by workplace, school or university. Since the social network doesn't change at runtime, unlike the infection tree, the network structure and layout only need to be calculated once.

4.5 ASSOCC Architecture

In this section we describe the overall architecture of the ASSOCC system and some of the salient implementation details that are important when trying to connect Unity to NetLogo.

4.5.1 Unity-NetLogo Communication Setup

Since the Unity display is supposed to mirror the NetLogo simulation, we first designed and built the necessary simulation elements in Unity and assembled them into the simulation world. Thus every agent in NetLogo has a corresponding entity in Unity. Similarly the homes, shops, hospitals, etc. of agents have counterparts in Unity. However, the agents in Unity do not duplicate the decision mechanism from the agents in NetLogo. Decisions are made in NetLogo agents and the results communicated to the Unity counterparts. Neither is the disease model replicated. This part of the process came with challenges that will be discussed in Sect. 4.5.4. However, in short, without input from NetLogo, the Unity world is "inert" and displays no behaviour of its own. It needs the input from NetLogo to drive the actions of the agents and indicate their social consequences. Since there is no easy way to get Unity to communicate directly with NetLogo, we took advantage of the Java API for NetLogo and wrote a small Java application that would send commands to and request data at runtime from the NetLogo simulation. We then used sockets to connect the Java application to the Unity application, effectively turning the Java app into a relay point between Unity and NetLogo. It took a bit of experimentation to figure out a good communication flow between the applications, and we eventually settled on the one presented in Fig. 4.15.

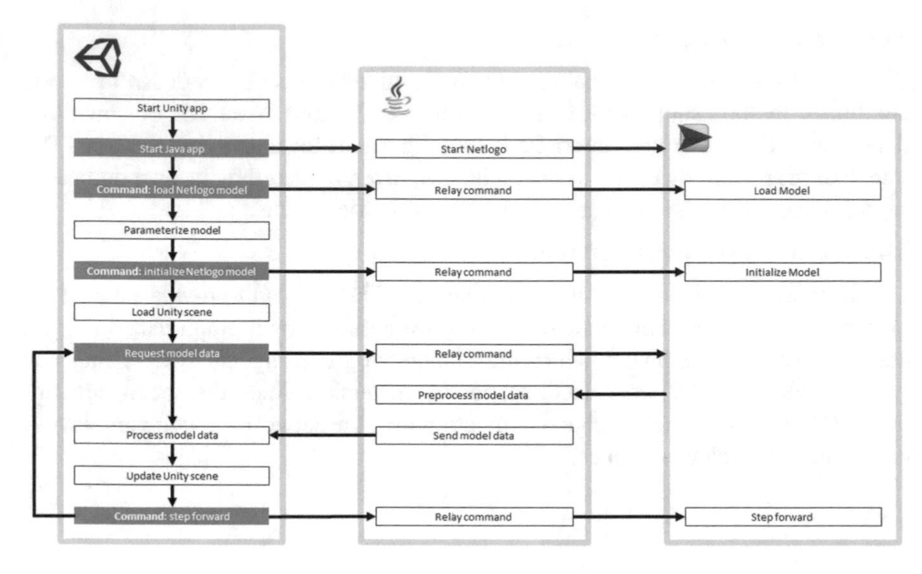

Fig. 4.15 Simplified call stack for the Unity-NetLogo connection

We faced two main challenges when deciding how the two applications should communicate. First, Unity and NetLogo do not work at the same "pace", and second, the amount of data being transferred between the two could slow down the process to such a degree as to render the final application unusable.

4.5.2 Syncing Unity and NetLogo

NetLogo's step is the simulation step. Unity's step, however, is the frame (as in animation frame) and a proper application that is visually pleasing (or, at the very least, not irritating) needs to run at 30–60 frames per second, which makes it unfeasible to match the two applications step-to-step. In order to bypass this issue, we let Unity drive the simulation. This means that Unity decides when NetLogo should move to the next simulation step and then commands NetLogo to do so. While NetLogo processes the next step, Unity has time to display and animate the data for the current step. This puts the Unity simulation world always slightly behind the NetLogo simulation world, but it makes no noticeable difference to our setup (or to our purpose). In our simulations the NetLogo simulation is still quick enough to provide Unity with a next step in the required frame rate. This renders this connection relatively straight forward to manage. This would become more of a problem if the NetLogo step starts to become too slow. Then more elaborate solutions are needed to keep the synchronisation working properly.

4.5.3 Data Bundling

Because the simulation is both large and complex, a lot of information is being passed from NetLogo to Unity through the Java relay, with varying degrees of processing at every step, and this can cause issues if the data flow isn't managed appropriately. In our case, the best solution is to have fewer calls, but larger data volumes being passed during those calls. As such, Unity never queries NetLogo for data on individual agents, instead asking for data on the whole collection of agents at once, every simulation step. This means that every time Unity displays data on request for an individual agent, that data was already present locally and no calls went out to NetLogo for it. The implementation is rather simple: we added reporters in the NetLogo code to provide the required data to the Java relay, the Java relay transforms the data it receives from NetLogo into a JSON format, then passes it to Unity and Unity maps the received JSON stream to the object structures it uses to create and display the simulation world.

4.5.4 Crisis Considerations

4.5.4.1 Specific Requirements

As time-efficient development tool, Unity is not very high on the list. Using Unity requires coding skills—preferably knowledge of C#—and scene building knowledge that is specific to Unity. Someone who lacks both, will have to climb a pretty steep learning curve if their first Unity project is an interface for a NetLogo model, especially a complex model. In our experience, coding skills are by far the more important of the two, so, in the absence of a Unity developer, the best next choice is a decent general programmer who wouldn't mind learning one more tool.

It might be convenient to hire an outside Unity developer, but this solution comes with a few caveats. First, and least important, game engine developers tend to specialise into code developers and visual assets/art developers, which means that a good Unity programmer might not be well versed in building a good looking visual interface. This is not the greatest of hurdles when the alternative is relying on the Netlogo interface as an alternative. Hiring a team of developers would easily deliver the best results, if cutting edge visuals are important to the project, but this solution is more costly. It is the second caveat that is harder to solve: the hired developer needs to continuously keep abreast of fast-paced changes in the model, sometimes with poor or incomplete communication to the modelling team, depending on how heavy or chaotic their workload becomes. Unless the developer has some basis in ABM and knowledge of Netlogo, updating the visual module in good time might be impossible to handle properly.

4.5.4.2 Assets

Using Unity properly also requires some knowledge of visual design, UI, UX, and possibly shaders, because visuals will do a lot of work in the final product. Luckily, this is a problem that can be solved by spending some money. Unity is a popular and well established game engine and its asset store can provide many useful bits to make building new applications easier, which is boon when time is of the essence.

We used Odin Inspector[6] and Peek[7] to speed up the workflow, GraphMaker[8] to make most charts, a couple packs of UI elements to build the UI (buttons, progress bars etc.), a couple packs of scene assets to flesh out the world (buildings, rocks, trees etc.) and DOTween[9] for animation (because, in our case, tweening was easier, less computationally intensive and less time consuming than animation). These are all paid assets, although some, like DOTween, have a free version too. There are plenty of free assets and asset packs for the financially frugal, but paid ones tend to be more comprehensive, or just better looking.

It is not impossible, or even difficult in many cases, to make your own assets using some elbow grease and other free software (such as Blender[10]), but it is invariably time consuming. Since we were working in crisis mode, getting as many ready-made elements as possible allowed us to focus on the actual code, visual and interaction development, and the workflow assets sped up the process considerably.

The one drawback to using paid assets is that we cannot share our full project freely and sharing it with the assets removed would be pointless since the final product wouldn't be functional anymore.

4.5.4.3 Crisis Architecture

The architecture is designed to keep as much separation between the simulation module and the visualisation module as possible. Given the crisis conditions under which the model was developed, fast paced changes were inevitable and keeping the two modules too closely intertwined would only have added to the effort and time required to keep them both functioning properly. Furthermore, we designed the architecture with a one-way dependency from Unity to Netlogo, which allowed the simulation module to evolve completely independently at its own pace, without being slowed down by considerations regarding visualisations. This independence did come with its own cost in terms of communication effort since the Unity side of the team needed to be kept appraised of the latest Netlogo developments, but, under the circumstances, it was a convenient trade-off.

[6] Asset Store: https://assetstore.unity.com/packages/tools/utilities/odin-inspector-and-serializer-89041.

[7] Asset Store: https://assetstore.unity.com/packages/tools/utilities/peek-editor-toolkit-149410.

[8] Asset Store: https://assetstore.unity.com/packages/tools/gui/graph-maker-11782.

[9] Asset Store: https://assetstore.unity.com/packages/tools/visual-scripting/dotween-pro-32416.

[10] Blender home: https://www.blender.org/.

4.5.4.4 Design Choices

Much of the visual design of the Unity app changed considerably from the initial stages to the current final state. We'll discuss some of these choices in this section.

Some of the changes in design came from our attempts to optimise and distribute computational load, but the main discussions happened around the issue of communication accuracy. The choice of spatial representation in the model does not map well onto a 3D space like the one used in the Unity visualisations. This led to a couple of instances where we had to pick between keeping the visualisation in line with the model data—and thus end up with uninspiring or confusing visuals—or "fill in the gaps" in spatial information for more compelling visuals—and risk giving users the wrong understanding of how the simulation works.

The map design is the main instance of this conflict and it went through a few iterations before we settled on the one presented here. Some of the issues had to do with optimising and distributing computational load (we cut out shadow rendering completely, for instance), but the bulk of the problem ended up coming from the scarcity of spatial information in the model. The agents gather in places where they are "physically" close enough to pass on the infection, but those places have no relation to one another. The closest accurate visualisation of this spatiality is probably a roiling cloud of location points the agents can move between and interact inside of. This means that our choice of map layout has no bearing on simulation results, which is both good and bad. It is good because we can choose a map layout that will be easy and fast to build, and time is always at a premium in a crisis. It is also bad because, by imposing a city layout over the nebulous cloud of locations the simulation actually uses, we might inadvertently end up sending users the message that geography matters. To make matters worse, geography does matter in a pandemic, it is just that our model accounts for only some of it. For instance, in real life, disease outbreaks sometimes tend to cluster geographically because people who work, study or shop together also tend to live in the same general area of a city. In our simulation, where people live has nothing to do with where they perform any of their other life activities, but the map layout may send the opposite message. Fortunately, our simulations are small enough in scale that the chaos of agents' movements around the map is not very noticeable.

Like the map design, the representation of the agents required a few iterations to reach its current final state. We decided fairly early on after a bit of experimentation that representing individual agents wasn't a very good use of computational resources because the agents' movements around the map are far to fast to add anything other than noise to the visual display. However, not representing them at all takes away one of the main points of the model, which is that agents do change location throughout the simulation, coming into contact with one another and spreading the infection. While the paths they take or the distances they cover are entirely irrelevant to the model, their destinations and the number of other agents at those destinations matters quite a lot. In the end, we settled on marking the agents' movements by drawing an arc between their starting and ending point, which completely ignores the fact that roads exist and emphasises the gathering in groups at various places during different simulation steps, bringing the visualisation more in line with the actual model concepts. It has the added benefit of saving us computational resources that would have otherwise gone to pathfinding. While this is a decent compromise as visualisations go, it is also a result of working in crisis mode. If we had more time to iterate through the modelling and development process, we might have added some space representation to the model to support the visualisation, found a cleverer way to convey agent movement or design the map layout.

4.5.4.5 Data

The data we choose to present is not all the data the simulation generates, just what the researchers in the team decided would yield the most insight in visual form. While useful and interesting, this approach fails to take full advantage of the communication opportunities this type of interactive visual interface can offer for possible stakeholders. Unfortunately, the crisis working conditions and pacing didn't allow much time for testing the setup "in the wild", but it is a direction worthy of more investigation in future work. It is very possible users would like to be able to see other data or see the data presented in different forms or have the option to desegregate certain data—all of which can lead to further insights both in what the simulation offers and in how different people interpret and relate to the simulation scenarios—but for now the current state reflects the mindset of the development team more than anything else.

Each scenario has its own most relevant data, and, in the beginning, we did keep them separated and gave them dedicated data visualisations. However, as the number of scenarios increased, we found it preferable to allow the user access to all data regardless of which scenario is being run. This resulted in a more complex representation of the world, which reinforces the idea that these simulations can generate a lot of data and keeps the sense of all scenarios happening in the same "world" (or sandbox). It also cuts down on redesign efforts (adding a visual element to one scenario often resulted in requests to add the same element to another one—adding everything everywhere seemed like the logical next step in the process). There is an argument to be made for highlighting relevant data depending on scenario, but the time constraints and the iteration speed on the NetLogo side made this option unfeasible for the time being.

4.6 Management Challenges

Given the relatively large team (around 15 persons) involved in this project, and its separation into groups working on vastly different modules (or different components of the same module), it was unavoidable that we'd deal with some management challenges. Most of them stemmed from insufficient communication, which, in turn, stemmed from the relatively small amounts of time we could afford to spend on this issue during the crisis. The string separation between modules helped mitigate many of the problems that might have otherwise arisen, since much of the work could be done within the bounds of modules without the need to involve someone unfamiliar with their inner workings, who would need to be brought up to speed before they could meaningfully contribute. It's not surprising, then, that there were no communication issues between people working on the visualisation and those working on the analysis module since the two are fully independent of each other. However, they are both dependent on the simulation module and that's where most if the issues appeared, as expected.

4.6.1 Netlogo

As the core of the system and source of data for the other two modules, the Netlogo module was the main driver of change in the system. Every model update or implementation change could ripple through visualisation or analysis and, thus, the Netlogo side of the team had the task of bringing the other members up to speed with the latest developments. Because of the fast-paced development, we agreed to keep much of this communication fairly high level and update the full documentation once we were reasonably sure we did not have to do it again within a week or less. It was far more time-efficient to leave it up to the Unity and R team members to bring up specific issues as they encountered them. This did result in a hilarious number of "what did you rename [variable] to" questions and might have caused way more issues if the team had been bigger or the modules more interconnected.

4.6.2 R

To make automated data processing worth the time and effort required to setup, the data in question needs to be standardised. Since the model kept evolving, a full stable standardisation was not forthcoming until later in the project. To cope with this, the R code is split into scripts that each deal with a single aspect of the data processing or analysis, which makes it easy to adapt and update as needed. Fortunately, many of the R developers were also involved in model development so there was minimal need for communication related to script-breaking changes.

4.6.3 Unity

The Unity team was rather firmly split between the Unity and the Netlogo side. This, together with the whole project operating under crisis conditions, inevitably led to some management challenges.

On the Unity side, the chief issue was the iteration speed on the Netlogo side, by virtue of its frustration causing potential. As the model became more and more complex and the Unity app started receiving more and more data from Netlogo, keeping track of changes became our main concern. Changes in the Netlogo code could mean Unity was no longer able to get the data without changes to the Java relay or use it without changes to its internal data structures. On a (mercifully) few occasions, changes in Netlogo didn't break the data pipeline, but rather changed the meaning of the data. The Unity side of the team spent quite a bit of time puzzling over new and inexplicable deviations in the app's behaviour caused by this type of miscommunication.

As a result, the Unity team resolved to keep the Unity app 1–2 weeks behind the Netlogo version and advance development with possible future Netlogo versions in mind. This did not fully eliminate the need to make changes to the Java relay in order to properly access the required data, but it considerably slowed down the pace at which they occurred. The data structures used by Unity were designed to be extended, rather than changed, which made for some classes with unused fields that hung around after outliving their usefulness, but it made for very fast mapping and remapping of the data from one Netlogo version to another without having to make significant code revisions on the Unity side. Given the rapid development pace, the Unity team also focused on modularity and reusability. Keeping as much separation as possible between data acquisition and processing modules and the rest of the code was, for us, not just general good practice, but a necessity.

4.6.4 Management Under Crisis

In order to meet the demands of working under crisis conditions, we had to prioritise fast prototyping and adaptation, which led to a number of choices that rippled out into the management of the team. Chief among these was the strong separation of the three modules, which led to the fragmentation of the team based on the primary focus of the members. Since the simulation module was the only truly independent one, this lead to some increase in the communication effort between the three groups regarding the latest developments in the model. These communication issues could have been overcome by implementing a documenting protocol that everyone can refer to as needed. However, given the chaotic nature of the process and the extremely fast pace of development, we decided it was less time consuming to address questions and comments as they came up and document everything once the project came to an end and no more changes were planned.

4.7 Conclusion

ASSOCC is ambitious in its goals and goes beyond the usual simulation methodologies to achieve them. Since there does not exist a single software solution that can fulfil all our requirements, we built three separate specialised modules that together cover all requirements. We chose NetLogo for the simulation module primarily for its fast-prototyping capabilities, which are of utmost importance when developing models under crisis conditions, but also kept in mind the widespread use of NetLogo in the social simulation community. The latter hopefully leads to other parties picking up the NetLogo model and use it (or parts of it) in their own simulations. We chose R for its wide spread use for statistical analysis and presentation use. We chose Unity for its vast graphical potential and strong developer ecosystem which provided us with numerous examples and out-of-the-box solutions for a modern, easily accessible user interface. By separating these three concerns into three different modules and keeping the modules relatively autonomous, the team was able to maintain a fast development pace throughout the development period.

One downside of working under crisis conditions is that speed of development and reaction to current situations takes precedence over full exploration of some of the potential features of the system. In this chapter, we discussed a few of the possibilities that are not realised yet. In Chap. 14 we will discuss future development of a social simulation platform for a resilient society in more detail. In Chap. 15 we will discuss some of the major issues that arise when developing a more general social simulation platform for these purposes.

Acknowledgements We would like to acknowledge the members of the ASSOCC team for their valuable contributions to this chapter. We wish to thank the team behind the development of NetLogo, and in particular Nicolas Payet, who assessed our code and provided some useful support for further optimising it. We also like to acknowledge the work of Tomas Sjöström who made substantial contributions to the visualizations in Unity.

Reference

1. Michael J. North et al., Complex adaptive systems modeling with repast simphony, in *Complex Adaptive Systems Modeling*, vol. 1, no. 1 (2013), pp. 1–26

Part II
Scenario's

In part I we have laid down the foundations and implementation of our simulation framework and argued that it is at least a good example of a simulation framework for simulations for crisis situations. In part II we will discuss a number of scenario's that have been run using the ASSOCC framework. They are intended to give an overview of the type of results that are interesting for a crisis situation. They also show how wide a range of results can be obtained and how these results can be combined when using a single, more abstract framework.

Chapter 5
The Effectiveness of Closing Schools and Working at Home During the COVID-19 Crisis

Mijke van den Hurk

Abstract In this chapter we show the results of simulations of two widely adopted measures that were taken in order to stop the spreading of the COVID-19 virus, namely *closing of schools* and *working from home*. We take these two measures together because in practice they are often instated together and at least parents with young children will have to stay at home if the children cannot go to school. We will simulate different scenarios in order to separately examine the effects of closed schools and people working from home on the number of infections, hospitalisations and social contacts, and the effect of the combination of the two measures. Although we expected a positive impact to come from people working from home, we see that closing of schools has the best results on decreasing the number of infected people. Working remotely has a negative effect as infections and hospitalisations are higher when people work from home. We will look into where and how many social contacts take place and how this results in the transmission of the virus. We will see that a decrease in physical social interaction is not enough to suppress infections by imposing these measures. The behaviour of people will change in such a way that smaller gatherings at busy locations cause almost as many infections as without the imposed measures.

5.1 Introduction

The closing of schools is one of the first measures that is taken when a country is facing an epidemic. This is because schools are places where both children and parents from a community gather, which potentially leads to further spreading of the virus. A lot of countries imposed this measure as soon as they got hit by the COVID-19 pandemic. The closure of schools has some major drawbacks. One of these drawbacks is the negative impact on the economy, in particular because people cannot go to work because they have to stay at home to take care of their children.

M. van den Hurk (✉)
Utrecht University, Princetonplein 5, 3584 CC Utrecht, The Netherlands
e-mail: m.vandenhurk@uu.nl

F. Dignum (ed.), *Social Simulation for a Crisis*, Computational Social Sciences,
https://doi.org/10.1007/978-3-030-76397-8_5

As nowadays a lot of households are dual earner families it is very common for grandparents to step in and look after their grand children. However, during this particular pandemic they were part of the group being at risk for the corona virus and were now recommended to keep social contact to a minimum. This meant that the closing of schools forced parents to stay at home, which put a burden on families where both of them are working. Some countries were therefore reluctant to close the schools, like Sweden and the Netherlands. In particular they were afraid that parents who worked in health care had to stay at home too, while they were urgently needed. Besides the potential dropout of essential workers, research at that time showed that children were less susceptible for the virus than adults and did not get sick or only showed moderate symptoms [1]. Thus, effects of the closing of schools on decreasing the number of infections and hospitalisations might be caused by parents staying at home, rather than the closure of schools itself.

At the same time working remotely from home was introduced as a separate measure aiming to prevent people of many age groups to get together at work and possibly infecting each other and subsequently the people in their own environment (family and friends). So, this restriction also resulted in adults staying at home. The digital age that we live in makes it relatively easy for some sectors to make employers switch to online meetings and working remotely. This measure reduced the number of social contacts with less or no gatherings of people at workplaces. At the same time travelling decreased which meant that clusters of infections where less likely to cross between different communities throughout the country.

The imposition of working from home makes it even harder to measure the effect of closing of schools as both of the measures implicate people staying at home. Since the closure of schools could potentially lead to a burden on vital sectors, such as health care, and the economy in general, it is particularly important to know if this measure is even effective.

We use the ASSOCC model to examine the effect of the *closing of schools* and *working at home* measures on the number of infections, hospitalisations and moments of contacts. We expect that both measures will lead to less social contacts. We also expect that the closing of schools itself will not have much effect on the number of infections or hospitalisations, as children are already less likely to spread the virus. However, the parents of children staying at home will have less contacts at work and thus put a hold on the spreading of the virus. We will therefore also take the *working from home* measure into account, to look at the effect of people staying at and working from home and children going to school. We expect that both of the measures will flatten the curve.

The next section will give some more background on the *closing of schools* measure in previous epidemics and pandemics, and the reluctance of some countries to impose this measure during the COVID-19 pandemic. Then we will describe the scenarios, the implementation specifics of the model regarding these scenarios, and the settings used during the simulations. Thereafter we will discuss the results of the simulations, followed by the discussion and conclusion.

5.2 Background and Context

The closing of schools is one of the first measures that is taken when a country is facing an epidemic, as schools are places where both children and parents from a community gather. This mixture of people from different age groups and families might potentially lead to further spreading of the virus. In case of influenza-like illnesses it has been proven that closing of schools is an effective and necessary measure, since children are highly susceptible for this disease [2] and also more vulnerable for this disease. An example of a pandemic where this measure was imposed was the Spanish flu in 1918. However, the COVID-19 pandemic is different, as it seems to be less contagious for children [1] and children under 12 years hardly show any symptoms and are thus not at risk for COVID-19. Early research already suggested that the spreading of the disease and the risk groups were different than for influenza-like diseases, like the Spanish flu. This means that the effect of closing schools might be smaller in case of the COVID-19 pandemic and also unnecessary.

Moreover, closing schools has some additional drawbacks [2]. First of all there is a potential economical loss, caused by parents who cannot work because they have to stay at home taking care of their children. Secondly, it has a negative impact on the economy in the long term, as children and students fall behind in their education [3].

These drawbacks are the reason that some countries were reluctant to impose this measure right away, like the Netherlands [4], or even kept the schools open, like Sweden [5]. For the Netherlands, the main concern of imposing this measure was the fact that the children would need some type of child care. Normally, families might use grandparents, but with the COVID-19 virus this age group was especially at high risk. Parents, who are mostly workers, are thus forced to stay at home, while some of them are needed in vital sectors. Their absence could lead potentially lead to a bigger burden on the health care sector. The schools were closed eventually, due to the fact that schools were struggling to stay open as too many teachers had to stay at home with corona-like symptoms. An exception was made for families with both parents having an essential job. Schools and day care were kept open for their children.

5.3 Purpose

We use the ASSOCC model to examine the direct and indirect effect on the spread of the virus when schools are closed. We expect that the main reason for the spread going down will be caused by people being forced to work at home to take care of their children. This will cause less interaction between adults, as they do not encounter each other at their workplaces. The question is, if closing schools will lead to the same expected results as for influenza like diseases, i.e. a flattened curve of infections, or if it has little or no effect at all, or even worse, intensifies the spreading of the virus. We will measure this effect not only by keeping track of the number of infections, but also by looking at the amount of patients in the hospitals and the number of

Table 5.1 Scenario descriptions

	Closing of schools	Working from home
Scenario 1	False	False
Scenario 2	True	False
Scenario 3	False	True
Scenario 4	True	True

social contacts for different locations. The number of hospitalisations tells us how much the healthcare system becomes overwhelmed by corona patients. The number of contacts gives us context on the behaviour of people and where infections take place.

5.4 Scenario Description

We will explore the effect on flattening the infection curve when schools are closed with and without people being advised to work at home. We will first run a baseline simulation, followed by a simulation with the measure *closing of schools*. Then we will run a simulation where only *working from home* is active. The results are compared with the final simulation, when both the schools and offices are closed. An overview of the four scenarios is shown in Table 5.1.

5.4.1 The Model

The timing of closing the schools influences the effect the measure has on the spread of the virus. A school can be closed once a student or teacher is found to be infected, or closure can be imposed in a more proactive manner, i.e. before an infection in a school is detected. In this case, the schools will be closed as soon as a certain ratio of the population of the city is infected, regardless of the number of infections at schools. When children are staying at home, at least one adult should be at home to take care of them. This adult will then, if possible, be working from home. Other workers will go to their office.

When the measure *working from home* is imposed, only the non-essential workers will be expected to do their job at home. All non-essential jobs are modelled as workers going to an office gathering point. When the *working from home* measure is simulated all offices are closed from the start of the simulation. People that work at the hospital, (non-essential) shop or school will be allowed to go to their workplace. This measure will be active throughout the whole simulation.

Table 5.2 Settings

Parameter	Value
Probability-hospital-personel	0.04
Probability-school-personel	0.03
Probability-university-personel	0.03
Probability-shopkeeper	0.04
Ratio-agent-infected-that-trigger-school-closing	0.02
All-self-isolate-for-35-days-when-first-hitting-2%-infected?	False
Productivity-at-home	0.75
R	5.75

5.4.2 Settings

In general, the settings are similar to the standard settings, see appendix B. This means that the scenarios are simulated with the national culture and household profiles set to Great Britain. This gives the best compatibility with the results in the other chapters with scenarios. Note, that it is also possible to run the same scenarios with profile of Sweden or The Netherlands. The number of households is set to 391 which results in 1126 agents. Differences relative to the standard settings can be obtained in Table 5.2. The variable R in the contagiousness function, see Sect. 3.5.1.3, is set to 5.75, i.e. 2 times lower than the default value. This is done to make sure that the spreading of the virus is limited. Because the scenario incurs only closing schools and working at home the number of infections would explode very quick with the default setting. This would distort the comparisons as almost all agents would be infected in a very short time and nothing could be deduced from the events after that. The probability of people that work at the hospital, school, university or a shop are set to 0.04, 0.03, 0.03 and 0.04 respectively. These percentages are roughly corresponding to reality in the UK, although the hospital workers in the simulation also represent the other health care workers. The rest of the working population works in offices (i.e. 86% of the workers go to the office in normal times). The probability for people to work at regular offices is very high, such that the effect of *working from home* has more impact and effects are more visible within the time of the simulation runs. The schools will be closed when a certain ratio of people has been infected. This ratio is set to 0.02. The variable *all-self-isolate-for-35-days-when-first-hitting-2%-infected* is set to false, as both *closing of schools* and *working from home* would be activated. The value of *productivity-at-home* is set to 0.75. This means productivity of the workplaces will go down. This is higher than the default value, such that shops do not end up with empty shelves. All scenarios are simulated with 10 repetitions and 1200 ticks, which represent 300 days, with random seeds between 1 and 10.

Table 5.3 Dependent variables. The different age groups are young, students, workers and retired. The different locations are homes, schools, universities, workplaces, essential shops, non-essential shops, public leisure, private leisure, hospitals, shared cars, public transportation and queuing

Dependent variables	Measure
#infected-{age group}	Count
Cumulative-infected-{age group}	Cumulative count
#hospitalisations-{age group}	Cumulative count
{age group}-contacts-in-{location}	Cumulative count
{age group}-infected-in-{location}	Cumulative count

First of all we will look at the total number of infected people at every time step. However, this by itself will not give us enough information about the effectiveness of the measurements. A child that gets infected will probably not get sick or only mildly, while an older person, i.e. retired in our model, might need to go to the hospital. The latter situation has a bigger impact on the healthcare system and is therefore more severe. Thus, we will also look at the number of infections within each age group and the number of hospitalisations per age group. In addition, we investigate the locations where infections took place and the number of contacts per location, so we can measure the difference in movements of people caused by the imposed measures. An overview of the measurements is given in Table 5.3.

5.5 Results

All four scenarios are run for 1200 ticks, which corresponds to 300 days, and 10 times with different seeds, which makes 40 simulations in total. The result of each scenario shown in the plots is the mean of the different simulations of that scenario.

In order to know if the measures lead to the desired outcome, i.e. less infections and unburdening the health care, we will first look into the number of infected people over time and total number of infections and hospitalisations. We compare the results from the four different scenarios to see if our expectations are met, i.e. the *closing of schools* will have no effect, while *working from home* will lead to less spreading of the virus. We will also look into the number of contacts within each scenario to see if the measures lead to less social interaction.

5.5.1 *Effect of Measures on Infections, Hospitalisations and Contacts*

Figure 5.1 shows the average number of infections per day for each of the four scenarios. The total number of infected people per day are plotted, where a person can be sick for more than one day. In case of the default scenario the peak of infections is reached on day 43, with an average of 315 infections. When the schools are closed the peak is reached earlier, namely on day 36, and the number of infections is lower, namely 191 infections. This is a decrease of 40%. Scenario 3, with open schools but people working from home, shows the highest peak of infection with a maximum of 333, which is reached on day 43. This is an increase of 5.5%. Finally, scenario 4 where both schools are closed and people work from home has a peak of infections of 167, a decrease of 47%, on day 43. This means that only *working from home* has a negative influence on flattening the curve. However, in combination with *closing of schools* it lowers the peak only more compared to the scenario where only the schools are closed.

In Fig. 5.2, the cumulative number of infections are plotted. The default scenario has the most people getting infected, with a total of 752. This number is reached after 300 days, which means the virus has not disappeared at this time. The *working-from-home* scenario has a total of 749 infections. It is almost the same amount, but in this case the maximum number is reached after 247 days. This means that the virus died out. Scenario 2 and 4 have a total of 650 and 637 infections respectively. In both cases the number of infections is still increasing after 300 days which means the virus is still being transmitted. Thus, scenario 4, where both measures are imposed,

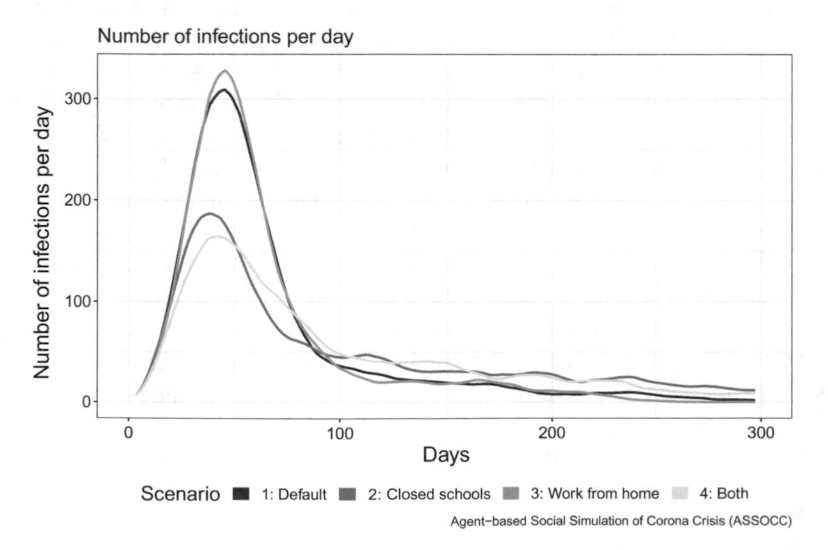

Fig. 5.1 The number of infections per day for each of the four scenarios

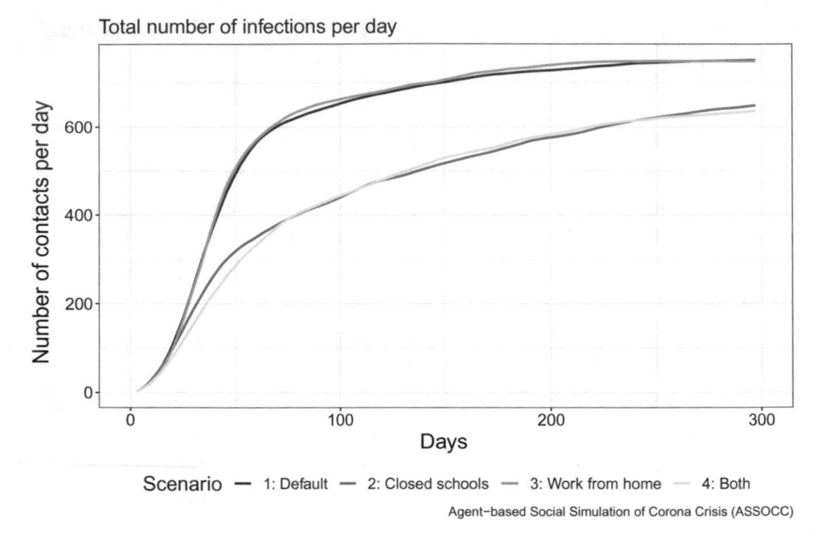

Fig. 5.2 The cumulative number of infections for each of the four scenarios

only has slightly less infections than scenario 2, although the peak is much lower. If the goal of the measures is to flatten the curve, both measures are better than only closing the schools. If the goal, however, is to limit the number of infections, the extra measure of letting people work at home does not have that much of an effect.

Next, we will look at the effect of these measures on the hospitalisations. The cumulative number of hospitalisations are shown in Fig. 5.3. The numbers reflect the effectiveness of the measures in the same way as the number of infections did. The measure *working from home* shows a slight decrease of the average number of people that have to go to the hospital, i.e. 163, versus the default scenario, i.e. 167. It should be noted that in case of working from home the number of hospitalisations stopped increasing after 242 days. If the simulations was run for a longer time the number of sever cases would become higher too. The lower average total number of hospitalisations in both the closed schools and default scenario corresponds to the lower number of infections, as seen in Fig. 5.2. In case of only closing schools an average total of 158 people go to the hospital. When both measures are active this number is 149. Again, we have seen that in these scenarios the virus was still active, so the total number of hospitalisations could end up higher. It is also interesting to see that the relative difference in the number of sever cases between scenario 2 and 4 is higher than the relative difference in infections. To understand this we have to look deeper into the results.

Finally, in Fig. 5.4 we plotted the average total number of contacts between people within each scenario. The default scenario has a total of around 17 million contacts, which means that there were 17 million moments in the simulations that the virus could have been transmitted. Intuitively, we would expect to see a decrease in the number of contacts for each of the measures. We see indeed that both the *closing of*

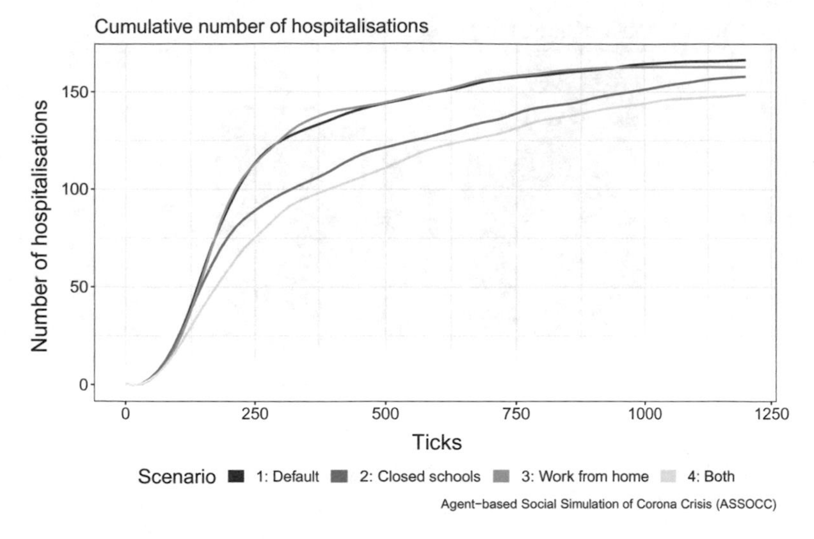

Fig. 5.3 The cumulative number of hospitalisations per scenario

schools and *working from home* measure reduce the number of social interactions, with a total of around 13 and 14.5 million. When both measures are imposed, this number becomes even lower, i.e. around 11.5 million. If we compare these results with the total number of people that get infected, we can conclude that a lower number of contacts does not necessarily decrease the spread of the virus. People working from home reduces social interaction compared to the case of people going to their workplaces. However, the number of infections and hospitalisations is higher. From these results we argue that reducing the moments of interaction is not enough to delay or stop the virus to spread. The places where social interaction take place seem to be important too.

The results so far are not as we expected. First of all, we expected that the closed schools would have no effect, as children are less susceptible for the corona virus, and therefore are less likely to spread the disease. We see however that this measure has a positive effect. Secondly, we see that, contrary to our expectations, *working from home* has a negative effect on the number of infections and hospitalisations. Furthermore, if we compare the number of contacts with the number of infections, we can conclude there is not a linear correlation between the two metrics. For example, the default scenario has more moments of contacts than the *working from home* scenario, but this is not reflected in the number of infections.

In order to understand these effects of the measures we have to look closer into the data, instead of looking at the totals. We will first look into the number of infections, hospitalisations and contacts per age group to see which age group is mostly effected by the measures. We will also examine the location of contacts to see where the increase in moments of interaction comes from in case of *working from home*. Finally,

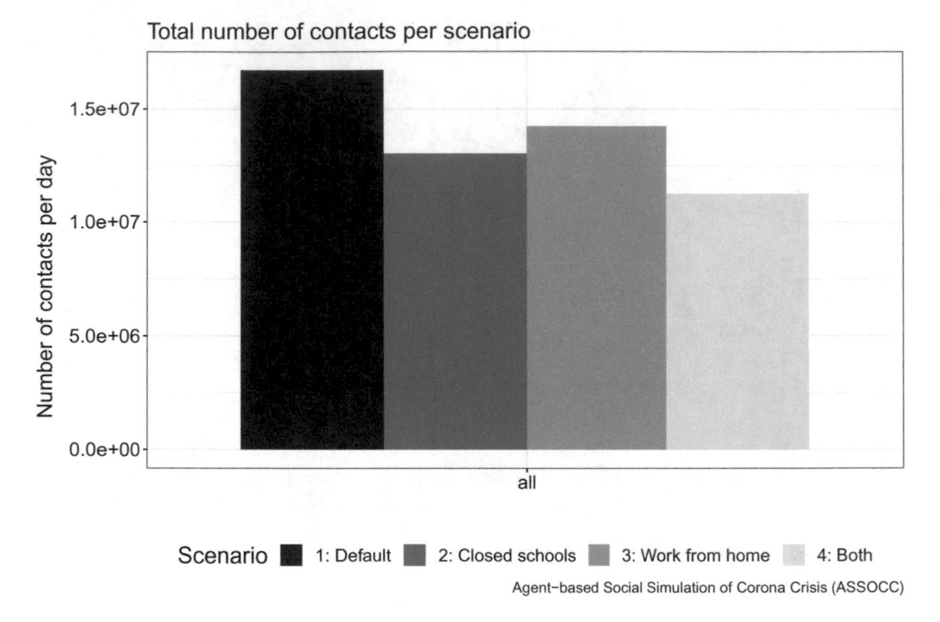

Fig. 5.4 The total number of contacts per scenario

we will also examine the location of infections in relation to location of contacts to understand why there is no linear correlation between moments of interaction and infections.

5.5.2 Infections and Hospitalisations Per Age Group

The first step we take is to check the infections per age group. This plot is shown in Fig. 5.5. We see that the number of infected children decreases when the schools are closed. The combination of the measures shows a similar decrease. The *working from home* measure seems to have a negative impact on the number of workers that get infected, which is counter intuitive. At the same do we see a decrease when schools are closed. Retired people get less infected in all scenarios compared to the default scenario, although the differences are minimal. We can conclude that the differences obtained in the total number of infections are mainly due to the effects of the measures on the children and the workers.

In Fig. 5.6 the amount of hospitalisations is split into the different age groups. We can see that most of the severe cases are among retired people. They are responsible for more than half of the hospitalisations. The total number of children or students that have to go to the hospital stays small in all four scenarios relatively, as they are less likely to get severely ill. For the workers the totals per scenario look similar to the totals in Fig. 5.5, i.e. the *working from home measure* causes more workers to be

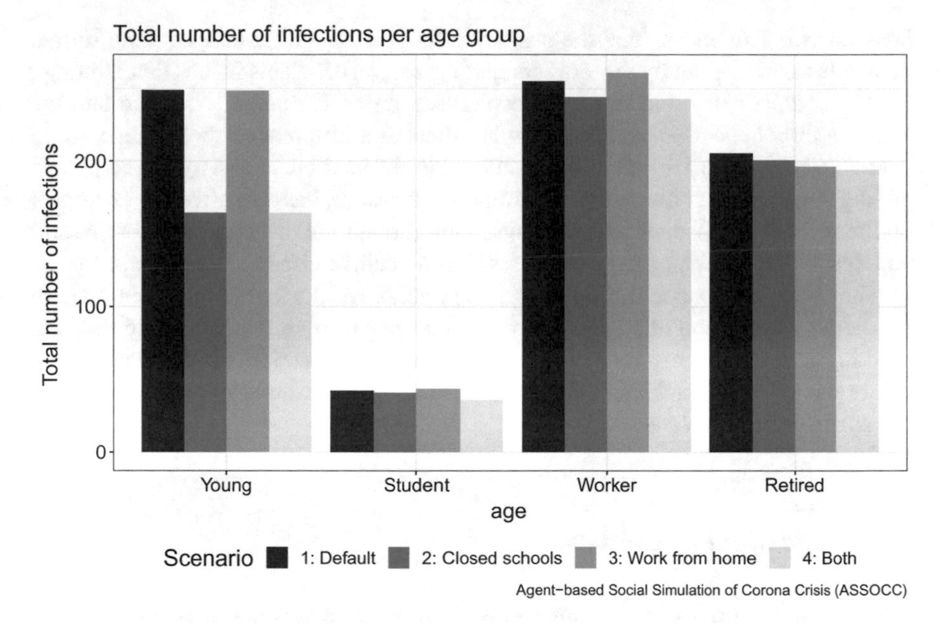

Fig. 5.5 The total number of infections per age group

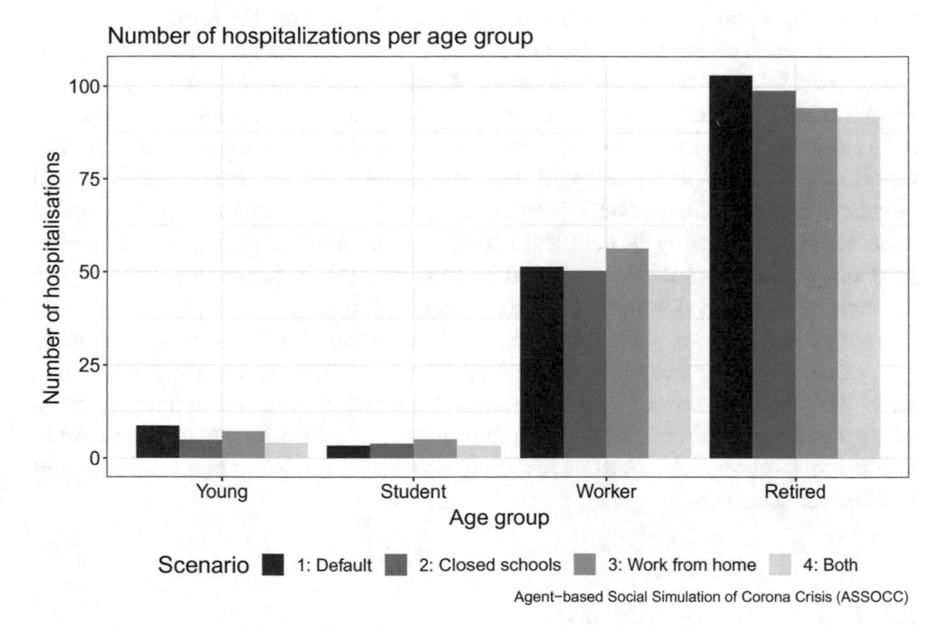

Fig. 5.6 The number of hospitalisations per age group

hospitalised. This means that the positive effect on the severe cases among retired people is cancelled out by the workers getting severely ill. This means that although the total number of infections and hospitalisations of the default scenario and the *working from home scenario* look similar, there is a difference in which age group is effected the most. We can also conclude that the small decrease in retired people getting sick, which comes with each imposed measure, has a relative major impact on the number of hospitalisations. Thus, even though the measures are not directly targeted at the high risk age group, it still has a positive effect on this group.

The above results confirm the surprising overall results, but do not seem to give any better explanation of it. The fact that workers get more infected when they are working at home seems counter-intuitive. In the next section we will investigate where the infections actually take place in order to see whether this can give us a more satisfactory explanation.

5.5.3 Location of Contacts

The number of contacts per location in each of the scenarios are plotted in Fig. 5.7. The average total number of contacts per scenario are now split according to the location where they took place. The result of closing of schools is obtained by far less social interaction at schools. The remaining contacts come from workers still going to the schools. We can explain this by teachers preparing and teaching digitally from school. Furthermore, we see no interaction at the workplaces when people are asked to work from home. However, the measures also have an impact on the number of contacts at other places. We will first compare the default scenario with school closures. This measure not only ensures an obvious decline in social interaction at schools, it also decreases the number of contacts in public transportation. This makes sense as children do not take the bus to go to school. At the same time, we only see a small increase in the number of social contacts at home. Looking at the *working from home* scenario, we see similarities with the *closing of schools* measure, namely a decrease in the number of people that take the bus. This decrease is smaller than when the schools are closed. We barely see an increase of more social interactions at home. Finally, we can see that both imposed measures makes the biggest decrease in public transportation. At the same time is the contact at home the highest, compared to the other scenarios.

5.5.4 Location of Infections

In order to understand why the number of contacts does not correlate with the number of infections, we will also look at the places where infections take place. Figure 5.8 shows a bar plot with the average sum of infections per location for each of the scenarios and Fig. 5.9 shows the same number but split for each of the age groups.

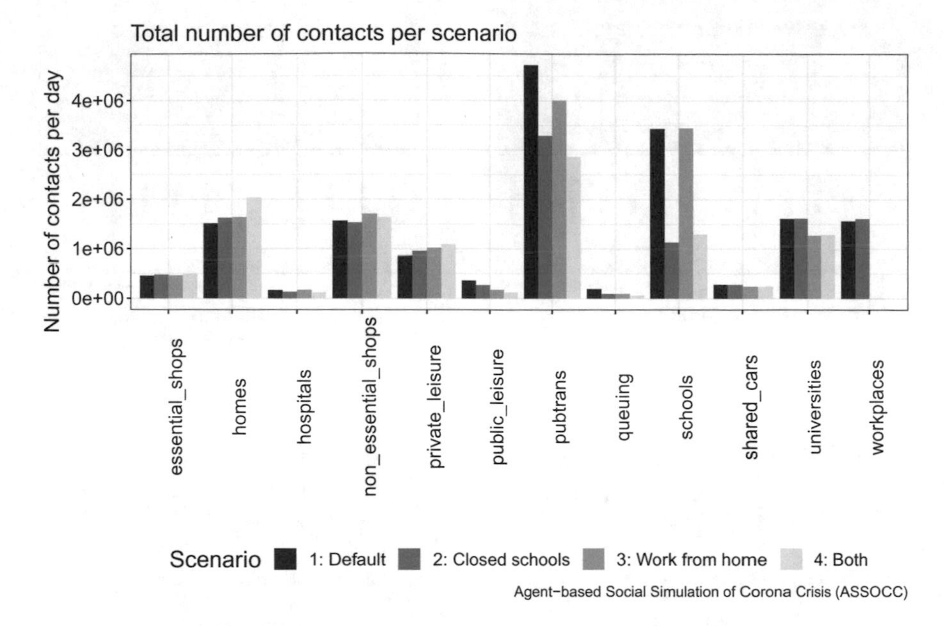

Fig. 5.7 The total number of contacts per location within each scenario

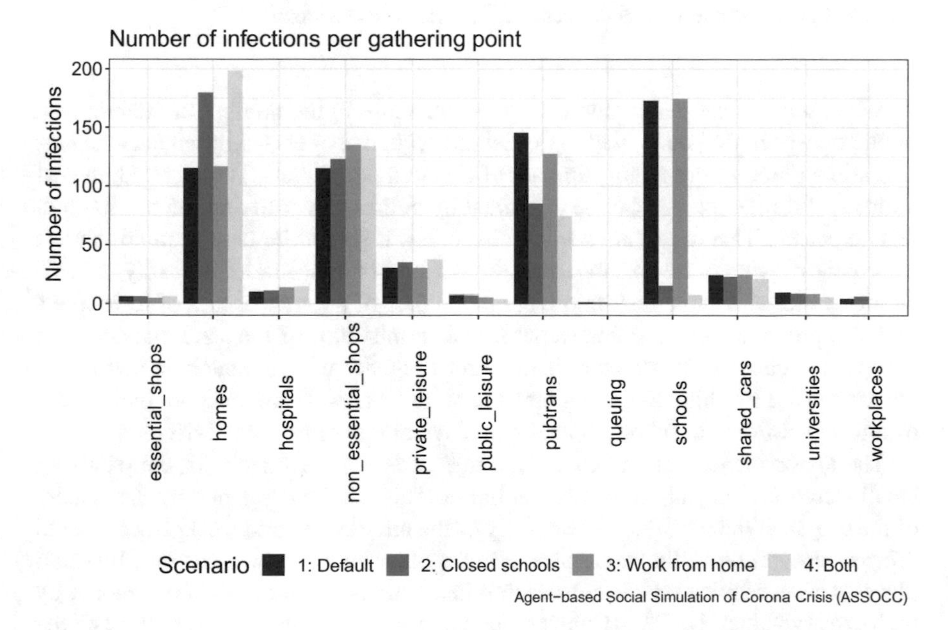

Fig. 5.8 The total number of infections per location

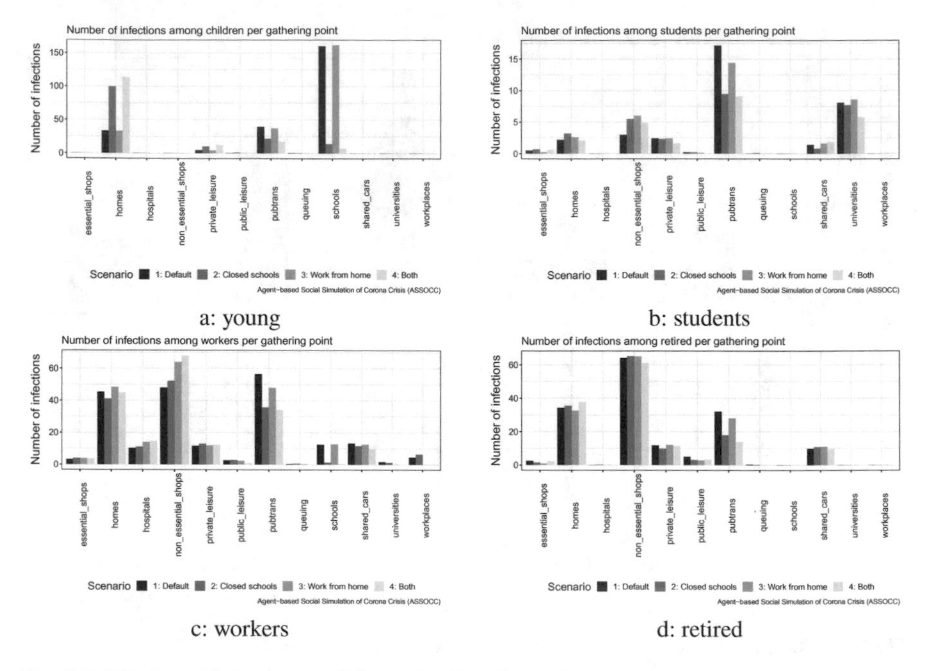

Fig. 5.9 Number of infections at different locations for each age group

When schools are being closed we see a decrease in the number of infections in public transportation compared to the default scenario, as less children take the bus. It is still a place where a large amount of children is infected. The fact that a lot of children still take the bus can be explained by children now taking the bus to go to leisure places. This might not seem realistic, but it should be kept in mind that we look at the *closing of schools* measure in isolation in this scenario. In reality, schools are never closed as sole restriction, but it is a part of a larger packet of restrictions that also prevents people going to pubs, restaurants, etc. At the same time, we see a major increase of infections at home. From Fig. 5.9 we can conclude that this is mostly caused by children getting infected at home now. If we compare the number of infected children at schools and at home, we obtain only a small decrease.

The major increase in infections at home does not correlate with the relatively small increase of social interactions at home. This can be explained by the density of the people at the locations. The density of the universities and workplaces is set to 0.2, as these are typically wide spaces. On the other hand, homes are typically small places where people stay close to each other. This is represented in our model by high density values, i.e. 0.95. Contacts are the result of possible interactions of all the people at the same location at the same time. When the density is high, the probability of actually being infected also becomes higher. The effect of contacts at a high density place is strongest at homes. The number of contacts in homes is only slightly higher in all scenarios but the default one. Looking at the number of infections, we see that when the schools are closed the number of infections increased substantially. This effect is even stronger when people also have to work from home. Thus, closing a relatively wide place, where big groups gather, results in small groups meeting at more high density places like homes. This will undo any effect of the closure.

Figure 5.2 showed that there was an increase in the number of infections when people had to work remotely, even though offices were closed. We can see in Fig. 5.8 that the number of infections at offices is relatively low. Closing them would not have that much of an impact. This effect, together with less workers being infected in public transportation, is undone by more workers now going to non-essential shops. The non-essential shops in our model are, just like the homes, places with a high density. This means that a small increase in contacts lead to a much higher increase in infections.

5.6 Discussion

The simulations in this chapter show how the measures *closing of schools* and/or *working from home* have an effect on the number of infections and hospitalisations in a city hit by the corona virus. We build a model to simulate these scenarios and expected that the closure of schools would not have that big of an impact on the number of infections, as children are less susceptible for this specific disease. Rather, a decrease in the number of infections and hospitalisations would be caused by parents that had to stay at home to take care of their children. This can become apparent when also introducing the measure of people working from home in our model.

However, results of the simulation show that *closing of schools* has a positive effect on the number of infections and hospitalisations, whereas *working from home* has even a negative effect on how many people get infected. Furthermore, only looking at the total number of infections within each scenarios does not yield enough information to understand the actual impact of the measures.

When looking at the total number of contacts between people, we see that *closing of schools* makes social contacts decline, which leads to no or far less children being infected in schools and public transportation. However, they are now more at home and have more free time. This results in asking friends over at home. The typically high density of people at those places make the probability of virus transmission higher. Workers also spend more time at home when they have to work from home. This causes a feeling of isolation, giving people the urge to also gather at each others homes or go to non-essential shops to meet with friends. This explains the increase in contacts at non-essential shops and leisure places. Another unexpected result is the number of hospitalisations in each of the scenarios. We expected less hospitalisations when offices were closed but the opposite is true. This is caused by more workers being infected and they are more likely to get severely ill.

From these results we can conclude that not all measures will automatically lead to less infections, although the total number of social interactions might decline. People will have the urge to meet other people and go outside. If these social interactions take place at smaller and more dense locations, like homes, the virus will still spread easily. Depending on what the desired outcome of the measures should be, i.e. less interaction between people, a decrease in infections or a decline in the number of hospitalisations, a measure can be successful or ineffective.

The way the ASSOCC model is implemented also leads to some limitations in the simulations that have some influence on the results. First of all, data from research showed that children are less likely to be infected and, if infected, show little or no symptoms. However, they defined children as being not older than 12 years. In our model children have ages between 4 and 18. Thus, we did not make a distinction between children being younger or older than twelve. Also the schools in our model represented both basic and high schools.

Furthermore, the model uses density settings to simulate people gathering in relatively cramped or spacious places. The non-essential shops have a high density setting and the probability of transmission of the virus is therefore higher compared to offices, which have a lower density setting. This is not seen through the number of contacts obtained from the simulations. The densities used in the simulation are a condensed representation for both the distance between the people in that location type and the duration of the contacts there. Research on the spread of the virus shows that the duration of contact and distance between people is an important indicator of getting infected or not. Therefore, an infection in a small shop is a bit more likely than in the office as people might stand in line for some time. At the same time, the office is a place where people sit apart from each other for a longer time. Except, of course, when they are in meetings or lunch breaks.

Also, the distribution of workers was quite arbitrary. The number of workers that had to go to an office was set to a high value in order to make the effect of *working from home* as high as possible. However, this number was not validated by data on labour distribution in a country like Great Britain.

Finally, the model has a limited number of locations in order to resemble a city during a pandemic. This model was developed during the early stages of the pandemic and the most relevant locations where chosen. Only after a few months, it became clear that nursing homes play an important role when looking at the burden of healthcare and the number of deaths [6]. Adding this location would give more insights on what measure will effectively protect this risk group.

Further research could be done for a better understanding of the effect of the measures *closing of schools* and *working from home*. For example, the simulations are run with the Great Britain settings. With different household settings and culture preferences the outcome of *closing of schools* or *working from home* can significantly differ from the results presented in this chapter. Great Britain is also modelled as a rich country. Thus, people have the money to go to non-essential shops. Other settings might result in less social gathering or at different locations. Furthermore, more measures are usually adopted at the same time, like closing non-essential shops, social distancing and isolation in case of an infected person in one's household. It would be interesting to include these measure as well and explore the effects of the combination and interaction of multiple imposed measures.

5.7 Conclusion

This chapter explored the effect on suppressing the spread of a virus by imposing two measures, namely *closing of schools* and *working from home*, within a city in a UK setting. We were particularly interested in the effect of the *closing of schools* measure. We predicted this would have little or no effect, as the coronavirus has been proven to be less contagious for children. Any effect would be rather caused by the parents, that now had to work at home to take care of their children. We therefore also took the measure of *working from home* into account. The effects were measured by looking at the number of infections, hospitalisations, and social contacts. First of all, we saw that *closing of schools* was effective, regardless if people were working from home or not. This had a positive effect on the number of hospitalisations. This effect was even bigger when both measures were activated. The main reason for the positive effect was that less retired people have been infected in public transportation. This shows that effects of restrictions are often not direct and also can have side effects that were not foreseen. At the same time, *working from home* was not effective on its own. The number of social interactions declined, but this effect was cancelled out by side effects, for example by people going to non-essential shops or meeting at each others' homes and being infected at that place. Because in this scenario we investigated the measures in isolation, the side effects might not be all realistic. However, what is important to conclude from the simulations is that people will look at alternative ways to satisfy their needs at some stage. Therefore one should be prepared to investigate these side effects of the measures. From the simulation we can see that the side effects can counter balance all positive effects of some measure! Finally, we can also conclude that the number of contacts does not correlate with the number of infections. This counter intuitive effect is caused by people spending more time together at high density places like their home, which accelerates the spread of the virus. Thus we see that the abstract and complex model gives us important information about the effects of a measure. It is not enough to just look at the result of restrictions on the number of infections, but rather look at where infections take place, in which age groups and how this might change as effect of the restriction. These insights can be used to compose more effective and focused combinations of measures that at the same time have less severe consequences for the whole society.

The ASSOCC model shows how the effects of different measures can be explored and explained. We can look behind the number of people getting infected and examine who gets infected at which place. This provides us with insights on the relation between the measures and the results, caused by an intuitive change in behaviour of people or unexpected side effects. Therefore, this model can serve as a tool for policy makers to understand the long term effects of different policies and facilitate customisation of measures in order to reach a desired outcome.

Acknowledgements We would like to acknowledge the members of the ASSOCC team for their valuable contributions to this chapter. We wish to thank Loïs Vanhée, Christian Kammler and René Mellema for having developed the model, code, and part of the conceptualisation and the analysis on which the results of this chapter are established. For the graphs of this chapter we would like to thank Maarten Jensen for providing the code base to build the graphs in this chapter with. Last but not least we would like to thank Christian Kammler and Frank Dignum for reviewing and discussing the chapter.

References

1. Q. Li et al., Early transmission dynamics in Wuhan, China, of novel coronavirus-infected pneumonia. New Engl. J. Med. (2020)
2. S. Cauchemez et al., Closure of schools during an influenza pandemic. Lancet Infect. Dis. **9**(8), 473–481 (2009)
3. H. Cooper et al., The effects of summer vacation on achievement test scores: a narrative and meta-analytic review. Rev. Educ. Res. **66**(3), 227–268 (1996)
4. A. Deutsch, S. van den Berg, Netherlands to close schools, restaurants in coronavirus fight (2020). https://www.reuters.com/article/us-health-coroavirus-netherlands/netherlands-to-close-schools-restaurants-in-coronavirus-fight-idUSKBN2120KG. Accessed 12 Apr 2020
5. Sweden has kept schools open during the pandemic despite spike in cases (2020). https://www.france24.com/en/20200917-sweden-has-kept-schools-open-during-the-pandemic-despite-spike-in-cases. Accessed 12 Apr 2020
6. A. Fallon et al., COVID-19 in nursing homes. QJM Int. J. Med. **113**(6), 391–392 (2020)

Chapter 6
Testing and Adaptive Testing During the COVID-19 Crisis

Christian Kammler and René Mellema

Abstract The scenario presented in this chapter is investigating the potential effects of different testing policies in combination with isolating households. In particular we will explore the effect of isolating the household of an infected member and giving priority in testing for healthcare and education workers. Assuming that we have more tests available than necessary for the healthcare and education workers, the effect of different strategies for the leftover tests, don't test youth, test only elderly with leftover tests, and test everyone with leftover tests are investigated. The results show that the combination of *no priority in testing + testing everyone with leftover tests + isolation of the household of an infected member* is the best combination to "flatten the curve". Furthermore, the amount of deaths, the impact on hospitals, and the effects on people in isolation are explored. This scenario has been developed on request of regional Italian authorities.

6.1 Introduction

Random testing of people can be a good start to identify infections, but since no treatment for COVID-19 is available at the point of this writing, just testing alone without any further action is not beneficial. To achieve a reduction in the spread of the virus, it is important that the infected person is also going into self-isolation. This can help to reduce the spread of the virus, because then the infected person can't spread the virus to other people outside. The most important benefit from random testing, however, is to have an early warning system that can trigger timely and possibly focused restrictions when the number of infections in a certain region or group of the population rise above a certain maximum.

C. Kammler (✉) · R. Mellema
Umeå University, Mit-huset, Campustogert 5, 901 87 Umeå, Sweden
e-mail: ckammler@cs.umu.se

R. Mellema
e-mail: renem@cs.umu.se

F. Dignum (ed.), *Social Simulation for a Crisis*, Computational Social Sciences,
https://doi.org/10.1007/978-3-030-76397-8_6

However, random testing has also disadvantages. One of the biggest ones comes from the design of random testing. Since people are only tested randomly, it can be possible that potential risk groups or other crucial people for reducing the spread of the virus are not tested (enough). This becomes even more problematic as there will be only a limited amount of tests and testing facilities available at a certain point in time. To make the most out of the testing limitations, it can be interesting to investigate what effects the testing of different groups of people has. For example: it has been identified that elderly are at greater risk [1] and therefore it can be worth to see if focusing on them during testing can have a positive effect on flattening the curve. Also, early research suggested that children are not as strongly affected by COVID-19 [2]. To investigate this, it can be interesting to exclude children from the testing process and test other groups more and see if that has an effect in terms of flattening the curve. Furthermore, groups can be not only formed along the age dimension, but can also be done along professions that are more likely to come into contact with infected people, such as education or hospital workers.

To investigate the potential effects, we will explore the scenario requested by regional Italian authorities in which they wanted to explore the possible effects of testing around 5% of the population, which was deemed to be the complete maximum of testing capacity per day if all resources would be stretched to the maximum. We will specifically prioritise healthcare workers and education workers. These two professions have been requested, because they have a high risk to get infected at their workplaces, as teachers get in contact with a lot of children and students, and hospital workers can get infected by the sick people admitted to the hospital. They can then infect co-workers, family members and vulnerable people they work for. In our simulation we only have health care workers working in hospitals. Education workers in this case means primary/middle/high school and university personnel. Priority testing for this scenario means that the available tests will first be used to test the groups that will be prioritised before the leftover tests will be used for other people. Since these two groups together do not make up the entire amount of available tests, the influence of testing different groups with the leftover tests is also investigated. In particular, we will explore only testing the elderly, not using tests to test young people, and testing everyone with leftover tests. Overall we assume that 5% of the population is tested daily. This can be quite a lot, but this percentage was specifically requested by the regional Italian authorities, as they assumed that it might be feasible.

The focus is on the number of infections, hospitalisations and deaths. In addition, we will look at hospital effectiveness, the amount of people in isolation and the people that break isolation. Furthermore, we will investigate if isolating the household of an infected member has a beneficial effect for reducing the spread of the virus compared to isolating only the infected person.

6.2 Settings

The national culture and household profiles have been set to Italy. These can be found in Appendix A and B.9. The probabilities for the different professions, i.e. being hospital personnel, school personnel, and university personnel, come from the Italian authorities and the other numbers are the general numbers from the simulation model scaled up to 400 households and therefore 1158 agents. These general numbers can be found in Appendix B. The aim is to test 5% of the population. Thus, the number of daily tests has been set to 58. Table 6.1 shows the used parameters which differ compared to the general set-up in Appendix B. The variables concerned with the amount of hospitals and universities are called *hospitals-gp* and *universities-gp* respectively. We had to adjust the amount of hospitals and university locations for this scenario compared to the standard settings, to ensure a sufficient amount of workers at each place, given the probabilities provided by the Italian authorities. The amount of both locations was halved.

The requested testing specifics have been grouped into three groups to make the analysis of the results easier. The first group is called *Isolation Policy*. Here a distinction is made between whether the household of an infected person is isolated or only the infected individual. The second group is called *Testing Regimes*. This group makes a distinction between the following three options: don't test youth, test only elderly with leftover tests, and test everyone with leftover tests. Finally, the third group is called *Prioritising Regimes*. This group consists of the different professions that have been prioritised during the simulation. The following options are part of this group: prioritise both (education and healthcare workers), prioritise only education workers, prioritise only healthcare workers, don't prioritise anyone in testing. Table 6.2 summarises these different groups and their members. All possible combinations have been tested in the simulation with twelve repetitions for each combination and a time limit of 1500 ticks and thus 375 days.

Given these settings, it is important to point out the following consequences. We can see in Table 6.1 that the amount of healthcare and education workers already takes up $25(= 11 + 12 + 2)$ out of the 58 tests that are available each day. Therefore, prioritising more professions leads to less available tests for random distribution over the rest of the population. This also has another big consequence. Since more tests are available than the amount of health care and education workers, these people will get tested every day when they get prioritised in testing. Therefore, only 33 (prioritise both)/47 (prioritise healthcare)/44 (prioritise education) tests will get distributed randomly over the remaining(!) population. Another important consequence can be seen when looking at the amount of hospitals and universities. Reducing the amount of university locations and hospitals present in the simulation has an effect on the amount of contacts and thus also the amount of possible infections, as potentially more contacts can happen at these locations. Furthermore, we want to mention that the settings resulted in only one teacher per school location and per university location. Once a school or university teacher gets infected and has to stay home to isolate, we don't close the school or university location. We assume that a solution will be found

Table 6.1 Scenario settings

Parameter	Value
#households	400
#children	316
#students	124
#workers	438
#retired	280
#hospitals-gp	2
#universities-gp	2
Probability-hospital-personnel	0.026
#hospitals personnel	11
Probability-school-personnel	0.028
#school personnel	12
Probability-university-personnel	0.005
#university personnel	2
#available-tests	58 (0 for no testing case)
When-is-daily-testing-applied?	Always (never for no testing case)
Food-delivered-to-isolators?	On

Table 6.2 Different options for testing

Group	Options
Isolation policy	a. Isolate household of infected membe
	b. Isolate infected individual only
Testing regime	a. Do not test youth
	b. Only test elderly with leftover tests after priority testing
	c. Test everyone with leftover tests
Prioritising regime	a. Prioritise both (education & healthcare workers)
	b. Prioritise only education workers
	c. Prioritise only healthcare workers
	d. No priority in testing.

and a replacement is found. Another consequence, although not from the settings for this scenario, is the time span an agent is sick. Given the disease and contagion model presented in Chap. 3, the sick time of an agent is around two weeks. It is very important to mention here that, in order to account for the no testing case, we set the number of available tests to zero and daily testing is never applied. This enables us to observe the effect of household isolation without testing. In addition, the *testing and prioritising regimes* in Table 6.2 have been set to false, for this specific case only.

6.3 Results

A large amount of results was obtained from the simulations. In order to gain access to the results and to not be overwhelmed, we will present them in two ways. First, a One Factor At a Time (OFAT) analysis [3] will be performed in order to identify the best value for each of the different dimensions. A dimension refers to one of the groups presented in Table 6.2. A One Factor At a Time analysis looks at one factor under investigation at a time. When the best value for this specific factor has been found, it will be kept fixed going forward and the next factor will be investigated. This process repeats over the three factors of interest that we have identified. This way of analysing each of the factors separately is possible as we assume that the three factors (Isolation policy, testing regime and prioritising regime) are (sufficiently) independent of each other.

The assumption of sufficient independence is based on the following observations. The *isolation policy* dimension is clearly independent, as it has no influence on the amount of tests. While the *prioritising regime* affects the amounts of tests available for the *testing regime*, we argue that these two dimension are still sufficiently independent enough, because they don't share common agents. Once an agent is part of the priority group, they will not be part in the leftover test group and vice versa.

After the OFAT results have been presented for each of the three factors, the complete picture with all the results combined will also be provided.

Overall, the results show that the following combination of options, from Table 6.2, is the most promising for flattening the curve of infections: *isolating the household of an infected member + no priority in testing + testing everyone with leftover tests*. While this combination doesn't result in the lowest amount of hospitalisations, as can be seen in Fig. 6.25 (comparing the dotted line of plot (c) with the red line of plot (j)), it flattens the curve of infections the best, as we will see in the following OFAT analysis and when looking at Fig. 6.22. Therefore, it reduces the stress on the hospitals the most, because it stretches out the hospitalisations over time. This is also reflected in the relation between the infections over time and cumulative infections, for example when comparing Figs. 6.1 and 6.2. In this case the isolation of the household of an infected member has an effect on lowering the peak of infections and spreading them out more over time. This is reflected in a lower cumulative infections curve. However, at a certain point (around tick 375) the infections over time (Fig. 6.1) are higher for the flatter curve compared to the steep curve. Therefore, the cumulative infection curves (Fig. 6.2) are getting closer together. This results in the observation that flattening the curve helps to spread out the infections over time but doesn't necessarily make a difference in the cumulative amount of infections. A reason for this could be that when isolating the household of an infected member, more people stay at home earlier resulting in less daily contacts overall. However, these isolated household members might be infected at home or still get infected later when the isolation is finished. Thus we claim that the infections rather get postponed rather than prevented.

Furthermore, the results (especially in the OFAT analysis) show that various testing strategies only have a small effect. Isolating the household of an infected member has the biggest effect.

6.3.1 OFAT

The OFAT analysis starts with the following baseline: no testing, no priority in testing, only isolate the infected person. This means that the first parameter under investigation will be compared against the results for the baseline setting. In this case, the first parameter refers to the *Isolation Policy* group.

Isolation Policy: Figure 6.1 shows that *isolating the household of an infected member* has a positive impact on flattening the curve when no testing is performed. While this is an intuitive and expected result, it is important to point it out and not take it as a given. When looking at the cumulative infections in Fig. 6.2, it is interesting to see that in the end about the same amount of agents got cumulatively infected. We argue that this might happen, because, due to household isolation, more people stay at home at the beginning and then go out afterwards and get possibly still infected. Nonetheless, the reduced peak of infections in Fig. 6.1 corresponds to the difference between the two lines in Fig. 6.2. Therefore, we argue that isolating the household of an infected member is an important measure to take in order to flatten the curve and reduce the stress on the healthcare system. Even though in the end, the amount of cumulative infections don't differ that much. Next, the results for the parameters of the group *Testing Regimes* will be analysed.

Testing Regimes: Figure 6.3 compares the effects of the various testing regimes on the number of infected people, with the focus here being on what is happening with the leftover tests. This figure shows that the flattest curve is testing everyone with leftover tests, although the differences are minimal. There is no observable difference between prioritising the elderly with leftover tests or excluding young people from testing. These small differences can be a result of the limited amount of agents in our simulation. It can be possible that the difference increases with more agents being present in the simulation.

When looking at the cumulative infections in Fig. 6.4, we can see that *testing everyone with leftover tests*, even though achieving the flattest curve, has a higher amount of cumulative infections compared to *only testing elderly with leftover tests*. We further investigate this looking at the box plots in Fig. 6.5, to see if the difference is meaningful or due to the inherent randomness in our model. While very sophisticated, our model still works with probabilities for the transmission of the disease, as shown in Chap. 3. The box plots indeed confirm this, because they are very similar. Therefore, the difference in the amount of cumulative infections is due to randomness and not meaningful and can be discarded. The argument that *testing everyone with leftover tests* for flattening the curve still holds.

We want to investigate a bit further what is actually happening during the spread of the virus by looking at the the contacts and infections at different locations.

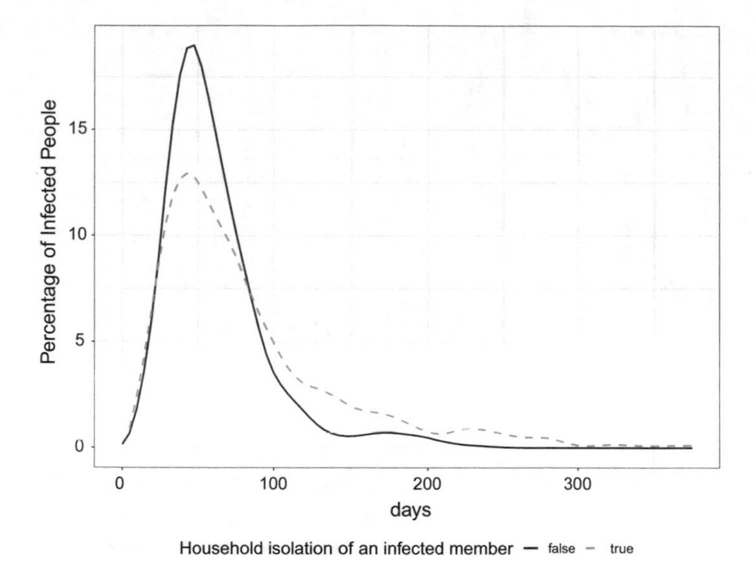

Fig. 6.1 The effects of isolation policies on infections when no testing is performed

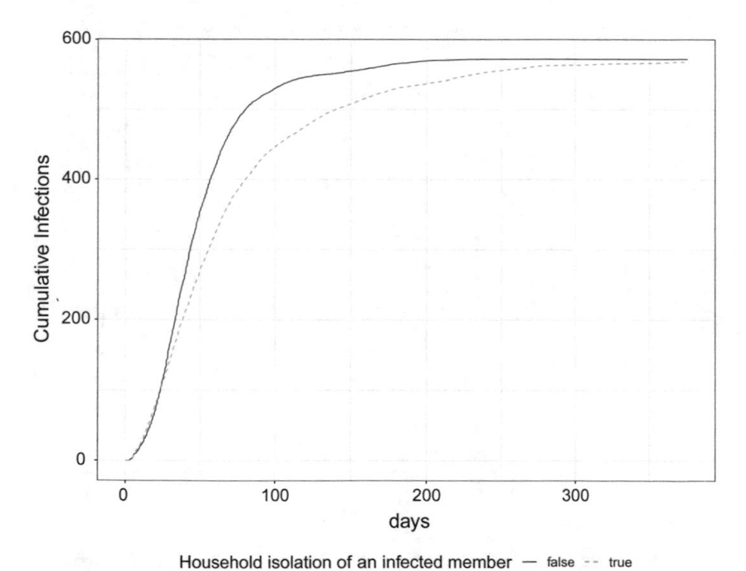

Fig. 6.2 The effects of isolation policies on the cumulative amount of infections when no testing is performed

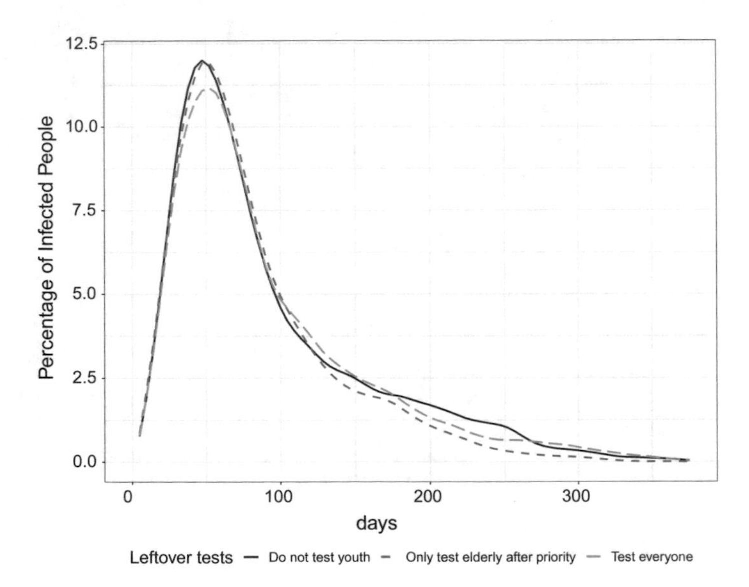

Fig. 6.3 The effects of testing regimes with isolating the household of an infected member

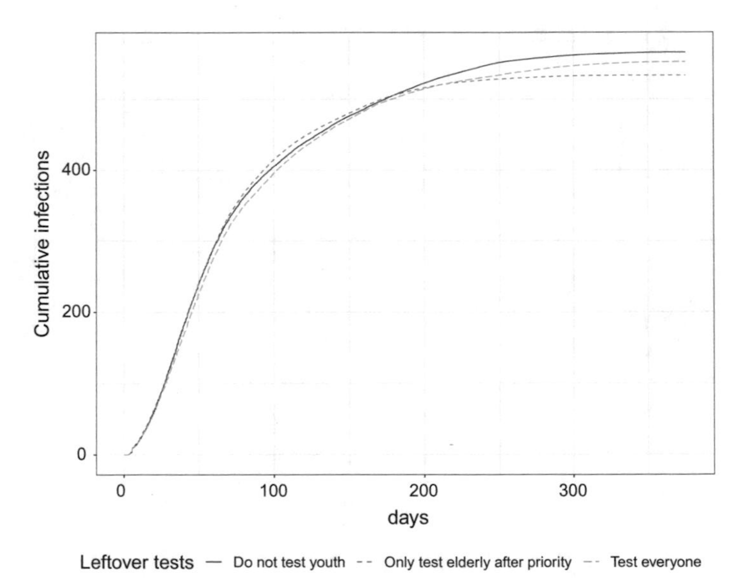

Fig. 6.4 The effects of testing regimes with isolating the household of an infected member on the cumulative amount of infections

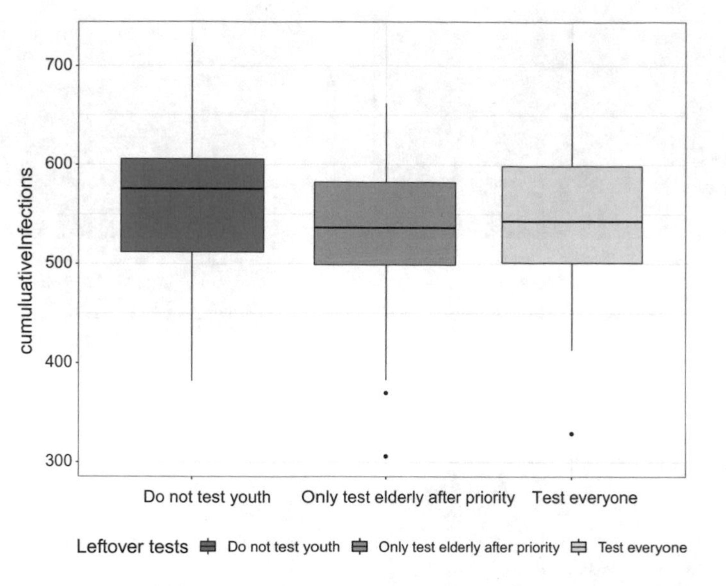

Fig. 6.5 Box plot of the amount of cumulative infections at the end of the simulations

Figure 6.12 shows that most infections happen at home, followed by schools and public transport. These places are among the top six in terms of contacts as can be see in Fig. 6.7. Furthermore, Fig. 6.7 also shows the effect of isolating the household of an infected member. It can be seen that the number of contacts drop noticeably around the time of the peak of the infections, as seen in 6.22. Corresponding to that, the amount of contacts at home increases visibly. This drop becomes immediately clear when looking at Fig. 6.8. Since the household of an infected member stays at home as well, there are almost no contacts while queuing (waiting for the bus) anymore, as enough busses are available to transport the agents at a certain point in time.

We can see in Fig. 6.12 that there are only small differences in the total amount of infections at the different locations for the different testing regimes. *Only testing elderly* shows a bit less infections at home, where most infections happen. However, at the two other worst locations, in terms of spreading the virus, schools and public transport, *only testing elderly* with leftover tests results in the highest amount of infections at these locations.

Furthermore, we can see in Fig. 6.6 that most contacts happen within the same age group of an agent, students get mostly in contact with other students and children get mostly in contact with other children for example. This also corresponds to the infections between the age groups, as shown in Fig. 6.11. Most infections happen within the same age groups, just like for the amount of contacts between the age groups in Fig. 6.6. It is interesting to note here that for adults (worker group agents) the ratios of whom is infecting them are a bit more spread out over children, adults, and retirees. However, we argue that this is due to them having to stay at home when

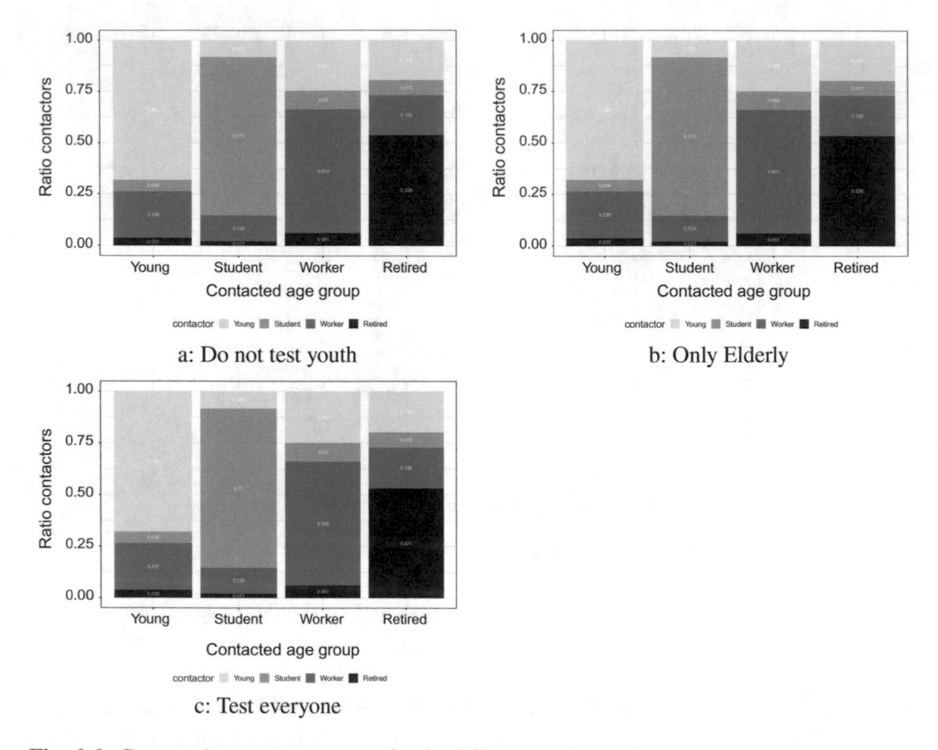

a: Do not test youth

b: Only Elderly

c: Test everyone

Fig. 6.6 Contacts between age groups for the different testing regimes

one of their children or retired agents in their household (given a multi-generational household) is infected. This corresponds to the stacked bar for children and retirees. We can't see a significant part of infections coming from adults. Here, the clear majority of infections are coming from their own age groups.

We can also observe in these two figures that the plots are looking similar for the different testing regimes. Based on this, we argue that no benefit is gained from *not testing youth* or *only testing elderly* compared to *testing everyone* with leftover tests in terms of contacts and infections between age groups.

This also transitions over to the contacts and infections at different locations, Figs. 6.7, 6.8, 6.9 and 6.10. All of the plots look similar to each other, within their own figure respectively. Furthermore, Figs. 6.9 and 6.10, the cumulative amount of infections at different locations, don't show any noticeable plateau. A plateau in these cumulative infection graphs means that no or only a few new infections happen at this specific location for a certain period of time. Therefore, this can be used as an indicator if certain measures have an effect on the amount of infections at specific locations. However, this is not the case here and thus we conclude that *not testing youth* or *only testing elderly* compared to *testing everyone* with left over tests has no benefit in terms of reducing the amount of infections at a certain location. One

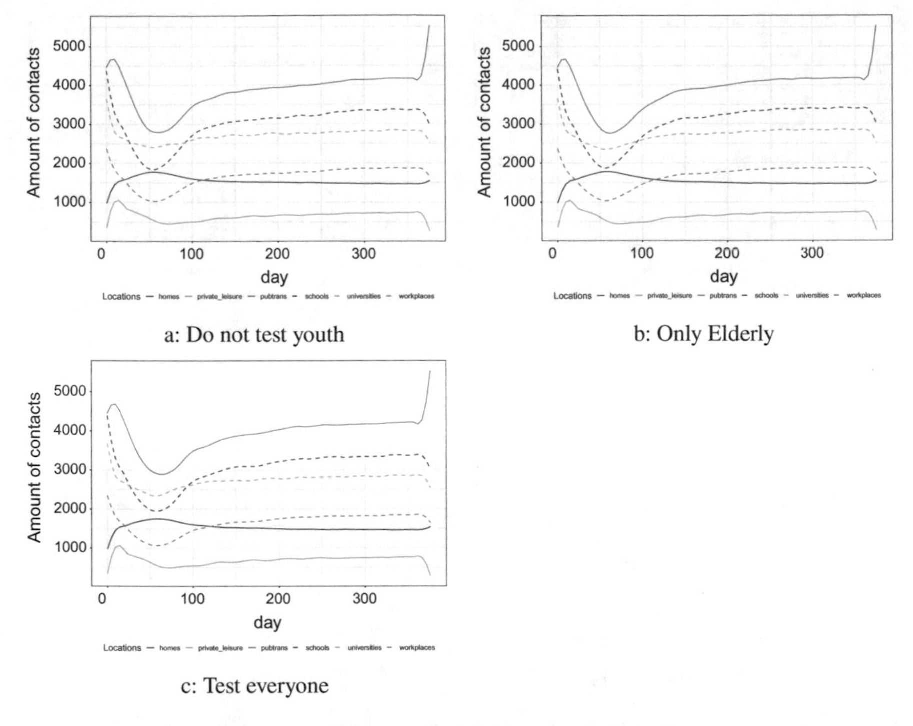

a: Do not test youth

b: Only Elderly

c: Test everyone

Fig. 6.7 Contacts at different locations over time for locations with most contacts

could argue that Fig. 6.10 shows a small plateau for queuing which corresponds to the aforementioned drop in contacts at queuing. Also here, the strongest effect can be observed when *testing everyone* with leftover tests.

As a result of all the above mentioned arguments, this we take *Test everyone with leftover tests* as the optimal choice for this factor and use it for the last factor. The last factor that has to be analysed now is the *Prioritising Regimes* factor.

Prioritising Regimes: Based on the results from the previous two factors, *isolating the household of an infected member* and *test everyone with leftover tests* were chosen. The results for the third factor, prioritising regimes, based on the optimal choices for the other two factors, are shown in Fig. 6.13. Prioritising different professions while testing everyone has small effects, similar to the results of Fig. 6.3. The flattest curve for the number of infections can be found for *no priority in testing*. However, it can bee seen that from around day 75 on wards, prioritising certain professions can have a positive effect on the total amount of infected people. Nonetheless, the focus here is on flattening the curve and thus, lowering the peak of infections and spreading them out more over time. With this in mind, *no priority in testing* achieves the most compelling result, as it spreads the amount of infections out the most over time.

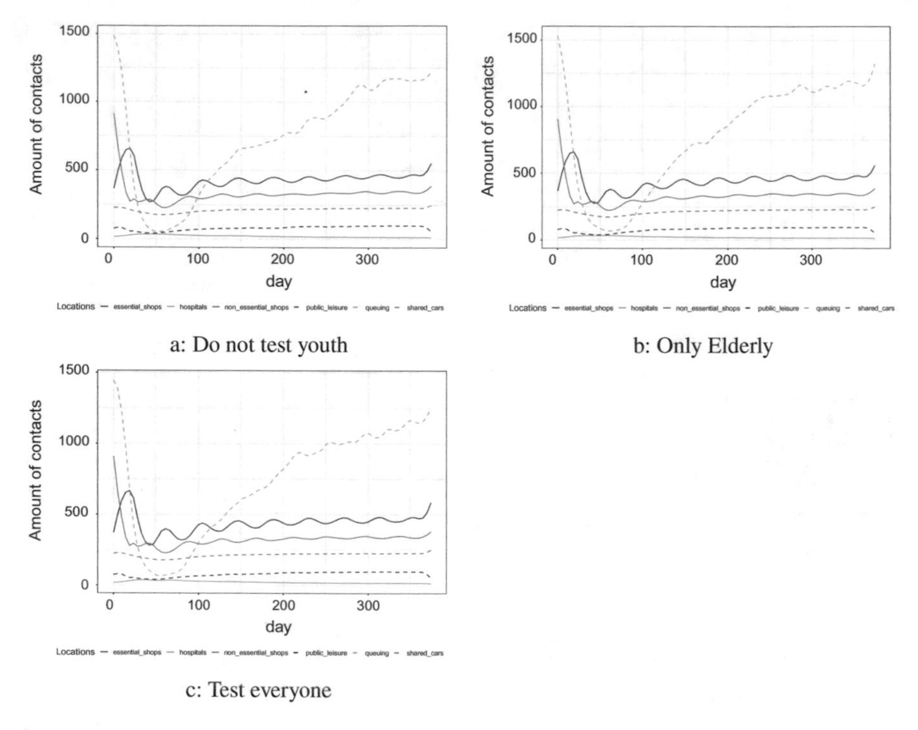

a: Do not test youth

b: Only Elderly

c: Test everyone

Fig. 6.8 Contacts at different locations over time for locations with least amount of contacts

Figure 6.14 confirms this choice, The graphs show that the lowest cumulative amount of infections can be found for *no priority in testing*. Reasons for the results in Figs. 6.13 and 6.14 can be found in the consequence of the settings, as described earlier in this chapter. With *no priority in testing*, the tests get spread out more over the total agent population.

We can also see in Fig. 6.15 that *no priority in testing* has the best results, with which we mean the lowest amount of infections, for most of the locations. When looking at the three locations with the most infections, we see homes, schools, and public transport are those locations. Only at schools does *prioritising education workers* have a slightly better effect for lowering the amount of infections, but this effect is really small. However, for the other two 'most infectious' locations, *no priority in testing* results in the lowest amount of infections, especially when looking at public transport. Here, the difference is comparatively large compared to all other prioritising regimes and not only compared to one.

When looking at the the cumulative amount of infections over time at the different locations in Figs. 6.16 and 6.17 we can see that none of the different prioritising regimes is resulting in any plateau and the graphs are looking similar. Therefore, as argued previously for the different testing regimes, we conclude that none of the other different prioritising regimes have a benefit for a specific location, compared to *no priority in testing*.

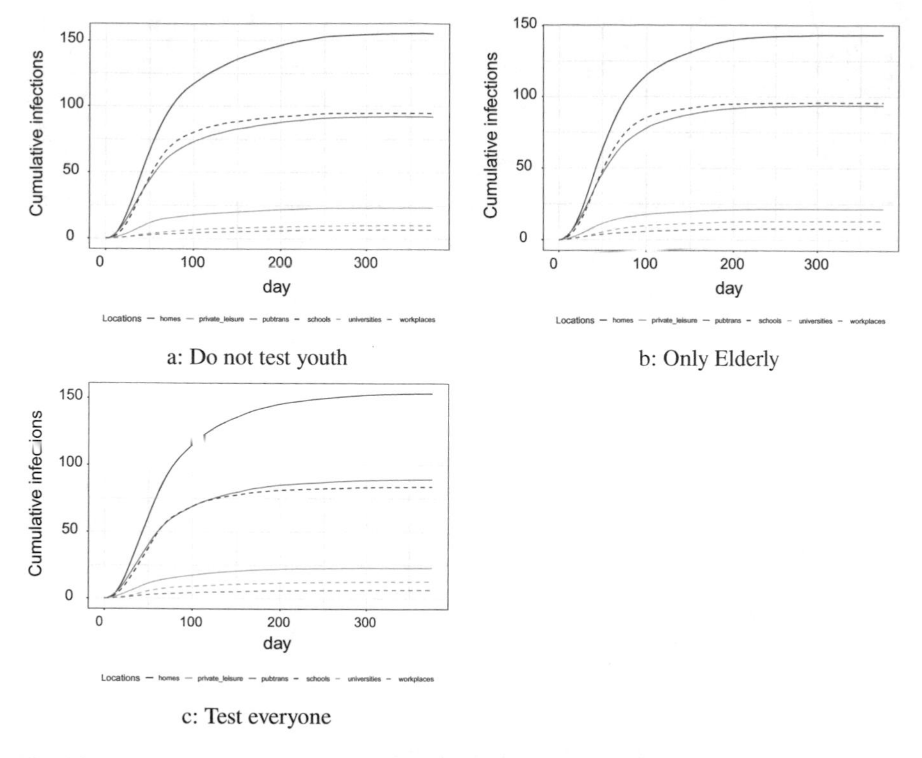

a: Do not test youth

b: Only Elderly

c: Test everyone

Fig. 6.9 Cumulative amount of infections at different locations over time

The other results about contacts between age groups, infections between age groups, and contacts at locations over time showed similar results as for the different testing regimes show in Figs. 6.6, 6.11, 6.7, and 6.8. Therefore, they have been omitted here to avoid redundancy.

Final result: Considering the analysis it can be concluded that the most compelling result was found when *testing everyone with leftover tests, in combination with household isolation and no priority in testing.* Fig. 6.18 shows the curve for this combination of regimes, and household isolation has a much bigger effect compared to not testing (Fig. 6.1). Furthermore, the flatter curve (black dotted line) only shows more infections over time for a short period of the simulation (around day 70 to day 150). Otherwise, the curve lays either below the red line or around it. This is in line with the findings in Fig. 6.19. Here, the amount of cumulative infections when isolating the household of an infected member is notably lower (black dotted line) compared to only isolating the infected person (red line). This also is the case after the flatter curve has more infections (mainly between around day 70 to day 150).

It can also be seen in Fig. 6.20 that hospital effectiveness goes down a bit. The effectiveness of a hospital goes down when there are less health care workers than needed at the hospital. However, given the wide spread of the infections over time, it still remains very high at about 90% at the lowest point.

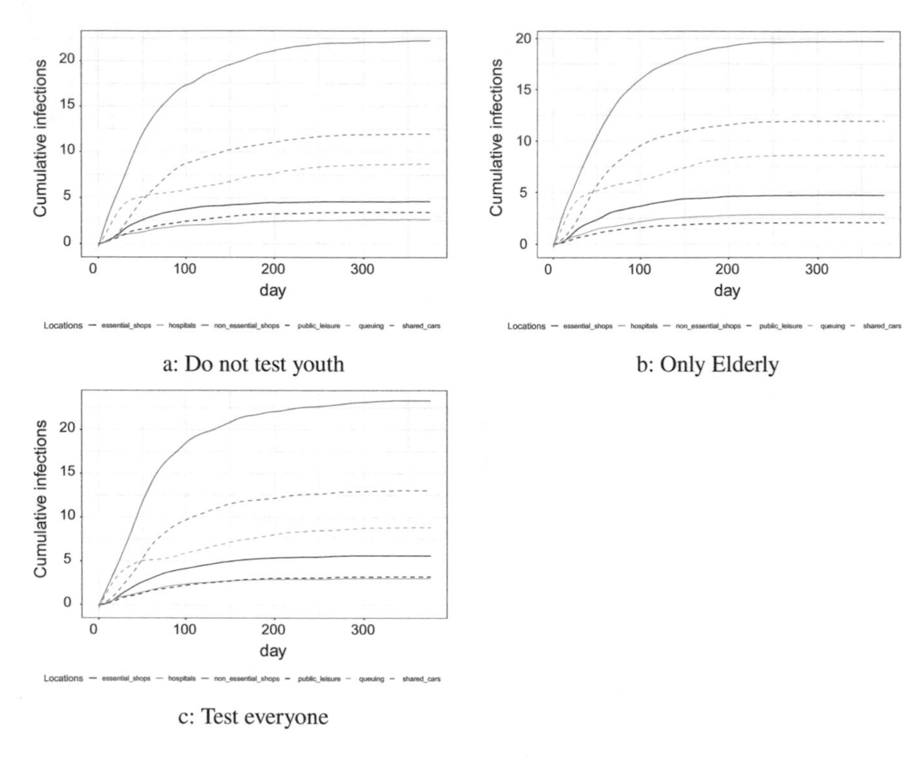

a: Do not test youth

b: Only Elderly

c: Test everyone

Fig. 6.10 Cumulative amount of infections at different locations

Figure 6.21 shows the amount of cumulative hospitalisations for the best combination identified. As can been seen in the figure, the amount of cumulative hospitalisations is 60 for the best combination (isolating the household of an infected person + test everyone with leftover tests + no priority in testing). While this amount sounds rather high it is easily accommodated by the number of beds available. There are eleven beds per hospital location and two hospitals, and the 60 patients are not in hospital at the same time. On average patients stay around two weeks in hospital. They are spread out over the duration of the simulation, which is around a year. The results showed that at maximum six people per day get hospitalised, during the peak of infections. Therefore, the risk of the hospitals running out of beds is rather low.

Finally, the same data showed that the mortality of the disease remained similar regardless of the testing or prioritising regime.

6.3.2 The Complete Picture

After presenting the accumulated version of the results using a One Factor at a Time analysis, the results of the cross product of choices and plots will be presented now. We mainly show these results to make sure there are no unexpected hidden depen-

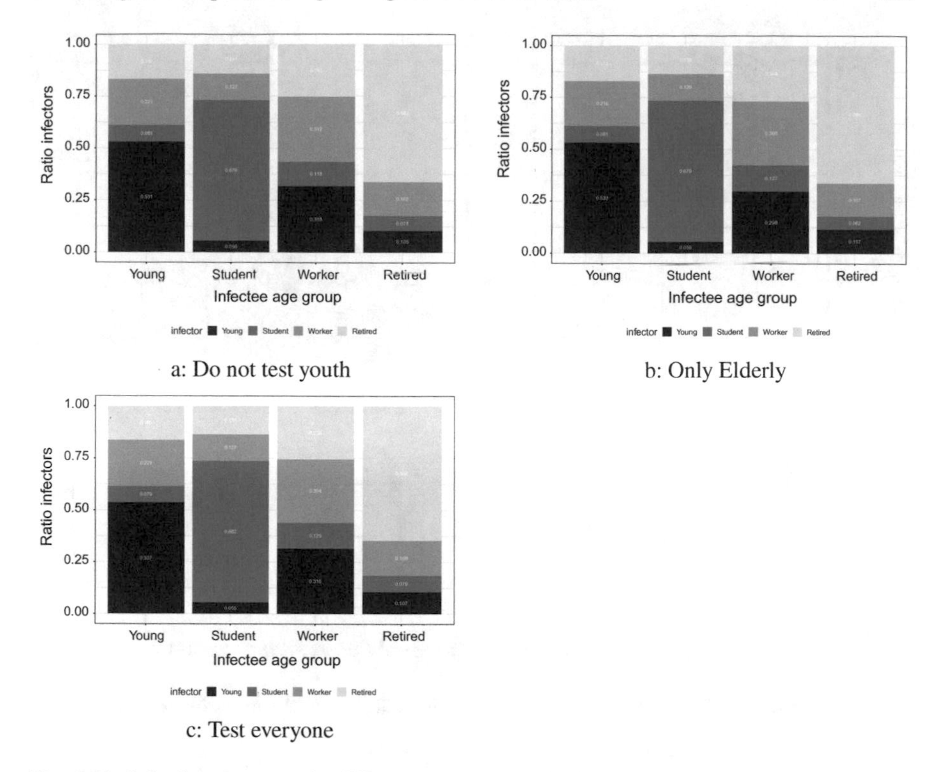

a: Do not test youth

b: Only Elderly

c: Test everyone

Fig. 6.11 Infections between the different age groups, who is infected by whom

dencies between the three factors that we investigated that might create other optimal combinations of choices that we have overlooked. The results are presented in the form of a plot matrix. The horizontal axis describes the different *testing regimes*. The vertical axis depicts the *prioritising regimes*. Figure 6.22 shows the complete results of the infections for the different combinations. Figure 6.23 shows the cumulative amount of infections. Figures 6.24 and 6.25 highlight the effect of the different testing strategies on hospital effectiveness and cumulative hospitalisations respectively. In addition, Fig. 6.26 shows the amount of people in isolation for the different possible combinations. As these are quite big plot matrices, labels have been added to each plot of each matrix, so they can be identified faster and easier. For example Fig. 6.22 plot (c) refers to the combination of *no priority in testing* and *test everyone* with leftover tests. Thus, this reflects the case for random testing.

Infections and Deaths: In general, it can be seen that the infection plot matrix and the cumulative infection plot matrix in Fig. 6.22 confirms the results of the OFAT analysis for the *isolation policy* that isolating the household of an infected member is beneficial for reducing the spread of the virus. Furthermore, plot (c) highlights the case of random testing, as everyone is tested with no priority testing. Comparing this plot (c) to the plot in (b) also reflects the finding in Fig. 6.13 that prioritising elderly in testing doesn't lead to a substantial benefit in reducing the spread of the virus.

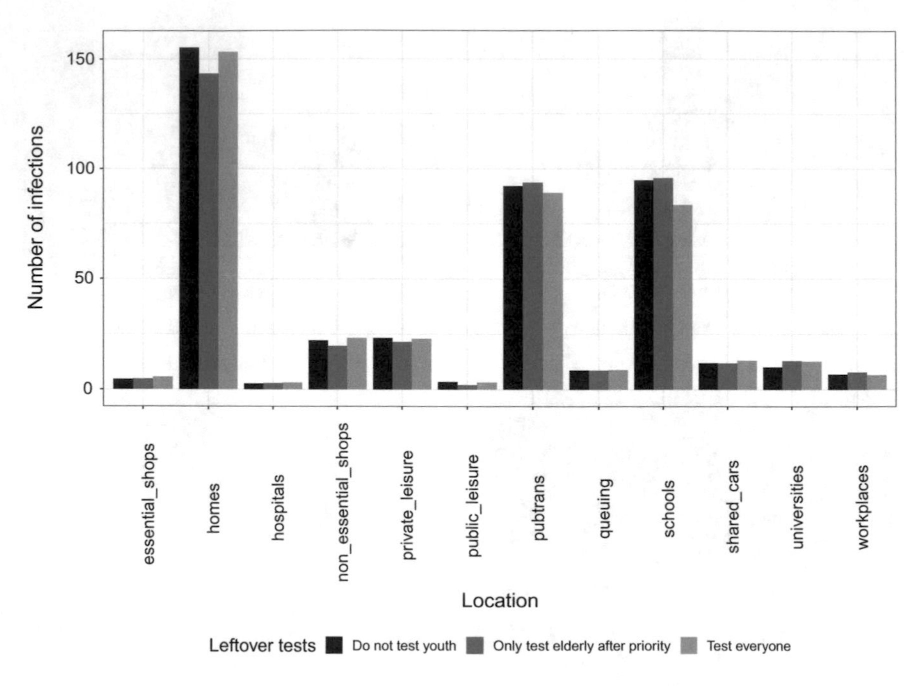

Fig. 6.12 Amount of infections at the different locations, for the different testing regimes

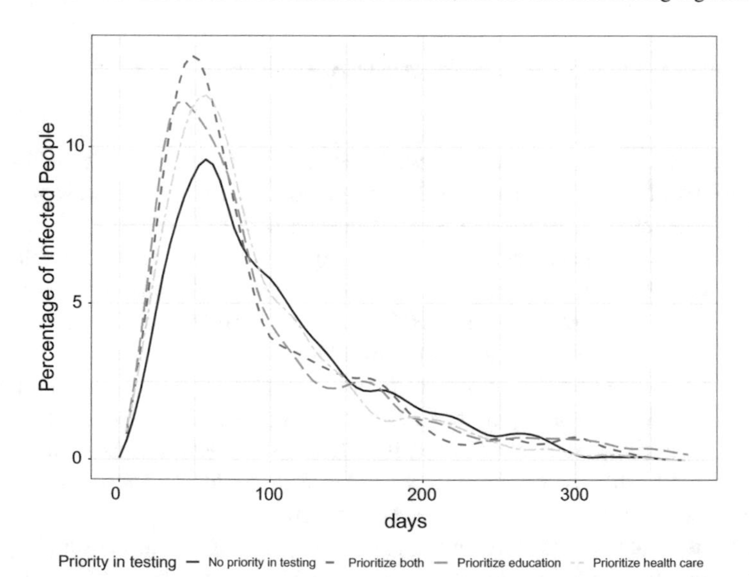

Fig. 6.13 The effects of prioritising regimes when testing everyone with leftover tests and households of an infected member are isolated

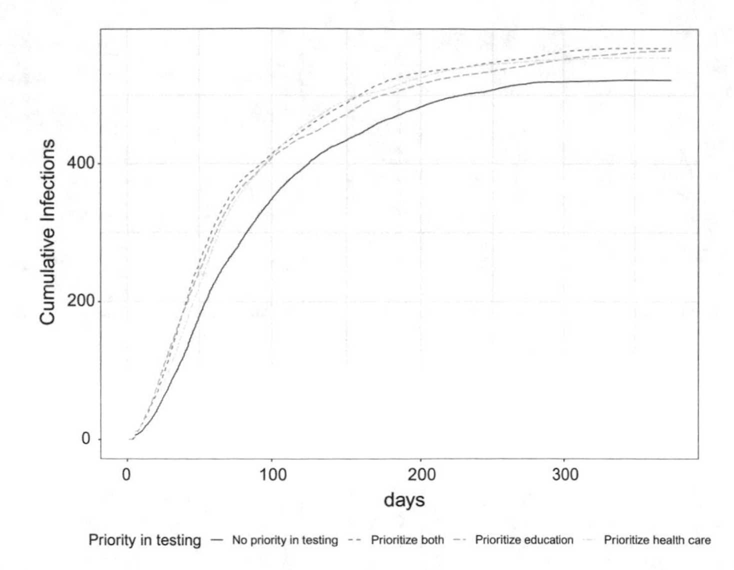

Fig. 6.14 The effects of prioritising regimes when testing everyone with leftover tests and households of an infected member are isolated on the amount of cumulative infections

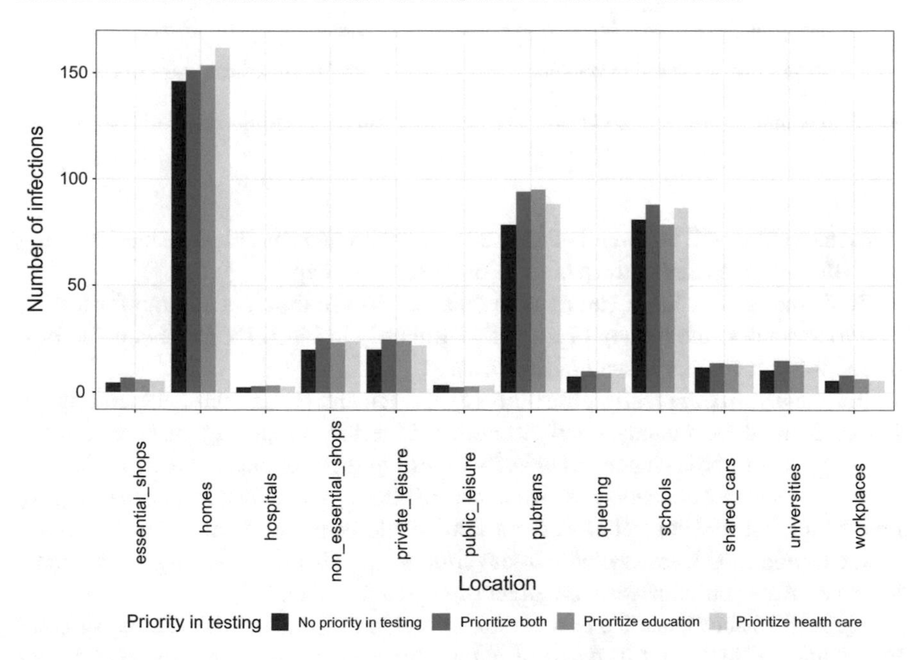

Fig. 6.15 Amount of infections at the different locations, for the different prioritising regimes

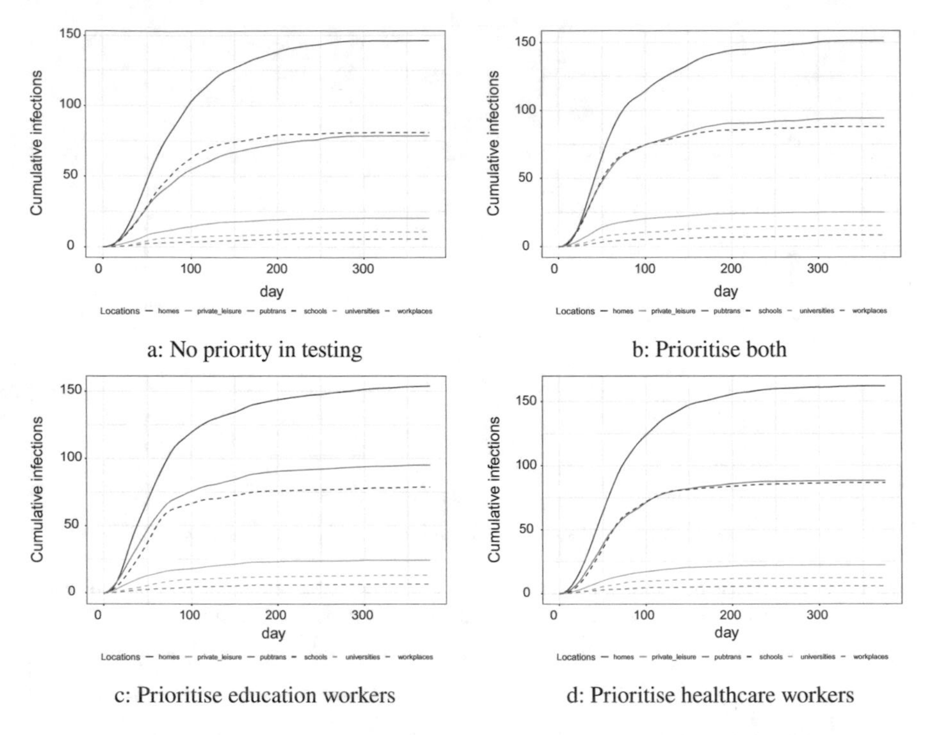

Fig. 6.16 Cumulative infections over time at different locations for the different prioritising regimes

Reasons for this will be given in the discussion, with one possible explanation being the different schedules of the agents as presented in Chap. 3.

To further generalise, it can be seen that the plots in the right column for testing everyone support the finding of the OFAT analysis in Fig. 6.13 that this is the best option for reducing the spread of the virus.

For the *prioritising regime*, the plots (c), (f), (i), and (l) also show the findings in Fig. 6.13 that there are only small differences in terms of reducing the spread of the virus. It can also be seen how *no priority in testing* is better than *prioritising healthcare* or *prioritising education* workers, and that these two are better than *prioritising both* healthcare and education workers. This could be because only a small number of agents and tests were available in the simulation, Therefore, having more agents and more tests could increase the effect of such a measure.

Figure 6.23 shows that the best combination identified in the OFAT analysis (plot (c)) is among the lowest in terms of cumulative infections. Furthermore, this plot matrix also highlights that there are no plateaus within any of the different graphs. This shows that there is no point in time during the simulation when the infections almost stopped. While other plots in this matrix such as (e) and (k) also show a similar low amount of cumulative infections, they are not best solutions in terms of flattening the curve, as can be seen in the corresponding plots (e) and (k) in

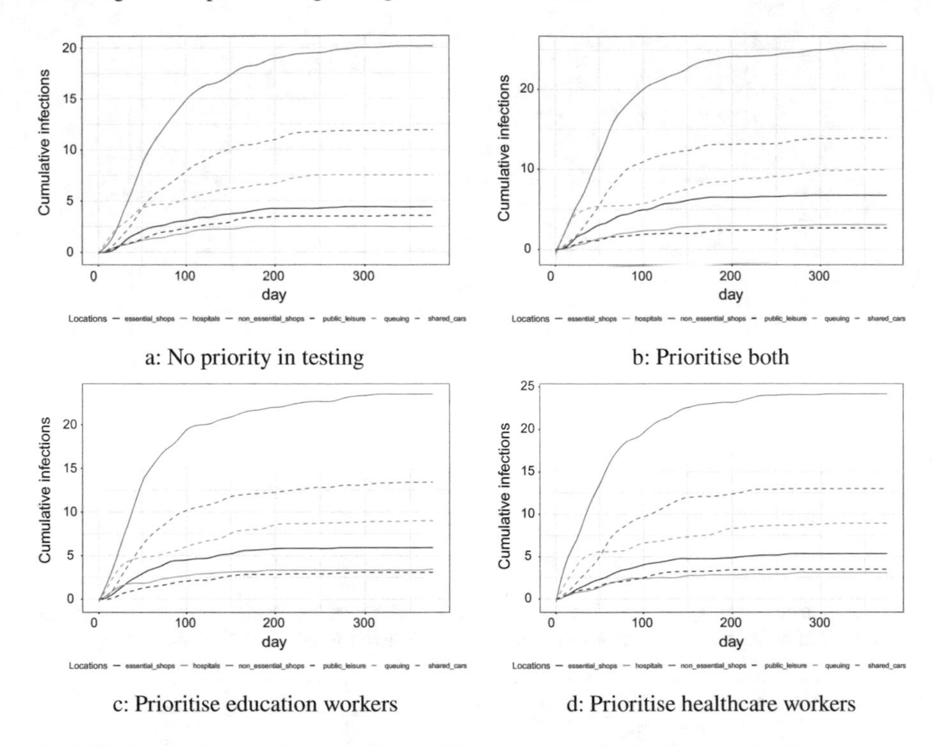

a: No priority in testing

b: Prioritise both

c: Prioritise education workers

d: Prioritise healthcare workers

Fig. 6.17 Cumulative infections over time at different locations for the different prioritising regimes

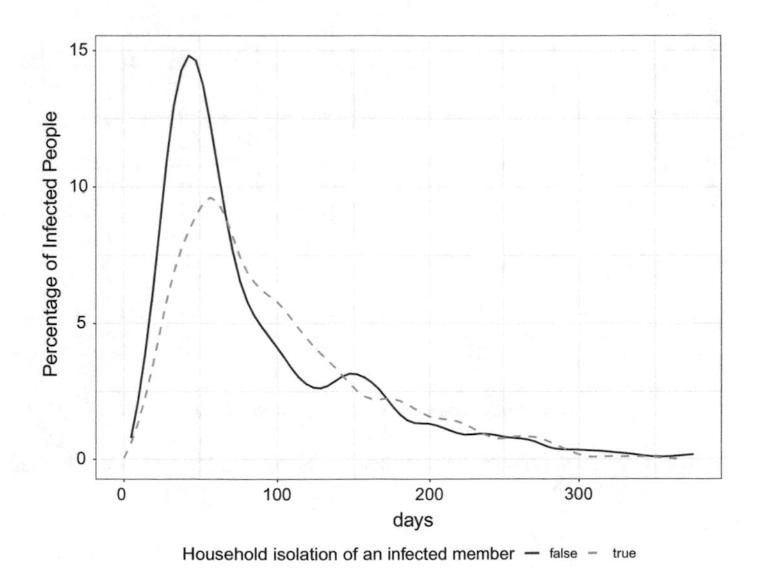

Fig. 6.18 Ratio of infections over time based on the previous OFAT analysis: isolating the household of an infected member + test everyone with leftover tests + no priority in testing)

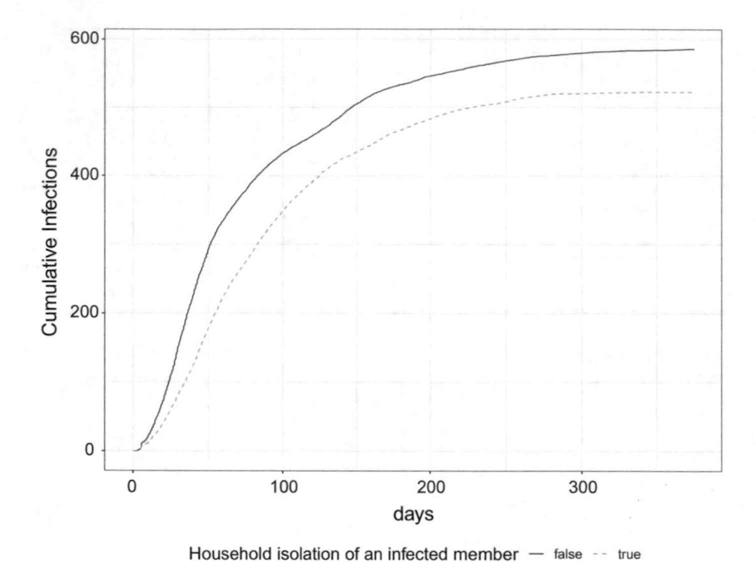

Fig. 6.19 Amount of cumulative infections based on the previous OFAT analysis: isolating the household of an infected member + test everyone with leftover tests + no priority in testing)

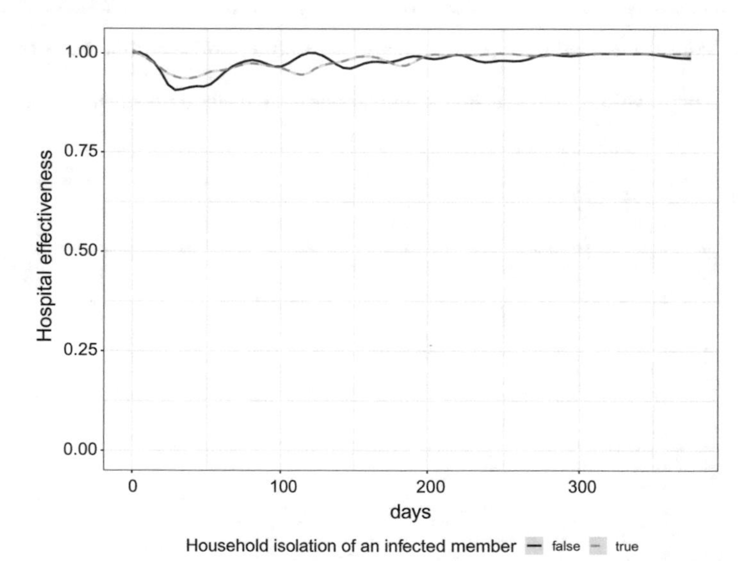

Fig. 6.20 Hospital effectiveness for the best regimes, prioritising healthcare workers in combination with testing everyone with leftover tests

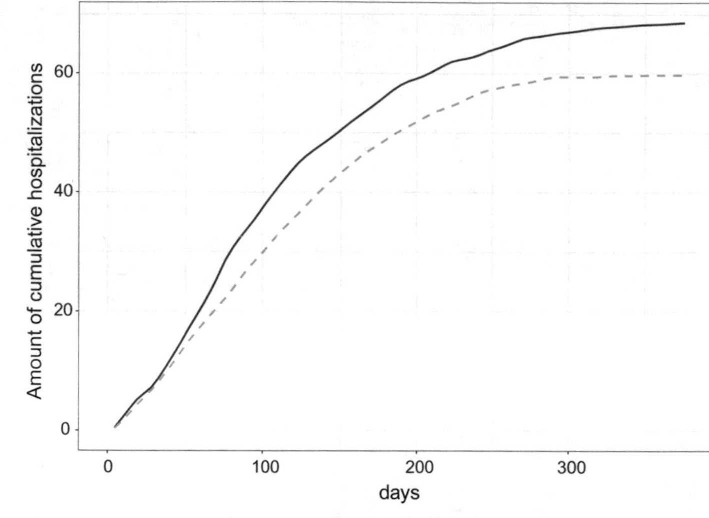

Fig. 6.21 Cumulative hospitalisations for the best regimes, prioritising healthcare workers in combination with testing everyone with leftover tests

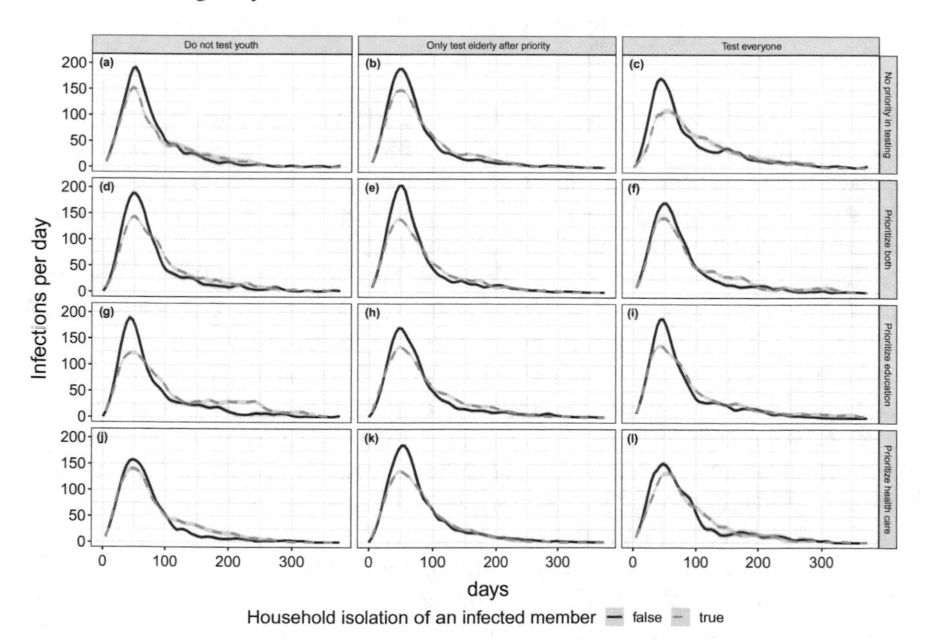

Fig. 6.22 The results for the amount of infections for the different testing strategies

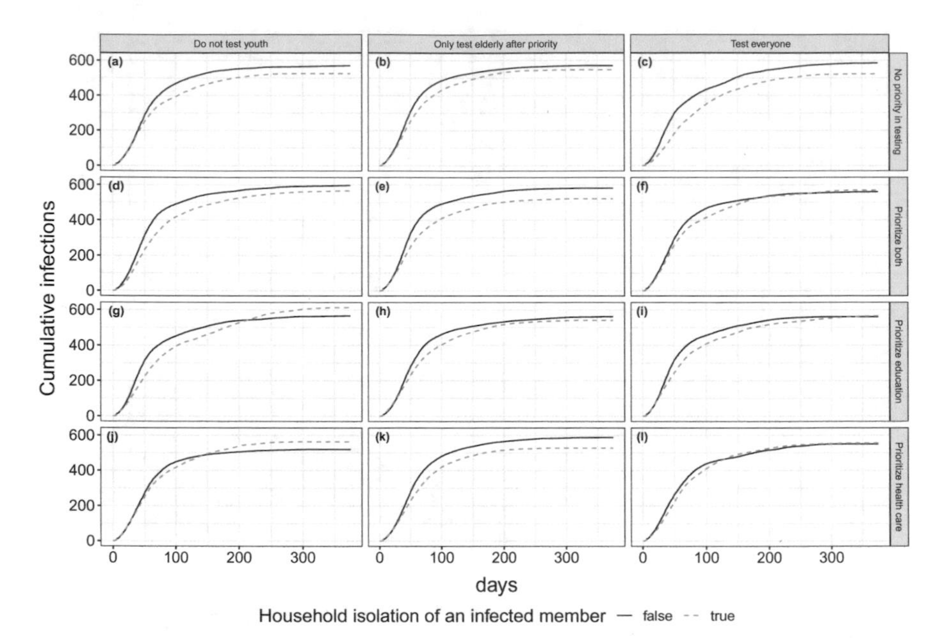

Fig. 6.23 The results for the cumulative amount of infections for the different testing strategies

Fig. 6.22. Prioritising in testing doesn't have an effect on the mortality of the virus. The only small effect can be found when the household of an infected member is isolated. Then one person less died compared to no household isolation. Notice that this effect is very small (around 0.1%), but still can amount to thousands of people when calculated for a population of 60 million people!

Hospital: Plot (l) in Fig. 6.24 confirms the finding in Fig. 6.20 that prioritising hospital personnel will have a small negative effect on the hospital effectiveness. A reason for this is that when they are prioritised in testing, they have to stay home earlier and thus less personnel is available to work at the hospital. In general, the plot matrix in Fig. 6.24 shows that the hospital effectiveness is not going down critically. This can be due to the limitations in our model. Since we have only a limited amount of agents in our simulation, the amount of simultaneously hospitalised agents never critically tackles the capacities of our hospitals. The maximum in our simulation was six simultaneously hospitalisations per tick.

Therefore, it is more interesting to look at the cumulative amount of hospitalisations. This can be seen in Fig. 6.25. We can see that in the end 60 people were hospitalised over the course of the simulation. The capacity of a hospital location was eleven beds. Having two hospital locations in the simulation with eleven beds each therefore means that in total 22 beds were available. Thus, only about three times more people were hospitalised in total than the amount of available beds. But given that not all hospitalisations happen at the same time, the hospitals were never in jeopardy. However, in reality countries hardly have more than 6 beds per 1000

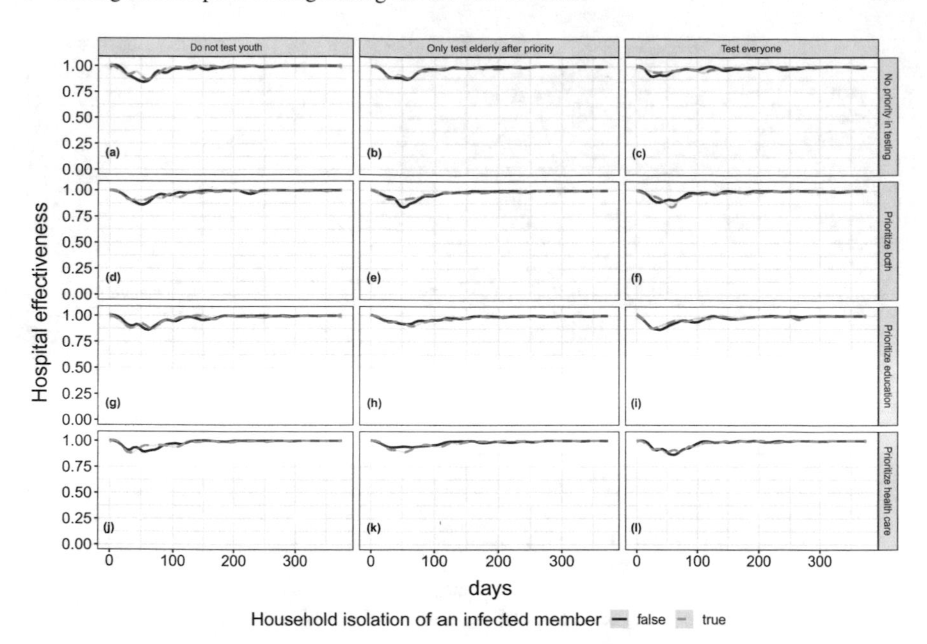

Fig. 6.24 The hospital effectiveness for the different testing strategies

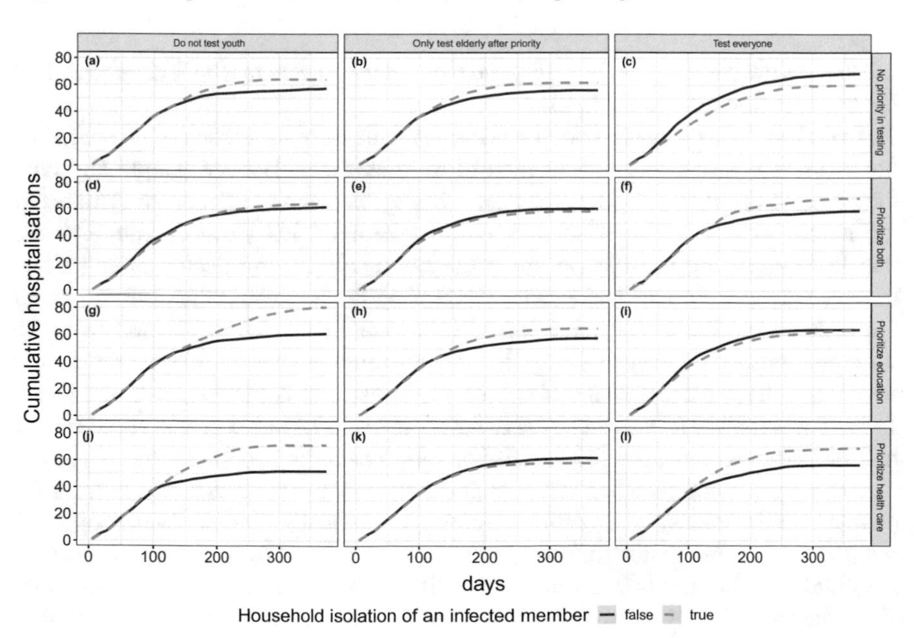

Fig. 6.25 The amount of cumulative hospitalisations for the different testing strategies

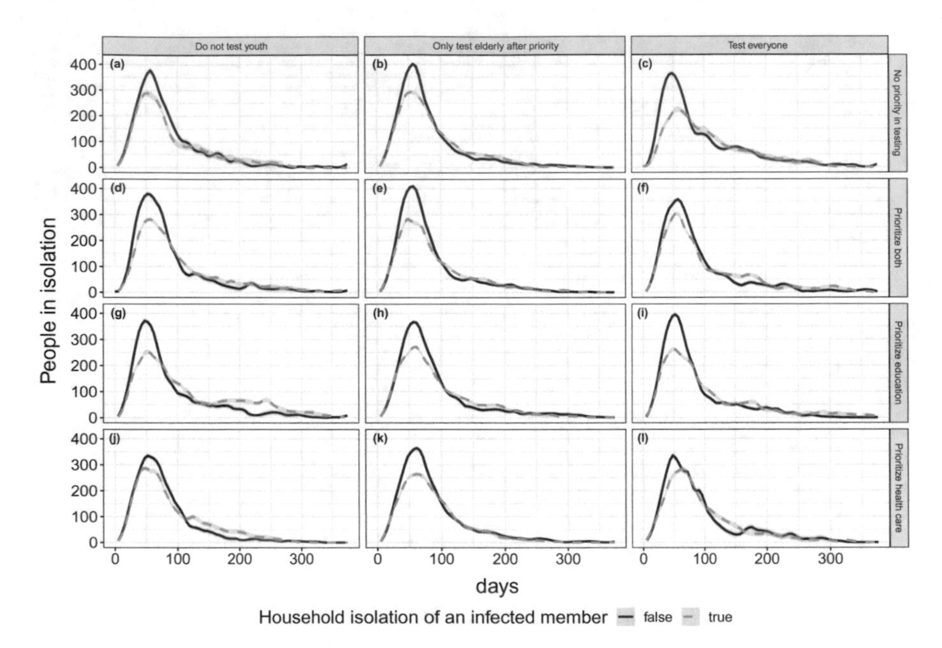

Fig. 6.26 The amount of people in isolation for the different testing strategies

inhabitants available.[1] Thus in reality the 6 patients on average would have stretched the hospital resources to the limit!

Isolation: When isolating people becomes a measure for reducing the spread of the virus, it is also interesting to look at the number of isolated people and how they obey to this measure. The plot matrix in Fig. 6.26 shows that for every infected agent, roughly two agents get isolated. While it seems on first glance counter intuitive that less people are in isolation when the household of an infected member has to isolate as well, it does make sense. Exploring the household distribution, given the settings specific to Italy, we can see that the agents are distributed over the different households in such a way that the resulting median of this distribution is two. Also, the average amount of people per household is between two and three, 1158 (#people)/400 (#households). However, it has to be noted here that more than half of the households, 242 (123 adult homes + 119 retired couples), are composed of two people living together. Given that for the rest of the households either four people (family homes) or six people (multi-generational living) are living together, the average is more distorted towards three than to two. Furthermore, we want to mention that if an agent infects a housemate, it has no effect on the people in isolation, since they are already in isolation. As a result, the number of people in isolation should be around the double amount compared to infections. This can be seen when comparing Fig. 6.22 with Fig. 6.26. Furthermore, the lines in the isolation plot matrix of Fig. 6.26 mimic the form of the lines in the infection plot matrix of Fig. 6.22.

[1] https://en.wikipedia.org/wiki/List_of_countries_by_hospital_beds.

Only one person on average broke isolation per tick. It has to be noted that this plot is not looking at unique isolation violations. Therefore, it can be possible that agents break their isolation on multiple occasions during the simulation run. This behaviour can be explained using the needs model, which has been presented in Chap. 3. Agents have the need of compliance which makes them comply to policies and also risk-avoidance which keeps them away from risky places. As a result, a lot of agents stay at home and comply to the requirement of isolation. On the other hand, people also have the need of autonomy and belonging. Belonging can be viewed as one the main reasons that leads people to break isolation. When people stay at home, they don't meet their friends and colleagues. Thus, their need satisfaction for belonging depletes. At a certain point in time, the need is so strongly depleted that the urge of satisfying it is way stronger than the need of complying to rules and policies.

6.4 Discussion

The results of the adaptive testing scenario show how advanced distinctions of groups of agents allows the exploration of different effects when applying different testing policies.

The importance of the division between agent groups based on various characteristics is strengthened by the results, as they show that single interventions have only a marginal effect in lowering the spread of the virus. Therefore, it is important to use a combination of different strategies to lower the spread most effectively. The ASSOCC model is useful here because it provides insight into the effect of various measures and highlight possible interdependencies between them. A fine-grained analysis can be done while exploring a wide array of small interventions which can then accumulate to a large effect. This effect is shown by the combination of small effects in Figs. 6.1, 6.2, 6.3, 6.4, 6.5, 6.6, 6.7, 6.8, 6.9, 6.10, 6.11, 6.12 and 6.13 leading to the bigger effect shown in Fig. 6.18.

The reason that *household isolation + no priority in testing + testing everyone with leftover tests* showed the most compelling results in terms of flattening the curve can be found in the consequences of the settings. Since more daily tests are available than education and healthcare workers are in the simulation, the same people get tested everyday when prioritising one or both of these two professions. Therefore, part of the tests would always get used for the same agents every day. This leaves less tests available for the rest of the population.

For healthcare workers this means that eleven out of the 58 tests are used for the hospital workers. These are the same agents every day and thus the probability for them getting tested is one. However, this is different for the rest of the population. Since the setting is that everyone is getting tested with leftover tests, the remaining 47 tests will be distributed randomly over the rest of the population every time that testing is done. Another important aspect to keep in mind is that people admitted to the hospital are already sick and infected. This limits the effect of prioritising hospital workers with testing, because no prevention in the spreading of the virus can happen at the workplace which is the hospital. The main focus is on protecting the contacts during transport and at the home of a hospital worker. However, the gain of this protection strongly depends on the method of transportation and the living situation of each hospital worker. If they own a car and live in a household with only one or two members, the possible effect of prioritising them in testing is even lower as they get in less contact with non-infected people, compared to other agents that use public transport and live in a household with three, four or even more members. An advantage of prioritising health care workers that we did not explore is the influence on the effectiveness of the hospitals. In case the hospital locations would have less beds and health care workers this aspect could be explored more fully.

This also applies to education workers and highlights why the two strategies that prioritise either education or healthcare workers are very similar in terms of their results. In total 14 people are considered education workers in the simulation, taking school and university personnel together. Similar to before, these 14 people get tested every day and the remaining 44 tests get randomly distributed over the population, given that everyone is tested with leftover tests. The small difference in the amount of personnel, eleven healthcare workers compared to 14 education workers, explains the very small difference shown in Fig. 6.13. However, the simulation contains only a limited number of agents and tests and therefore it could be possible that these effects will be stronger in a larger scale simulation where the numbers could be more statistically representative of the real world. For this specific scenario the probabilities for people being education or healthcare workers have been given by Italian authorities. However, health care workers also includes people working in elderly care homes, in home care, etc. Because we do not have these roles in our simulation the figures are only weak approximations of the real values.

As mentioned in the introduction, one of the reasons that these professions have been proposed and investigated is that they have a high risk to get infected at their workplaces, as teachers get in contact with a lot of children and students and hospital workers can get infected by the sick people admitted to the hospital. These people can then infect others again, but also lead to decreased efficiency if staying at home afterwards. However this is not necessarily the case. Figure 6.12 shows that hospitals contribute very little to the amount of infections and therefore, not many new infections occur at the hospital. Furthermore, while schools are among the top three locations where most infections take place, prioritising teachers in testing only has a small effect on reducing the amount of infections. Schools still remained among the top three spreading locations, as can be seen in Fig. 6.15. Even more so, when prioritising education workers, more infections happen in public transport compared to other prioritising regimes. Therefore, we argue that, while it is natural to think that these professions have a high risk of getting infected at their respective workplaces, this assumption is not supported by our results. Furthermore, Fig. 6.16 doesn't show any noticeable plateau that would indicate a stop or strong reduction of infections at schools and universities at any point in time.

Given these results, we argue that education and hospital workers are no potential super spreaders. Therefore prioritising and testing them every day, as in our simulation, can lead to a 'waste' of tests which can otherwise be used for other people. As we argued earlier, being able to test a wider array of people is more beneficial for flattening the curve. Furthermore, if tests are 'wasted', the risk that potential super spreaders remain unnoticed and keep spreading the disease is increased.

Nonetheless, it could also be interesting to observe what happens when prioritised groups are tested on a regular basis with longer intervals between the tests, for example every other day or every Friday afternoon. This is one area of potential future work.

By keeping families of sick agents isolated, the people that have a greater chance to be infectious are also kept inside, and thus cannot spread the disease further to other families. The ASSOCC model is able to provide insight into the strength of this measure since different cultures can be represented, including their respective different household distributions of: families, adults living together, elderly couples living together, and multi-generational households. In this case, the settings have been adapted to Italy and its distribution of households.

As a surprising result, testing elderly people alone did not lead to a substantial benefit (Fig. 6.3). This can be explained by the schedules of the agents. As shown in Chap. 3, retired agents have a rather empty schedule with a lot of free time. Thus, they can do their trips to shops or other places during the day and are not restricted to do that in the evening or at the weekend. Contrasting this to the schedules of the other age groups can then explain why retired agents get mostly in contact with other retired agents, as can be seen in Fig. 6.6. The different plots in this figure look almost identical. The changes are only marginal. Therefore, testing only elderly doesn't lead to a substantial benefit, as the amount of contacts does not change. Children have most of their contacts with other children as they meet in the school and students have most of their contacts to other students as they meet at the university. Furthermore, children are taken care of by their parents and thus they don't have to go to shops. Therefore, the chance of meeting agents from other age groups is even more reduced. These arguments are also supported when looking at the plots in Figs. 6.7, 6.8, and 6.12, as these plots are also the same for each of the different options.

The results presented in this chapter and the effectiveness of the testing regimes may vary for different countries and cultures. This is one option for further research. Furthermore, the simulation was run with 1158 agents. Therefore, it can be possible that effects of measures can not be distinguished so clearly. The number of agents in each age group and professions is limited. This means several of the tested policies were only applied to a limited number of agents, limiting their effects. Given more agents, it can be possible that the small differences will become larger and more noticeable. In addition, only 58 people were tested daily, which can also limit the effectiveness of the measures.

While the effectiveness of these measures can change, the results presented in this chapter show the value of the ASSOCC model as a tool and how it can help to explore different interventions for the COVID-19 crisis and how to get through it.

6.5 Conclusion

The presented scenario focuses on adaptive testing techniques for Italy. The simulation outcomes show that testing is needed in combination with isolating the household if one of its members is infected. The most compelling results in terms of flattening the curve was achieved for *isolating the household of an infected member + no priority in testing + test everyone with leftover tests*. Furthermore, the death rate stays the same for the different strategies and most of the people adhere to stay in isolation. Given the schedule and the behaviour of the agents, only testing elderly with leftover test didn't lead to a noticeable benefit.

The ASSOCC project is a valuable tool for decision-makers to gain insights into the effect of different policies on the population. Given the broad scope of the ASSOCC model, different possibilities for future work, which still addresses adaptive testing, exist. Other professions, such as supermarket personnel, or different age groups could be prioritised. Different countries and cultures can be investigated, given the implementation of cultural diversity. In addition, people who have been in contact with sick colleagues or people who are just about to leave self-isolation could be tested.

As general future work, the ASSOCC model is being migrated to a Repast simulation, which eases the design of even more advanced agents and enables us to run simulations with many more agents.

Acknowledgements We would like to acknowledge the members of the ASSOCC team for their valuable contributions to this chapter. Furthermore, we would like to thank Annet Onnes for creating the basic R code that has been used in the OFAT analysis. Additionally, we would like thank Loïs Vanhée, Harko Verhagen for initial support with the OFAT analysis, and Mijke van den Hurk for proof reading and her valuable feedback for this chapter. The simulations were enabled by resources provided by the Swedish National Infrastructure for Computing (SNIC) at Umeå partially funded by the Swedish Research Council through grant agreement no. 2018-05973. This research was conducted using the resources of High Performance Computing Center North (HPC2N). This research was partially supported by the Wallenberg AI, Autonomous Systems and Software Program (WASP) and WASP - Humanities and Society (WASP-HS) funded by the Knut and Alice Wallenberg Foundation.

References

1. J.-M. Jin et al., Gender differences in patients with COVID-19: focus on severity and mortality. Front. Public Health **8**, 152 (2020)
2. Q. Li et al., Early transmission dynamics in Wuhan, China, of novel coronavirus-infected pneumonia. New Engl. J. Med. (2020)
3. V. Czitrom, One-factor-at-a-time versus designed experiments. The Am. Stat. **53**(2), 126–131 (1999). https://doi.org/10.1080/00031305.1999.10474445. https://www.tandfonline.com/doi/abs/10.1080/00031305.1999.10474445

Chapter 7
Deployment and Effects of an App for Tracking and Tracing Contacts During the COVID-19 Crisis

Maarten Jensen, Fabian Lorig, Loïs Vanhée, and Frank Dignum

Abstract The general idea of tracking and tracing apps is that they track the contacts of users so that in case a user tests positive for COVID-19, all the other users that she has been in contact with get a warning signal that they have potentially been in contact with the COVID-19 virus. This is, to quarantine potential carriers of the virus even before they show symptoms. We set up a scenario in which we test the effects the introduction of such an app has on the dynamics of infection with varying amounts of app users. Running the experiments resulted in a slightly lower peak of infections for higher app usages and the total amount of infected individuals over the course of the whole run decreased not more than 10% in any case. The app seems mainly effective in decreasing contacts and infections in public spaces (except hospitals) while increasing the contacts and infections at home.

7.1 Introduction

One of the standard procedures to control the spreading of a virus is to track and trace all persons an infected person has been in contact with from the likely moment the person became contagious until the moment of the positive test. Traditionally, this is done by hand and thus very labour-intensive. Additionally, the time between a person becoming contagious and the symptoms becoming apparent can be up to 7 days for the corona virus. Consequently the risk that contacts are missed during the

M. Jensen (✉) · L. Vanhée · F. Dignum
Umeå University, Mit-huset, Campustorget 5, 901 87 Umeå, Sweden
e-mail: maartenj@cs.umu.se

L. Vanhée
e-mail: lois.vanhee@umu.se

F. Dignum
e-mail: dignum@cs.umu.se

F. Lorig
Department of Computer Science and Media Technology, Internet of Things and People Research Center, Malmö University, 205 06 Malmö, Sweden
e-mail: fabian.lorig@mau.se

F. Dignum (ed.), *Social Simulation for a Crisis*, Computational Social Sciences,
https://doi.org/10.1007/978-3-030-76397-8_7

tracing is quite high, while contacts also can have spread the virus during the time the tracing is taking place (which can take several days). This means, that effective manual track and tracing is difficult. Therefore, worldwide, tracking and tracing apps (TTAs) are introduced to assist the tracking and tracing process.

The basic idea of all the apps is to keep track where a person has been or with whom a person has been in contact, such that if that person has been tested positive, all persons on the accumulated list can be alerted, self isolate, and get tested. In the Netherlands, the app is installed on mobiles and makes contact with all other app users when they get close enough for the Bluetooth signal to connect. In this way, it keeps a list of other persons someone has been in close contact with and could possibly infect. Once a person gets symptoms and gets tested she can put that in the app on her mobile, which then notifies all the people in the list she has been in contact with. Based on the simulation reported in [1] (which we refer to as the Oxford model), the Dutch government stated that if 60% of the people would use a TTA it would be possible to keep the spread of the corona virus under control.[1] We are interested in the effects of this TTA on the dynamics of the pandemic. Therefore, we pose the following research question: *How do different numbers of users with the tracking and tracing app influence the corona virus spread?*

We want to evaluate the effects of introducing a TTA by analysing the number of infected individuals and where they get infected. The first goal will be to check if the introduction of the app influences the height of the infection peak and the amount of people infected. However besides this main result (of which we will see that it deviates from some of the more commonly cited results) we also will identify potential reasons for these results by evaluating where people meet, how the app influences behaviour and which people get tested due to the usage of the app.

The next section will describe literature on TTAs used over the world. As many countries have an app, they are implemented in different ways. Then we describe the scenario that we use for our implementation of the TTA in the simulation. This is followed up by results and analysis which will elaborate the type of people that get notified, how the TTA influences people meeting others and where most infections take place.

7.2 Background and Context

For controlling the spread of a disease that can be transferred via airborne transmission, the isolation of infected individuals seems a promising intervention. In case of the COVID-19 pandemic, however, the timely identification of infected individuals is challenging. Even though individuals are not showing symptoms, they might still be capable of transmitting the virus to others. This characteristic of COVID-19 has already been identified at an early stage of the pandemic. Hence, only isolating individuals that already show symptoms is not sufficient. The standard practice in

[1]https://nos.nl/artikel/2332235-meerderheid-zou-veilige-corona-app-installeren.html.

epidemiology to curb the spread is to reconstruct contact-networks of individuals that have been diagnosed with COVID-19 and to quarantine other individuals that they have encountered during the last 7 days and potentially infected. Similar approaches have also been successfully conducted for other diseases such as SARS, MERS, or influenza [2].

For local health authorities, this process of manually tracking contacts is time- and resource-consuming and quickly becomes impracticable with a large number of infections. Hence, supporting the tracking and tracing of COVID-19 contact-networks by means of technical means such as smartphone apps was seen as a very promising solution. By August 2020, approximately 9 months after the initial outbreak of COVID-19, [3] identified 17 different apps that were used for contact tracing in case of COVID-19. The introduction of these apps, however, has been subject to debates in several countries. They included both the epidemic benefits of such apps as well as ethical concerns regarding the collection of personal data. As examples for ethical concerns that might be discussed prior to the implementation of app-based contact tracing and epidemic surveillance, [4] outline the type and extent of collected data, accessibility to the data, and the extent to which large information technology companies contribute to the development and provision of such apps. From an engineering perspective, the implementation, data management, security, vulnerability, and the ability to correctly identify contacts with others have been discussed [5]. Technically, there are two different approaches for app-based contact tracing: centralised and decentralised. Once two individuals meet that are both using the app, their smartphones will exchange keys (pseudonyms) via Bluetooth in case the distance between the devices is smaller than a certain threshold (e.g., 1.5 m) for a certain period of time (e.g., 15 min). For decentralised approaches, the user of the app might choose to upload its history of pseudonyms once he or she receives a positive test result. This list of published pseudonyms can then (automatically) be downloaded by other app users and be compared against the locally stored list of encountered pseudonyms to identify a potential encounter with an infected person. In the centralised approach, all app-data and health authority data is uploaded to a central server, which then informs users about potential encounters with persons tested positive for the virus [6].

Singapore was the first country to implement the use of apps for contact tracing in the COVID-19 pandemic, releasing TraceTogether on March 20th, 2020, regardless of the prototypical status of the app [7]. A major concern from other countries regarding developing and relying on a tracing app for containing the spread of COVID-19 was the uncertainty about the amount of people that must actively use the app in order to be effective. This is, due to different compositions of the population, e.g., a larger share of elderly people that might not even possess a smartphone for the installation of the app, as well as different configurations of the app and the respective tracing process. To evaluate potential epidemic benefits as well as the acceptability of TTAs for specific populations, the use of simulation is reasonable. In a simulation the differences in population and app use can be tried and thus better predictions for specific countries given. In Germany, for instance, the Corona-Warn-App that has been released on June 16th, 2020 relies on the Exposure Notification APIs provided by Apple and Google. Contract histories are stored on smartphones for 14 days and reporting infections to the app is optional via a decentralised approach [8]. As of September 22nd, 2020, 100 days after the release of the app, it had 18.4 Million downloads (ca. 22% of the population) and ca. 5000 users have reported a positive COVID-19 test via the app. As a side note it should be said that as of now there has been no country where the adoption rate of the app has exceeded the 25%. (see https://www.statista.com/statistics/1134669/share-populations-adopted-covid-contact-tracing-apps-countries/ for statistics in July 2020).

An example of a simulation for the configuration of digital contact tracing apps has been presented in [9]. A population of 1 million individuals is simulated, which represents the typical size of an organisational unit (NHS trust) within the National Health Service in England. The authors use an individual-based model, where each simulated person has a demographic profile, interaction networks, and a disease status. From the simulations run on this model, the authors conclude that the epidemic can be suppressed in case 80% of all smartphone users use the app, but even with much lower percentages of app use the results on the spread of the virus will be significant. In this chapter we aim to check the effectiveness of the app based on our behavioural model. We will see that our model predicts a much smaller effectiveness of the app. In Chap. 12 we will make a detailed comparison between the two models in order to validate our model and show where the differences in the results originate from.

7.3 Scenario Description

This section describes the setup and implementation of our TTA scenario. The reason we provide more technical details than some other scenario chapters is because the TTA implementation is not just a setting in the model, but also required the addition of the use of the app in the model. For example closing of the schools is just a setting that can be turned on or off. While the TTA is an actual addition that goes on top of the model described in Chap. 3. This allows for many variations in the implementation, for example we had to choose who are notified when a person get positive. This could just be the contacts, but can also be the contacts of the contacts and even recursively expending further than that. We first give an introduction of the TTA and follow this up with a clear and precise explanation of how we implemented the TTA in our model.

7.3.1 Implementation of the Tracking and Tracing App

To implement the TTA, we had to make some abstractions from the real world. In our simulation a day is divided into for ticks representing the morning, afternoon, evening and night. This leads to interactions always being a tick in time and agents interact with all the other agents at the same location. In the real world you could have interactions of only a few seconds, when you for instance greet someone on the way, but also longer interactions when for example talking for hours. The real-world TTA would in the greeting example not register and log the contact but would do this in the talking for hours example. In general the app tries to register contacts taking more than 15 min. In our implementation, the app of an agent captures agents that are at the same location for a tick. This may seem too crude, but this is compensated by the configuration of the locations. We created e.g., a number of shops for a population that would lead to an average amount of person being in the same shop that is in the same range as the amount of people one would normally be in contact with in a shop for some more time. I.e. taking into account waiting in lines, interacting with personnel, etc.

The mechanism of the app is assumed to work perfectly, i.e., every contact will be tracked correctly. The tests are also done immediately and their results known and people immediately go into quarantine when one of their contacts has been tested positive. Thus we take a very optimistic and positive perspective on the functioning of the app. In real life this ideal situation will not be reached. The app will not always register all contacts correctly, people might not register the fact that they are tested positive in the app, or not do that straight away. Some people will go home when receiving a message from the app, but some might stay at work and finish the job for the day, before going into quarantine or even not listen to the app at all and continuing to work the days after as well. We took this positive perspective on the working of the app in order to get the most positive estimate on the effectiveness of the app. I.e. if the app is not effective under these assumptions it will only be worse when some of the assumptions are slackened. We also adopted parts of the Oxford model [9]. They start the TTA at the end of a 35 day lockdown. This lockdown is initiated when 2% of the population is infected. The app started collecting data 7 days before the end of the lockdown.

Figure 7.1 shows the steps in our TTA implementation. The first step is to distribute app users which is done at the start of the simulation, a certain number of agents randomly become app users (dependent on *ratio-of-people-using-the-tracking-app*). Agents that have a higher anxiety avoidance are first selected as users of the app (dependent on *ratio-of-anxiety-avoidance-tracing-app-users*, as we assume that people who are more risk avoiding will more likely download such an app. The second step is registering contacts which is started 7 days before the end of the global quarantine until the end of the simulation run. It will register every agent at the location who is also an app user and stores this for 7 days (this can be varied by changing *#days-recording-tracing*). The third step is to report positively tested users. App using agents that get symptoms will record this immediately in the app and every

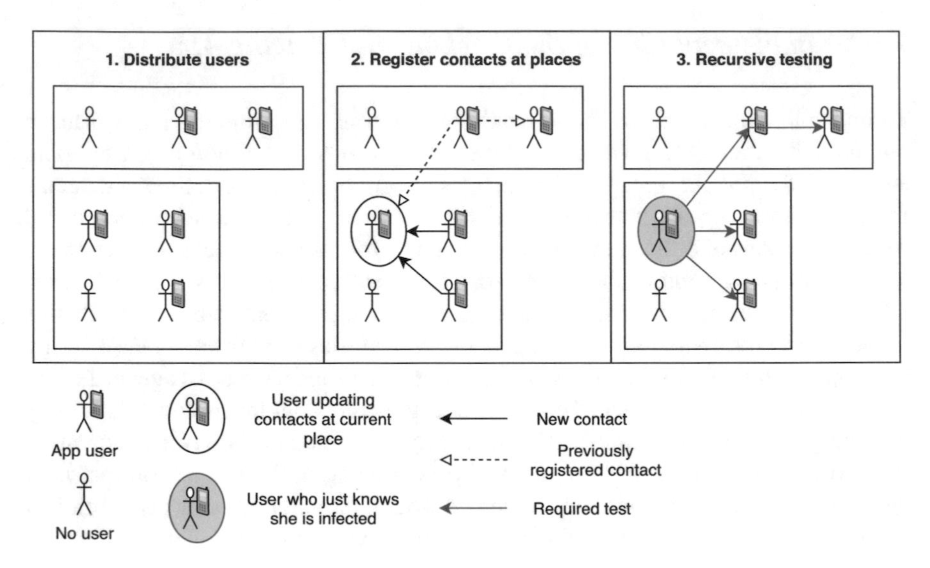

Fig. 7.1 Diagram of the implemented TTA

agent in the contacts list will get tested. When one of them is positively tested all their contacts get tested, and when recursive testing is enabled (dependent on *is-tracking-app-testing-recursive?*) all the contacts' contacts get tested until everyone indirectly connected to the positively tested agent is tested.

The specific parts of the implementation of the app are explained in more technical detail in the next sections. Readers who are not interested can skip to the scenario settings section.

7.3.2 Selecting the Tracking and Tracing App Users

Selecting users for the TTA goes through a couple of steps which are represented in Fig. 7.2. At the start of the simulation first the phone users are determined. All students and workers have a phone, as data shows that the majority uses mobile phones [9]. We implemented a variable for the amount of phones that young (*ratio-young-with-phones*) and elderly (*ratio-retired-with-phones*) have, as the data shows that for age groups 0–20 and 70+ less than half of the people have a phone. These parameters allow us to set the phone users and thus the app users to a more realistic amount.

After setting the mobile phone users, the app users are selected. First the total number of agents that should use the app is calculated by multiplying the *ratio-of-people-using-the-tracking-app* variable and the total number of phone users. In Fig. 7.2 as an example we have ten phone users, so with a 0.6 *ratio-of-people-using-the-tracking-app* this means the function should end up with six app users. Then we

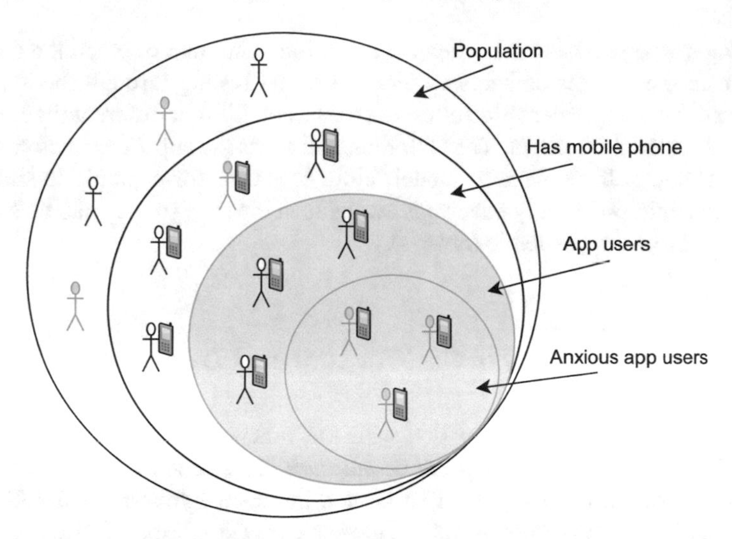

Fig. 7.2 An example for the selection of users for the TTA. Here we have a *ratio-of-people-using-the-tracking-app* of 0.6 and *ratio-of-anxiety-avoidance-tracing-app-users* of 0.5. The red agents are more anxiety avoiding. The green are indicates the app users

calculate the number of agents that should be selected based on anxiety avoidance, this is done by multiplying the total number of app users with *ratio-of-anxiety-avoidance-tracing-app-users*. When these numbers are known, agents are selected to become app users based on weighting of *risk-avoidance* and *compliance* needs, here higher needs lead to a higher chance of becoming an app user. The figure shows the agents with high *risk-avoidance* and *compliance* in yellow, three of those agents are selected since there is 0.5 *ratio-of-anxiety-avoidance-tracing-app-users*. The remaining required app users are selected at random from the remaining agents who have a mobile phone. Since the *ratio-of-anxiety-avoidance-tracing-app-users* can limit the amount of app users selected based on their needs we still see a yellow agent that has not been selected as app user. These settings lead to six app users in total. Note that a 0.6 *ratio-of-people-using-the-tracking-app* does not mean 60% of the population, but rather 60% of the phone users.

7.3.3 Enabling the Tracking and Tracing App

The TTA is started for all the agents who have the TTA when the variable *is-tracing-app-active?* is set to true. This is dependent on the *when-is-tracing-app-active?* variable which has one of the following settings.

- Never
- 7-days-before-end-of-global-quarantine
- at-end-of-global-quarantine
- Always

is-tracing-app-active? is false when the condition is not met or when the condition is never. In these cases contacts are not saved and testing through the app is not performed. This implementation allows us to start the TTA at different times to more realistically model the effect. The Oxford model starts the app 7 days before the end of the lockdown. If we want to model future scenarios for example to simulate a future pandemic we already have apps available at the start so they can be started at the beginning of the pandemic (always).

7.3.4 Saving Contacts in the Tracking and Tracing App

The contacts are updated every tick with the function *update-contacts-of-a-set-of-people-in-contact-tracing-app-public-measure*. This function will only activate when the condition for using the TTA is met. For each agent that is a TTA user it adds a list of other app users that are at the same location. This function thus saves every app users at the same location. This is an abstraction from real apps as they could have more specific indications of distance and time between users, while we take everyone at the same place at the same tick. This tracking of app users is done without any error which is of course better than real life where there could be errors in the software or people could even have forgotten their phone.

Implementing saving contacts in this way was the most straightforward with our implementation of locations where agents remain for just one tick. If we wanted to implement more advanced contact saving we would also have to adjust the base model, for example with explicit interactions or a finer granularity in ticks such that agents could vary how long they stay at a location.

7.3.5 Testing and Reporting with the Tracking and Tracing App

The app users in our simulation can get tested in two ways:

1. When a user is experiencing symptoms (which can come from COVID, but also from having a cold).
2. When one of the contacts of a user is tested positive. This can also happen with contacts from a contact when recursive testing is enabled.

The testing for corona in this tracking and tracing scenario is only performed through the TTA. There are no random tests performed when we evaluate the effects of the TTA. The agents that are no app users do not get tested as they already 'in most cases' stay in quarantine upon getting symptoms. The testing for corona is always 100% valid and all app users will report their infection once aware of it. This makes our implementation of the TTA more optimistic than real life, where there is a small error margin in the tests and people could forget or not be willing to indicate they are positive on the COVID-19 virus. Because of this implementation we can expect that the TTA will lead to slightly better results in our simulation than in real life.

Table 7.1 Track and tracing app scenario settings for NetLogo

Parameter	Value
General	
Preset-scenario	Scenario-6-default
Household-profiles	Great Britain
#random-seed	1 2 3 ... 48 49 50
infected-start-tick	112
#available-tests	10.000
#bus-per-timeslot	30
Tracking and tracing app	
Ratio-of-people-using-the-tracking-app	0 0.2 0.4 0.6 0.8 1
When-is-tracing-app-active?	7-days-before-end-of-global-quarantine
When-is-daily-testing-applied?	Never
Is-tracking-app-testing-recursive?	True
Ratio-young-with-phones	0.42
Ratio-retired-with-phones	0.44
#days-recording-tracing	7
Rratio-of-anxiety-avoidance-tracing-app-users	1
Is-quarantining-for-14-days-people-in-contact-with-a-sick-person-track-and-trace?	True
Global quarantine and Self-isolation	
Ratio-infected-to-start-global-quarantine	0.02
All-self-isolate-for-35-days-when-first-hitting-2%-infected?	True
Food-delivered-to-isolators?	True
Keep-retired-quarantined-forever-if-global-quarantine-is-fired-global-measure?	False
Is-infected-and-their-families-requested-to-stay-at-home?	True
Ratio-self-quarantining-when-symptomatic	0.80
Ratio-self-quarantining-when-a-family-member-is-symptomatic	0.80

7.4 Scenario Settings

Table 7.1 shows the settings relevant for the scenario of this chapter. The default settings of the simulation can be seen in (Appendix B), the variables indicated in this chapter overwrite the default ones. This preset scenario *scenario-6-default* sets the parameters for this scenario, the household profile *Great Britain* is used as later in the book we compare this specific scenario with a model that simulates Great Britain and the use of a TTA [9]. The *random seed* is manually set and creates a different initialisation for each random seed. Having manually set random seeds enables us to rerun individual runs whenever we want to check the results more in depth. The *infected-start-tick* variable is the tick at which the first three individuals are infected, 112 corresponds to 4 weeks, which should be enough time to balance out the initialisation period. We have set the number of *#available-tests* to 10.000 per day, such that everyone who has to get tested can get tested. The *#bus-per-timeslot* is set to get a more realistic amount of number of individuals in the bus.

Most of the Tracking and Tracing App variables are explained in the previous sections. The *Independent variable* in our experiment is the *ratio-of-people-using-the-tracking-app* which sets the amount of users of the TTA. Then there are a couple of quarantining and isolation variables. The global quarantine is fired when the first 2% of individuals are infected, this means the individuals stay in quarantine (self-isolate) for 35 days. Food is delivered to those who stay in self-isolation. After the global quarantine also retired individuals are released, as this seems more realistic then keeping them in self-isolation for the remainder of the run (1 year). When an individual is infected the household is requested to stay home. The ratio for going into quarantine when becoming symptomatic is 0.80. This seems realistic as the majority of individuals will stay home when becoming sick, however still a decent portion keeps on going out.

7.5 Results and Analysis

To answer the research question: *How do different numbers of users with the tracking and tracing app influence the corona virus spread?* we have set up an experiment where we vary the ratio of app users, i.e. 0, 0.2, 0.4, 0.6, 0.8 and 1. Per setting 50 runs were done. The total number of agents is 1126 of which 312 are children, 115 on average are students, 425 on average are workers and 274 are retired. There are in total 798 agents that have a phone, the total amount of app users in a run will be a percentage of the the total amount of phone users. Across all runs on average the global quarantine started at tick 140 which is during day 35 and it ended at tick 280 which is during day 70. In this section we will discuss the results of app usage on the current infections, cumulative infections, tests performed, contacts at location types and infected per age group.

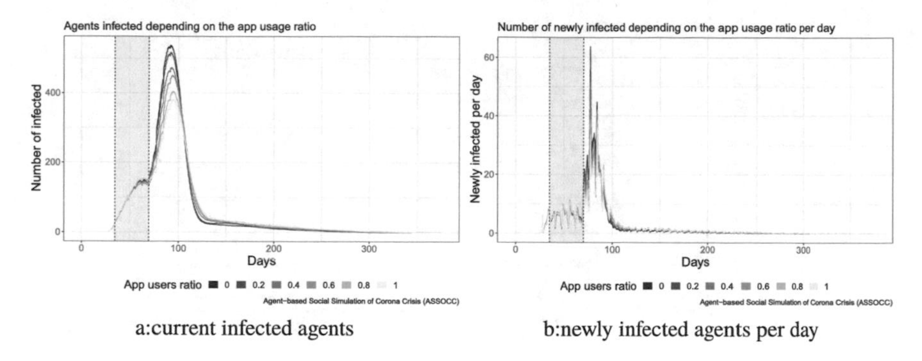

a:current infected agents b:newly infected agents per day

Fig. 7.3 The effect of the TTA on the currently infected agents and the cumulative infected agents. The red vertical bar in the plots indicates the global quarantine period, the starting day is 35 and the ending day is 70

7.5.1 Effect of App Usage on the Spread of the Virus

In Fig. 7.3a the main result of this chapter is shown. It shows the current number of infected individuals over time for the different ratios of app use. In Fig. 7.3b the daily new cases are shown for different ratios of app use. The period of lockdown is indicated by the red shaded area. The number of infections increases at the start of the simulation, around day 35 the lockdown is initiated which slows the spread of the virus, but when the lockdown is ended the virus spreads rapidly. Figure 7.3a may be not so clear in showing the diminishing effect of the lockdown as it shows the total number of infections, which still increases (but at a slower pace), however Fig. 7.3b shows that at the end of the lockdown (at day 70) there are only a couple of newly infected agents per day. From the moment the lockdown is released large peaks follow a few days later. After the large wave happened the amount of currently infected agents and amount of daily new infected agents diminish steadily until the end of the run.

We can see that the highest peak of infections is at around day 95. For 0 app usage ratio about 48% of the agents are infected at the peak (on average 538). With higher app usage ratios, we see that the peak is lower. The number of infected agents at the peaks for 0, 0.2, 0.4, 0.6, 0.8 and 1 app usage are respectively 538, 518, 473, 452, 405 and 382. The difference between 0 and 0.2 app user ratios is small, only a 3.7% decrease. It gets somewhat bigger when comparing 0 and 0.8 app user ratios, with a decrease of 24.7% of the peak. Given that in most countries the app usage ratio is less than 0.2 we can already see that even at the peak the difference in infections per day by using the app is not more than 5%.

However, if we want to investigate the overall effect of the app usage we should check the cumulative number of infected agents over time. When we look at the cumulative infections (Fig. 7.4), there is hardly any difference when using the TTA. Here the cumulative amount of infected agents at the end of the run for app usage ratios of 0, 0.2, 0.4, 0.6, 0.8 and 1 are respectively 945, 935, 922, 911, 894 and 884.

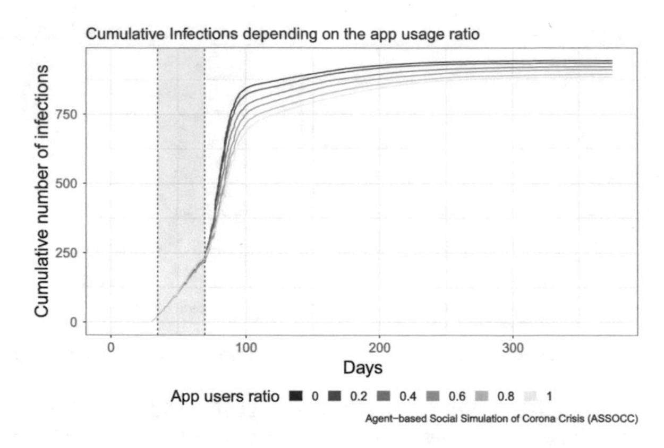

Fig. 7.4 The cumulative amount of infected agents by app ratio

Again, the dark blue line (no one uses the app) rises quickest and highest. However after a year one can see that the accumulated number of infected agents when no one uses the app is around 84% of the total population. With 0.2 app usage ratio this decreases to 83%, while with 0.8 app usage ratio the accumulated number of infected agents gets to 79%. We should reiterate that these infection numbers are much higher than can be expected in reality (in order to make results more clear). However, the shape of the curves and relative positions of the curves are what we expect to happen in reality as well. Thus the overall positive effect of using the app is never bigger than 5%!

The assumption is that using the app will lead to quicker detection of possibly infected people that can be isolated and thus preventing more infections. Why does the effect of this simple theory seem to be much less than expected? For this we will analyse how the usage of the app affects the infection rate. Are there less contacts? Or are people staying in quarantine more? Where do infections arise mainly and how does this change when the app is used? Could it be that some infections arise in places where the use of an app does not really change the situation? Additionally, we will also investigate the consequence of app usage on how many tests can be expected when many people use the app. Would there be enough test capacity in reality? In the next sections we will discuss these questions in order to get a better insight in why the use of an app seems to make so little difference in the infection rate.

7.5.2 Number of Contacts Per Location Type

To evaluate which locations are contributing most to the spreading of the virus we investigate the average number of contacts per day per agent and where they take place. Figure 7.5 shows that during the lockdown phase the number of contacts goes down, while it increases again after the lockdown since many agents try to get back to

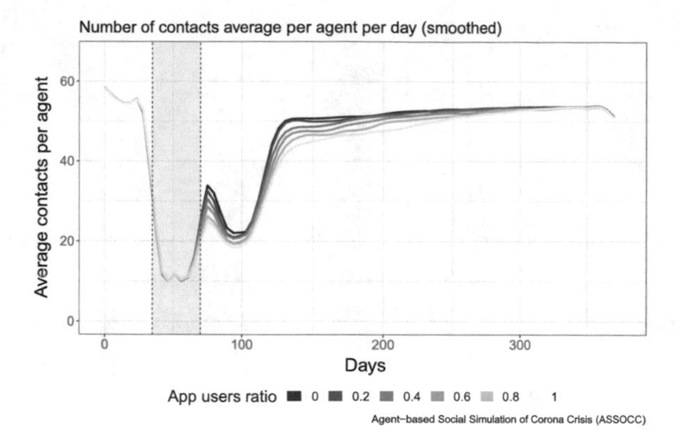

Fig. 7.5 The average number of contacts per day per agent

their normal live. We can see that at day 75, the peak of number of contacts is higher for 0 app usage than for higher app usage ratios. However, this peak is followed up by a trough since many agents get infected. Only when the total amount of infected agents starts to decrease the number of contacts starts to go up again. For lower app usage settings it goes to 'normal' a bit quicker than for higher app usage settings.

Since there is some difference in average number of contacts per day depending on the app usage ratio we want to see whether this is a consequence of more people sitting at home or whether contacts at other locations also change. Figure 7.6 shows the cumulative number of contacts per location type per app usage setting. Having more app users slightly decreases contacts at public transport, schools, universities and workplaces. The decrease of contacts in essential shops and private leisure is almost negligible as they have a small amount of cumulative contacts. It however slightly increases the contacts at homes. This makes sense as more people get notified through the app and get tested, which will make them go into quarantine (home) sooner keeping them away from other location types. This leads to one of the first reasons why the app might not be very effective. If the people that get warned through the app are more likely to be one's housemates, (who already know you are infected and cannot really avoid you) the app does not have much effect on the number of infected agents.

7.5.3 Infections Per Location Type

It is clear by Fig. 7.7 that the impact of the app on where infections take place is the largest in homes, public transport, schools and non essential shops. These changes are more or less exponential with higher app usage giving proportionally more change. For example there is hardly any change between 0 and 0.2 app usage ratios while the difference between 0.8 and 1 app usage ratios is quite notable. Figure 7.6 showed

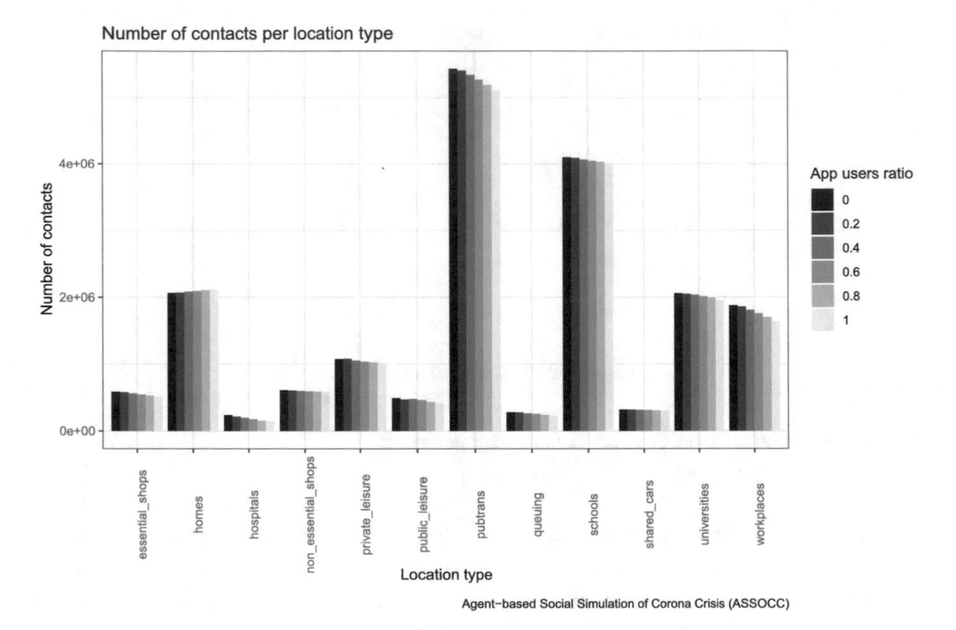

Fig. 7.6 Cumulative number of contacts per location type per app users ratio

us that the contacts at homes, public transport, schools, universities and work places have a varying amount of contacts for different app usage ratios and therefore it seems logical that the number of infected agents changes as well. A small decrease or increase in number of contacts could lead to respectively a larger decrease or increase in infections. As this change in contacts is due to app users staying home because of notifications, these are potentially infectious agents and keeping them home more often influences the infections at other locations.

The number of infected agents decreases most at public transport with schools, shared cars, non essential shops and private leisure having a lower decrease. The infections at homes increases quite a lot, which was expected as there were more contacts at homes. The other locations have no decrease or such a small decrease that it hardly has any effect on the cumulative infections. A small detail at the non essential shops is the higher number of infected agents for 0.2 app usage ratio. This may have to do with agents going less to work since they are supposed to be in quarantine, however they want to fulfil their belonging need and instead go to the non essential shop. Only with a very low amount of app usage ratio this is apparent as with higher amounts enough agents stay home to decrease the number of infected agents at non essential shops. At 0.6 app usage ratio we see slightly higher number of infections at schools however this could be an effect of the random runs, which would diminish if we would do more runs per setting.

Remember that we have called pubs and restaurants private leisure places. They are the places where you meet with a relative small number of people in a confined space. Thus again, when there are no restrictions these are the places where you

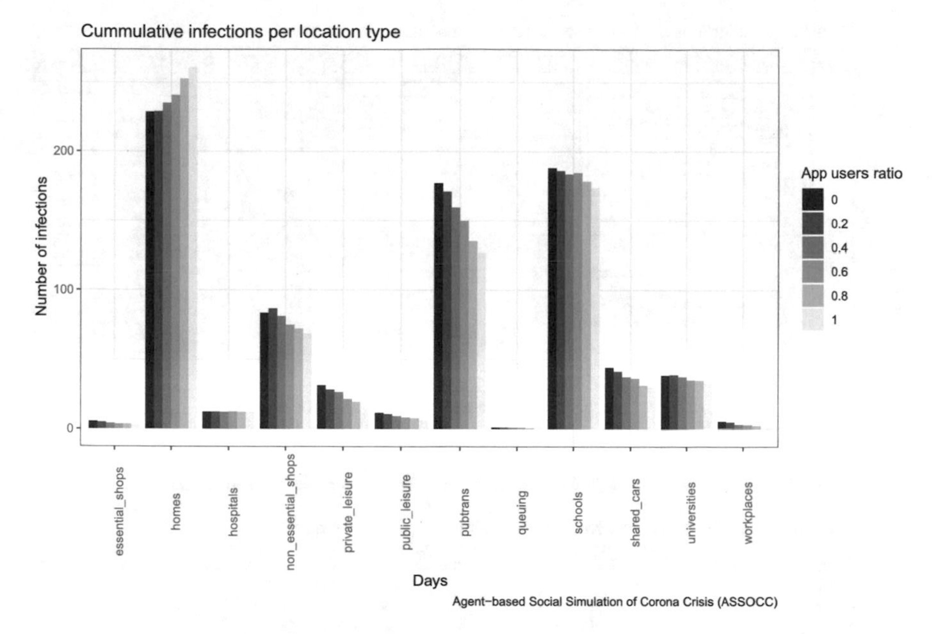

Fig. 7.7 Cumulative number of infected per location type per app users ratio

expect more infections to take place and also spread over different groups of people that meet in different constellations on different days and places. This is confirmed by the graph, but it also shows that the positive effect of the app usage is small due to the relative small number of people meeting in these places (compared to e.g., work and school) (Fig. 7.7).

Figure 7.8 shows the accumulated number of infections per location type over time. I.e. the two graphs show the contribution of each location type to the total number of infections over time. If no one uses the app (top-left figures, 0 app usage ratio) the number of infections first rises most at home while a bit later the infections in public transport rises. We can explain this by the fact that people go home when they feel ill or are tested and stay in quarantine at home. Thus, the first place where infected people are in long contact with other people is at home. Therefore, the chance that those get infected is high.

If no one uses an app, the most obvious places where many people meet strangers in many different constellations is the public transport. Thus, this will be indeed a place where many infections occur. This decreases with the use of the app. This is also explainable as the app targets exactly the public spaces where many strangers meet and can warn all the other passengers if someone happened to be infected. The fact that the cumulative amount of infections does not decrease more with the use of the app can be explained from the fact that often the warning of infections comes

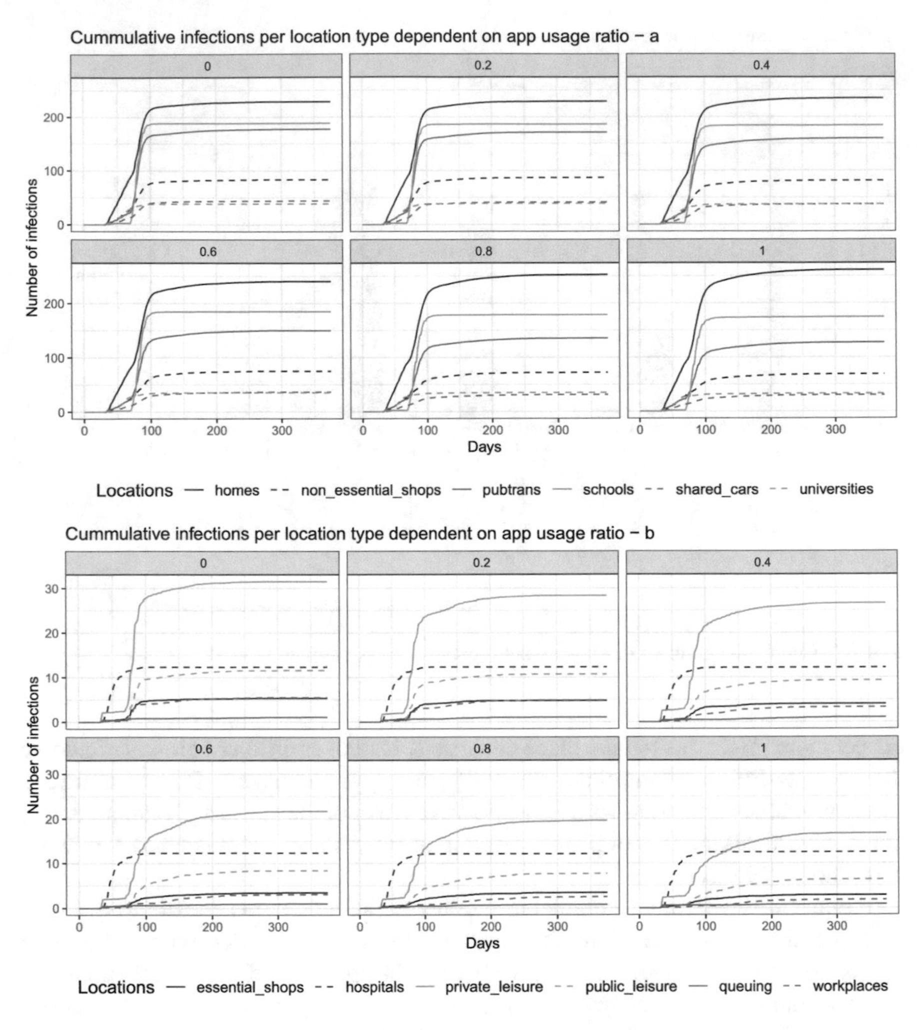

Fig. 7.8 The cumulative number of infections per gathering point. Notice the difference in the vertical axis, i.e. the top plot consists of the locations with the highest infections, while the bottom plot consists of the locations with the lowest infections

too late. In the days between an infection and the first symptoms arising the agents can already have infected others in the public transport as well.

The main result from the analysis on where infections seem to take place and how they are affected by the use of an app is that most infections take place at home. These infections are not much affected by the use of an app as they are kind of unavoidable. People can infect their housemates also when the housemates know they have an infected housemate.

7.5.4 *The Different Age Groups Infecting Each Other*

The next item to check is whether we can see which agents infect which other agents most and whether this could be influenced by using an app. Figure 7.9 shows the proportion of an age group getting infected by specific other age groups. Per column in each figure, one sees by which groups that age group is infected relative to the total number of infected in that age group. Thus, young people are mostly infected by other young people (more than 50% of column 1 is light green). Students are mostly infected by other students (more than 50% of column 2 is orange). Workers are, relative to the other groups, infected by the most diverse groups. We can explain this by noticing that those people tend to work, while also having children at home and having parents that are retired. Thus, they come in contact in a natural way with all other age groups. Finally, retired people are mainly infected by other retired people (also more than 50% of column 4 is dark red). They are more vulnerable and thus infection spreads quick in care homes (as can also be seen in the actual world).

From the figures it is clear that the app use does not change anything in the relative infection by other age groups. Thus, the places where the app has effect, like in public transport are not the only or even main places where the people of different age groups meet each other. The figure would probably look different if e.g., universities would be closed as that is the main place for students to meet each other. We could in that case expect that the ratio of students infecting each other (the large orange bar) becomes smaller.

While the ratio of age groups infecting each other does not change, we can see some changes in the number of cumulative infected agents separated by age group. Table 7.2 shows that workers are the ones that get most affected by having agents using the TTA. In that age group the total amount of infected agents decreases from 319 (with app usage ratio 0) to 278 when every agent, that has phone, uses the app. This may have to do with workers visiting most types of places. They are in contact with their partner and their children at home. They work at a workplace and come in contact with other workers, however they may also work at a hospital, shop, school or university which broadens their contacts. In their leisure time they can go to shops, private leisure and public leisure as well. This variety of places means a variety of contacts, which adds them to more TTAs lists. Therefore they may get tested more. The elderly only get slightly affected going from 231 (no app usage) to 218 (1 app usage). The young and student age groups are hardly effected by higher TTA usage.

7.5.5 *Hospitalisations Based on App Usage Ratio*

Finally, we want to see if there is a difference in hospitalisations. Because there were slightly less cumulative infected people but these seem mainly to be workers, which have a smaller chance to become severely ill than retired people. In Fig. 7.10, we can see that both the number of workers and retired that are hospitalised decreases slightly

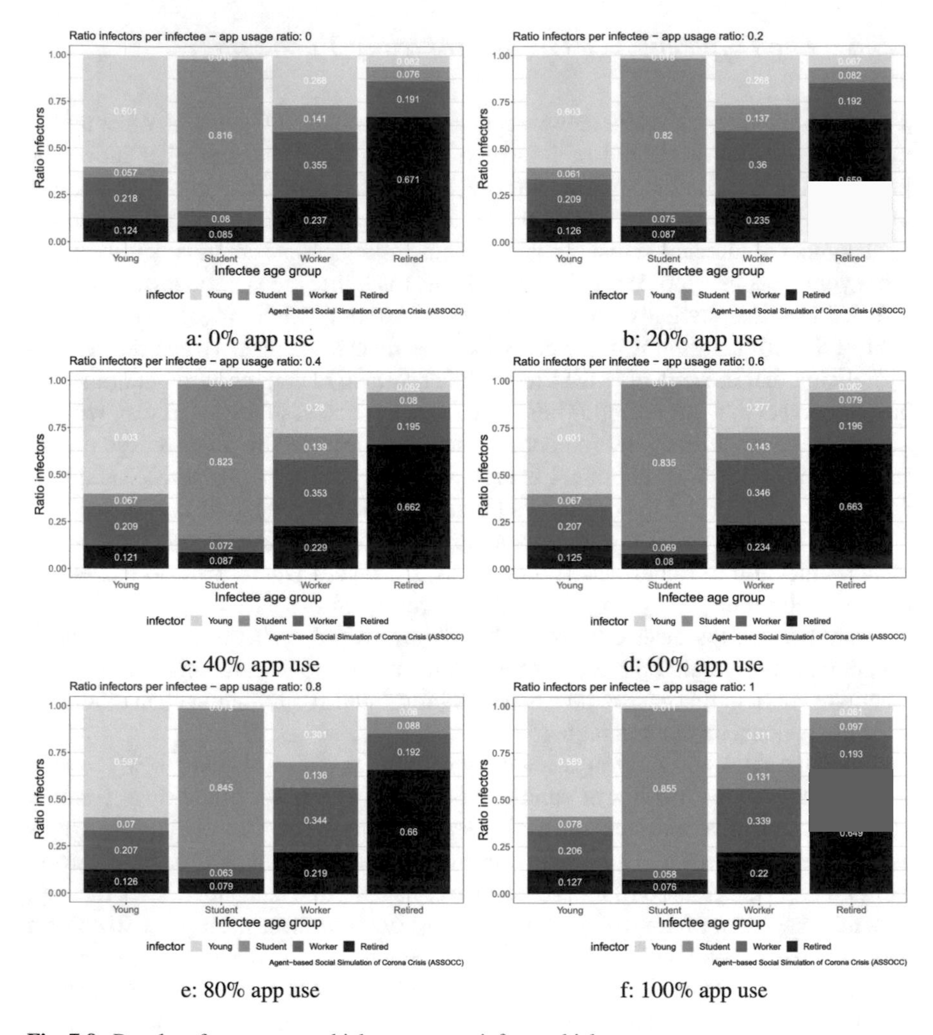

Fig. 7.9 Results of app use on which app groups infects which age group

Table 7.2 Cumulative infected per age group

App usage	Young (312)	Student (115)	Worker (425)	Retired (274)
0	301	93	319	231
0.2	299	95	312	229
0.4	298	93	303	228
0.6	298	93	295	225
0.8	296	93	285	220
1	295	92	278	218

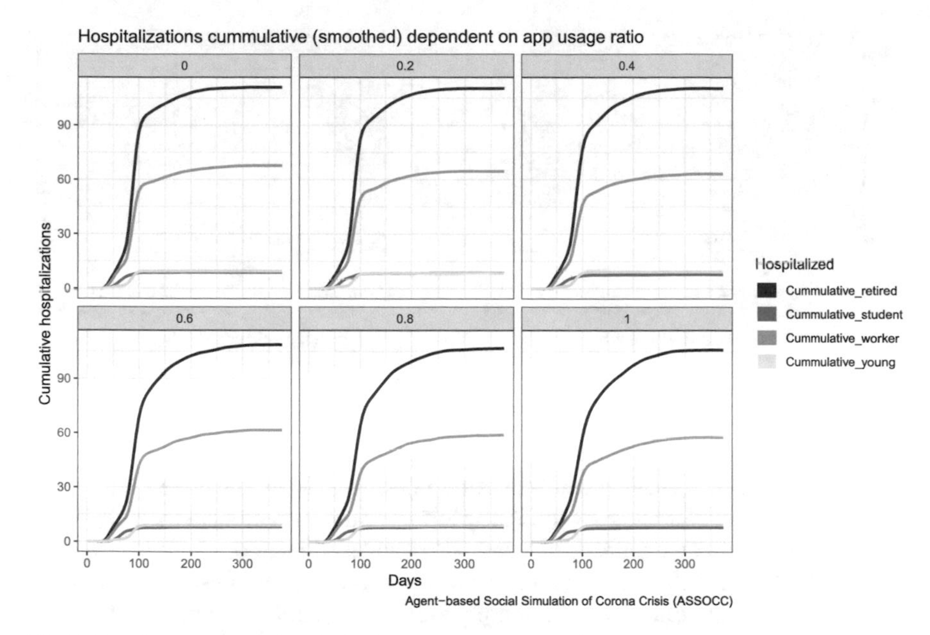

Fig. 7.10 The cumulative hospitalisations per age group and app usage ratios

for higher app user ratios. However, the decrease is in the range of 5%, which is not very substantial. In Table 7.2 we saw the cumulative infected workers being most diminished by app usage ratio, compared to retired. The figure here shows almost the same decrease in hospitalisations among those groups. This is due to the increased severity of the disease for older people. Elderly have a higher chance to get severely symptomatic and thus end up in the hospital. Thus a decrease of 5% in elderly getting infected could lead to the same amount of decrease in hospitalisation while a decrease of 10% infected workers only leads to 5% decrease in hospitalisations as well. The amount of hospitalised young and students does not change with higher amount of app use ratios. The minor changes are probably part of random perturbation.

7.5.6 Effects of App Usage on Tests Needed

In the previous sections we have discussed the effect of using the TTA on the spread of the virus. However, another effect of using the app is that all people that have been warned through the app who have been in contact with someone who tested positive should go in isolation and being tested. What does this mean for the number of tests that are needed when many people use the app? Fig. 7.11a shows the cumulative tests per app usage ratio and Fig. 7.11b the tests per day. As expected, the number of tests increases exponentially with an increasing app usage ratio. This is a direct consequence of the fact that if the ratio of app users goes up the chance that people an app user is in contact with also use the app. Thus the number of contacts that is

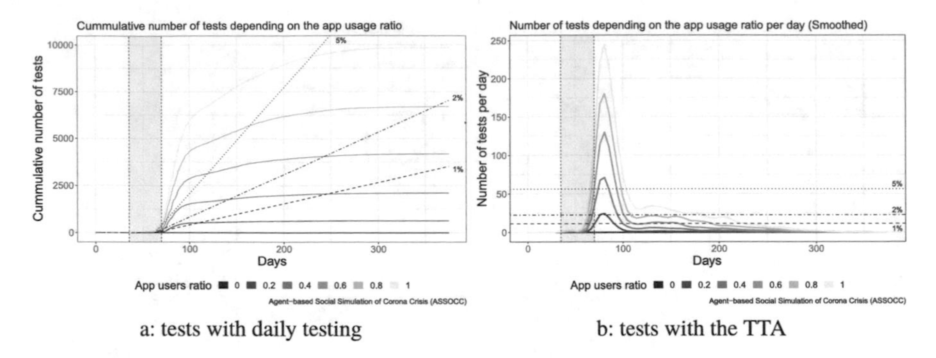

a: tests with daily testing

b: tests with the TTA

Fig. 7.11 Number of tests required with different ratios of app usage

registered by the app goes up and thus if one person is tested positive the number of contacts that are warned goes up. And if the warning is recursive the number of warned people goes up exponentially and with that also the amount of tests needed. In Fig. 7.11b we see that the amount of tests per day that are needed follows the same curve as the infections. It rises sharply after more people become infected after the lockdown finishes and it decreases again after most vulnerable people have been infected and the pandemic dies down slowly. To give a sense of how many tests are required per day we have included lines to indicate how many tests would be performed if a certain percentage of the population was tested.

The cumulative tests lines for different app usage ratios are much steeper than the percentage lines. Especially in Fig. 7.11b, which shows the tests required per day, we can see that the peaks far exceed the random testing lines. We see that even if only 20% of the people use the app there is a period between day 70 and 90 where more than 1% of the people require a test every day. In most countries the test capacity is less than this 1% per day. Thus many more tests would be required due to the app use than available! The only solution available is that all the people who cannot get tested should stay in quarantine. If we take as a target that 60% of the people use the app it means that in the top demand time around 10% cannot be tested and should go in quarantine per day(!). Effectively this means that in a few days the country would be in a lockdown again.

7.6 Discussion

We performed a range of simulations to evaluate the effect of a TTA on the dynamics of the pandemic. Based on the results, we can answer the research question both with yes and no. Yes, since there are some effects apparent with higher number of app users. These effects are that the curve of infected people is lower and there is a slight decrease in total number of infected people. More app usage reduces contacts at public transports, schools, university workplaces and hospitals, likewise the number of people infected at those places. The number of infected children and workers decreases slightly and the number of hospitalisations of workers and elderly decreases slightly. At some places, the app has less effect or a detrimental effect. A negative effect, which is expected, is that the number of contacts and infections in households increases. This is expected as people get into quarantine more often when the app does its job. Furthermore, most of the positive effects need a higher app usage ratio (for example 0.6 or 0.8) before they become really significant. In the last section of this chapter we have shown that the consequence of a higher ratio of app usage requires an excessive amount of tests available. Since these amounts of tests are not available in most countries, this will lead to effective lockdowns as people that are warned, but cannot be tested should stay in quarantine.

In reality mobile phone usage is not equal for people of different ages. For example young children or elderly have a lower percentage of mobile phone users than other age groups [9]. Our model allows for changing the proportion of children and elderly that have a mobile phone with the use of two parameters. This setting can be used in the future, when we want more specific results. Some preliminary exploits in changing these parameters made the effect of the app usage even a bit smaller, but the effects are minimal due to the fact that app usage already affects the spread of the virus among young and retired agents very little.

We simplified the implementation of the app. In our model, the intensity and duration of interactions is always assumed to be equal, while in reality when you go to a place you will have more contact with specific people and often no contact with some others. Since in real life the app, if implemented with a 15 min timer, can for example mark you when you talked with a person behind a screen for 30 min. However, the interaction will not be tracked when you talk with a person up close for 12 min.

With regards to the TTA, we do not take into account that people could do the tracking and tracing by themselves without the use of an app. If a person gets a positive corona test she is likely to inform her friends, family and colleagues. Certainly not every person will do this and if a person wants to do it, he/she does not have all the contacts of people randomly met in a shop or on the street. However, it could influence the results and soften the number of infected for 0% app users as people would inform each other.

In practice, in all countries that have started to use the TTA, not more than 20% of the people actively use the app. With these low percentages, the app barely has any effect on the dynamics of the pandemic.

7.7 Conclusion

Higher numbers of TTA users lead to a decrease of peak infections, but only a very small decrease in cumulative infected over the complete run. The number of contacts at public transport, schools, universities and workplaces decreases. While contacts at homes increase. In terms of total infections at locations there are less infected agents in public transport and slightly less infected agents in schools, shared cars, non essential shops and private leisure. However due to more people getting infected at home these effects get somewhat undone. The total amount of workers infected and total amount of retired agents infected decreased slightly while the amount of infected young and student agents stayed relatively the same. There is a small decrease in hospitalisations of retired agents and workers. The number of tests required a day increases exponentially with higher app usage ratios. Especially the peak of tests performed per day goes far beyond what is realistically possible, which requires that many of the app users have to stay in quarantine.

To conclude, a higher number of TTA users seems to have a positive effect on decreasing the corona virus spread. However these positive effects seem to be almost non existing with low number of TTA users $< 20\%$, and only become significant when app usage is $> 60\%$ (which is unrealistic in the real world). While the number of contacts and infections at most public places such as schools, workplaces, leisure places and shops go down, the number of contacts and infections at homes increases. The TTA could have some benefit, however the benefit stays marginal and only becomes apparent at higher app usage percentages. When we also take into account the tests required for higher app usage percentages, it seems that the app leads to a partial lock down as all users that cannot be tested have to stay home.

Acknowledgements We would like to acknowledge the members of the ASSOCC team for their valuable contributions to this chapter. This research was partially supported by the Wallenberg AI, Autonomous Systems and Software Program (WASP) funded by the Knut and Alice Wallenberg Foundation. The simulations were enabled by resources provided by the Swedish National Infrastructure for Computing (SNIC) at Umeå partially funded by the Swedish Research Council through grant agreement no. 2018-05973.

References

1. L. Ferretti et al., Quantifying SARS-CoV-2 transmission suggests epidemic control with digital contact tracing, in *Science* (2020). ISSN: 0036- 8075. https://doi.org/10.1126/science.abb6936. eprint: https://science.sciencemag.org/content/early/2020/04/09/science.abb6936.full.pdf. https://science.sciencemag.org/content/early/2020/04/09/science.abb6936
2. I. Braithwaite et al., Automated and partly automated contact tracing: a systematic review to inform the control of COVID-19, in *The Lancet Digital Health* (2020)
3. Samira Davalbhakta et al., A systematic review of smartphone applications available for corona virus disease 2019 (COVID19) and the assessment of their quality using the mobile application rating scale (MARS). J. Med. Syst. **44**(9), 1–15 (2020)
4. F. Lucivero et al., Covid-19 and Contact Tracing Apps: Ethical challenges for a social experiment on a global scale. J. Bioeth. Inq. 1–5 (2020)
5. Nadeem Ahmed et al., A survey of covid-19 contact tracing apps. IEEE Access **8**, 134577–134601 (2020)
6. M. Zastrow, Coronavirus contact-tracing apps: Can they slow the spread of COVID-19? Nature (2020)
7. G. Goggin, <?covid19?> COVID-19 apps in Singapore and Australia: reimagining healthy nations with digital technology. Media Int. Aust. 1329878X20949770 (2020)
8. J.H. Reelfs, O. Hohlfeld, I. Poese, Corona-Warn-App: tracing the start of the official COVID-19 exposure notification app for Germany (2020). arXiv:2008.07370
9. R. Hinch et al., Effective configurations of a digital contact tracing app: a report to NHSX. (Apr. 2020). https://github.com/BDI-pathogens/covid-19_instant_tracing/blob/master/Report (2020)

Chapter 8
Studying the Influence of Culture on the Effective Management of the COVID-19 Crisis

Amineh Ghorbani, Bart de Bruin, and Kurt Kreulen

Abstract In this chapter we investigate the influence of culture on the effective management of the COVID-19 crisis. In order to study culture we first describe cultures in terms of values. These values are connected to the needs of the agents, giving them a certain default priority, which differs across cultures. Then we will show how culture actually influences how people react to certain types of measurements and the effect this has on the effective management of the COVID-19 crisis.

8.1 Introduction

The outbreak of the novel Corona Virus Disease in 2019 ('COVID-19') continues to have a tremendous impact on the daily lives of people around the globe. Although the scale at which SARS-CoV-2 (the virus that causes COVID-19) has been able to spread across the globe is unprecedented, this is certainly not the first pandemic the world has witnessed [1]. Such global epidemics are inevitable and are expected to occur more often given our increasingly connected lives. To illustrate, in our modern connected and urbanized world the outbreak of an infectious disease can move from a remote village to a major city on the other side of the world in less than 36 h [2]. This highlights the importance of finding ways that facilitate the effective management (read: mitigation) of the impact of such disease outbreaks.

While, at the time of writing this article, enormous efforts are geared towards mass vaccination of populations, altering the behavior of individuals and the consequent patterns of social interaction remains the main approach to underpin the virus transmission. For instance, increasing the frequency of hand washing, wearing face masks in public and maintaining a sufficient physical distance from others all help to exert a downward pressure on the transmission potential of the virus. Crucially, the readiness of people to comply to these public health related measures is dependent upon prevailing cultural beliefs. Gelfand et al. [3] already show that nations with

A. Ghorbani (✉) · B. de Bruin · K. Kreulen
Faculty of Technology, Policy & Management, TU Delft, Jaffalaan 5, 2628 BX Delft, The Netherlands
e-mail: a.ghorbani@tudelft.nl

F. Dignum (ed.), *Social Simulation for a Crisis*, Computational Social Sciences,
https://doi.org/10.1007/978-3-030-76397-8_8

'tight', rather than 'loose' cultures have been most effective at limiting the spread of SARS-CoV-2 during the first 'wave'[1] of COVID-19 infections. Also, [4] show that cultural beliefs played an important role in the transmission dynamics of Ebola outbreaks in West Africa. Moreover, [5] suggest that collectivistic cultures may hold an edge over more individualistic cultures in terms of being able to slow the spread of the virus through non-pharmaceutical measures. These postulations are supported by [6] who show that citizens of countries that endorse pro-social, rather than ego-centric, cultural values tend to display a stronger willingness to adopt preventative public health behaviors.

All in all, the cumulative evidence suggests that strategies aimed at limiting the spread of any contagious virus ought to account for the prevailing cultural context within which those strategies are formulated and subsequently put into practice. Hence, gaining a better understanding of how culture influences the transmission of a virus as well as the effectiveness of policy measures aimed at managing a pandemic is a valuable objective to pursue.

The goal of this research is to use a theory exposition model [7] to explain how culture influences the transmission of a virus and the effectiveness of policies in a population of individuals. The model simulates the relationship between the cultural profile of societies and the acceptance of, and compliance with public measures aimed at limiting the spread and mortality of the COVID-19 pandemic. The outcomes of in-silico experiments are interpreted and discussed in light of empirical data reported by cultures regarding the spread and mortality of COVID-19 within their populations.

The current paper starts off with a description of our theoretical representation of how culture influences behavior during a pandemic. Next, we describe the research methodology and a description of the model. The paper subsequently describes the model experiments and concludes by discussing the experimental findings in light of empirical data and presenting avenues for further investigation.

8.2 Theoretical Framework

In this section we describe the theories that are, in combination with each other, used to model culturally-influenced decision-making behavior during a pandemic.

8.2.1 Values

Humans are fundamentally motivated to extract meaning from, and make sense of reality in order to resolve ambiguities, reduce complexity, and avoid feelings of

[1]The notion of 'infection-waves' refers to the characteristic 'wave-shaped' development of cases of infection over time which tends to transit through phases of growth, stabilization and decay, respectively.

confusion and anxiety [8, 9]. Humans do so by cognitively transmuting the necessities inherent in existence into higher-order guiding principles (i.e. values) that can be communicated effectively. Values help humans cope with the reality of living in a complex social context [10, 11] by providing an authoritative justification for norms that dictate how one is expected to act [9]. In doing so, values guarantee some kind of predictability and stability of the behaviour of individuals and society as a whole [12].

As explained comprehensively in Chap. 2, we use the Basic Value Theory (BVT) [13] that distinguishes ten values that are universally present within the value systems of humans. Note that although the nature and structure of the Schwartz BVT values may be universal, individuals and groups differ substantially in the relative importance they ascribe to each value. Notwithstanding the differences between humans in their relative prioritization of values, it is shown that they exhibit a reliable and characteristic correlational pattern. For instance, people that indicate they find the accumulation of material wealth to be important in life are more likely to give a lower rating of the importance of fairness in survey studies. These inter-value correlations have been tested extensively and shown to be consistent across nations, cultures, genders and age-groups [11, 14–16]. Thus knowing that someone ascribes special importance to a particular value—or set of values—enables one to draw reliable inferences with regards to the structure of that person's value system as a whole. The BVT summarizes these findings by specifying the structure and dynamism of the relations between each of the values within the *value circumplex model* (see Fig. 8.3 Values placed close to one another in the circumplex model are considered *mutualistic*, and values placed further away from one another become increasingly *antagonistic*. The circumplex model implies that actions in pursuit of any particular value will have consequences for a person's ability to cater to the fulfillment of the other values [13]. Hence, antagonistic values are generally in conflict with one another, while mutualistic values are those whose prescribed actions tend to harmonize with one another. The notions of antagonistic and mutualistic values help to understand how people's value systems can be logically structured.

8.2.2 Needs

Values refer to desirable end-states of reality on the basis of which *goals* are formulated that in turn motivate actions [13]. Motivation theory provides a means of conceptualizing how values may drive behavior. Two strands of motivation theory are distinguished, namely: *content* versus *process* theories [17]. Content theories specify the factors within a person that energize, direct, sustain and stop behavior [17]. Process theories, on the other hand, focus on elucidating the process by which behavior comes about [18]. The current section builds on content-based motivation theory to formulate a conceptualization of what drives of behavior. Section 8.2.4 applies insights derived from process-based motivation theory to explicate how behavior is driven.

Table 8.1 Connecting values, needs and actions

Construct	Thought processes
Values [Beliefs]	How should the world be? What state(s) of reality is/are desirable?
Needs [Desires]	What discrepancies do I currently perceive between how the world is and how it should be?
Actions [Intentions]	What can I do at this moment within the constraints posed by my current contextual circumstances that bring the world as it appears closer to how I think it should be?

In line with Maslow's Hierarchy of Needs (HON) that is explained in Chap. 2, we presume that the satisfaction of needs is what drives behavior. Moreover, we propose that values inform the prioritization of needs. In doing so, our conceptualization is based on the Beliefs-Desires-Intention (BDI) model of agency [19]. Table 8.1 explicates how values, needs and actions are connected through an individual's thought processes which are represented as questions one may ask him-/herself during decision-making.

Building on the concept of *value trees* [20, 21], each Schwartz value can be characterized by a set of needs whose satisfaction promotes a particular set of values (see Table 8.2). We assume there are no interactions taking place between the satisfaction of various needs in the fulfillment of a particular value [22].

A 'Maslow-based' categorization of needs is applied whereby needs are classified as being *physiological* or *psychological*. The physiological needs are grouped within the Survival category (see Table 8.3), whereas the psychological needs are to be found in the other categories (i.e. Self-Esteem, Belonging and Safety). The current study omits the Self-Actualization and Self-Transcendence categories of Maslow, since it is presumed that the impact of a pandemic on these types of needs is non-existent [23].

The physiological needs (i.e. biological drifts) are the same for all humans, which implies that the differential prioritization of tending to particular psychological needs is what distinguishes humans from one another in terms of their decision-making behavior. In line with Self-Determination Theory (SDT) [25], it is presumed that a person's values exerts an influence solely on the psychological set of needs.[2]

In accordance with Maslow's HON theory, it is proposed that some basic satisfaction level of physiological needs must be met before an individual becomes concerned with meeting her psychological needs [22]. Thus, when an individual is hungry, she eats regardless of the status of any of her psychological needs. Although Maslow proposes that the sub-classes of psychological needs are also hierarchically structured, a decision is made not to integrate this proposition within the current model. This is because empirical evidence for the specific, linear hierarchical order of psy-

[2]The decision to satisfy a particular psychological need is deemed to be 'good' or the 'right' thing to do by a particular individual because of the values he/she cherishes. In this way, values serve to justify the outcomes of a decision-making process [26].

Table 8.2 Linkages between values and needs

Value	Description of needs	Needs
Achievement (ACH)	Needs that relate to the attainment of personal success, social approval, and competence	Autonomy
Power (POW)	Needs that relate to the attainment of social status and prestige, control or dominance over people and resources	Financial Stability, Luxury
Hedonism (HED)	Needs that relate to the attainment of pleasure and sensuous gratification for oneself	Luxury, Leisure
Simulation (STM)	Needs that relate to the attainment of excitement, novelty, and challenging oneself	Leisure
Self-Direction (SD)	Needs that relate to the attainment of independent thought and action; choosing, creating, and exploring	Autonomy
Universalism[a] (UNI)	Needs that relate to experiencing appreciation, tolerance and protection of the welfare of all people and for nature	N/A
Benevolence (BEN)	Needs that relate to the preservation and enhancement of the welfare of those with whom one is in frequent personal contact (the 'in-group')	Belonging
Conformity & Tradition (CT)	Needs that relate to maintaining respect of and commitment to social norms and one's cultural heritage	Compliance, Conformity, Belonging
Security (SEC)	Needs that relate to the attainment of safety, harmony, and stability of society, of relationships, and of the self	Risk Avoidance, Compliance

[a]The value of Universalism is presumed to be related to the self-actualization and self-transcendence needs which are not included within the current model

Table 8.3 A description of needs and their Maslow-based categorization

Type	Category	Needs	Description
Psychological	Self-Esteem	Leisure, Luxury, Autonomy	Humans cherish the desire for both self-esteem and for the esteem a person obtains from others (i.e. social recognition) [24]
Psychological	Belonging	Belonging	People seek to overcome feelings of loneliness and alienation. This involves both giving and receiving love, affection and experiencing a sense of belonging [24]
Psychological	Safety	Risk Avoidance, Compliance, Conformity, Financial Safety	While adults have little awareness of their security needs except in times of emergency or periods of disorganization in the social structure (such as a pandemic), children often display the signs of insecurity and the need to be safe [24]
Physiological	Survival	Food, Health, Sleep, Financial Survival	These are biological needs which consist of the need for oxygen, food, water, and a relatively constant body temperature. They are the strongest needs because if a person were deprived of all needs, these are the ones that would come first in the person's search for satisfaction [24]

chological needs that Maslow proposed is sparse [27]. Furthermore, [28] asserts that the hierarchy proposed by Maslow suffers from Western-oriented ethnocentrism. In doing so, [28] implies that the hierarchical structure of needs is culturally-dependent. By coupling values with needs and value systems to culture, the current conceptual model ensures that people's hierarchy of needs is culturally sensitive.

8.2.3 Culture

Culture can be thought of as a macro-level, or 'collective' value system; that is, the value system of a group of people, a population, rather than that of an individual [29]. Culture is defined as "*the collective programming of the mind which distinguishes the members of one group or category of people from those of another*" [30]. It has been argued that national cultures differ in particular at the level of values held by a majority of the population [31]. Culture is therefore considered to constitute the blueprint upon which the value systems of individuals are constructed. In a similar vein, [32] describes culture as "*the press to which individuals are exposed by virtue of living in particular social systems*". Conceiving of culture in this way helps to understand how it orients the formation of an individual's value system so that it comes to resemble the 'cultural standard', which is a function of a population's *cultural profile*. A cultural standard can be thought of as the set of collectively shared notions within a population of what is considered to be acceptable (i.e. norms) or desirable (i.e. values) under various contextual settings [33]. If one could construct a collective value system from all the individual value systems present within a population, it represent the cultural standard (see Fig. 8.1). A cultural profile is defined here as a set of features that summarises a culture's key characteristics. Cultural profiles provide a means to compare different cultures to one another [34]. To summarize, it is proposed that the top-down influence of culture on individual value systems is what differentiates populations from one another in their propensity to accept and comply to public health related policy measures.

It is important to note that culture informs and is formed by the content and structure of individual value systems. The slow but steady accumulation of changes in people's value systems is what lays the foundation for broader cultural change. Cultural change is a process that unfolds over the course of several decades [35, 36], it is therefore not considered to be relevant within the scope of the current research.

Another important dimension of cross-cultural variation is the degree of *looseness versus tightness* that characterizes a population's cultural profile. Individuals living in 'tight-culture' societies tend to exhibit lower levels of psychological differentiation [37] than individuals living within societies that hold loose cultures. Moreover, social norms are expressed more clearly and unambiguously in tight cultures [37]. Based on these findings and on reasoning presented by [38], we propose that the magnitude by which individual value systems deviate from the cultural standard is proportional to a nation's cultural tightness. Specifically, the tighter the culture, the higher the consensus among the structure of the value systems of a nation's people [39]. Figure 8.2 helps

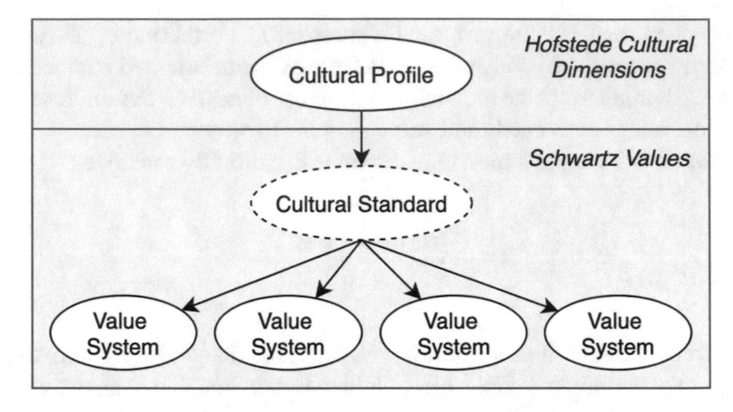

Fig. 8.1 Cultural profile, cultural standard and value systems

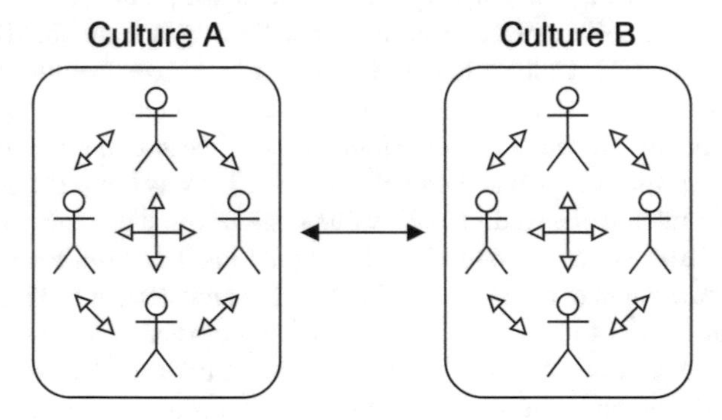

Fig. 8.2 Within (Intra) and Between (Inter) cultural variation

to visualize how cultural tightness modulates the degree of intra-cultural variation (as indicated by the white-headed arrows), whereas differences in cultural profiles determine the inter-cultural variation (as indicated by the black-headed arrows). The current study utilizes the Hofstede Cultural Dimensions (HCDs),[3] or the Hofstede 6D-Model, in order formalize nations' cultural profiles. The construction of the six HCDs is grounded in factor analysis of empirical (survey) data [40]. Nations can be scored on the HCDs, thereby providing a quantitative representation of a nation's cultural profile. The 6 HCDs are: (1) Power Distance vs Egalitarianism (PDI), (2) Individualism vs Collectivism (IDV), (3) Uncertainty Avoidance vs Uncertainty Tolerance (UAI), (4) Masculinity vs Femininity (MAS), (5) Long-term vs Short-term Orientation (LTO), and (6) Indulgence vs Restraint (IVR). The Hofstede framework is one of several ways in which culture may be quantified, see e.g. [41]. One alternative formalization of culture that competes with the HCD-framework is presented by Schwartz's notion of Cultural Value Orientations (CVOs) [42]. The HCDs are

[3]For a detailed description of the six Hofstede dimensions, see: https://hi.hofstede-insights.com/national-culture.

favoured over the CVOs because of the extensive availability of up-to-date empirical data on the HCDs. Moreover, the HCD-framework enjoys a unique position of earned respect within the field of cross-cultural research [41]. Lastly, one may argue on the basis of the principle of *triangulation* that combining several independent research methods for studying a particular phenomenon adds to its validity [43]. Since the Schwartz BVT and CVO both build on the same theoretical foundations [38], it is probably wise to apply the HCD-framework in order to diversify the risk of potentially building on invalid conceptual underpinnings. For a comparison of the HCDs with Schwartz's CVOs the reader is referred to [44–47].

Coupling the Schwartz BVT values with the Hofstede dimensions is done through the notion of Dominant Cultural Correlates (DCCs). A DCC denotes the cultural (i.e. macro-level) concept that shows strongest fit with the defining goal of a particular value at the micro-level. Specifically, each Schwartz value is assigned one or more DCCs on the basis of theoretical descriptions provided by [13, 48]. Note that a DCC can be positively (DCC^{+}) or negatively correlated (DCC^{-}) with a given Schwartz value. The following items present argumentation for the theoretical linkages depicted in Table 8.4:

- **Power Distance (PDI)**: High-PDI cultures prescribe decision-making power to be concentrated in the hands of figures of authority rather than to be distributed equally across the members of a society as is the case in low-PDI cultures [49]. High-PDI cultures are designed around aristocratic principles that promote social exclusivity. Hence, individual value systems within high-PDI cultures will be biased towards ascribing a high importance to the Power value of Schwartz which is characterized by a motivation to pursue and obtain social status and a proprietary control over resources and decision-making power [13]. Therefore, the DCC^{+} of the Power value is PDI. Conversely, PDI constitutes the DCC^{-} of Universalism (see Table 8.4). This is because Universalism is concerned with promoting egalitarianism [11].
- **Masculinity (MAS)**: High-MAS cultures are organized around meritocratic principles and prescribe assertiveness, mastery, toughness and competition [49, 50].

Table 8.4 Overview of conceptual linkages between the Schwartz BVT values and hofstede cultural dimensions

Value	DCC^{+}	DCC^{-}
Hedonism (HED)	IVR	–
Stimulation (STM)	–	UAI
Self-Direction (SD)	IDV	–
Universalism (UNI)	–	MAS, PDI
Benevolence (BEN)	–	MAS
Conformity & Tradition (CT)	PDI	IDV, LTO, IVR
Security (SEC)	UAI	–
Power (POW)	PDI, MAS	–
Achievement (ACH)	MAS	–

Low-MAS cultures are characterized by promoting consensus-building, compromise, modesty, compassion and social equality [49, 50]. Based on these characteristics it is proposed that individuals in high-MAS (vs low-MAS) societies tend to value Achievement and Power. This is because Achievement promotes the obtainment and exhibition of success, and Power subscribes to the importance of gaining social status and prestige. Conversely, high-MAS societies suppress the valuation of Benevolence and Universalism since these two values combine to promote cooperation and social equality [11]. MAS therefore constitutes the DCC^+ of the Achievement and Power values and the DCC^- of Benevolence and Universalism (see Table 8.4).

- **Uncertainty Avoidance Index (UAI)**: The UAI-HCD is about how societies cope with the unknowable [49]. Low-UAI cultures are intolerant and/or dismissive of things that challenge the status quo. Low-UAI cultures tend to have strict rules and rituals that suppress change by enhancing predictability and order [49]. On the contrary, high-UAI cultures promote curiosity, exploration and experimentation [49]. It is hypothesized that the Security and Stimulation values are the ones that are most closely related to he UAI-HCD. People that value Security seek to promote stability, order and safety [11]. People that value Stimulation find it important to live a life that is exciting and loaded with novel experiences. High-UAI societies are therefore presumed to promote Security and demote Stimulation. Thus, UAI forms the DCC^+ of Security and the DCC^- of Stimulation.
- **Long-Term Orientation (LTO)**: This dimension is about the extent to which a society looks forward to the future rather than resorting to the past to solve problems [49]. High-LTO cultures promote planning, foresight and perseverance [49]. Low-LTO cultures, on the other hand, prescribe time-honoured traditions and are suspicious of societal change. Stability is important to low-LTO societies, leading them to stick to conventions that uphold the status quo. With regards to the LTO-HCD, it seems that the value Conformity & Tradition is most relevant. People that value Conformity & Tradition find it important to obey to social norms, to respect a society's traditions and orient oneself to the past to obtain the information to deal with problems occurring in the present and future [11]. Based on these descriptions, it is hypothesized that low-LTO cultures promote the valuation of Conformity & Tradition, whereas high-LTO societies will tend to suppress the importance ascribed to this value. The LTO-HCD therefore forms the DCC^- of Conformity & Tradition.
- **Individualism (IDV)**: This dimension is essentially about affiliation [49]. Low-IDV societies view humans as fixed members of a single well-defined group in which all members are interdependent [48]. Low-IDV cultures promote loyalty, harmony and a general promotion of the collective over the individual. In contrast, high-IDV societies promote self-sufficiency, self-actualization and self-expression [49]. We propose that the Self-Direction and Conformity & Tradition values relate most closely to the IDV-HCD. Self-Direction is concerned with the valuation of individual freedom, independent thought and autonomy. It is hypothesized that people in high-IDV societies will tend to ascribe high importance to Self-Direction. On the contrary, high-IDV cultures will demote Conformity & Tradition, since this

value prescribes the restraint of individual freedom for the sake of the collective. Thus, IDV forms the DCC^+ of Self-Direction and the DCC^- of Conformity & Tradition.

- **Indulgence (IVR)**: High-IVR societies allow a relatively free gratification of basic and natural human drives related to enjoying life and having fun [48]. Low-IVR cultures suppress instant gratification and promote tight regulation of individual behavior by means of strict social norms. We propose that the characteristics of high-IVR societies harmonize with the goals promoted by the Hedonism value. Hedonism prescribes pleasure and sensuous gratification for oneself [11]. Low-IVR societies, on the other hand, promote the value of Conformity & Tradition as it prescribes self-restraint and discipline. The IVR-HCD therefore constitutes the DCC^+ of Hedonism and the DCC^- of Conformity & Tradition.

8.2.4 Decision-Making Behavior

Values have a reliable, albeit weak, effect on behavior [51]. The effect of values on behavior is generally indirect and/or contingent upon contextual factors [51, 52]. In the current study, the effect of values on behavior is influenced by amongst others the satisfaction level of needs, the actions of peers within one's personal social network, and contextual factors such as where an individual is situated and the time of day it is during a instance of decision-making. In doing so, values are conceived of as "*abstract fixed points that actions over many contexts can be traced back to*" [51].

At any given moment in time, the relative importance an individual ascribes to its values determines the priority assigned towards satisfying particular needs. Thus, needs are assigned weights according to the structure of one's value system. These weights remain static over the course of the simulation.[4] What varies is the satisfaction level of needs. The dynamism of satisfaction levels is modelled using the water tank approach as presented by [54] and modelled by [21, 55]. In short, needs are represented as tanks filled with water (i.e. satisfaction). As time passes these tanks gradually leak 'satisfaction'. Individuals are driven to keep their tanks filled up to some extent. High-priority needs are represented by tanks that are especially important to keep filled up. Hence, individuals are more sensitive to fluctuations in the satisfaction level of high-priority needs than those of low-priority needs.

Satisfying needs (i.e. the process of filling up one's water tanks) is done through engaging in particular activities; that is, the actions that an individual performs alter the satisfaction levels of its needs. During the act of decision-making, individuals form expectations with regards to how much satisfaction will be gained by performing certain actions. This expectation-based activity selection builds on the process-based motivation theory of Vroom [56]. People are expected to choose among alternative courses of action in a manner that maximizes the potential satisfaction to be gained

[4]Although an individual's values may change over the course of a lifetime[53], this process lies beyond the scope of this study.

from executing the action. Importantly, people may choose to act in a certain way in the short-term as to increase the probability of realizing a particular outcome in the longer term. This tactical ('short-to-medium term') or strategic ('short-to-long term') behavior is excluded from the current model since modelling such behavior adds enormous complexity, much of which is deemed unnecessary given the scope of the current study.

To summarise, individuals choose to engage in activities that generate the highest expected level of satisfaction for the most *urgent* needs. The urgency of a need is defined as a function of that need's satisfaction level (dynamic component) and its value-based priority-weight (static component).

8.3 Data Sources

Data is obtained from academic literature and publicly available databases such as the European Social Survey (ESS) [57] and World Value Survey (WVS) [58] for data on people's values, the Our World in Data (OWID) [59] for data on demographics and COVID-19 statistics, and the Hofstede Insights database for data on nations' cultural dimensions [48]. Data on the cultural tightness of nations is obtained from [39, 60].

8.4 Model Description

8.4.1 Formal Description of Model Procedures

Ensuring Logical Consistency of Agent Value Systems: Modelling the value circumplex of the BVT [13] is done on the basis of a procedure presented by [21]. Specifically, a set of $Values = \{V_1 \dots V_9\}$ is defined, where V_1 = Hedonism (HED), V_2 = Stimulation (STM), V_3 = Self-Direction (SD), V_4 = Universalism (UNI), V_5 = Benevolence (BEN), V_6 = Conformity & Tradition (CT),[5] V_7 = Security (SEC), V_8 = Power (POW), and V_9 = Achievement (ACH). The second set relevant to this procedure contains the importance levels (Val_i) tied to each value (V_i). Suppose we define a function (f) that takes as a value index as input and outputs the importance level of that value such that $Val_i = f(V_i)$. If $Val_i = 100$, then the agent in question will ascribe maximum priority to satisfying needs related to the fulfillment of V_i during the act of decision-making. If $Val_i = 0$, then the fulfillment of V_i is assigned the lowest possible priority by an agent during decision-making.

[5]Originally, conformity and tradition are considered to distinct values; the latter forming a more extreme version of conformity. For the sake of simplicity, these values are currently thought of as constituting a mild—i.e. conformity—and a more extreme—i.e. tradition—case of the same type of value.

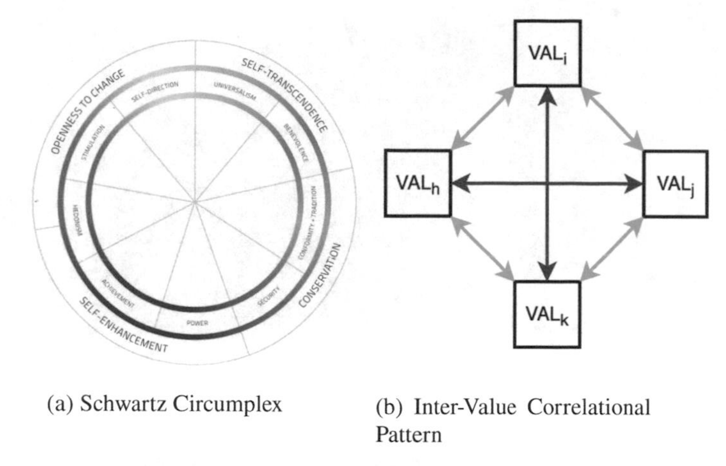

(a) Schwartz Circumplex

(b) Inter-Value Correlational Pattern

Fig. 8.3 Logical consistency of agent value systems

The following condition ensures that any two instances of $Values$ whose indices are sufficiently close to one another hold similar importance levels: $\forall i, j \in \{1 \ldots 9\}$ and $0 \leq |Val_i - Val_j| \leq UB_{i,j}$. Where $UB_{i,j}$ serves as an upper boundary to the dissimilarity between the importance levels of V_i and V_j (see Eq. 8.1). This way, values placed close to one another within the circumplex will be constrained in their dissimilarity (i.e. their importance levels will be similar), whereas values placed further apart will be less constrained in their dissimilarity. The parameter c in Eq. 8.1 is an integer ranging from [1, 100]. The c parameter is set to 20 by default.

$$UB_{i,j} = \begin{cases} |i - j| * c & \text{if } |i - j| \leq 5 \\ (9 - |i - j|) * c & \text{if } |i - j| > 5 \end{cases} \tag{8.1}$$

Figure 8.3 depicts the Schwartz circumplex model (a) and an abstract representation of the current procedure (b). The right-hand sub-figure visualizes how values placed close to one another on the circular continuum are positive correlated (as indicated by the green bidirectional arrows). Values that are placed further apart—e.g. on opposites sides of the circular structure—are negatively correlated as visualized using red bidirectional arrows. A consequence, the structure of agent value systems exhibit a form of homeostasis.

Age and Structure of Agent Value Systems: Across cultures, older people tend to exhibit a lower endorsement of agentic personal values and See (Fig. 8.4) higher endorsement of communal personal values than did younger people [61]. To account for the effect of age on the structure of agent value systems, a procedure is implemented that performs a linear transformation of an agent's values according to a set of coefficients taken from [62]. By default, the magnitude of this transformation decreases as agents get older, this implies that the value systems of younger agents is affected most by the current way the procedure is designed (see description

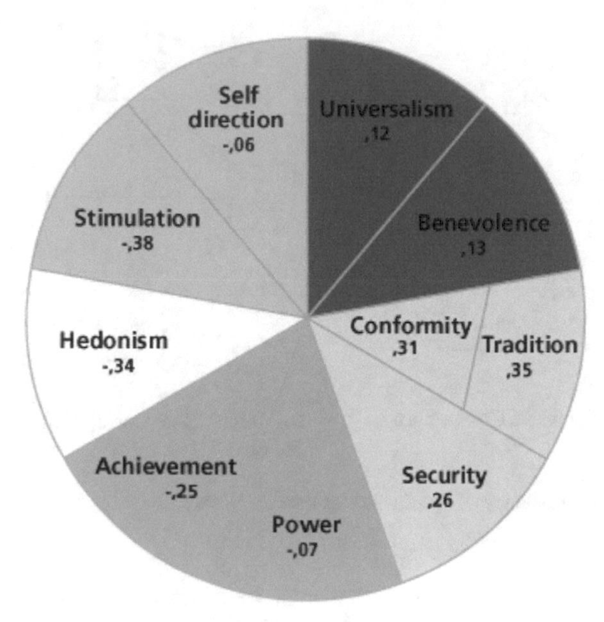

Fig. 8.4 Influence of age on values (taken from [57])

of Algorithm 1). The global parameter INFLUENCE- OF- AGE- ON- VALUE- SYSTEM modulates the strength of the age-dependent transformation; it is set to 5 by default. The age-group dependent weights are denoted as ω_{Age}, and the value-specific coefficients are symbolized as C_i. The re-calibrated value importance levels (Val_i') are computed by applying Eq. 8.2 $\forall Age \in AgeGroups$ and $\forall i \in \{1 \ldots 9\}$, where $AgeGroups = \{Young, Student, Worker, Retired\}$.

$$Val_i' = Val_i - (C_i \cdot \omega_{Age}) \tag{8.2}$$

Mapping Cultural Profiles to Agent Value Systems: The relationship between the HCD's and the values of individual agents is modelled by a procedure that computes the population means of the BVT values ($\mu[Val_i]$) as linear combinations of HCD scores (see description of Algorithm 2). The values of agents are drawn from a normal distribution with $\mu[Val_i]$ as the mean, and the global parameter VALUE- STD- DEV as the standard-deviation. The global parameter HOFSTEDE- SCHWARTZ- MAPPING- MODE specifies the way in which HCD's are mapped onto a particular $\mu[Val_i]$. The default setting is "*empirical & theoretical*", which means that only those mappings that are empirically supported (see Figs. A.1 and A.2) are included within the procedure (see Table 8.4 for the hypothesized relationships between HCDs and Schwartz BVT Values). Those that are not backed by empirical findings are instead calibrated by drawing a random decimal, denoted as **X**, from a uniform probability distribution with a range of [40, 60].

Cultural tightness is implemented by mapping a nation's cultural tightness score (CLT), which ranges from [0, 100], to the standard-deviation ($\sigma[Val_i]$) that is used to draw agent values from their respective normal distributions. Note that the following

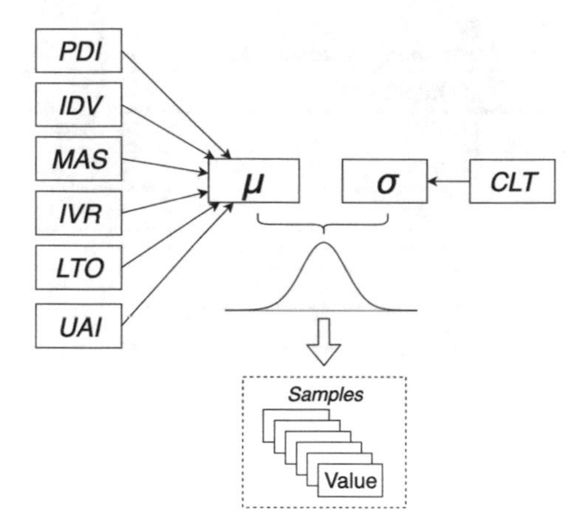

Fig. 8.5 Linking cultural variables with Schwartz BVT values

condition holds: $5 \leq \sigma[Val_i] \leq 15$. This implies that the higher a nation's cultural tightness score, the smaller the standard-deviation that is used during the sampling of agent values (see Eq. 8.3). The terms in the denominator of Eq. 8.3 are $\max(CLT)$ and $\max(\sigma[Val_i])$ which represent the upper limits of CLT (= 100) and $\sigma[Val_i]$ (= 15), respectively.

$$\sigma[Val_i] = \max(\sigma[Val_i]) - \frac{CLT}{\frac{\max(CLT)}{\max(\sigma[Val_i])}} \tag{8.3}$$

Figure 8.5 provides a visual representation of how the cultural parameters (a nation's HCDs and CLT index) combine to determine the value systems of agents. The sample distributions of agent values within nations with cultural profiles that are characterized by low cultural tightness and extreme HCD scores may become skewed due to the fact that the importance levels of agent values are capped to fall within a range of [0,100]. To prevent samples from piling up towards extreme ends of the value importance-level spectrum, a function is deployed that pulls extreme values for $\mu[Val_i]$ towards the median of the interval, which is 50. Extreme values are defined as: $\mu[Val_i] \leq 10$ or $\mu[Val_i] \geq 90$. Extreme values of $\mu[Val_i]$ are modified according to Eq. 8.4; these modified entities are denoted as $\mu[Val_i]'$.

$$\mu[Val_i]' = \frac{100}{1 + e^X} \tag{8.4}$$

The X parameter in Eq. 8.5, and particularly the β parameter contained within it, determines the degree to which the logistic function depicted in Eq. 8.6 distorts extreme values of $\mu[Val_i]$. Lower values of β lead to more rigorous transformations of extreme values of $\mu[Val_i]$ into values that lie closer to the median. As can be seen in Eq. 8.6, β is dependent upon the magnitude of $\sigma[Val_i]$, and a global parameter coined CULTURAL- TIGHTNESS- FUNCTION- MODIFIER, which is set to 0.01 by default and is denoted as λ within Eq. 8.6. The higher the values of $\sigma[Val_i]$ and/or λ, the lower the β.

Table 8.5 Cultural tightness function input-output table

Input	Output	Δ
0	6.01	6.01
5	7.76	2.76
10	9.98	−0.02
90	90.02	0.02
95	92.24	−2.76
100	93.99	−6.01

$$X = (50 \cdot \beta) - (\mu[Val_i] \cdot \beta) \quad (8.5)$$

$$\beta = 0.055 - (\sigma[Val_i] \cdot \lambda) \quad (8.6)$$

By designing Eqs. 8.5 and 8.6 in this fashion the following properties are ensured: nations with $CLT = 100$ will have a $\sigma[Val_i] = 15$, nations with $CLT = 0$ will have a $\sigma[Val_i] = 5$. Any individual nation's $\mu[Val_i]$ that falls below 10 or that exceeds 90 will be transformed so that it takes on a more moderate value which is denoted as $\mu[V_i]'$. Table 8.5 provides insight into the workings of Eq. 8.6 by presenting a function input-output table.

Figure 8.6 depicts the population-level distributions of agent values that result from the way in which the procedure described in this section is designed. Note how nations with a higher CLT index exhibit more homogeneity with respect to agents' value systems; i.e. there exists a lower spread in the distribution of the importance levels agents ascribe to their values. In doing so, agents in tight cultures will, in general, think and act more similar than their counterparts in loose cultures.

Construction of Value-Based Social Network Topology: The settings of the social network determine how agents, after being spawned, link up with one another to form a particular type of social network. There are two global parameters that change the topology of the network: (i) the number of peer links each agent makes PEER- GROUP- LINKS, which is set to 7, and (ii) the proportion of agents that makes a random link with another agent in the population RANDOM- LINKS, which is set to 0.15. The default settings for these global parameters are decided upon by means of a visual inspection of the type of social network topologies they generate. The desired network topology is based on the principle of *preferential attachment* and should exhibit a realistic combination of weak ties, social clusters, and wide bridges [63]. Preferential attachment refers to the condition that any given node in the network is incentivized to link up with other node that exhibit similar characteristics. In doing so, the network is built on the principle of *homophily*, which describes how people tend to befriend like-minded others [64]. Upon initiation of the model, each agent is asked to look for other agents that hold similar value systems and to subsequently link up with them. The similarity of value systems—referred to as the *social distance* that exists between two agents—is determined by computing the Euclidian Distance

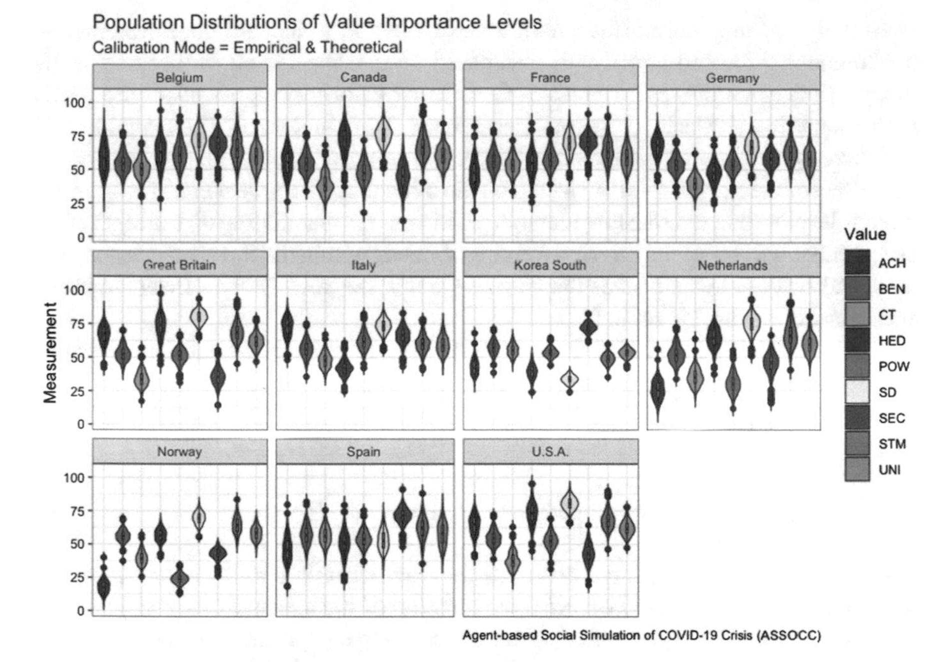

Fig. 8.6 Country-level distributions of value importance levels

between a targeted pair of value systems; i.e. value system X and value system Y (see Eq. 8.7).

$$SocialDistance_{X,Y} = \sqrt{\sum_{i=1}^{9}(X_i - Y_i)^2} \tag{8.7}$$

The parameter PEER- GROUP- LINKS specifies the number of relationships an agent makes on the basis of Eq. 8.7. The RANDOM- LINKS determines the proportion of agents that randomly link up with another agent. These random links ensure that the network contains 'bridges' between clusters of like-minded agents, similar to what is observed in the real world [63].

Mapping Agent Values to Needs: As described in Sect. 8.2.2, an agent's values inform the priority-weights ascribed to its 12 needs. Agent needs are indexed as η, where $\eta \in \{1 \ldots 12\}$. The priority-weight of a need is denoted as ω_η. After having computed the priority-weight for all needs, they are normalized to ensure that their sum equals 1 (see Eq. 8.8). Normalized priority-weights are denoted by ω'_η, where $\sum_{\eta=1}^{12} \omega'_\eta = 1$.

$$\omega'_\eta = \frac{\omega_\eta}{\sum_{\eta=1}^{12} \omega_\eta} \tag{8.8}$$

Optionally, one may activate the MASLOW- MULTIPLIER, which introduces a hierarchical component to need priority-weights based on the Maslow HON theory. Specifically, when MASLOW- MULTIPLIER > 0, then the priority-weights of all the needs within a particular Maslow category (see Table 8.3) are increased. The magnitude of this increase depends on the Maslow category to which a group of needs is allocated. The effect of the MASLOW- MULTIPLIER on the other categories weakens in proportion to their level within the Maslow hierarchy. This means that dialing up the MASLOW-MULTIPLIER will boost the priority-weights of needs within the Survival category the most and Esteem-related needs the least. Note that the MASLOW- MULTIPLIER is not active by default as it is set to 0.

Decision-Making and Satisfaction of Needs: During an instance of decision-making, agents are presented a set of activities ($Activities$) from which they are prompted to select the activity (A) that grants them the highest total satisfaction gain (α), where there are N activities within ($Activities$) (see Eq. 8.9).

$$\alpha = \max\{TSG_A : A = 1 \ldots N\} \tag{8.9}$$

The types of activities within $Activities$ that are accessible to the agent is dependent upon the time of day ($Period_i$), the location of the agent ($Location_i$) and the agent's properties ($Agent_i$) at the instant it is making a decision on what to do next, such that $Activities \Leftarrow \{Location_i, Period_i, Agent_i\}$. Figure 8.7 summarises the needs-based activity selection procedure in a conceptual scheme. An action's total satisfaction gain (TSG_A) is computed as the sum of the satisfaction gains of that action for all of the agent's needs (see Eq. 8.10).

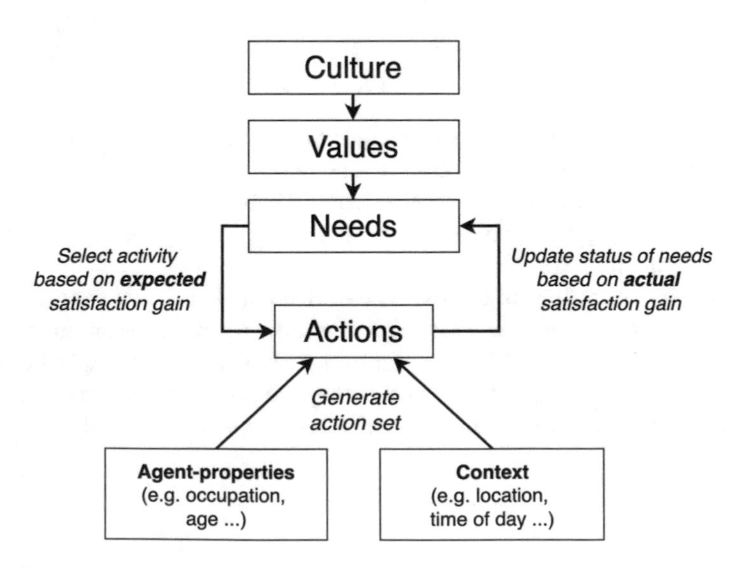

Fig. 8.7 Needs-based activity selection

Table 8.6 Satisfaction gains quadrant

Satisfaction gains	Expected	Actual
Undiscounted	ESG = raw satisfaction gained from performing an activity	ASG = the raw actual satisfaction gained from performing an activity (generally the same as ESG)
Discounted	ESG' = the satisfaction gained from an activity weighted by the urgency of the need(s) addressed by that particular activity	ASG' = the actual satisfaction gained from performing an activity corrected for mistakes in agent's expectations

$$TSG_A = \sum_{\eta=1}^{12} ESG'_\eta \tag{8.10}$$

A distinction is made between four different types of satisfaction gains, namely: *expected* versus *actual* and *undiscounted* versus *discounted* satisfaction gains. Expected Satisfaction Gains (ESG) are those the agent expects to incur *before* performing an activity, whereas the Actual Satisfaction Gains (ASG) represent the satisfaction gained from an action *after* it has occurred. Undiscounted Satisfaction Gains (SG) can be conceived of as 'raw' estimates of the satisfaction potential of an activity. An SG is used to compute a Discounted Satisfaction Gain (SG') by incorporating information about the status of an agent's needs in terms of their respective priority ranking and satisfaction depletion status (i.e. the degree to which a need's satisfaction level is depleted). Table 8.6 presents a quadrant that shows how these various satisfaction gains relate to one another.

Assume a function (τ) is defined that takes as input an activity (A) and returns the Undiscounted Expected Satisfaction Gain (ESG) for each of the agent's needs (η), $\tau(A) : Activity \rightarrow ESG_\eta$. The ESG_η is then multiplied by the urgency ($Urgency_\eta$) of that need to obtain ESG'_η (see Eq. 8.12). As can be seen in Eq. 8.11, the urgency of a need is a function of that need's satisfaction level (dynamic) and priority-weight (fixed). This implies that needs with a high urgency will have a relatively large influence on the action that is selected by an agent during decision-making. Thus, those activities that present a satisfaction gain for needs that are ascribed a high priority-weight and whose satisfaction levels are depleted, are most likely to be picked from the activity set during decision-making.

$$Urgency^t_\eta = \omega_\eta \cdot (1 - SatLevel^t_\eta) \tag{8.11}$$

$$ESG'_\eta = \tau(A) \cdot (\omega_\eta \cdot Urgency^t_\eta) \tag{8.12}$$

As time passes, the satisfaction level of needs diminishes until one of them reaches a status of depletion that becomes so critical that it comes to dominate the decision-

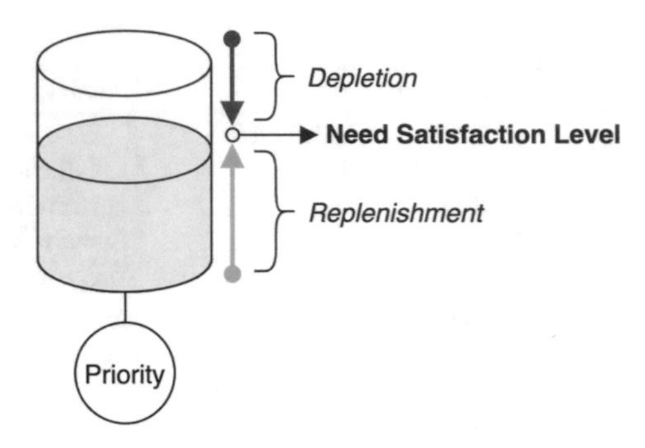

Fig. 8.8 Water tank model of need satisfaction

making process. The magnitude by which $SatLevel_\eta$ decreases is modulated by the decay rate (which is static but differs across needs), which is denoted as r and is subject to the condition: $r < 1$. Note that the multiplicative effect of the decay rate in Eq. 8.13 leads full water tanks to leak relatively higher quantities of satisfaction than emptier tanks.

$$SatLevel_\eta^t = (r \cdot SatLevel_\eta^{t-1}) + ASG_\eta' \tag{8.13}$$

Although the raw or undiscounted satisfaction gains tied to activities are static and the same for all agents, the subjective urgency of catering to a need ensures that agents hold differing views on the attractiveness of engaging in a particular activity at given point in time. This implicitly represents the concept of diminishing marginal utility which states that each incremental effort aimed at satisfying a need yields a lower marginal satisfaction gain or 'utility' in economics jargon [65]. Thus, as the satisfaction level of a need increases, it becomes increasingly unappealing to seek to obtain an even higher level of satisfaction. Moreover, despite the fact that the need-specific decay rates are the same for all agents, agents do exhibit a differential sensitivity to the depletion of any particular need. Specifically, the salience of the depletion of a particular need increases in proportion to the priority ascribed to that need by a given agent (for more information see Sect. 2.3.5). Figure 8.8 presents a visualization of the mechanisms at play in the water tank model of need satisfaction. Agents within our model do not only base their decision-making on their internal states, they also look outward to see what their peers are doing. In this way, the ASSOCC model offers a simple representation of 'social decision-making' [66]. Agents register the actions of peers at t and may obtain higher satisfaction gains from pursuing similar actions at $t + 1$. The 'boost' in the satisfaction gains from performing similar actions to peers is moderated by the priority ascribed to the Conformity need. That is, agents that prioritize Conformity tend to base their decision-making process more on what others are doing than agents that ascribe a lower priority to this need.

Figure 8.9 summarises the 'value-based needs' and 'needs-based activity' models into a single visual conceptualization.

Modelling Social Distancing Behaviour: Each agents has a *social-distancing-profile* (SDP) which dictates the degree to which it feels the urge to be in physical

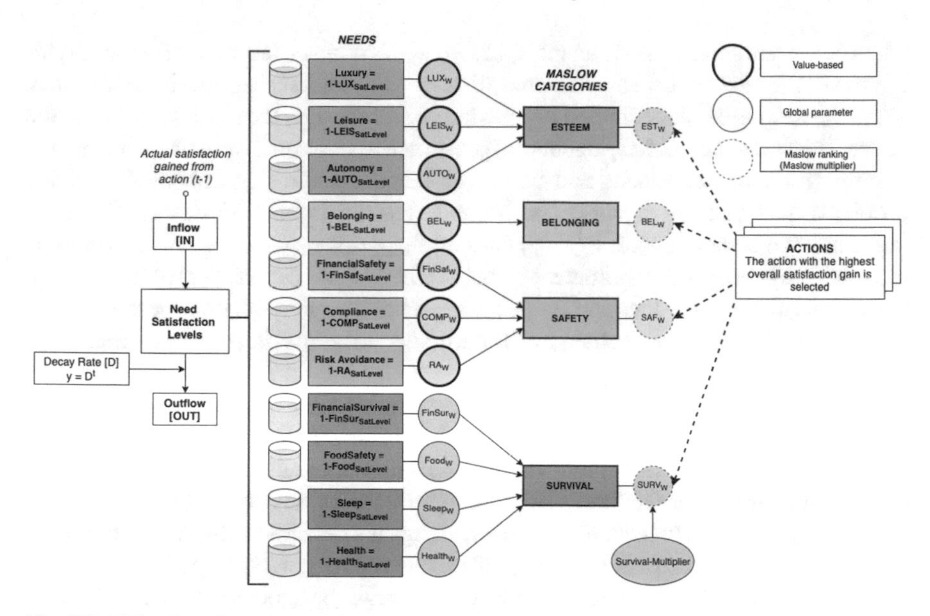

Fig. 8.9 Value-based needs model

contact with other agents. It is presumed that agents that are (highly) extrovert will have more difficulty maintaining physical distance from their friends and family [67]. Hence, it is proposed that extroverts will find it more difficult to comply to norms prescribing the restraint of direct or proximate physical contact. It is shown that one's personality type and the structure of one's value system are closely related to one another [68]. On this basis, we propose that agents that ascribe a relatively high importance to Achievement, Stimulation and Hedonism will be extrovert, whereas agents scoring high on Conformity & Tradition and Security are considered to be introvert [68]. The description of Algorithm 3 shows how agent SDP's are calculated. Agent SDP's are drawn from a normal distribution with a mean that is higher for extroverts than for introverts. The standard deviation used is set by the global parameter STD- DEV- SOCIAL- DISTANCING- PROFILE, which is denoted as $\sigma[SDP]$ and is set to 0.1 by default. Agent SDP's are capped to fall within a range of [0, 1].

If the social distancing measure is activated, then the *Activities* set presented to agents during each time-step contains two variants of each activity: a 'non-social-distancing' and a 'social-distancing' variant. When an agents chooses to perform the social-distancing variant of an action, it will engage in the chosen activity whilst maintaining a proper physical distance to others engaging in the same activity at the same location. The decision of agent's to engage in social distancing or not affects the transmission potential of the virus. Formally, the probability that a given agent X is infected by agent Y is expressed as:

$$P(Y\,infects\,X) = Y_{cont} \cdot X_{suscept} \cdot X_{sd-status} \cdot Location_{density-factor} \tag{8.14}$$

where the contagiousness of agent Y ($Y_{contagiouness}$) since the moment that Y was infected is modelled using a Gamma distribution. The susceptibility of agent X ($X_{susceptibility}$) is dependent upon its age; the higher the age, the more susceptible the agent. This is because elderly agents are presumed to have relatively weaker immune systems (immunosenescence) and are more likely to be suffering from health-related problems than younger agents. The decrease in P(Y infects X) due to X engaging in social distancing—i.e. $P(Y\ infects\ X | X_{social-distancing-status} = Active)$—is modulated by the global parameter SOCIAL- DISTANCING- DENSITY- FACTOR. Lastly, the density-factor of a location, which represents, amongst others, the number of people per m^2 of space available at a particular location, is positively related to P(Y infects X).

Agents with a relatively low SDP experience higher satisfaction gains from engaging in their activities whilst complying to social distancing measures than agents with a high SDP. This implies that high-SDP agents will be more likely to choose the 'non-social-distancing' over the 'social-distancing' variant of an activity when the social distancing measure is activated. Moreover, an agent's decision between the two variants is also dependent upon the status of its needs during t. Specifically, the status of the needs for Belonging and Autonomy may incite the agent to elicit a preference for the non-social-distancing variants. Whereas the urgency status of the needs for Risk-Avoidance, Health and Compliance do so for the social-distancing variants. In doing so, the SDP of an agent represents a static proclivity to (dis)obey to the measure of social distancing, whereas the status of the aforementioned needs induce a dynamic component to the decision to comply to social distancing or not.

8.4.1.1 Implementation Details: Software, Initialisation and Input Data

The ASSOCC model is programmed in Netlogo 6.1.1 and the data analysis is performed in RStudio. Where possible the calibration of model parameters is done using empirical data. In case empirical data was not available, calibration of model parameters is done on the basis of a series of model tests where theoretically valid outcomes were obtained mainly through a process of trial & error. With regards to the cultural sub-model, the HCDs are calibrated according to their country-specific scores.[6] Moreover, the theoretical linkages made between the HCDs and the Schwartz BVT Values are tested for their empirical validity by analyzing data from the World Value Survey (WVS)[7] and European Social Survey (ESS).[8] Specifically, country-level data on the HCDs was regressed on country-level survey data for the Schwartz BVT Values using multiple Ordinary Least Squares (OLS) regressions (see Eq. 8.15).

$$\mu[Val_i] = \beta_0 + \beta_1 PDI + \beta_2 IDV + \beta_3 MAS + \beta_4 UAI + \beta_5 LTO + \beta_6 IVR + \varepsilon \tag{8.15}$$

[6]These scores are obtained from: https://www.hofstede-insights.com/product/compare-countries/.
[7]http://www.worldvaluessurvey.org/wvs.jsp.
[8]https://www.europeansocialsurvey.org/.

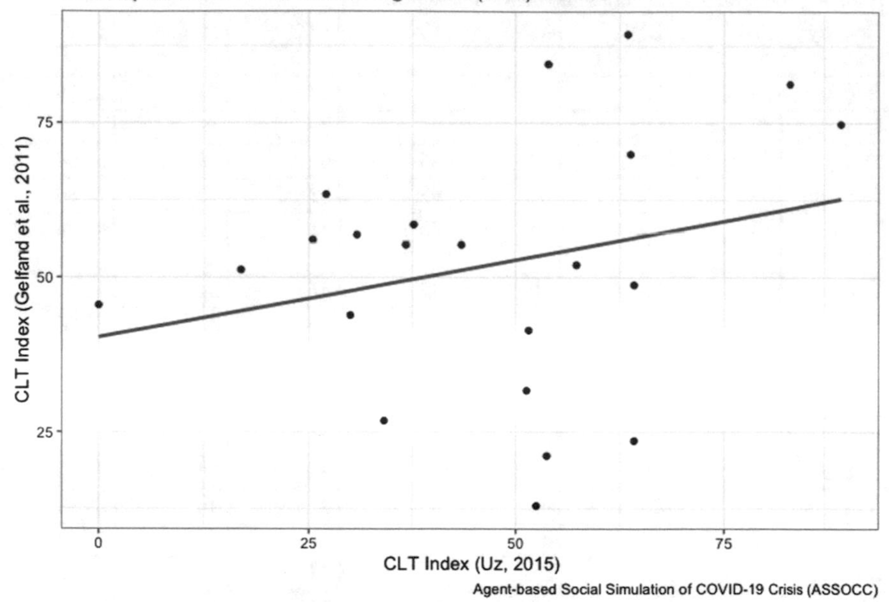

Fig. 8.10 Scatter-plot of cultural looseness-tightness indices

The β-coefficients are analyzed to see which ones correspond to the hypothesized relationships. If a β-coefficient is statistically significant ($p < 0.05$) and its direction (i.e. whether it is positive or negative) corresponds to a hypothesized relationship between a HCD and Schwartz BVT Value (Table 8.4), then it is incorporated into the mapping procedure (see Algorithm 4 and Figs. A.1 and A.2). Thus, by default, only those hypothesized relationships that are supported by the results of the statistical analysis are used to inform the mapping of HCDs onto Schwartz BVT values.

With regards to cultural tightness, empirical data was obtained from [39] and [60]. Nation's cultural tightness scores from both data sets are scaled to fall within a range of [0, 100], where 100 represents maximal tightness and 0 stands for maximal looseness. As can be seen in Fig. 8.10, the CLT indices relate positively to one another which justifies the act of combining them into a single score for each nation present within both data sets. A total of 13 countries remained after merging data sets and subsequently filtering out countries with missing data. Table 8.7 provides an overview of the scores for each of the 13 countries on the HCDs and CLT.

Table 8.7 Overview of countries and their cultural profiles

Country name	PDI	IDV	MAS	UAI	LTO	IVR	CLT
Belgium	65	75	54	94	82	57	22.76
Canada	39	80	52	48	36	68	29.38
France	68	71	43	86	63	48	34.04
Germany	35	67	66	65	83	40	43.86
Great Britain	35	89	66	35	51	69	40.78
Italy	50	76	70	75	61	30	49.35
South Korea	60	18	39	85	100	29	82.26
The Netherlands	38	80	14	53	67	68	30.48
Norway	31	69	8	50	35	55	77.24
Singapore	74	20	48	8	72	46	69.24
Spain	57	51	42	86	48	44	36.93
Sweden	31	71	5	29	53	78	26.63
The United States	40	91	62	46	26	68	46.52
Mean	47.92	66	43.77	58.46	59.77	53.85	45.34
Std. Dev.	15.15	23.15	22.16	26.03	21.30	15.80	19.42

8.5 Model Evaluation

Where the process of model verification concerns formulating an answer to the question of whether one has "built the *thing right*", model validation involves checking whether one "built the *right thing*" [69]. The ASSOCC model is the result of extensive collaboration between a group of researchers, making sure that *unit tests*[9] were performed each time the model was expanded upon. Hence, this paper will focus primarily on assessing the validity of the ASSOCC model. Specifically, the model's quality is evaluated by (A) investigating the robustness of the model's behavior to perturbations in the input parameter settings, (B) reflecting on the empirical underpinnings of the theoretical assumptions embedded in the model, and (C) assessing how well the model output resembles real-world data. With regards to point A, it is important to understand how the various assumptions embedded within the model

[9]Performing unit tests involves taking individual units or components of the model and providing them with well-defined inputs as to assess whether the consequent outputs are in agreement with the a-priori expectations of the modeller(s). In other words, each model component is tested to verify whether what it does is the same as what it should do.

each contribute to the variability in simulation outcomes [70]. Gaining a solid understanding of this sorts helps to reduce the risk of drawing conclusions on the basis of *modelling artefacts*.[10] Regarding point B, we aim to base our modelling assumptions on empirical data where possible. Section 8.4.1.1 describes what assumptions are calibrated on the basis of empirical findings and what procedure is used to realize this. With respect to point C, Sect. 8.7.3 reflects on the (dis)similarities between model outcomes and real-world data so as to assess the external validity of the ASSOCC model.

8.5.1 Global Sensitivity Analysis: Time-Dependent Sobol Variance Decomposition

The evaluation of ASSOCC[11] comprises of an exploration and quantification of the sensitivity of model outputs with respect to marginal changes in the settings of model input parameters (henceforth referred to as *factors*). In doing so, we assess the robustness of emergent properties to changes in *core* and/or *ancillary* assumptions embedded within the model [71]. The approach implemented to reach this objective is termed *Global Sensitivity Analysis* (GSA). In contrast to a *One-Factor-at-a-Time* (OFAT) approach where one particular factor is varied at a time whilst keeping all other factors fixed, the GSA method involves varying all selected factors simultaneously and analyzing the consequent model behaviour. This is done by drawing a large number of random samples of factor settings from the *factor space*. The factor space is an abstract representation of the collection of settings that each distinct factor is allowed to take on as specified by their sampling ranges.

Factor settings are sampled from the factor space on the basis of a uniform probability distribution as to avoid imposing unnecessary restrictions on the sampling process. The samples are drawn using a statistical technique called Latin Hypercube Sampling (LHS). The reason for this is that LHS guarantees a uniform sampling of factor settings given a Y dimensional factor space constrained to a limit of X samples [72]. The selection of factors included in the GSA is limited to the ancillary parameters that fall within the cultural sub-model, which are henceforth denoted as X_i'. The factors and their sampling ranges are depicted in Table 8.8.

Factor samples are fed into the ASSOCC model to execute simulations with. The variance in simulation outcomes is then used to analyze the model's sensitivity to changes in factor settings. Specifically, the model's sensitivity to a factor is then measured as the proportion of variance of model output variance that can be attributed to changes (i.e. variance) in that factor [71, 73]. The advantage of GSA over OFAT is that the former is able to cover a much wider range of the factor space and takes into

[10] A modelling artefact occurs when there is a mismatch between the set of assumptions that causes the occurrence of a certain phenomenon and the set of assumptions that the modeller believes is responsible for producing that phenomenon [70].

[11] Of the cultural sub-model in particular.

account the complex interactions between factors in determining emergent properties. Broadly, one may choose to perform a parametric (e.g. regression-based) or a non-parametric (i.e. 'model-free' or 'variance-based') GSA. We currently adhere to the non-parametric approach as it circumvents the assumptions of linearity inherent in regression-based approaches that are ill-suited for dynamic complex system modeling [73, 74]. Variance-based GSA partitions the variance (V) of model output (Y) that is attributed to changes in k factors represented individually as V_i and in combinations with an increasing level of dimensionality [74] (see Eq. 8.16). For instance, V_{ij} represents the sensitivity of model outputs to the interaction between factors X_i and X_j.

$$V = \sum_{i} V_i + \sum_{i<j} V_{ij} + \ldots + \sum_{i<\ldots<k} Val_{i<\ldots<k} \tag{8.16}$$

These variance statistics are used to compute first-order (Si) indices for every factor (see Eq. 8.17). An Si quantifies the partial contribution of a factor to the variance in Y independently from the other $k - 1$ factors [74]. The higher the Si of a factor relative to others, the more influential it is in determining a model's behaviour. The numerator in Eq. 8.17 can be interpreted as follows: it is the variance (V) in the expected value of Y conditional upon X_I which is denoted as $E(Y|X_i)$. An instance of $E(Y|X_i = x_i)$, where x_i represents a particular setting of X_i, is computed by varying all factors but X_i ($X_{\sim i}$) and computing the average value for Y across the drawn samples of $X_{\sim i}$. This procedure is repeated many times for many unique settings of X_i so that at some point we obtain a credible estimate of $V_{X_i}[E_{X_{\sim i}}(Y|X_i)]$.

$$S_i = \frac{V_i}{V} = \frac{V_{X_i}[E_{X_{\sim i}}(Y|X_i)]}{V(Y)} \tag{8.17}$$

Table 8.8 GSA factor settings

ID	Factor	Default value	Min	Max
X'_1	Contagion-factor	10	5	15
X'_2	Social-distancing-density-factor	0.45	0.2	0.7
X'_3	Std-dev-social-distancing-profile	0.1	0.05	0.15
X'_4	Min-random-value-generator	20	20	40
X'_5	Max-random-value-generator	60	60	80
X'_6	Survival-multiplier	2.5	1.25	3.75
X'_7	Maslow-multiplier	0	0	1

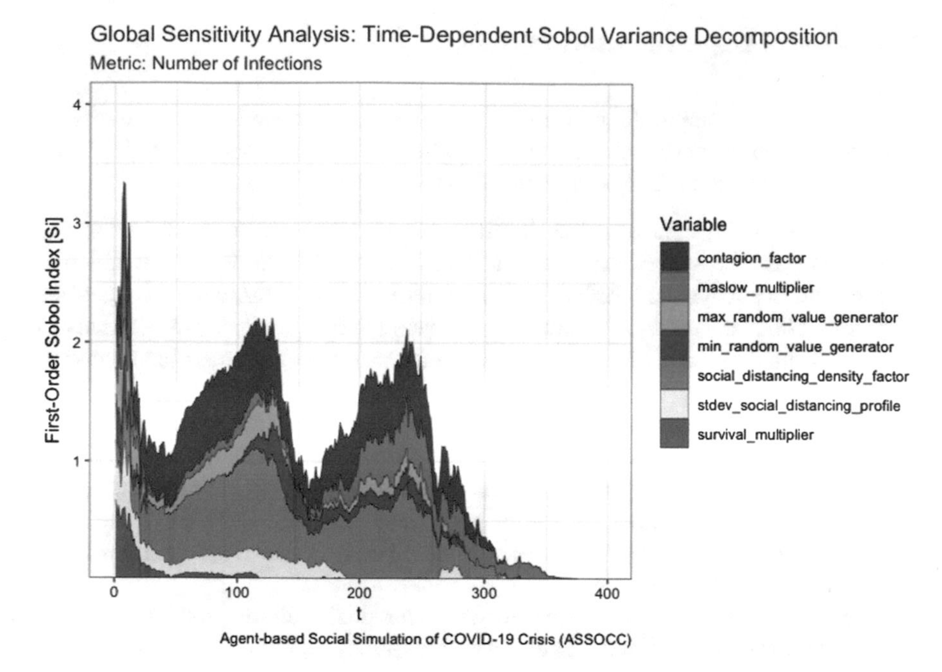

Fig. 8.11 Global sensitivity analysis of infections

We compute a Si statistic for each factor k for each time-step t of an executed model simulation. This approach is coined *Time-Dependent Sobol Variance Decomposition* (t-SVD) [74]. Note that the outcome of the sum of the Si of all factors corresponds to the percentage of output variability attributed to variability in the factors whilst treating them independently from one another. The remainder—that which makes the output variability sum up to 100%—is the proportion of output variance that must be attributed to the interactions among factors [74]. By applying t-SVD we gain a deeper understanding of how the sensitivity of the model to factor variability evolves over the course of a simulation.

As can be seen in Fig. 8.11, the model tends to produce three 'waves' of infections under default model settings (see Table A.1). Figure 8.11 depicts the relative contribution of each X_i' to this emergent outcome over the course of a simulation. Notably, X_1' and X_5' are shown to have the greatest influence on the variability of infections. The influence of the X_2' seems to swell during the third wave of infections. The influence of X_3' and X_4' is negligible, which is desirable since the sole purpose of these parameters is to fill in the gaps left behind by unsupported Hofstede-Schwartz linkages (see Figs. A.1 and/or A.2). The conclusion drawn on the basis of this GSA is that the parameters X_1' and X_5' are the most important non-core parameters within our model. Hence, the empirical calibration of these parameters would have the strongest positive impact on the external validity of our model.

8.6 Model Experimentation

Three model experiments are performed to explore the emergent patterns of disease transmission. All experiments are run for 500 ticks, which amounts to 125 d in our simulated world and each experimental run is replicated 10 times.

1. The first experiment investigates the effect of each HCD and of CLT on the spread of the virus. The first experiment consists of a t-SVD that aims to elicit which of the cultural parameters (HCDs and CLT) contributes the most to the spread of the virus without any policy measures implemented. Subsequently, an OFAT analysis is performed to point out *how* each cultural parameter influences the spread of the virus. Note that where t-SVD helps to gauge the relative *magnitude* of the investigated relationships, OFAT provides an indication of their *direction*.
2. The second experiment focuses on analyzing the effect of the cultural parameters on the spread of the virus under the implementation of different policy measures. For the second experiment, we only performed an OFAT analysis as we were unable to execute the number of model runs needed to obtain sufficiently reliable t-SVD results [75].
3. The third experiment shows how different national cultural profiles affect the spread and mortality of the virus under various policy scenarios. We keep country-

Table 8.9 Overview of experimental input parameter settings

Input parameter	Exp 1	Exp 2	Exp 3
POLICY- SCENARIOS*	PS1	PS2, PS3, PS4, PS5	PS2, PS3, PS4, PS5
COUNTRY- SPECIFIC- SETTINGS	World	World	World
CONTAGION- FACTOR	2	7	7
SOCIAL- DISTANCING- DENSITY- FACTOR	1	0.4	0.4
COUNTRY- HOFSTEDE- SCORES	–	–	Netherlands, South Korea
PDI	2.5, 50, 97.5 | [0, 100]	2.5, 50, 97.5	–
IDV	2.5, 50, 97.5 | [0, 100]	2.5, 50, 97.5	–
MAS	2.5, 50, 97.5 | [0, 100]	2.5, 50, 97.5	–
UAI	2.5, 50, 97.5 | [0, 100]	2.5, 50, 97.5	–
LTO	2.5, 50, 97.5 | [0, 100]	2.5, 50, 97.5	–
IVR	2.5, 50, 97.5 | [0, 100]	2.5, 50, 97.5	–
CLT	2.5, 50, 97.5 | [0, 100]	2.5, 50, 97.5	–
Replications	10	10	10
Ticks	500	500	500
Analytical Approach	OFAT*** | t-SVD**	OFAT***	Full Factorial

*PS = Policy Scenario, PS1 = no policy measures, PS2 = only social distancing, PS3 = soft lockdown, PS4 = hard lockdown, PS5 = tracking, tracing, testing, isolating, **t-SVD = Time-Dependent Sobol Variance Decomposition, ***OFAT = One-Factor-at-a-Time level demographic data fixed to control for their confounding effects on simulation outcomes. Here, the analytical approach adhered to is termed 'full factorial', which involves executing simulations for all possible combinations of prespecified experimental parameter settings [72].

For the first two experiments we executed three OFAT analyses, each with a varying baseline. The baselines are set to [2.5 | 50 | 97.5], which corresponds to slicing the range of the cultural parameters, which is [0,100], into three equal parts. For each OFAT analysis, the setting of 2.5 is labelled as 'Low', and 97.5 is labelled as 'High'. During the execution of the OFAT analysis, each cultural parameter is manipulated sequentially whilst the others remain fixed in the prespecified baseline state. The OFAT-based results are presented in two ways: (i) for each baseline ('Disaggregated') and (ii) as LOESS regressions ('Aggregated') (see Sect. 8.7). For the t-SVD, random samples are drawn for each cultural parameter from the range of [0,100] using LHS (see Sect. 8.5.1 for a detailed description of the t-SVD procedure). The experimental parameter settings are depicted in Table 8.9.

8.7 Experimental Results

8.7.1 Experiment 1

Figure 8.12 depicts the results of the first experiment and shows that varying Uncertainty Avoidance (UAI) and Indulgence (IVR) tend to have the most notable impact on the number of infections; this conclusion is drawn by looking at what cultural parameters account for the largest area under the infection curve. This finding can be

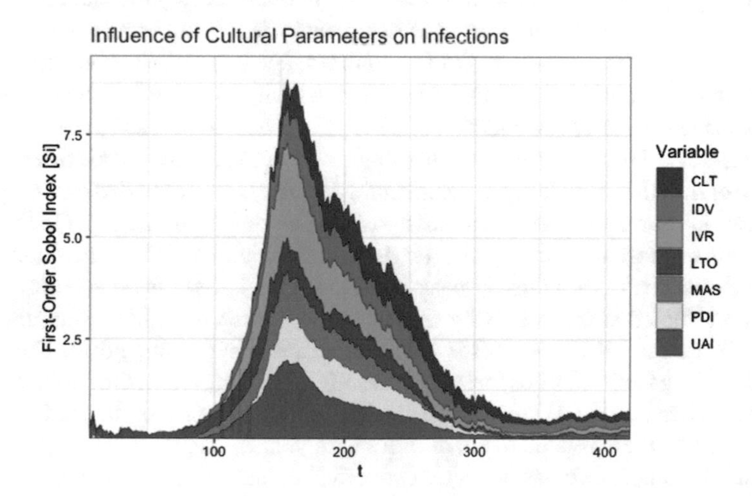

Fig. 8.12 Figures/SVD results of experiment 1

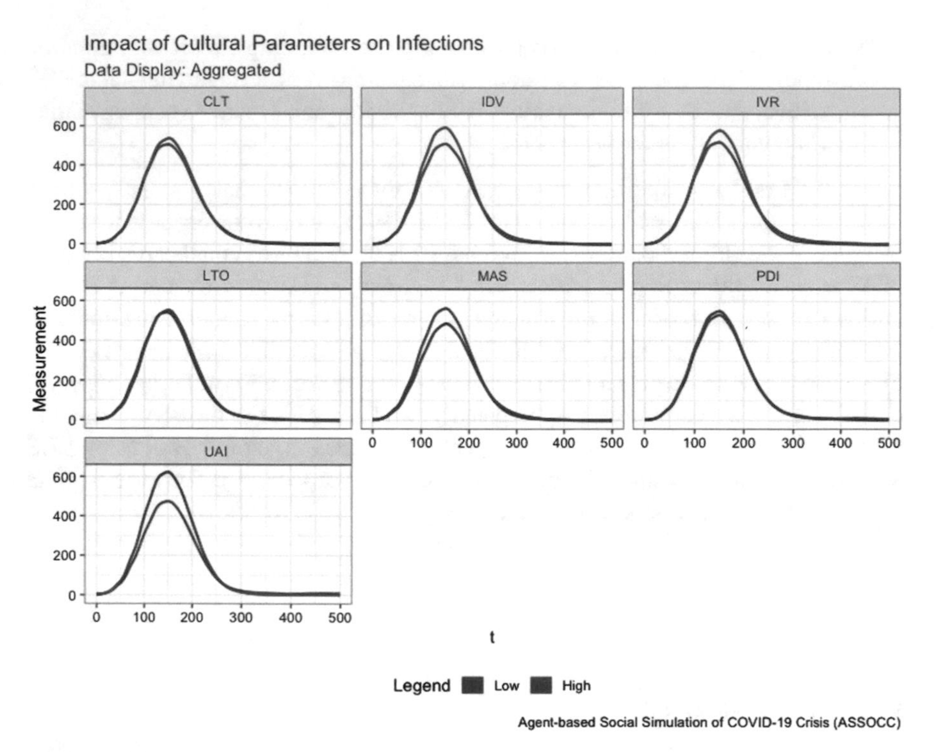

Fig. 8.13 OFAT results of experiment 1

explained by the fact that UAI and IVR are tied to Security and Hedonism, respectively. Security and Hedonism are in turn related to the needs for Risk Avoidance, Leisure and Luxury. The priority ascribed to these needs by agents has a notable effect on their propensity to interact with one another. To illustrate, agents that ascribe high priority to Leisure will be inclined to meet up with friends and family despite of the increased risk of contamination. On the contrary, agents that allocate high priority to Risk Avoidance will be reluctant to do such a thing. Figure 8.13 shows that high settings for Uncertainty Avoidance (UAI) tend to exert downward pressure on the infections-peak. This is because agents in high-UAI cultures tend to prioritize the satisfaction of the Risk Avoidance need, meaning that they are inclined to eschew places and situations that exposes them to an increased risk of contamination. On the other hand, high settings for Masculinity (MAS), Individualism (IDV) and Indulgence (IVR) tend to push the infection-peak upwards. This is because these dimensions map onto values that are positively related to the priority-weights of the needs for Autonomy, Luxury and Leisure (see Algorithm 2 and 4 in the appendix). The effects of manipulating Cultural Looseness-Tightness (CLT), Long-Term Orientation (LTO) and Power-Distance (PDI) are not as well-defined. This can be explained by the fact that LTO and PDI are related to the need for Compliance. However, in Experiment 1 the need for Compliance has no significant effect on the behavior of agents because there are no active policy measures that may work to alter its satisfaction level.

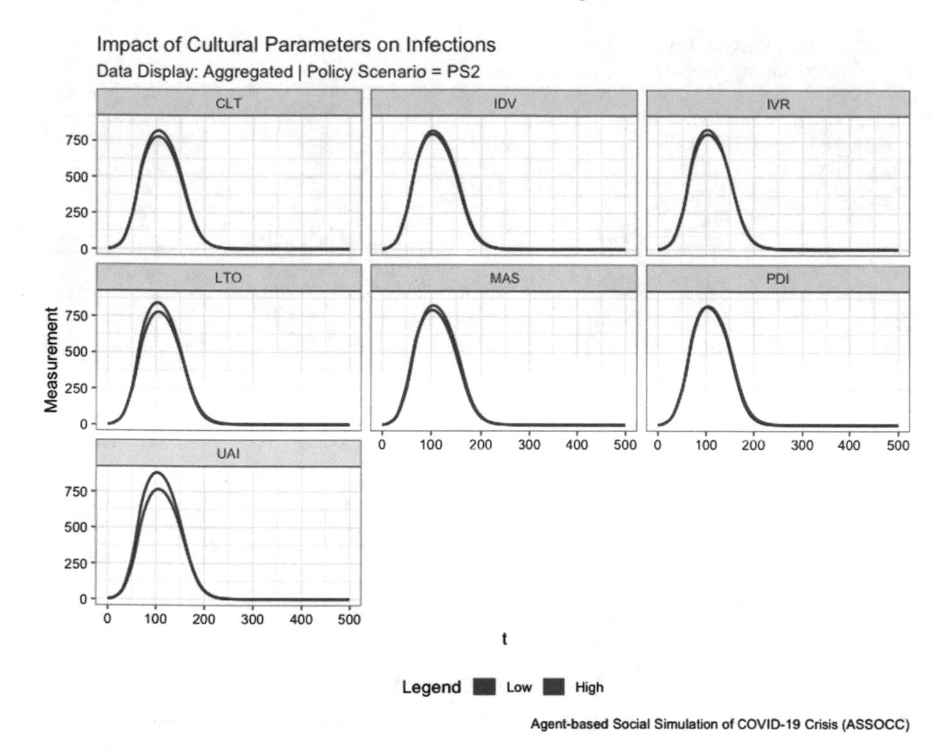

Fig. 8.14 Results of experiment 2: policy scenario: only-social-distancing

8.7.2 *Experiment 2*

Figure 8.14 shows that varying the cultural factors for PS_2 (only social distancing) has limited to no effect on the shapes of the infection curves. Only the manipulation of Uncertainty Avoidance (UAI) shows consistent variation between low and high settings, where lower settings of UAI tend to generate higher infection-peaks. This is because engaging in social distancing increases the satisfaction gains of the need for Risk Avoidance. Figure 8.15 shows the differences for varying the cultural factors under the activation of PS_3 (soft-lockdown). The infection curve trajectories show that high settings for Masculinity (MAS), Individualism (IDV), Indulgence (IVR) and low settings for Uncertainty Avoidance (UAI), Cultural Tightness (CLT) and Power Distance (PDI) tend to push the peak of infections upwards. Compared to the first two scenarios (PS_1 & PS_2) one observes a larger variation between low and high settings for UAI, PDI and IDV. An explanation for this is that these three dimensions are related to the need for Compliance (see Algorithm 2 and 4 in the appendix),

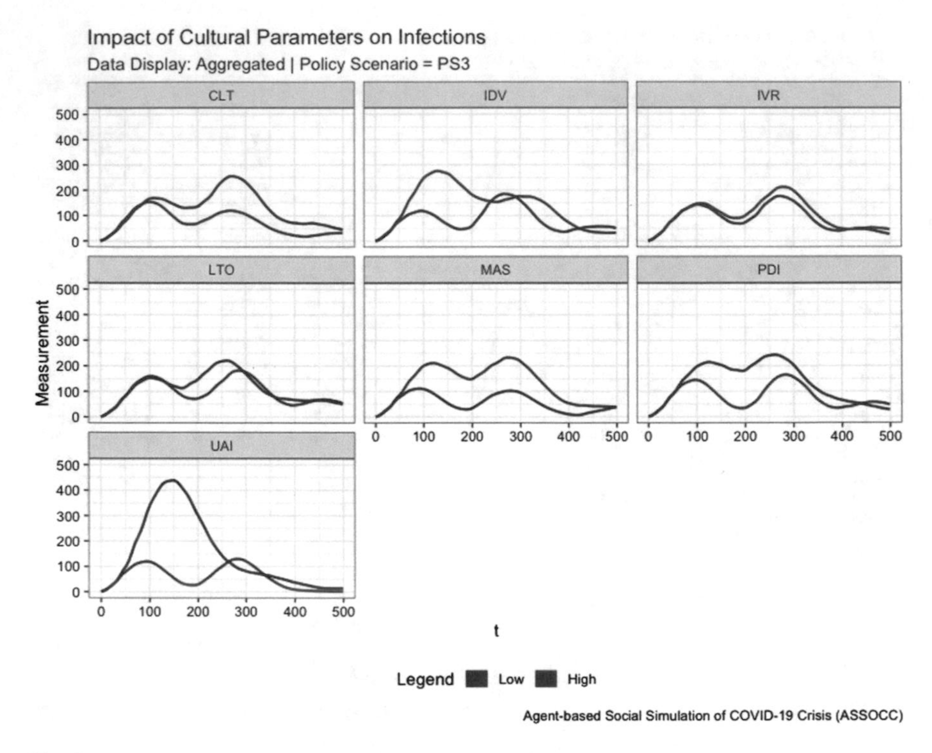

Fig. 8.15 Results of experiment 2: policy scenario: soft-lockdown

which influences the propensity of agents to comply to 'soft' public health related policy measures.[12]

Figure 8.16 shows that for PS_4 (hard-lockdown) higher levels of Indulgence (IVR) and Masculinity (MAS) result in higher infection-peaks. Moreover, it stands out that the second wave of infections is much larger than the first one, regardless of the cultural factor being manipulated. This emergent property can be explained as a consequence of the status of agent needs. Specifically, the satisfaction levels of agent needs become largely depleted after having been restricted by the hard-lockdown in their ability to freely choose the activities that provide them with the most bountiful satisfaction gains. Once restrictions are lifted, agents swiftly seek to replenish the satisfaction of the most depleted needs, which primarily concerns the needs that have to do with e.g., getting together with friends to enjoy leisure time and going shopping. This leads to an increase in social interactions taking place between agents that do not constitute a household, which drives up the transmission potential of the virus. It can be seen that IVR and MAS affect the spread of the virus in the most consistent manner; when set to 'High' both dimensions lead to a higher number

[12] 'Soft' policy measures are considered to be non-coercive strategies aimed at limiting the spread of SARS-CoV-2 and mortality of COVID-19. 'Hard' measures, on the other hand, are coercive. An example of a 'hard' measure is to fine or incarcerate people for not engaging in social distancing.

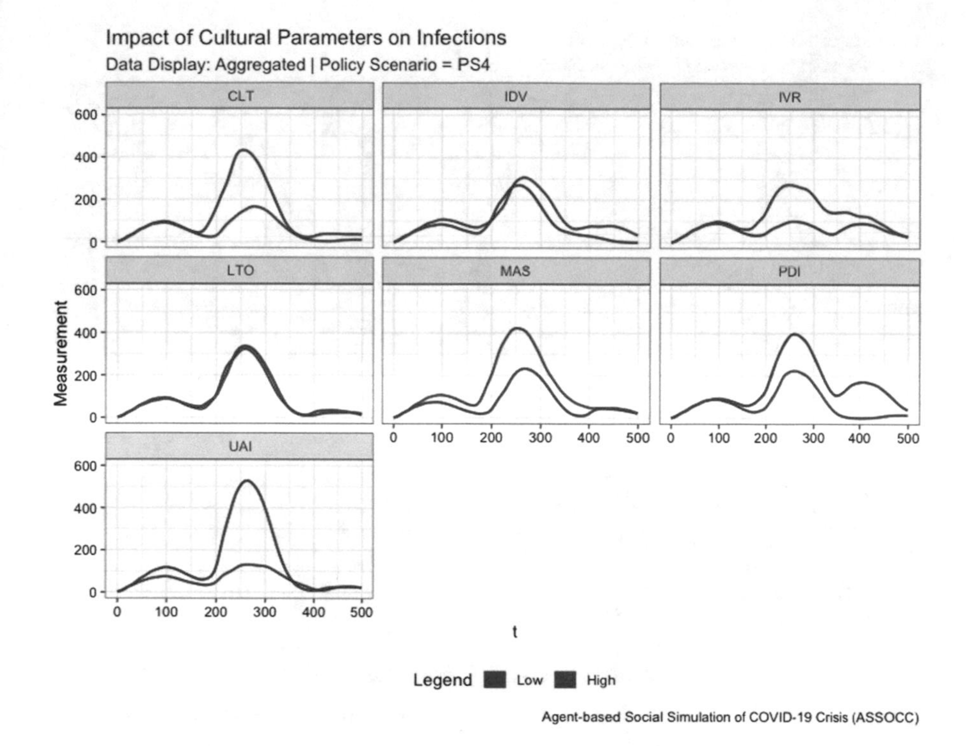

Fig. 8.16 Results of experiment 2: hard-lockdown

of infections relative to when they are set to 'Low'. Figure 8.17 shows the differences in the trajectory of infections when varying cultural factors for PS_5 (track & trace, testing and isolation). It can be seen that low settings for Uncertainty Avoidance (UAI) lead to lower infections curves, whereas high settings for Masculinity (MAS), Indulgence (IVR) and Individualism (IDV) result in higher infection-peaks. Manipulating Cultural Looseness-Tightness (CLT), Long-Term Orientation (LTO) and Power-Distance (PDI) does not appear to result in consistent differences in the shape of the infection curves.

It can be seen that the infection curves are similar to the ones observed in PS_2. The only major difference is the relatively lower infection-peak for PS_5, which is due to having an adequately functioning testing, tracking and tracing infrastructure in place. A second or third wave, which was visible in the soft- (PS_3) and hard-lockdown (PS_4) scenarios, is absent within this scenario as the policy measures remain active during the simulation.

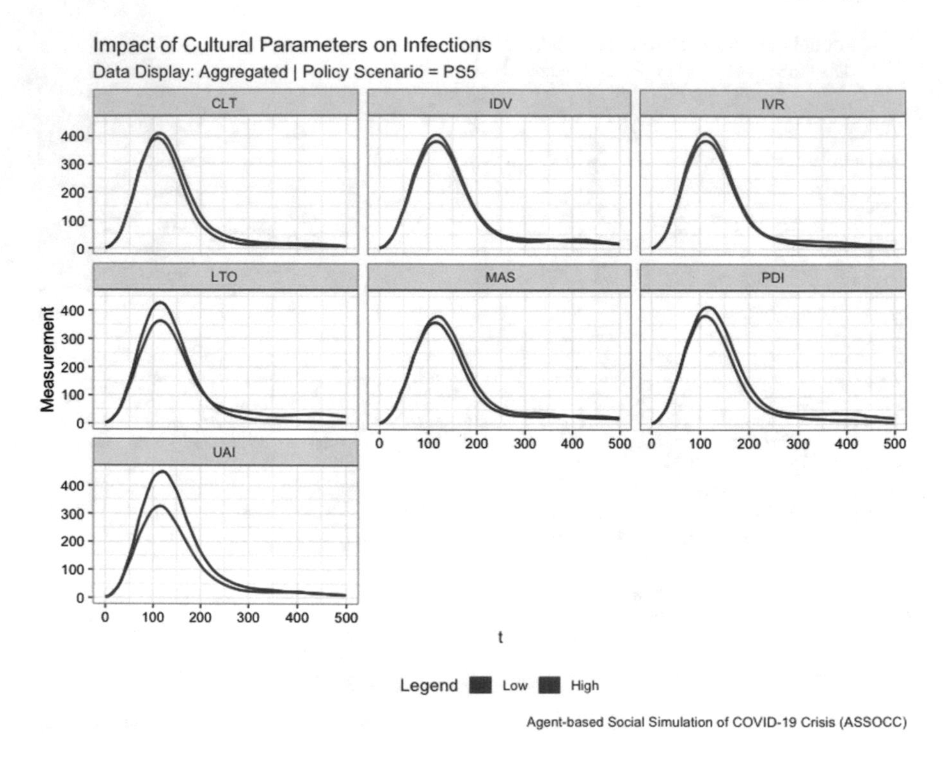

Fig. 8.17 Results of experiment 2: tracking-tracing-testing-isolating

8.7.3 Experiment 3

Figure 8.18 shows simulated time series data on (A) R_0, (B) infections, (C) cumulative infections and (D) deaths generated on the basis of the settings of cultural parameters for the Netherlands and South Korea. We compare these specific countries because their cultural profiles are highly dissimilar, as is their success at managing the spread of the virus thus far (see Fig. 8.18). Regarding Exhibit A, it stands out that the R_0 exceeds the critical threshold of 1 more often in the Netherlands, indicating lower success at limiting the spread of the virus in the Netherlands than in South Korea. Looking at Exhibit A in Fig. 8.18, it can also be noted that the infection waves in the Netherlands tend to peak at higher measurement levels than in South Korea. Moreover, the second waves of infections in the Netherlands are significantly larger than the first waves. In South Korea, it can be seen that this is not the case. Exhibit C shows that the Netherlands performs worse than South Korea across all policy scenarios. Interestingly, while the peak of infections for PS_5 is much higher in the Netherlands than in South Korea, the cumulative infections for PS_5 do not differ all that much, which indicates the importance of also taking into account the 'area under the curve' (AUC). However, one could argue that to effectively manage the COVID-19 crisis, it is required that governments prioritize lowering the infection-

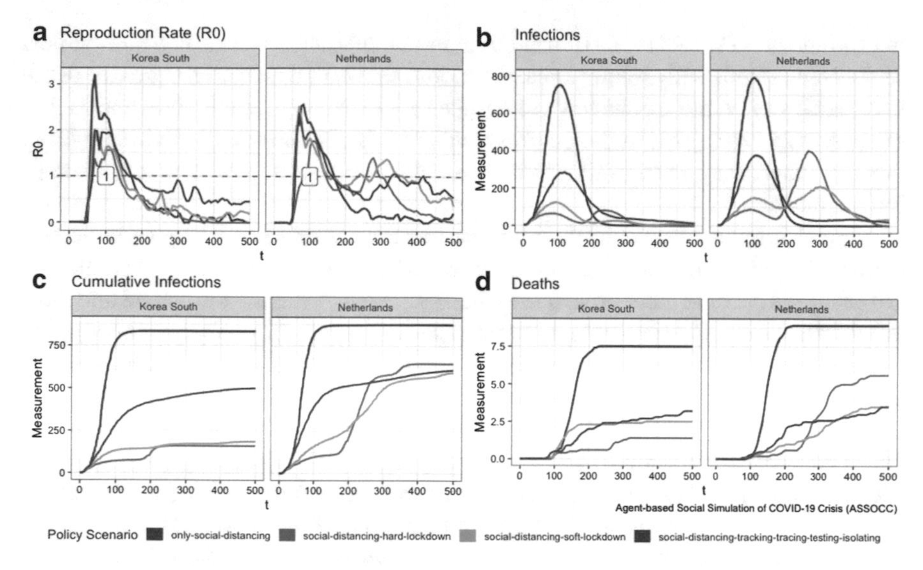

Fig. 8.18 Differences between the cultural parameters of the Netherlands and Korea South on reproduction rate (**a**), infections (**b**), cumulative infections (**c**) and deaths (**d**)

peak over minimizing the AUC. The reason for this is that the height of infection-peaks determine whether hospitals are able to accommodate the influx of patients suffering from severe COVID-19 symptoms. As a consequence, the mortality of the disease could increase drastically if infection-peaks become too high. It is therefore not surprising to see that the number of deaths is higher in the Netherlands across all policy scenarios (see Exhibit D in Fig. 8.18). On the basis of these findings, it can be concluded that the cultural profile of the Netherlands is less conducive to the effective management of the COVID-19 crisis than that of South Korea.

8.8 Discussion and Conclusion

8.8.1 Interpretations and Implications

The current study set out to construct a modelling representation of a candidate explanation for the cross-cultural differences observed in the effectiveness of policies implemented to manage the COVID-19 pandemic.

Model experimentation pointed out that coupling cultural variables at the country-level with values at the individual level leads to emergent properties that show resemblance to data observed in the real world. It is shown that nations whose cultural profiles are characterized by high levels of Uncertainty Avoidance and low levels of Individualism, Indulgence and/or Masculinity tend to be more successful at managing the spread of the COVID-19 pandemic in our simulated world. In a comparison of the Netherlands with South Korea, the model shows that the cultural profile of the latter seems to be more conducive to dampening the spread of the virus and lowering mortality rates. This finding corresponds to what is observed in the real world, thereby subscribing to the value of incorporating cultural forces within epidemiological models.

Several insights are obtained with regard to the cultural sensitivity of the effectiveness of public health related policy measures. Firstly, the model shows that as governments become more proactive in their efforts at managing the spread and mortality of the virus, the consequential response of the general public exhibits higher variability across cultural contexts. To illustrate, the effect of cultural parameters on the spread of the virus under the implementation of policy measures such as a soft- or hard-lockdown and tracking-tracing-testing-isolating shows higher variability than when governments decide to adopt a more passive stance (e.g. by implementing social distancing only). This suggests that as the complexity[13] of the implemented policy measures increases, so does the influence of culture on the effectiveness of these measures to limit the transmission of the virus. In this regard, it stands out that the cultural dimensions of Individualism, Indulgence, Masculinity and Uncertainty Avoidance exhibit the most consistent effects on the transmission dynamics of SARS-CoV-2. Findings point out that a binary manipulation of these cultural dimensions leads to predictable changes in the development of infections; e.g. increasing Masculinity, whilst controlling for the other dimensions, leads to a reliable increase of peak-infections, regardless of the type of policy measures implemented. On the other hand, dimensions such as Cultural Looseness-Tightness, Long-Term Orientation and Power-Distance are shown to affect infections in more volatile and less consistent ways. An explanation for this could be that these parameters affect the spread of the virus only under certain (specific) settings of other cultural factors or simply not at all; more elaborate analysis is needed to provide conclusiveness here.

Another insightful finding generated by the model is that it seems like implementing a soft-lockdown tends to lead to infection waves whose characteristics are similar (i.e. similar heights of peaks and symmetry of dispersion), whereas implementing a hard-lockdown tends to leads to quite drastic difference between the first and second infection waves (i.e. second waves tend to exhibit much higher peaks). A remarkable outlier in this regard is the Uncertainty Avoidance cultural dimension, which leads to a dramatic first wave when calibrated to its minimum setting under the activation of soft-lockdown policy scenario. Based on this finding, one might conclude that it is sensible of countries with low levels of Uncertainty Avoidance to implement rigorous rather than mild measures at the onset of a pandemic.

---

[13]The notion of complexity refers here to the quantity of measures concurrently implemented and the quality of these measures in terms of their managerial demands such as the need to coordinate and monitor processes of cooperation among a multitude of stakeholders over varying spatio-temporal scales. For instance, solely advising a population to maintain a safe physical distance from one another whilst carrying out one's daily activities is defined by a lower degree of complexity than organising and managing the implementation of a hard-lockdown.

Besides these experimental findings, the current study provides a coherent conceptual framework that explains how culture affects the behavior of individuals. It is proposed that cultural dimensions shape the formation of people's values, which, together with biological drifts, exert an influence on people's behavior during a pandemic. The current conceptualization may inform future studies aimed at incorporating the effect of culture on individual decision-making processes within social simulation models.

Lastly, the current study subscribes to the usefulness of applying an agent-based modelling approach when studying social phenomena. Statistical or equation-based epidemiological models are limited in their explanatory power since they do not explicate the causal processes that generate the data on which they are parameterized and calibrated. On the contrary, agent-based models, such as the one presented in this paper, present explicitly causal (as opposed to merely correlational) mechanisms to explain the emergence of certain social phenomena. Once the causal mechanisms embedded within the model are considered to be valid, it becomes possible to perform experiments that would otherwise have been impossible to conduct simply because the nature of reality does not allow for it. In doing so, the current study was able to study the effect of culture on the management of the COVID-19 pandemic whilst singling out the confounding effects of other country-level variables such as demographic composition, geographical & infrastructural features, political climate and institutional systems.

8.8.2 Limitations and Recommendations

The current study is characterized by several limitations that deserve attention. Firstly, the current theoretical framework contains no explicit notion of *norms*. While value-based models, such as the one presented in this paper, are well suited to portray cultural differences, they do not serve as well to capture cultural dynamics [76]. These dynamics refer to the micro-level processes that explain how individuals acquire, utilize and mutate their cultural assumptions and habits, as well as the macro-level processes that describe how cultural practices and institutions spread and change over time [76]. In sum, modelling norms, such as is done by e.g. [51, 77], enables a more context-dependent, contingency-based and dynamic approach to representing the influence of culture on the behavior of individuals during a pandemic.

Secondly, the current theoretical framework presumes that each time persons are engaged in decision-making, they enjoy unobstructed access to and overview of the contents and structural properties of their value systems. Moreover, decision-makers are presumed to know, in advance, to a high degree of fidelity the satisfaction to be gained from performing a broad range of activities. Decision-makers are then presumed to consistently select the activity that holds the maximum level of satisfaction to be gained. A large body of research points out that humans do not make decisions in such a rational manner [78, 79]. Incorporating representations of bounded rationality and cognitive biases into the decision-making process would therefore enhance the external validity of our theoretical framework.

Third, the current model does not include any notion of super-spreading individuals or events. For SARS-CoV-2, it has been shown that ~80% of secondary transmissions have been caused by a small fraction of infectious individuals (~10%) [80]. Future studies could choose to expand the current epidemiological model with super-spreaders through stochastically attributing individual agents with increased levels of contagiousness or by increasing the density-factor of particular gathering-points as to model the occurrence of super-spreading events.

Fourth, the model assumes that the government has continuous access to a highly accurate and timely indication of the virus's transmission potential (i.e. the R_0). In reality, this is not the case since there exist many factors that obstruct a government's efforts at monitoring the transmission of the virus within its nation's borders. Furthermore, our model does not allow for modelling policy measures related to suspending tourism or other forms of 'people-flows' across national or jurisdictional (e.g. regional, municipal) borders. These country-level differences may well be a function of cultural differences, which makes it an interesting avenue for further research.

Finally, the execution time of our model is rather slow due to the computational complexity that characterizes it. Due to limitations in the availability of computing power disposable to us, it was not possible to run simulations with populations of larger than approximately 1,200 agents. This 'ceiling' in the number of agents that could be simulated, combined with the assumptions that an agent becomes immune once it recovers from the virus, often led to the situation wherein the virus 'burnt through all of the available wood' during model executions. That is, at some point either all agents are immune or have passed away, leaving none of them susceptible to become infected anymore. As a consequence, the current NetLogo-based version of the model is not able to simulate more than three waves of infections. Moreover, it was not possible to run an adequate t-SVD analysis for the second experiment in this study as it required a infeasible number of simulations to be performed. A faster model, ideally combined with an access to more computational resources, would have enabled us to execute a more elaborate analysis by running more and lengthier simulations.

Acknowledgements We would like to acknowledge the members of the ASSOCC team for their valuable contributions to this chapter. The simulations were enabled by resources provided by the Swedish National Infrastructure for Computing (SNIC) at Umeå partially funded by the Swedish Research Council through grant agreement no. 2018-05973. This research was conducted using the resources of High Performance Computing Center North (HPC2N).

References

1. D. Huremovic, Brief history of pandemics (pandemics throughout history), in *Psychiatry of Pandemics* (Springer, 2019), pp. 7–35
2. P. Sands et al., *Outbreak Readiness and Business Impact: Protecting Lives and Livelihoods Across the Global Economy* (Switzerland, Technical Report (Geneva, 2019)
3. M. Gelfand et al., Cultural and Institutional Factors Predicting the Infection Rate and Mortality Likelihood of the COVID-19 Pandemic (2020)
4. F.B. Agusto, M.I. Teboh-Ewungkem, A.B, Gumel, Mathematical assessment of the effect of traditional beliefs and customs on the transmission dynamics of the 2014 Ebola outbreaks. BMC Med. **13**(1), 96 (2015)
5. C.L. Fincher et al., Pathogen prevalence predicts human cross-cultural variability in individualism/collectivism. Proc. R. Soc. B Biol. Sci. **275**(1640), 1279–1285 (2008)
6. S. Dryhurst et al., Risk perceptions of COVID-19 around the world. J. Risk Res. 1–13 (2020)
7. B. Edmonds, Different modelling purposes, in *Simulating Social Complexity* (Springer, 2017), pp. 39–58
8. R.F. Baumeister, E.J. Masicampo, Conscious thought is for facilitating social and cultural interactions: How mental simulations serve the animal-culture interface, in Psychol. Rev. **117**(3), 945 (2010)
9. M.J. Gelfand, J.C. Jackson, From one mind to many: the emerging science of cultural norms. Curr. Opin. Psychol. **8**, 175–181 (2016)
10. R. Boyd, P.J. Richerson, Culture and the evolution of human cooperation. Philos. Trans. R. Soc. B Biol. Sci. **364**(1533), 3281–3288 (2009)
11. S.H. Schwartz, Are there universal aspects in the structure and contents of human values? J. Soc. Issues **50**(4), 19–45 (1994)
12. E. Kostrova, The "Ought"-dimension in value theory: the concept of the desirable in John Dewey's definition of value and its significance for the social sciences. *Philosophy of Science* (Springer, 2018), pp. 171–185
13. S.H. Schwartz, An overview of the Schwartz theory of basic values. Online Read. Psychol. Cult. **2**(1), 2307–0919 (2012)
14. W. Bilsky, M. Janik, S.H. Schwartz, The structural organization of human values-evidence from three rounds of the European Social Survey (ESS). J. Cross-Cult. Psychol. **42**(5), 759–776 (2011)
15. E. Davidov, P. Schmidt, S.H. Schwartz, Bringing values back in: the adequacy of the European Social Survey to measure values in 20 countries. Public Opin. Q. **72**(3), 420–445 (2008)
16. J.R.J. Fontaine et al., Structural equivalence of the values domain across cultures: distinguishing sampling fluctuations from meaningful variation. J. Cross-Cult. Psychol. **39**(4), 345–365 (2008)
17. P.A. Gambrel, R. Cianci, Maslow's hierarchy of needs: Does it apply in a collectivist culture. J. Appl. Manag. Entrep. **8**(2), 143 (2003)
18. R.M. Steers, R.T. Mowday, D.L. Shapiro, The future of work motivation theory. Acad. Manag. Rev. **29**(3), 379–387 (2004)
19. T. Balke, N. Gilbert, How do agents make decisions? A survey. J. Artif. Soc. Soc. Simul. **17**(4), 13 (2014)
20. T.L. van der Weide, Arguing to motivate decisions. Ph.D. thesis. Utrecht University, 2011
21. S. Heidari, M. Jensen, F. Dignum, Simulations with values, in *Advances in Social Simulation* (Springer, 2020), pp. 201–215
22. L. Tay, E. Diener, Needs and subjective well-being around the world. J. Personal. Soc. Psychol. **101**(2), 354 (2011)
23. B.J. Ryan et al., COVID-19 community stabilization and sustainability framework: an integration of the Maslow hierarchy of needs and social determinants of health, in *Disaster Medicine and Public Health Preparedness* (2020), pp. 1–7
24. N. Jerome, Application of the Maslow's hierarchy of need theory; impacts and implications on organizational culture, human resource and employee's performance. Int. J. Bus. Manag. Inven. **2**(3), 39–45 (2013)

25. E.L. Deci, R.M. Ryan, *Handbook of Self-determination Research* (University Rochester Press, 2004)
26. F. Dechesne et al., No smoking here: values, norms and culture in multi-agent systems. Artif. Intell. Law **21**(1), 79–107 (2013)
27. M.A. Wahba, L.G. Bridwell, Maslow reconsidered: a review of research on the need Hierarchy theory, in *Academy of Management Proceedings*, vol. 1973.1. (Academy of Management Briarcliff Manor, NY 10510, 1973), pp. 514–520
28. G. Hofstede, The cultural relativity of the quality of life concept. Acad. Manag. Rev. **9**(3), 389–398 (1984)
29. A. Knafo, S. Roccas, L. Sagiv, The value of values in crosscultural research: a special issue in honor of Shalom Schwartz (2011)
30. G. Hofstede, Management scientists are human. Manag. Sci. **40**(1), 4–13 (1994)
31. G. Hofstede, *Culture's Consequences: Comparing Values* (Institutions and Organizations Across Nations (Sage publications, Behaviors, 2001)
32. S.H. Schwartz, National culture as value orientations: consequences of value differences and cultural distance, in *Handbook of the Economics of Art and Culture*, vol. 2. (Elsevier, 2014), pp. 547–586
33. A. Thomas, E.-U. Kinast, S. Schroll-Machl, *Handbook of Intercultural Communication and Cooperation. Basics and Areas of Application* (Vandenhoeck & Ruprecht, 2010)
34. B.J. Hurn, B. Tomalin, *Cross-Cultural Communication* (Springer, 2013)
35. R. Inglehart, W.E. Baker, Modernization, cultural change, and the persistence of traditional values. Am. Sociol. Rev. 19–51 (2000)
36. C. Welzel, R. Inglehart, H.-D. Kligemann, The theory of human development: a cross-cultural analysis. Eur. J. Polit. Res. **42**(3), 341–379 (2003)
37. M.J. Gelfand, L.H. Nishii, J.L. Raver, On the nature and importance of cultural tightness-looseness. J. Appl. Psychol. **91**(6), 1225 (2006)
38. S.H. Schwartz, Rethinking the concept and measurement of societal culture in light of empirical findings. J. Cross-Cult. Psychol. **45**(1), 5–13 (2014)
39. I. Uz, The index of cultural tightness and looseness among 68 countries. J. Cross-Cult. Psychol. **46**(3), 319–335 (2015)
40. K. Chudzikowski et al., The evolution of Hofstede's doctrine. Cross Cult. Manag. Int. J. (2011)
41. V. Taras, J. Rowney, P. Steel, Half a century of measuring culture: Review of approaches, challenges, and limitations based on the analysis of 121 instruments for quantifying culture. J. Int. Manag. **15**(4), 357–373 (2009)
42. S.H. Schwartz, Values: cultural and individual (2011)
43. R. Heale, D. Forbes, Understanding triangulation in research. Evid. Based Nurs. **16**(4), 98–98 (2013)
44. B. Tekes et al., The relationship between Hofstede's cultural dimensions, Schwartz's cultural values, and obesity. Psychol. Rep. **122**(3), 968–987 (2019)
45. R. Fischer et al., Are individual-level and country-level value structures different? Testing Hofstede's legacy with the Schwartz Value Survey. J. Cross-Cult. Psychol. **41**(2), 135–151 (2010)
46. S.I. Ng, J.A. Lee, G.N. Soutar, Are Hofstede's and Schwartz's Value Frameworks Congruent? Int. Mark. Rev. (2007)
47. S.I. Ng, X.J. Lim, Are Hofstede's and Schwartz's values frameworks equally predictive across contexts? Revista Brasileira de Gestão de Negócios **21**(1), 33–47 (2019)
48. G. Hofstede, The 6D model of national culture. https://geerthofstede.com/culture-geert-hofstede-gert-jan-hofstede/6d-model-of-national-culture/ (Mar. 2020)
49. G.J. Hofstede, C.M. Jonker, T. Verwaart, Cultural differentiation of negotiating agents. Group Decis. Negot. **21**(1), 79–98 (2012)
50. L. Vanhée, F. Dignum, Explaining the emerging influence of culture, from individual influences to collective phenomena. J. Artif. Soc. Soc. Simul. **21**(4), 11 (2018)
51. R. Mercuur, V. Dignum, C. Jonker, The value of values and norms in social simulation. J. Artif. Soc. Soc. Simul. **22**(1) (2019)

52. S. Hitlin, J.A. Piliavin, Values: reviving a dormant concept. Annu. Rev. Sociol. **30**, 359–393 (2004)
53. I. van de Poel, Design for value change. Ethics Inf. Technol. 1–5 (2018)
54. D. Dörner et al., *A Simulation of Cognitive and Emotional Effects of Overcrowding, in Proceedings of the Seventh International Conference on Cognitive Modeling* (Italy, Edizioni Goliardiche Triest, 2006), pp. 92–98
55. G. Di Tosto, F. Dignum, Simulating social behaviour implementing agents endowed with values and drives, in *International Workshop on Multi-Agent Systems and Agent-Based Simulation* (Springer, 2012), pp. 1–12
56. J.B. Miner, *Organizational Behavior: Essential Theories of Motivation and Leadership. One*, vol. 1. (ME Sharpe, 2005)
57. ESS, The European Social Survey (ESS) (2020). https://web.archive.org/web/20201101015551/, https://www.europeansocialsurvey.org/. Accessed 01 Sept 2020
58. WVS, The World Value Survey (WVS) (2020). https://web.archive.org/web/20201106095007/, https://www.worldvaluessurvey.org/wvs.jsp. Accessed 01 Sept 2020
59. OWID, Our World in Data-Coronavirus Source Data. https://web.archive.org/web/20201106095104/, https://ourworldindata.org/coronavirus-source-data. (2020). Accessed 01 Nov 2020
60. M.J. Gelfand et al., Differences between tight and loose cultures: a 33-nation study. Science **332**(6033), 1100–1104 (2011)
61. H.H. Fung et al., Age differences in personal values: universal or cultural specific? Psychol. Aging **31**(3), 274 (2016)
62. S.H. Schwartz, How age influences values (2020). https://web.archive.org/web/20200216043738/, http://essedunet.nsd.uib.no/cms/topics/1/2/2.html
63. D. Centola, The social origins of networks and diffusion. Am. J. Sociol. **120**(5), 1295–1338 (2015)
64. M. McPherson, L. Smith-Lovin, J.M. Cook, Birds of a feather: homophily in social networks. Annu. Rev. Sociol. **27**(1), 415–444 (2001)
65. M.K. Goetz, The paradox of value: water rates and the law of diminishing marginal utility. J. Am. Water Works Assoc. **105**(9), 57–59 (2013)
66. J.K. Rilling, A.G. Sanfey, The neuroscience of social decisionmaking. Annu. Rev. Psychol. **62**, 23–48 (2011)
67. J.H. Park, Introversion and human-contaminant disgust sensitivity predict personal space. Personal. Individ. Differ. **82**, 185–187 (2015)
68. S. Roccas et al., The big five personality factors and personal values. Personal. Soc. Psychol. Bull. **28**(6), 789–801 (2002)
69. M. Calder et al., Computational modelling for decision-making: Where, why, what, who and how. R. Soc. Open Sci. **5**(6), 172096 (2018)
70. J.M. Galán et al., Errors and artefacts in agent-based modelling. J. Artif. Soc. Soc. Simul. **12**(1), 1 (2009)
71. G. Ten Broeke, G. Van Voorn, A. Ligtenberg, Which sensitivity analysis method should i use for my agent-based model? J. Artif. Soc. Soc. Simul. **19**(1) (2016)
72. K.H. Van Dam, I. Nikolic, Z. Lukszo. *Agent-Based Modelling of Socio-Technical Systems*, vol. 9. (Springer Science & Business Media, 2012)
73. A. Ligmann-Zielinska et al., Using uncertainty and sensitivity analyses in socioecological agent-based models to improve their analytical performance and policy relevance. PloS One **9**(10), e109779 (2014)
74. A. Ligmann-Zielinska, L. Sun, Applying time-dependent variancebased global sensitivity analysis to represent the dynamics of an agent-based model of land use change. Int. J. Geograph. Inf. Sci. **24**(12), 1829–1850 (2010)
75. A. Saltelli et al., *Global Sensitivity Analysis: The Primer* (Wiley, 2008)
76. M.W. Morris et al., Normology: integrating insights about social norms to understand cultural dynamics. Organ. Behav. Human Decis. Process. **129**, 1–13 (2015)

77. J. Mc Breen et al., Linking norms and culture, in *2011 Second International Conference on Culture and Computing* (IEEE, 2011), pp. 9–14
78. H.A. Simon, *Models of Bounded Rationality: Empirically Grounded Economic Reason*, vol. 3 (MIT Press, 1997)
79. G.A. Klein et al., *Decision Making in Action: Models and Methods* (Ablex Norwood, NJ, 1993)
80. A. Endo et al., Estimating the overdispersion in COVID-19 transmission using outbreak sizes outside China. Wellcome Open Res. **5**(67), 67 (2020)

Chapter 9
Economics During the COVID-19 Crisis: Consumer Economics and Basic Supply Chains

Alexander Melchior

Abstract The COVID-19 pandemic has other perspectives than just the epidemiological one that impacts human lives. In this chapter, we look at the economic perspective by modelling a consumer economic system with a basic supply chain and a very basic government role to create a circular economic system. People have to work at shops or workplaces to earn money to buy food or other items. These items are sold by shops who in turn buy them from workplaces. We devise multiple scenarios to compare the effects of the pandemic, measures to lessen the epidemiological effects and additional economic effects. The results show us that we can: (1) create an useful economic model in the complex ASSOCC context, (2) that from an economic perspective repeated lockdowns are more harmful than not taking any action at all, and (3) economic measures do support economic well being of the population. While it is very clear that the real world is a lot more complex than how we have modelled it, the modelling process helps us pinpoint where next steps of policy investigation, model improvement and research could be performed.

9.1 Why Read This Chapter?

In this chapter, we discuss various economic aspects of a pandemic and the COVID-19 pandemic in particular. In particular, we focus on consumer economics and a basic supply chain. This allows us to see the consequences of lockdowns and restrictions for different types of persons and how these consequences influence the behaviour and reactions to further restrictions. This perspective on the effects of a pandemic comes with its own challenges, both for the economic sub-model and the ASSOCC model as a whole. We discuss the motivation behind adding this perspective and its scenarios to the ASSOCC model. After this, we present the conceptual model and

A. Melchior (✉)
Department of Information and Computing Sciences, Utrecht University, Princetonplein 5,3584 Utrecht, The Netherlands

Ministry of Economic Affairs and Climate Policy and Ministry of Agriculture, Nature and Food Quality, Bezuidenhoutseweg 73, 2594 AC Den Haag, The Netherlands
e-mail: a.t.melchior@uu.nl

F. Dignum (ed.), *Social Simulation for a Crisis*, Computational Social Sciences,
https://doi.org/10.1007/978-3-030-76397-8_9

the chosen fundamental basis for this model. This proved to be non-trivial as we could not just an economic model to the rest of the ASSOCC framework, but had to fit the economic consequences of actions and decisions to the model. So, we started of with some very generic economic principles and fitted the specific details of the economic model around these principles.

The translation from conceptual model to the implementation in the bigger ASSOCC model, comes with its own challenges. In this chapter, we will discuss this extra dimension of complexity of using the resulting economic model as a sub-model. In order to show the influence of economics on society during the COVID-19 crisis we will look at four different economic scenario's: no pandemic, a pandemic with no measures, a pandemic with a lockdown and a pandemic with a lockdown and an economic measure that pays wages. We will discuss our results by analysing a number of key metrics and why we have chosen these specific key metrics.

9.2 Introduction and Motivation

The COVID-19 pandemic has severe direct medical consequences. A relative high number of people get hospitalised or even die after contracting COVID-19. This alone has various direct consequences on other aspects of life. For example; a deceased parent can no longer provide care for his or her children. People, and mainly governments, try to reduce or even prevent the spread of the virus with various measures, such as closing schools, restaurants and even complete lockdowns. This shows that the COVID-19 pandemic also has *indirect* effects on our societies. These indirect effects cause *second-order effects* on the society as a whole, which in turn affects the spread of the disease.

Some of the most prevalent second-order effects are the effects on the economy. Economic recessions were expected early into the pandemic. In the Netherlands, by the end of March 2020 and one month into the local pandemic, the economy was predicted to shrink by 7.7%[1] which would be a bigger recession than the banking crisis recession from 2009. While this sounds severe, only five months later it became clear that this was still a conservative estimation as the economy shrunk by 8.5%.[2] The economy of the EU as a whole has shrunk by 7.8% in November 2020 and predictions made at the time indicate that the economy will not recover very quickly.

These figures are more than just numbers: unemployment rises, people have less income and self worth, government tax income decreases whilst government expenditure on social security rises. Given these severe (long-term) effects of the pandemic on the economy, it is imperative to have a better understanding how measures regarding the pandemic affect the economy. Incorporating the economic aspect in the

[1] https://nos.nl/artikel/2328385-cpb-vreest-diepe-recessie-economie-kan-krimpen-met-7-7-procent.html.

[2] https://nos.nl/artikel/2344026-catastrofale-krimp-van-de-nederlandse-economie-min-van-8-5-procent.html.

ASSOCC model will allow evaluation of various trade-offs between economic, social and medical consequences of government interventions. E.g. should only restaurants and pubs be closed or should all shops be closed (except for limited access to supermarkets). Complete lockdowns can provide a quicker decrease of infections, but also might have much bigger economic consequences.

Undoubtedly there are many economic effects of the COVID crisis. Many of these effects are still unknown and will only unfold in the years after the pandemic has subsided. In ASSOCC, the main actors are individuals, consumers, and their daily lives. Thus in our model we focus on the economic aspects of their daily lives. With "our economy" we focus on the consumer economics system and the directly related other agents. We will investigate the following three issues in this chapter:

1. Investigate the effect of a pandemic on the local economy.
2. Investigate the effects of measures taken by government to stop the pandemic on the economy.
3. Compare economic measures that could mitigate the negative effects stemming from the pandemic.

To investigate these issues in a structured manner we have structured the chapter as follows. We first describe the components of the economic model and their implementation in ASSOCC in Sect. 9.3. Once the fundamentals are described, we discuss various scenario's in Sect. 9.4 that help us gain insight in the three issues. Finally we show the results in Sect. 9.5 and discuss them in Sect. 9.6. In Sect. 9.7, we draw our final conclusions on the topic of modelling an economy in ASSOCC and indicate options for further work.

9.3 An Economic Model for ASSOCC

In the ASSOCC model, we are interested in the decision making and actions of individuals. Thus, an economic model for ASSOCC should be based on the behaviour of individual people. This requirement leads us in the direction of micro-economic models, with the challenge of finding an appropriate one. Interesting enough, this starts us off with a contrast when we look at the "newspaper headlines" that are supposed to inform us about the economic effects of COVID-19 in the real world. These, at least in the beginning of COVID-19, mainly reported on macro-economic effects and figures such as unemployment, GDP or bankruptcies. The beauty of ABM, in theory, is that we can easily extract such macro-economic metrics from a micro-economic ABM by measuring aggregates over groups of agents.

We also see this reflected in our brief literature review[3] when looking for feasible economic ABM models to use. Most of the found ABM literature investigates a specific type of economic model or an implementation of them and how this could

[3]Recall that ASSOCC has been created in a crisis situation.

replicate reality. Herbert Dawid and Domenico Delli Gatti give an overview of macro-economic models in [1]. The models themselves are not validated against real world data or mechanisms, but these are supposed to be given by the economic theories. In [2], on the validation of economic ABMs, Fagiolo et al. write in their concluding remarks on page 782 that validation of economic ABM is hardly possible:

> Furthermore, validation of ABMs will never tell whether a model is a correct description of the complex, unknown and non-understandable real-world data generating process. However, in a Popperian fashion, ABM validation techniques should eventually allow researchers to understand whether a model is a bad description of it.

Thus, rather than basing our model on a complete economic agent based model we base our model on some basic economic principles underlying all micro-economic models. On the Wikipedia page for *economic system*[4] we find the following definition:

> An economic system, or economic order, is a system of production, resource allocation and distribution of goods and services within a society or a given geographic area. It includes the combination of the various institutions, agencies, entities, decision-making processes and patterns of consumption that comprise the economic structure of a given community.

Given this definition we can scope our model in the following way: *the economic model consists of the economic value transition resulting from interactions between individual agents, such as consumers, government, shops and workplaces and their resulting state.* In order for the economic model to give any insights in economic consequences of the COVID crisis it should to some extend be a closed model. I.e. neither should we constantly drain money from nor insert money into the simulation. If this would have to be done the results of a simulation would mostly depend on how these leaks would be regulated rather than a result of the interactions within the simulation. As a consequence we have to somehow *close* the economy such that money more or less gets preserved and we get better sight on how it gets redistributed. In order to achieve this we use an artificial *government* agent to serve as a buffer and a means to absorb some money while also able to create money and subsidise agents during a crisis. Governments already have this function in the real world, but we have made this explicit and simplified many aspects of this role as well. We would like to stress that we exclude macro-economic aspects in this chapter (readers that are interested in possible macro-economic systems can visit Chap. 15 on Challenges) because we would need a far larger model where all economic sectors would be represented in the right proportion. This would be too complex in times of crisis and would require also far more agents than we could at present accommodate in the simulation.

Our model is based on the simple economic model of circular flow of income [4]. This model is based on the economic relationship between individuals and businesses, as depicted in Fig. 9.1. Individuals supply labour to businesses and get an income in the form of wages in return. The business can pay the wages with the the expenditure

[4] Wikipedia (https://en.wikipedia.org/wiki/Economic_system https://en.wikipedia.org/wiki/Economic__system).

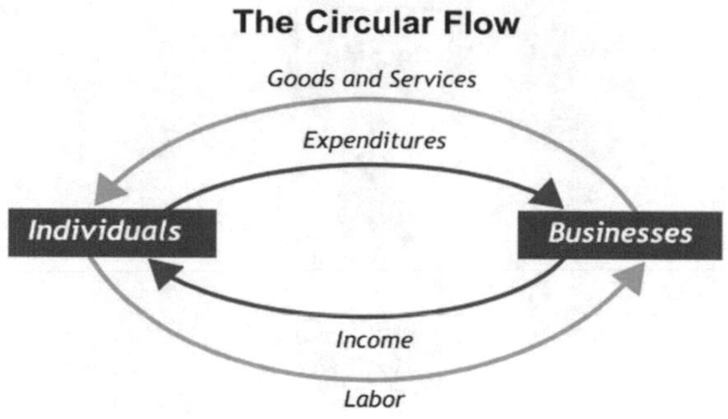

Fig. 9.1 The circular flow of income, figure from [3]

of (other) individuals at the business. The individuals get goods or services from the business in return.

The circular flow of income model is generally regarded as a macro-economic model as individuals and businesses are grouped together. To create an ABM, we don't have to group them together but can actually implement the individuals. We can use this model in the following way:

- Goods, Services and Labour are *resources* an agent (e.g. consumers, shops or workplaces) can provide to others.
- Income and expenditure is the exchange of capital between these agents in return for an action or good.

With this, we can take the idea of circular flow and use it for a conceptual micro-economic model with a grounding in accepted economic theory.

Given the basic principles of our model (based on general economic standards), we can direct our attention to making these principles actually work. Just like a machine, a working economy needs moving parts. In our case these moving parts are the agents that perform their actions, showing certain *behaviour*. In [1], Herbert Dawid and Domenico Delli Gatti make an interesting remark is made regarding behaviour and economic models:

> The design of behavioural rules determining such locally constructive actions is a crucial aspect of developing an agent-based macroeconomic model. The lack of an accepted precise common conceptional or axiomatic basis for the modeling of bounded rational behavior has raised concerns about the "wilderness of bounded rationality" (Sims (1980)), however agent-based modelers have become increasingly aware of this issue providing different foundations for their approaches to model individual behavior.

In the ASSOCC model, the foundation for agents is based on *needs* as explained in Chap. 2 and made concrete in 3. To fulfil some of their needs, agents perform actions, like acquiring food to eat and live, for which the agents need capital to be able to execute them. Agents also want to have some capital in reserve so they can survive for a while if for some reason they no longer have any income. This enables us to reuse principles of the ASSOCC framework that are already in place: the homeostatic model that is used to model the various needs of agents (see Sect. 2.3.5). The minimum reserve of capital is the threshold, income increases the homeostatic

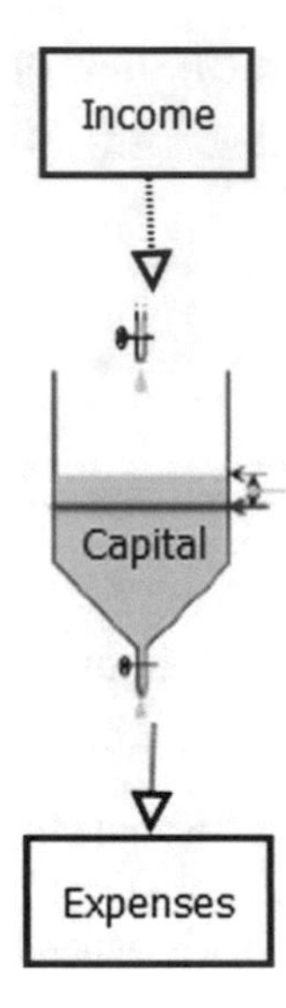

Fig. 9.2 Homeostatic model for capital based on [5]

level and the expenditure drains it, as depicted in Fig. 9.2. The use of the homeostatic model reflects that the amount of capital available to an individual should be more or less balanced over time. To provide for basic needs such as food and shelter agents have continuous expenses that drain the capital tank over time. Income fills the tank but might not come in the same frequency or time as it linked to different actions. If the tank is drained to a level below the threshold the agent will have a salient *need* to refill the tank. In general it holds that when the level of capital decreases the need to gain capital increases. This is very similar to the continuous use of energy by the body during the day and the replenishment of energy by eating food and drinking water at certain times a day.

9.3.1 From Fundamentals to a Real Economic System

In this section we describe how the fundamentals can be put to work in our model. Agents can have jobs to perform work at shops, workplaces, hospitals, universities, etc. to make money and provide labour in return. When they are not working, they can go to shops and spend money to satisfy their needs. In our model we assume that not everybody is able to work, like students and retired people. They receive scholarships and pensions from government in the form of monetary benefits or a subsidy. We also assume healthcare and education are paid for with taxes, so the universities and hospitals do not make money from the services they provide to agents, but get their income from the government. The tax component is modelled by creating the government as an agent in the ASSOCC model. The government makes money by levying taxes on things like sales and spends money by paying agents working in education and healthcare, next to paying the pensions and scholarships. These

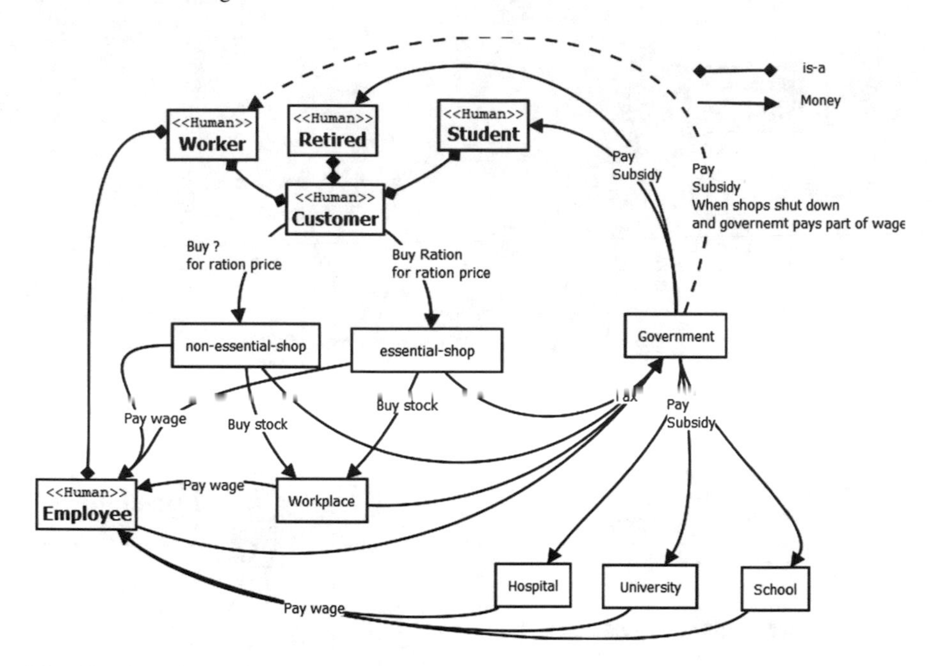

Fig. 9.3 Flow of capital

assumption are based on western European societies, which is also the perspective on how other things work in the model.

A generalised model is depicted in Fig. 9.3. In this diagram we see three of the four ASSOCC agent types. Children are not considered to be active economic parties. They can go the shops and restaurants, but the money they spend comes from their parents. The *worker* agents function as employees in the supply side of the model, while they are customers in the demand side of the model. Since students and retired agents do not work or produce resources for others they are only customers in this model. For the sake of simplicity, we have chosen not to model a full supply chain of goods and manufacturing. Thus, workplaces include all kinds of businesses, ranging from factories to wholesalers and from consultants to banks. Workplaces do not buy goods but create goods to supply to the shops when workers perform work at workplaces. Workplaces are generic and offer generic goods for all our shops, both essential and non-essential. Each of the entities (rectangles) in the model contains its own capital tank. Thus a shop has capital which gets replenished by customers paying for goods or services, while the capital diminishes by having to buy products from suppliers, pay tax and having to pay wages to employees (including the owner of the shop). Hospitals, schools and universities are not paid by their "customers", but get money from government that comes from the tax income as mentioned before. This income from government is completely used for paying wages of employees of these institutions. Using this very simple model affords us to concentrate on the micro-economic aspects related to a basic supply chain of both labour and goods.

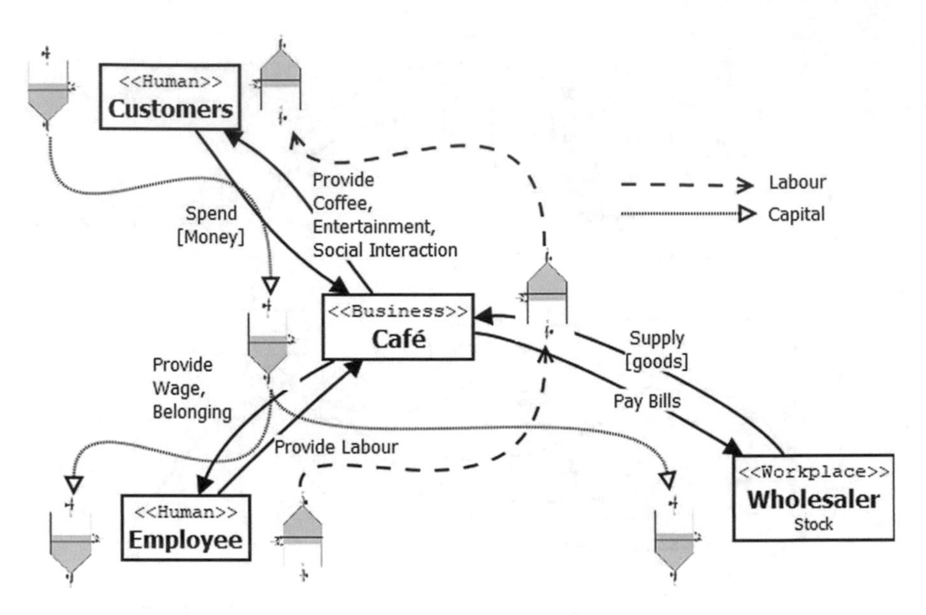

Fig. 9.4 A visualisation of the café example

Example—Home Town Café

We use one example to illustrate the combination of a homeostatic model and a circular economy. In this example, capital is exchanged for other things.

Imagine a café in your local town or city. This café offers various services to customers, from just a quick cup of coffee on-the-go to a complete experience with really comfy seating and great service. As a customer, you receive these goods, services and experiences that fulfil one or more of your needs in exchange for capital. In exchange you pay the café and drain capital from your *capital tank*, which in turn fills the café's capital tank. At the same time, an employee (or owner) at the café provides a service to you, labour. For this, the café pays the employee a wage and drains its own capital tank to fill the capital tank of the employee.

In turn, all these employees of the café are also customers at other places, just like you. This way, the capital, or money, keeps circulating from one person, shop or agent to another and back again. Thus "Money makes the world go round" while you visit a café and enjoy your drink.

An illustration of the scenario is shown in Fig. 9.4, where also the role of workplaces (e.g. wholesalers) is included to buy and sell stock for the café to operate.

9.3.2 Needs and Economic Activity

The ASSOCC model is based on needs of people to guide their behaviour as explained in Chaps. 2 and 3. The economic activities that people perform are also driven by the needs system. The most basic needs related to economic factors are *food-safety*, *financial-survival* and *Luxury*. As discussed in Sect. 3.7.4, working satisfies the financial-survival need as people gain money (wages) by performing this action. Food-safety and luxury are satisfiable by buying goods at essential and non-essential shops. Buying at non-essential-shops also gives a sense of *belonging*. On the other hand *risk-avoidance*, *financial-stability*, *autonomy* and *financial-survival* satisfaction are decreased by buying things at (non-)essential-shops.

The model consists of four daily time-steps and has a morning-afternoon-evening-night cycle. During this cycle, different actions are available to agents depending on the time of day. This also means that different needs are satisfiable at different points in time. The situations described in the previous paragraph are valid in the mornings. In the afternoon, evening and night, people can also *relax* at a non-essential-shop. At night people, can not gain *belonging* at a non-essential-shop as they are closed. More details on satisfiable needs can be found in Appendix C.

The needs model uses stochastic distributions in assigning the strength of needs to our population, different agents value the same things things differently. Some agents will value financial-safety highly, while other really like to live in luxury without having concerns for their financial well being. Thus, our agents will make different decision regarding how and when they will spend their money, even when they are in the same situation.

Note that it would be possible to use the combined needs of an agent as a kind of utility function that would allow for the agent to determine how much it would be willing to pay for resources or services. This could then lead to market mechanisms that drive prices for each resource. However, although prices for certain resources changed during the COVID crisis, this is not the focus of this chapter. Most of the consumer products that we are considering in this chapter are food and luxury things like clothes, etc. The changes in prices of these products have not changed significant enough to influence the whole economic condition of the society. This is influenced far more by the lockdowns and closures of shops, restaurants and pubs. Thus we decided to keep the model simple and not incorporate the market making mechanisms.

9.3.3 Supply Chain

One economic relevant aspect that got impacted by COVID-19 is the supply chain. Initially, Chinese factories closed down which led to shortages of goods in Europe. How a supply chain works is simple in concept but in reality these chains can be very complex. To get from a raw resource to a final product can take hundreds of steps in the complete supply chain.

In our model, we have taken a very basic but practical view on this part of the economic system. *Workplaces* produce goods, primarily for the local market. The goods from a workplace are bought by *essential shops* and *non-essential shops* to replenish their stocks. If shops run out of stock they can no longer provide services or goods to people. If workplaces should run out of goods, shops can no longer buy new stock. Thus, if workplaces have no workers producing goods (e.g. due to lockdowns or many workers being ill), people can no longer buy food once reserves run out. This gives us the basic supply chain of workers → workplaces → shops → consumers. This minimal supply chain gives us at least an impression how shops can suffer from both a lack of customers and a lack of supply due to the crisis. Also, if workplaces have excess production that isn't bought locally they can sell this outside the system, but mostly for a reduced price.

9.4 Economic Scenarios

With the economic model in place, we can develop various scenarios to analyse the consequences of a pandemic and measures from the economic perspective. In this section, we will discuss the various scenarios, their goals and how these are configured within the ASSOCC model.

For the analysis of the economic effects we have devised four scenarios:

1. No pandemic, life is as usual and provides a baseline for analysis.
2. A pandemic without any measures taken.
3. A pandemic with a lockdown to deal with the epidemiological consequences.
4. A pandemic with a lockdown and an unemployment subsidy from the government.

In the following sections we first discuss the key metrics that we use for the analysis. Then we will go into more detail to see what the different scenario's entail and set expectations on what we expect to see with regard to the key metrics.

9.4.1 Metrics for Analysis

For each of the scenarios, we can measure general things such as *infections* and the *Quality of Life*.[5] For the economic perspective, we can look at additional effects, such as the *financial position* of people and shops. This financial position entails to "can agents and shops buy everything that they need?". We can check this through the amount of people being in *poverty* or *shops being out of capital*. To measure possible positive effects we also look at the *capital of shops* and *capital of people* as it might increase more during a pandemic for certain groups. Do note that there are no ways for people or businesses to go into debt in our simulation.

[5] See Sect. 3.12 for the description of Quality of Life.

If we look at the whole economic system, we can use the *velocity of money* to give us an idea of the health of the current economy. With the velocity of money we look at how often money is changing ownership in an economy. The velocity of money is defined (as in [6]) as follows:

$$V_T = \frac{VT_T}{M_T}$$

where in a certain time-frame T, V_T is the velocity of money, VT_T is the total value of transactions and M_T is the amount of money in circulation. In this definition, we ignore economic phenomena such as inflation, because we have no inflation in our model due to the relatively short time scale of our simulations. Generally, an economy is seen as more healthy if V_T is higher. I.e. if there are more economic interactions preformed with the same overall amount of money.

We expect to see the velocity of money decreasing when a lockdown is in effect. People have less options to spend money while the amount of capital in the system stays effectively the same. It will be interesting to see how the velocity changes once a lockdown is lifted or the pandemic dies out.

Other aggregate metrics that are of interest are *goods produced in the system* and the *government capital reserve*. the government is the only agent that can go into debt, which is a reflection of reality in the model.

This gives us the following list of key metrics:

1. Number of infections.
2. Quality of Life.
3. Capital of people.
4. People in poverty.
5. Capital of shops.
6. Shops and workplaces out of capital.
7. Velocity of money.
8. Goods produced in the system.
9. Government capital.

9.4.2 Calibrating a Baseline

With our economic scenarios, we want to gain more insight in the effect of a pandemic and related measures on the economy. We do this by comparing our key metrics between different scenarios. This requires a stable and *realistic* baseline scenario. With realistic we basically mean that the baseline should adhere to basic observations from the real world as it functions in every day life, such as:

- People always have enough money to survive.
- Shops, workplaces, schools, etc. have enough money to pay wages.
- Shops are supplied with enough goods to operate.

- The government operates roughly break-even.
- Shops and workplaces don't have a monopoly and cause other shops or workplaces to go bankrupt.
- People go to their workplace to work.

Thus, in short, we assume there is a kind of balance in the economy that lets every agent survive even when they can have temporary shortages or slowly accumulate some capital. An other important aspect is the choice not to implement a real market making mechanic with supply and demand influencing prices; in principle, all prices are static.

This does leave us with a challenge: *money sinks*. These sinks are formed by agents, e.g. people or shops, that have more capital flowing in than out by being good at saving money or making profit. Shops have no motivation to spend this extra money and as a result hoard the money, basically extracting it from the system. People are driven by needs, thus, agents that have a high need to own money prioritise *not* spending money over spending money. Conversely, we see that some people really like to spend money, basically living payday to payday and are considered to be in poverty often.

During our balancing work, we learned that the sinks can drain most of the money from the system and lead to a complete stand-still in economic activity, simply because there is no money available to exchange. We have been calibrating the settings (see the next section for the particular settings) of the baseline scenario to avoid this effect of these sinks. We have also implemented three mechanics to reduce the impact of these sinks as it was both unfeasible and unrealistic to remove them completely:

1. The government is able to go into debt, thus providing infinite money with subsidies.
2. If people have more money, they spend more money at shops.
3. Workplaces can sell excess production outside the local system and add extra capital to the system.

Each of these mechanisms is realistic as they represent mechanics from the real world, where economies are never completely closed or the amount of available money being static. Mechanism one replicates the real world of government bonds and loans to acquire money to spend. Number two also reflects what happens in the real world where we generally see that people spend more in shops when they have more money available. The main motivation for mechanism three is that we saw that the workplaces seemed to be particularly sensitive to the sinks hoarding all the money as they were "at the end of the line" regarding the capital chain. With this mechanism they can have a small constant influx of money which makes them less dependent on the behaviour of the sinks in the system. In the real world, we can compare this to transactions with a neighbouring town or country which are not part of the local economy.

9.4.3 Generic Scenario Settings

With the assumptions and mechanics explained we can take a further look at our general settings for all the economic scenario's. All the settings in Table 9.1 that are related to prices and production are the result from the calibration based on the discussion in the previous section. As such the prices have no resemblance to prices in the real world. Basically we only model a part of the costs and expenses that people normally have and thus the wages and daily expenses are much tighter related in our model than in real life. The calibration is also heavily influenced by the choice of culture and household profile, see Chap. 8 for more on different cultures. We chose the United Kingdom for this as this profile has been thoroughly investigated with the work on the Oxford model, see Sect. 12.3 for more information. The UK profile comes with around 1000 agents in the simulation that nicely divide over the different age groups and households. The other generic settings for the scenarios found in Table 9.1 follow from these principles.

9.4.4 Scenario Descriptions and Expectations

Now that we have explained the fundamentals of all the scenario's we can describe the four scenarios and what they aim to show.

1. Baseline—no pandemic or infections
    a. This scenario is the baseline scenario for our analysis: no pandemic and life goes on as normal.
    b. Economy is stable and continues to operate within certain bounds.
2. Infections, no measures
    a. Infections occur.
    b. No measures from the government.
    c. Agents do not adapt to the pandemic.
    d. We expect a bad epidemiological situation.
    e. We expect the economy to stagnate when many agents get ill and/or die.
3. Infections, lockdown
    a. Infections occur
    b. Government enforces the a "10-5 lockdown". A strong lockdown is enforced once 10% of the population is infected and lasts until the numbers are down to 5% infected. This results in closing workplaces and non-essential-shops, among others. Workers get paid by workplaces, that still have money, if they can't work due to lockdown.
    c. The lockdown cuts most cycles of income, as depicted in Fig. 9.5.

Table 9.1 General settings for the economy scenario's

Description	Value
Culture profile	United Kingdom
Household profile	United Kingdom
Population size	~1000 people
Population distribution	28% children
	10% students
	38% workers
	24% retired
Workplaces	23 workplaces
Essential-shops	9 essential-shops
Non-essential-shops	9 non-essential-shops
Income	Worker wages: 12.5 per tick worked
	Retired pensions: 3.5 per tick
	Student scholarship: 3 per tick
Essential-shop rations base price	2.8
Non-essential-shop rations base price	4
Max stock at shop	500
Unit price of goods from workplace	2.9
Value modifier of exported surplus workplace goods	10% of price (= 0.29 per good sold)
Tax	Workers: 42%
	Essential-shops: 76%
	Non-essential-shops: 85%
	Workplaces: 60%
Starting capital	Workers: 60 retired: 50 students: 40
Goods produced per "work performed"	12
Productivity at home	50%
Percent of a wage paid by government	80%

  d. We expect agents that continue to have an income to get richer during the lockdown as they can't spend it on things they want.
  e. Agents without income as well as closed shops and workplaces might go bankrupt.
  f. Once a lockdown is lifted we expect a massive shopping spree as people have saved up money and want to satisfy their needs.

4. Infections, lockdown and wage subsidy

  a. Same as scenario 3.
  b. Government pays 80% of the wages for the workers of closed shops and workplaces once they are unable to.

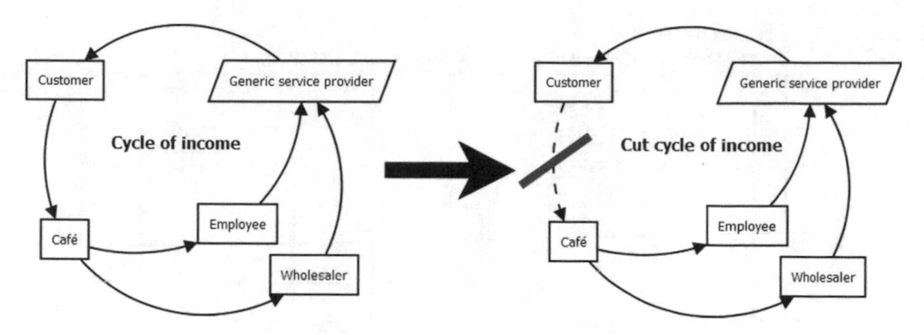

Fig. 9.5 The cycle of income cut by a lockdown

c. We expect less shops to go bankrupt and possibly quicker recovery after the crisis finishes.
d. We expect a bigger shopping spree after a lockdown as more people will continue to have an income without the options to spend it.

9.5 Results

We will present the results of the four scenarios by discussing the key metrics as defined in Sect. 9.4.1 for each scenario. All the scenarios have been run nine times to minimise any influence of some incidental values of stochastic parameters on the overall results. Where appropriate, we also analyse the variance between runs with standard deviations or min/max run values, this will be represented in our graphs as areas around the smoothed lines. We have used the same 9 seeds for each scenario, this way we have identical starting conditions in all four scenarios. The runs lasted for 420 ticks, which is equal to 105 days or 15 weeks. Figures 9.6 through 9.25 display the graphs of the key metrics over time in our runs of the simulation. These graphs are a subset of all the graphs that we created and have been selected as they are deemed to be the most informative. Because the results can best be understood when we compare them between the four scenarios, we present the results ordered on the key metrics rather than per scenario. We will first give the brief results of the four scenarios and discuss the results in more depth in Sect. 9.6.

9.5.1 Number of Infections

1. **Baseline**
   In Fig. 9.6, we can see the flat line representing zero infections (as it should be) based on the scenario definition.

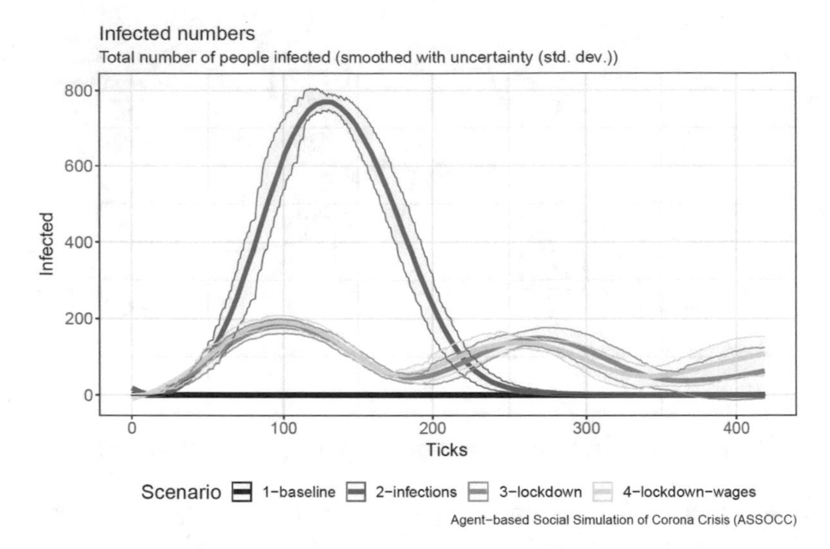

Fig. 9.6 Number of infected in the four scenario's

2. **Infections—no measures**
   We see a steep rise in infections after the first infections occur. Once everyone is infected and has progressed to either being immune or dead, the infections steeply drop. We see the assumption that people are immune after surviving an infection clearly expressed in this line as no reinfections occur.
3. **Infections—lockdown**
   Compared to the no-lockdown scenario, we see that the first spike in infections is about $\frac{1}{4}$th lower. But on the other hand we see a wave pattern in the number of infections. This is due to the repeated lockdowns that are in effect when the the number of infections rises and are lifted when the number is low enough.
4. **Infections—lockdown and subsidy**
   We see the same wave pattern as in the lockdown scenario. The biggest difference is that after the first lockdown it seems that the infection phases are a bit shorter.

9.5.2 Quality of Life

1. **Baseline**
   In Fig. 9.7, the QoL indicator stays very stable and high throughout the whole simulation indicating that most agents can get a satisfaction level overall of around 80% in normal life. Also, the difference between people with the highest and the lowest QoL stays stable.
2. **Infections—no measures**
   We see a major decrease in QoL during the peak of the amount of infections.

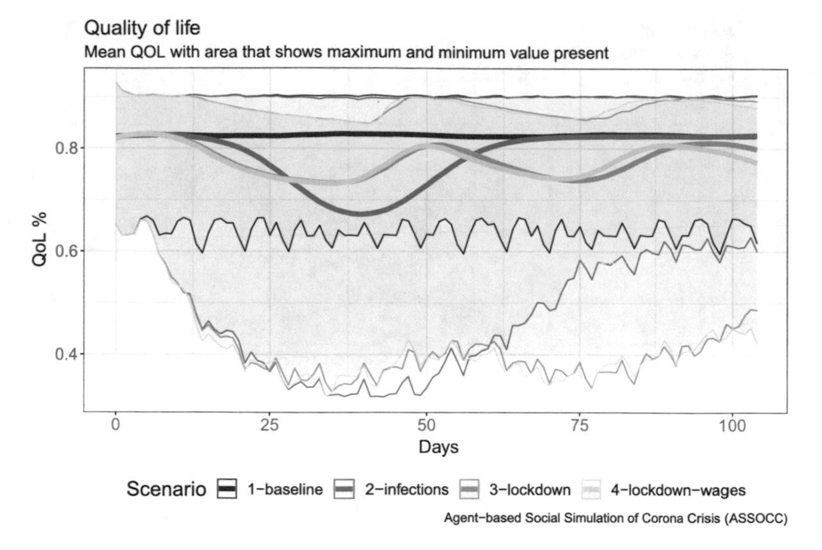

Fig. 9.7 Quality of life

During this period the difference between people's QoL increases a lot as well: the highest QoL level doesn't decrease at all while the lowest levels goes down from 0.65% to 0.35%. After the infections the QoL slowly returns to the same level as before the pandemic.

3. **Infections—lockdown**
   The quality of life during a lockdown is lower than in the baseline scenario. We also see the average QoL decreasing earlier than in the no-lockdown scenario. The lowest point of the average QoL during a lockdown is not as low as during the no-lockdown scenario, but the QoL of the people with the lowest QoL is roughly the same as in the no-lockdown scenario. We also see that the people with the highest QoL are experiencing a lower QoL compared to both the baseline and the no-lockdown scenario's. This shows that everyone is impacted by the lockdown, while in the no-lockdown scenario some people are not affected in their QoL.
4. **Infections—lockdown and subsidy**
   The QoL matches the lockdown scenario both in min and max and in the average. We also see shorter lockdown periods reflected in the QoL graph.

9.5.3 *Capital of People*

1. **Baseline**
   In the graphs in Figs. 9.8, 9.10 and 9.11 we see that the baseline scenario is stable. At most, we see that workers (Fig. 9.11) get steadily richer.

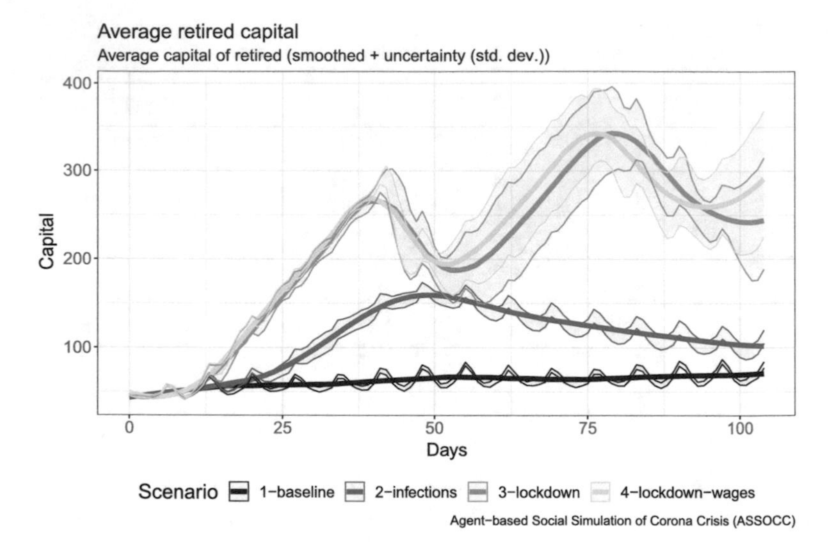

Fig. 9.8 Average retired capital

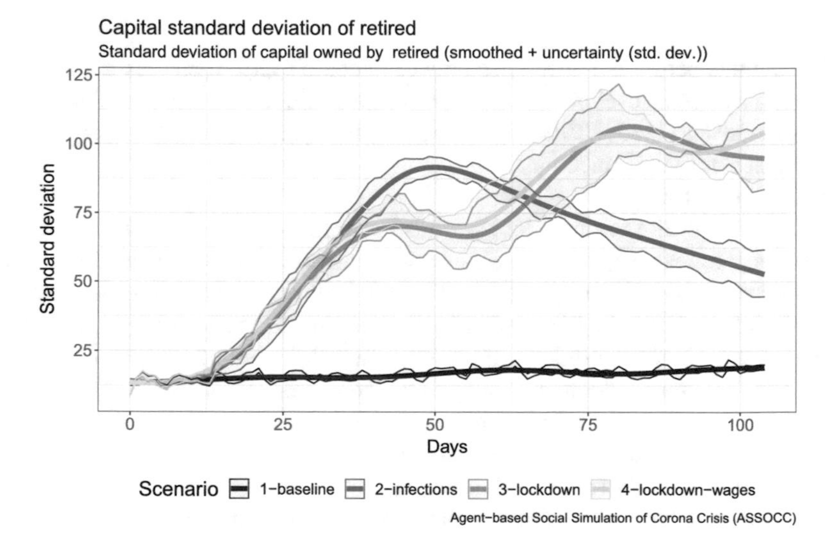

Fig. 9.9 The standard deviation of the retired's capital

2. **Infections—no measures**
   Compared to the baseline scenario we see that during the infection period people get richer on average. This is due to loss of options to spend money. If you are sick in the hospital, or at home, you can not go out to do shopping and spend your money. As retired and students have a steady income this effect is very visible in these groups, as seen in Figs. 9.8 and 9.10. We also see the inequality in the society increase where some agents get richer while others don't. This is

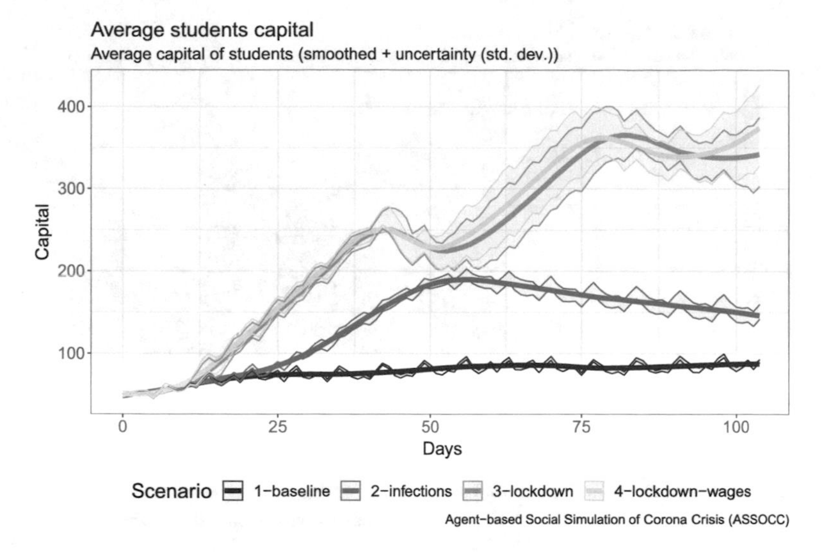

Fig. 9.10 Average student capital

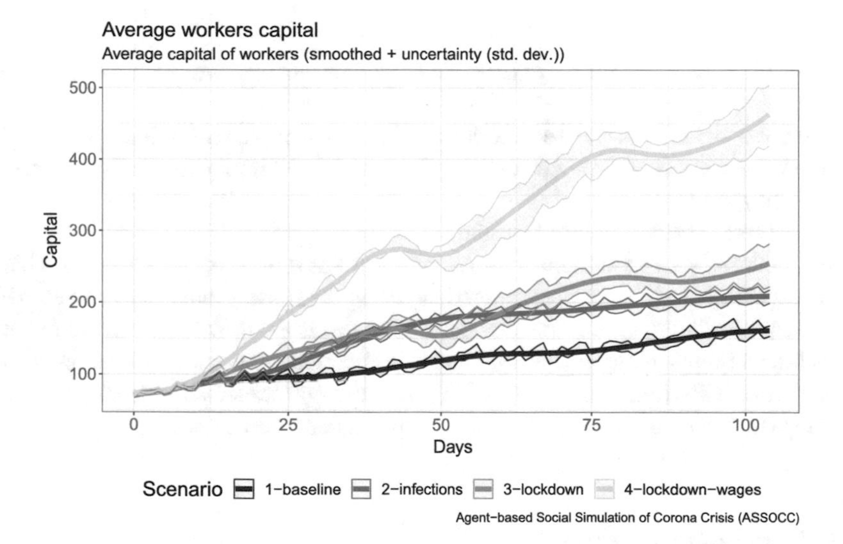

Fig. 9.11 Average worker capital

especially prevalent within the group of retired people, as shown in Fig. 9.9. In this figure we see the increase of the standard deviation during the infection period and then slowly going back to the same levels as before the pandemic.

3. **Infections—lockdown**
   In this scenario we see a clear effect of the lockdowns in the same sort of way as the pandemic has in the infection scenario. People can no longer exchange their money for various needs as non-essential shops are closed, despite their income

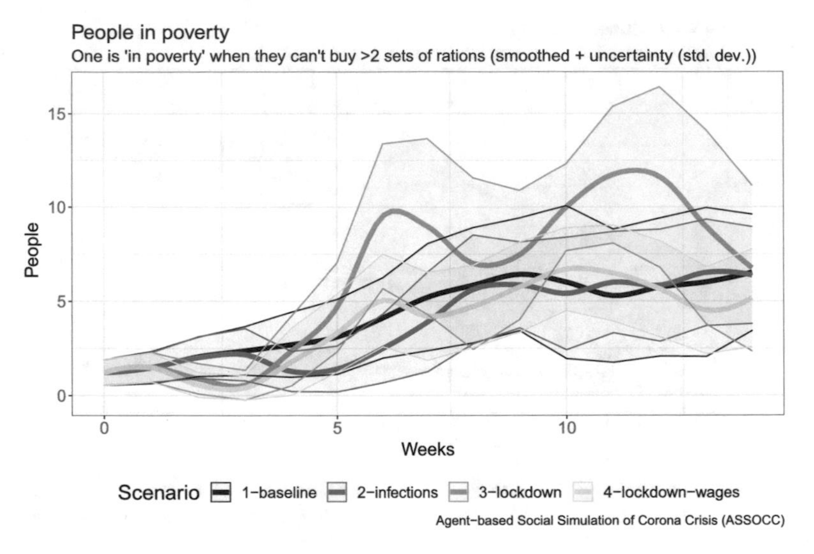

Fig. 9.12 Poverty of people

remaining the same for retired and students. For workers we see the same effect, but being less strong.
After the lockdowns we see a sudden decrease in capital. This reflects that people, when the non-essential shops open again, spend their money to satisfy their needs that they couldn't before.

4. **Infections—lockdown and subsidy**
For both retired and students this scenario mimics the lockdown scenario. This is as expected as we have not changed anything for these groups. We do see a big change for the workers. The average capital develops in the same wave pattern, but the amount of capital gained is a lot higher when compared to the lockdown scenario. This is a clear effect of the government subsidy for workplaces and shops that can no longer pay the wages of workers themselves.

9.5.4 People in Poverty

1. **Baseline**
In Fig. 9.12, during the first $\frac{2}{3}$ of the simulation, we see that the amount of people in poverty slowly rises to ∼0.5% of the population. This is due to the fact that all agents of the same type start with the same amount of capital. This leads to agents that have an above average need for luxury to spend more money and balancing themselves around the poverty line. This stabilises after some time due to the influence of their other needs.

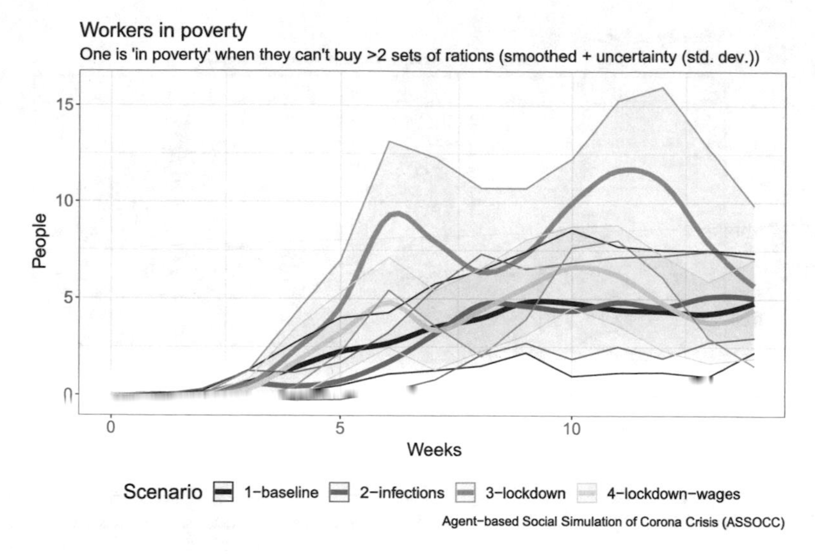

Fig. 9.13 Poverty of workers

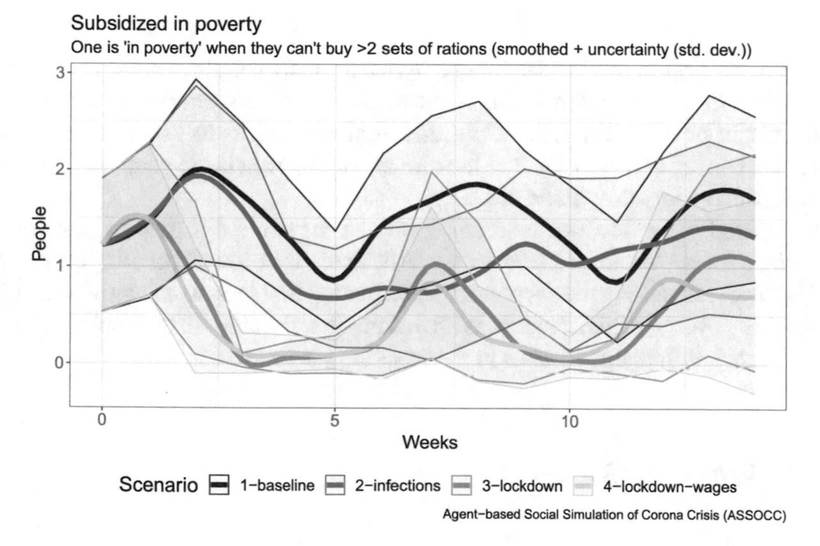

Fig. 9.14 Poverty of subsidized (students and retired)

2. **Infections—no measures**
   Poverty develops comparable to the baseline scenario to ~0.5% of the population. We do see a difference during the peak of the amount of infections where the poverty is going down or staying stable.
3. **Infections—lockdown**
   During the lockdown period we see more workers in poverty compared to the baseline and no-lockdown scenario's, but actually less students and retired people

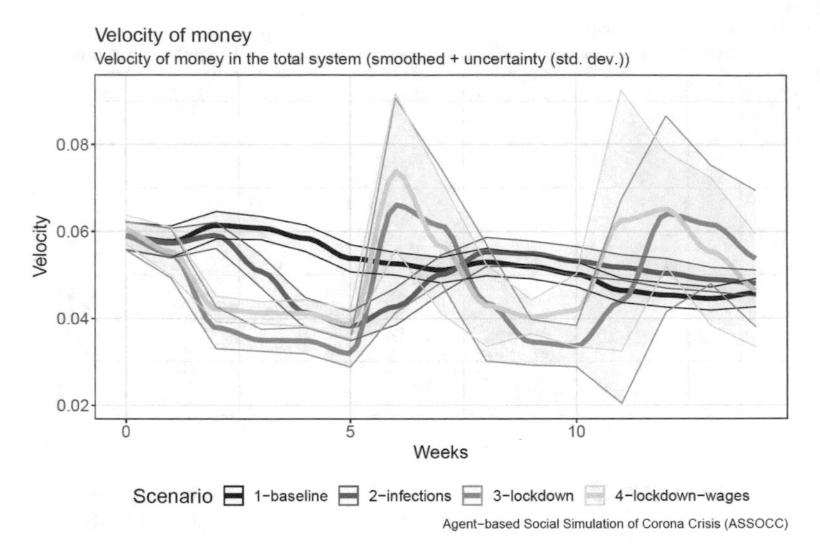

Fig. 9.15 Velocity of money

in poverty (Fig. 9.13). We see roughly twice as many people in poverty *right after the first lockdown is lifted* compared to the baseline scenario. The amount of people in poverty decreases a bit after that and seems to go back to baseline levels. During and after the 2nd lockdown we see even more people in poverty.

4. **Infections—lockdown and subsidy**
   Compared to the lockdown scenario we see the same dynamic in poverty, but being a lot more damped (Fig. 9.14). Where the lockdown scenario has roughly 1% in poverty during the shopping spree after the first lockdown, we have roughly 0.5% in poverty in this scenario. With this number the amount of people in poverty is roughly at the same level as in the baseline scenario.

9.5.5 *Velocity of Money*

1. **Baseline**
   In Fig. 9.15, we see the velocity of money slowly decline during the run from ∼0.06 to ∼0.05.
2. **Infections—no measures**
   The velocity of money seems to have the inverse trend of the infection rate. When the infections go up, the velocity goes down and the decrease peaks at ∼30% of the baseline velocity. After the pandemic the velocity of money restores itself to the level of the baseline and continues with the baseline trend.
3. **Infections—lockdown**
   The velocity of money seems to have a direct relationship with the lockdowns. The

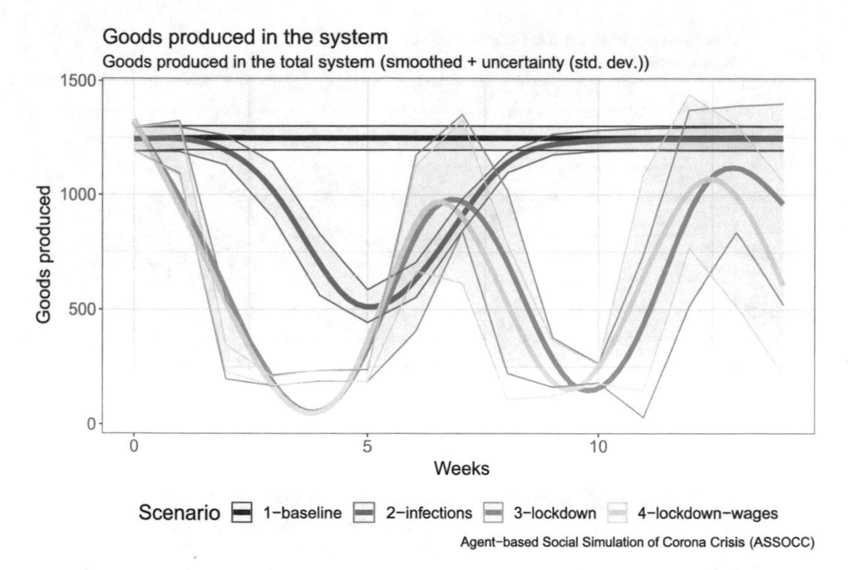

Fig. 9.16 Goods produced in the system per week

velocity instantly goes down at the beginning of a lockdown and stays relatively constant throughout the lockdown, this in contrast to the clear peak in the no-lockdown scenario. The level is lower than the lowest velocity during the no-lockdown scenario, roughly at ~50% of the baseline scenario velocity. Once a lockdown is lifted we see a sudden increase in the velocity, roughly up to ~135% of the baseline velocity, but the increase quickly declines after that. The next lockdown comes into effect before the velocity has stabilised. During the 2nd lockdown the velocity stabilises at the same level as the first lockdown, followed by the same sudden increase. We do see a less sudden change in during the 2nd lockdown, which might be explained by the different timings of the follow up lockdowns of the various runs.

4. **Infections—lockdown and subsidy**
   The velocity has the same trend as the lockdown scenario but stays about 15% higher during the lockdown period. With this it is at same velocity as the no-lockdown scenario is at its lowest. Also, the peaks after the lockdown is higher compared to the no-wage-subsidy lockdown scenario.

9.5.6 *Goods Produced and Stock in the System*

1. **Baseline**
   The goods production stays stable as seen in Fig. 9.16. In Fig. 9.17 we see that the stock for non-essential shops stays stable. This also holds for essential shops.

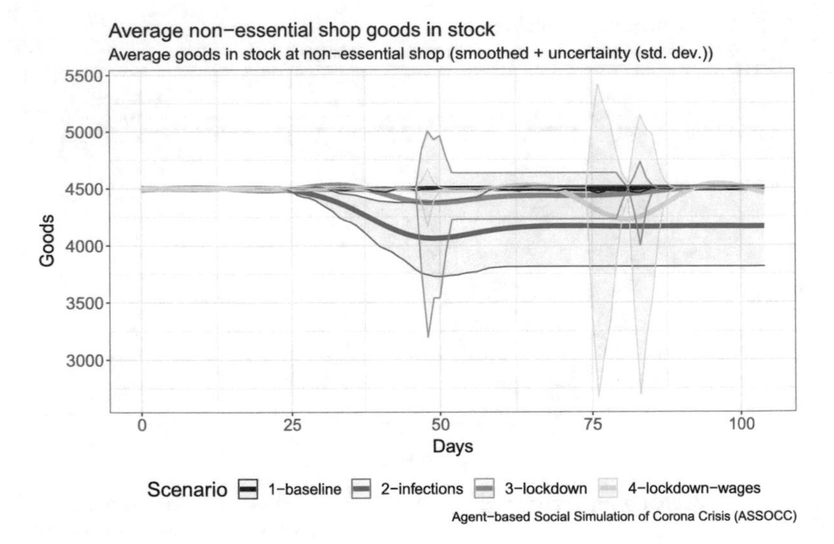

Fig. 9.17 Non-essential shops goods in stock

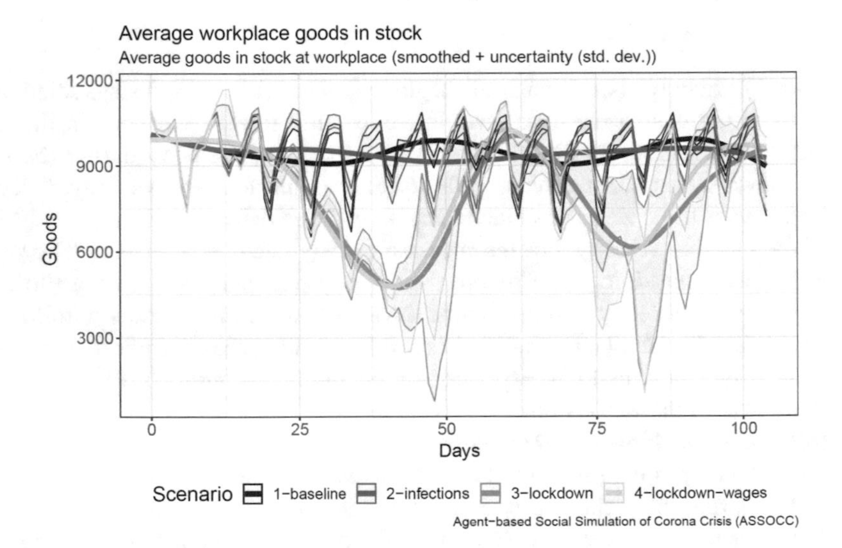

Fig. 9.18 Workplaces in stock

In Fig. 9.18 we can see the weekly cycle of people shopping in the weekends. As they shop more in the weekends, the workplaces need to supply more goods from their warehouses. In the days after the weekend the reserves in the warehouses of the workplaces is replenished.

2. **Infections—no measures**
   This graph is very comparable to the velocity of money: production goes own ~40% during infection period, but is restored to pre-infection numbers after that.

We do see that one some non-essential shops did not survive the pandemic in Fig. 9.17 as they no longer restock and cause the total stock to go down. The pandemic seems to have no real impact on the amount of stored goods at workplaces, indicating that the amount of sold goods is roughly the same as the decrease in production.

3. **Infections—lockdown**
   The production is heavily influenced by the lockdowns. People work from home and have less productivity when working from home, thus reducing the production. In combination with all the people getting sick it goes down to about 25% of the normal production. In Fig. 9.18 we see that in some scenario's the average stock of workplaces almost reaching 0. This most likely means that some warehouses actually have no stock left and can not supply the shops.
   This explains the sudden decrease in stock in Fig. 9.17 for some runs around day 50: workplaces can not supply then new stock, so they run out of their own stock. The timing of this spike around day 50 also coincides with the lifting of lockdown and the sudden increase of sales. Once the lockdown is lifted we see that production is quickly restored to the baseline level of production. The warehouses of the workplaces need some time to be replenished, but the shops quickly go back to healthy levels of stock.
4. **Infections—lockdown and subsidy**
   We see no difference in the production of goods compared to the lockdown scenario, other than the earlier mentioned stretching of the waves. For the level of stocks for non-essential shops we the amount of sales increasing around days 75–80, thus also the amount of stocks declining for a few days.

9.5.7 Capital of Shops

1. **Baseline**
   In Figs. 9.19, 9.20 and 9.21, we see that shops and workplaces are doing good on average and make decent profit. The amount of capital keeps rising as the shops and workplaces have no other options than to use this capital other than pay their workers, pay taxes and buy goods. They have no options to invest or otherwise use their capital.
2. **Infections—no measures**
   Both essential and non-essential shops start on the same trajectory as the baseline scenario. During the infection phase the trajectory of the essential shops becomes a bit steeper: the shops make more profit. This new trajectory remains for the rest of the simulation, thus the essential shops continue making more profit than in the the baseline scenario. This can be explained by the fact that people have more capital on average in the infection scenario. As they have more capital, they spend more money at the shops, increasing the profit of the shops. The non-essential shops get hit during the infection phase and see their available capital going down.

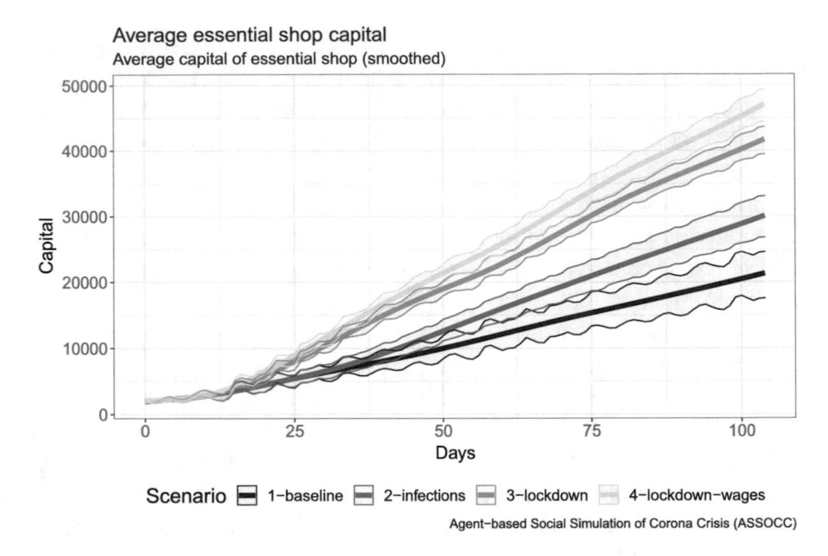

Fig. 9.19 Essential shops capital

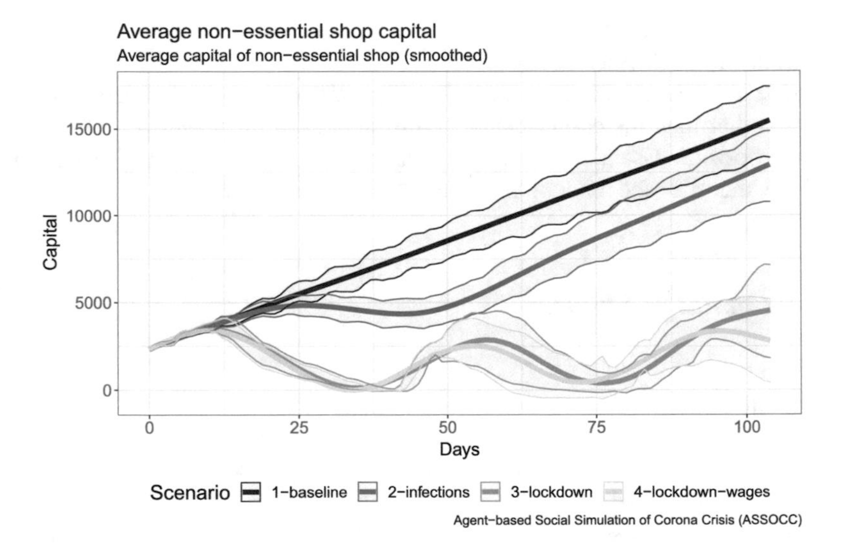

Fig. 9.20 Non-essential shops capital

This is due to the fact that agents try to shop less to minimise risk of infection and satisfy a minimal need for luxury.

3. **Infections—lockdown**
   Essential shops stay open and make more profit than the baseline and no-lockdown scenario. Non-essential shops are closed and lose money as they continue to pay wages to their employees. Once the lockdown is lifted they suddenly have an immense amount of sales as people have money saved up and have a need to buy

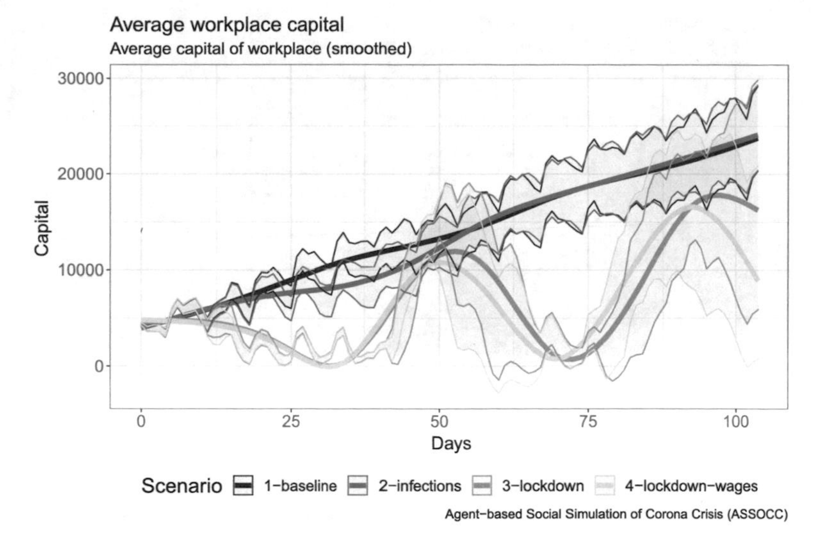

Fig. 9.21 Workplaces capital

things from the shops. This pattern repeats itself over the next lockdowns, but the variation in the amount of capital that they have increases a lot with more lockdowns.

For the workplaces we roughly see the same as they produce a lot less during lockdowns but still pay their workers. But where the non-essential shops don't seem to fully recover after a lockdown, the workplaces do seem to do this. This can be explained by the sudden higher demand and the better margins when selling locally vs selling globally.

4. **Infections—lockdown and subsidy**
   We see almost no difference other than the essential shops doing even better than in the lockdown scenario. The non-essential shops behave the same as in the lockdown scenario as they need to page wages until they can't.

9.5.8 Shops and Workplaces Out of Capital

1. **Baseline**
   In Figs. 9.23 and 9.24 we see that non-essential shops and workplaces do not run out of capital Essential shops seem to be having a harder time as on average ∼0.5 of them run out of capital as seen in Fig. 9.22. These figures are using weekly averages but when we look at day numbers we see that during the weekends shops always make enough money to build up some capital. We see a big difference, up to 3 times in amount of essential-shops out of capital, between the average

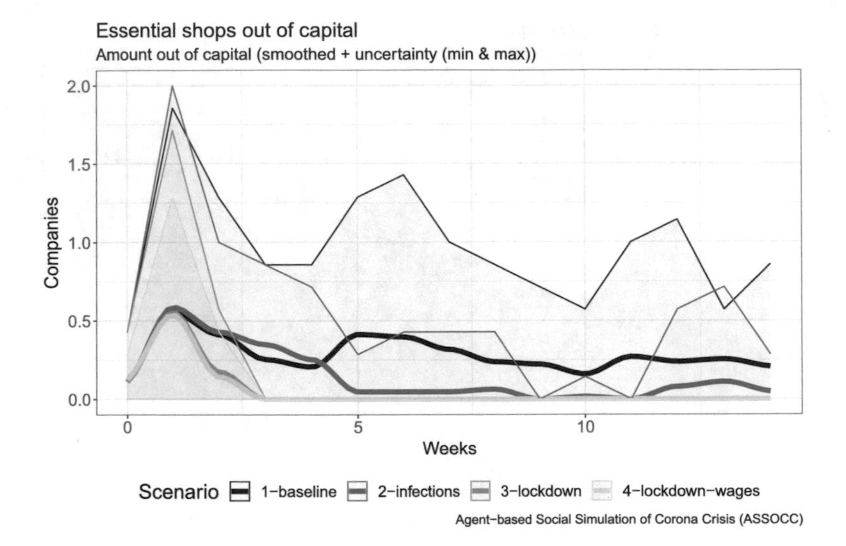

Fig. 9.22 Essential shops out of capital

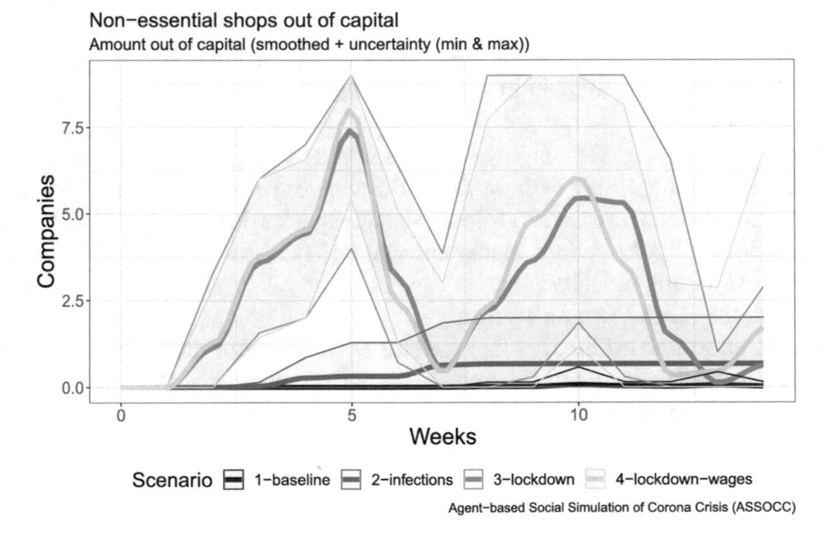

Fig. 9.23 Non-essential shops out of capital

and the maximum amount of essential shops out of capital. We also saw a high standard deviation, thus there seems to be a high variance in this metric between runs.

2. **Infections—no measures**
The trends we see here are comparable to the capital situation in the previous section. The essential shops are doing better than in the baseline scenario, having less shops being out of capital. The non-essential shops are a bit different as by

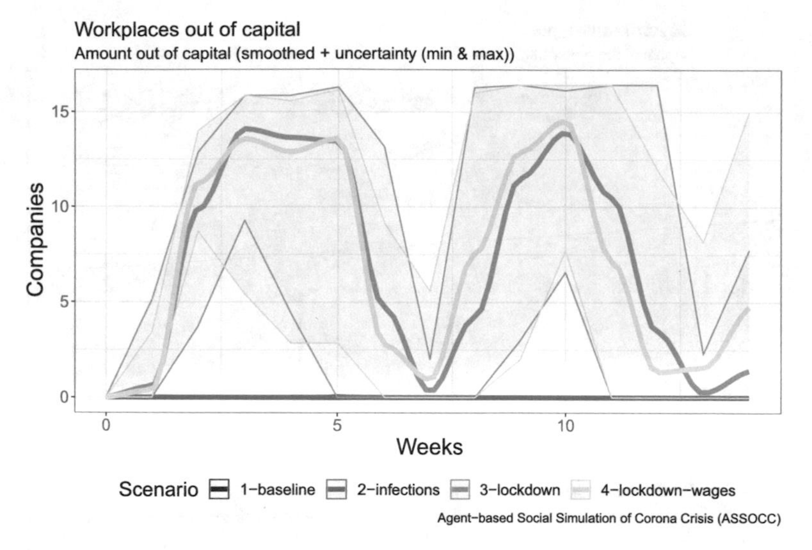

Fig. 9.24 Workplaces out of capital

the end of the infection phase one or two shops are permanently out of capital and remain closed.

3. **Infections—lockdown**
   Essential shops are booming and none of them run out of capital during or after the lockdowns. As said the non-essential shops have a harder time during the lockdown as most of them (we have 9 in total, see Table 9.1) run out of capital by then end of the lockdown. It also shows in Fig. 9.23 that on average the 2nd lockdown has roughly the same effect, but the "max cases" line shows that in some runs we have a lot more non-essential shops out of capital. Some non-essential shops didn't have enough reserve to survive the lockdown and remain closed after the lockdowns.
4. **Infections—lockdown and subsidy**
   We see the same behaviour as in the lockdown scenario for both the shops and the workplaces.

9.5.9 Government Capital

1. **Baseline**
   In Fig. 9.25, we see that the government is operating with a small deficit and is slowly losing money.
2. **Infections—no measures**
   During the infection phase the government income decreased, but after this it stabilises again and seems to slowly increase.

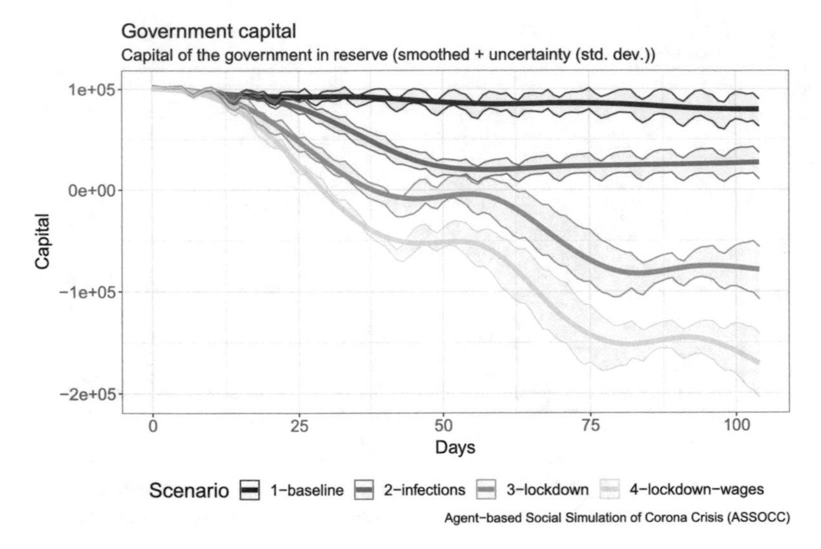

Fig. 9.25 Government capital in the four scenario's

3. **Infections—lockdown**
   During the lockdown the government is running with a bigger deficit than during the no-lockdown scenario. After the lockdown it is recovering a bit, but before it has time to show a clear trend the next lockdown is enacted. At the end of the 2nd lockdown we see the same pattern of slowly recovering but going going down again as another lockdown is enacted.
4. **Infections—lockdown and subsidy**
   The government capital has the same dynamic as in the lockdown scenario where we see a big deficit during the lockdowns. The only difference is that deficit is bigger than the deficit during the lockdown periods. At the end of the simulation the difference between the baseline the lockdown + wages scenario is about 3.5 times bigger than between the baseline and the no-lockdown scenario.

9.6 Discussion

Now that we have discussed the separate metrics we can look at the overall picture of our scenario's.

The first result is the baseline scenario itself. This is a functioning economic system that continues to function during the simulation runs. Workers work, receive salary, spend this to buy food, shops use the money to buy new stocks from workplaces and workplaces pay wages to workers that work. The expected behaviour of sinks and avid spenders is especially visible with the workers where the standard deviation in average worker capital continues to increase during the simulation. It is interesting to note that we do not see real change in the QoL indicator during the simulation, indicating that the increase of the amount of people in poverty does not affect their QoL.

In the second scenario people get infected while no measures are taken. Here we see the biggest effect of people who are sick. They don't work, hence the decrease in production, and they stay home unless they need to buy food to survive. When people are sick they continue to get paid their wage or allowance but don't spend it. This results in people having more money after the infection phasc and, as a result, spend more money in shops. This explains shops making more profit after a pandemic in our model.

We do see that this the reserves people have build up go down over time. If we would run this simulation for a longer period we expect the profit to return to the baseline trajectory. One big effect that we see is that some non-essential shops don't have enough capital in reserve to survive the infection phase. They remain open and pay full wages but sell less goods, thus running at a deficit. This results in a situation where some of them end up spending all their capital on paying wages without being able to buy new stock to sell after the pandemic, and going out of business completely. Also, the government takes a one-time hit due to loss of taxes while expenses stay the same. In reality this could be more severe as we would expect that the government needs to spend more money on healthcare during and after a pandemic, for example on vaccines.

The third scenario has multiple lockdowns, as we can clearly see in Fig. 9.6. This causes a wave pattern in most of our metrics. The wave seems to be getting weaker for every next wave, most likely as the amount of people that are immune is increased with every resurgence of the virus.

These lockdowns create an interesting economic dynamic. People can not spend money at non-essential shops during a lockdown and are forced to save up money as they have way to spend money to fulfil some of their needs. But once a lockdown is lifted they have two things: (1.) highly unfulfilled needs, and (2.) a lot of available money. We see that consumers start throwing money at the non-essential shops to fulfil these needs which gives the non-essential shops a big boost in revenue. Also, when a lockdown is lifted the need for food-safety is, relative to the other needs, low as essential shops, with food, have remained open. Agents choose to prioritise to satisfy their other needs over food-safety, thus spending most of their money. This causes for more people to be seen as "in poverty" due to way we measure poverty: if an agent doesn't have enough money to buy two sets of rations it is seen as being in poverty. So, due to the lockdowns, we first see people getting richer, but once the lockdowns are lifted they (temporarily) become impoverished.

This dynamic also affects the supply chain. As workplaces produce less and essential shops sell more, the stock reserves of the workplaces is going down. In our current simulation the workplaces have enough in stock to be able to supply the essential shops during the lockdowns, but the workplace reserves are more than 50% depleted. So if the lockdown would last twice as long, or the initial stock in reserve would be halved, the essential shops would run out of goods to sell, possibly causing them to go bankrupt and maybe even a famine in the town.

The point of economic reserves is also relevant for the non-essential shops and their workers: while they still have money they can pay the wages during the lockdown, but once they run out of capital the wages stop being paid. We see this reflected in the higher amount of workers being in poverty, despite the average workers capital being higher than the baseline and no-lockdown scenario. Non-essential shops don't go bankrupt as their stock is frozen during the lockdown and they can instantly resume operating once a lockdown is lifted in our current model. People spending a lot of money to satisfy their needs helps the non-essential shops to kick-start them and restore some of their financial reserves to pay wages and buy new stock.

In the last scenario, we primarily see a difference in the workers capital, essential shop profit, the velocity of money and government reserves when we compare it to the lockdown with no subsidy scenario. This difference show that workers have more money available and thus spend more. The other parts of the system behave the same as in the basic lockdown scenario.

9.6.1 Research Questions

Above, we have discussed the results for each scenario in relative isolation but our main goals are of a broader sense. Let's recall our issues to investigate in this chapter:

1. Investigate the effect of a pandemic on our economy.
2. Investigate the effects of measures taken to stop the pandemic on our economy.
3. Test various economic measures that could mitigate the negative effects stemming from the pandemic.

In the rest of this section we will structure the discussion based on these general goals. In order to discuss these goals we first want to get back to some of the basic principles of the ASSOCC model. The economy is driven by the needs of our population. Both performing work and buying things is driven by the needs, not a static rule that people have to work and buy food. Where they spend their money is also driven by the needs system. As discussed in our baseline scenario, we see that this sets the whole consumer economic system in motion as they need money and/or require goods to perform the actions. If peoples' needs change, e.g. during a pandemic, so does the way the economy behaves.

Another aspect of the ASSOCC model that is important for the economy is the supply chain that can have a profound effect on the ability of the economic system to satisfy needs. As discussed while analysing the third scenario in Sect. 9.6, when workplaces have less workers, or workers working from home and being less productive, they produce less goods to sell to shops. If shops keep selling more goods than the workplaces produce they will run out of stock, and after a while the shops will do the same. When this happens people can no longer satisfy their needs by shopping and, among other, their food-safety and luxury are unsatisfied. Thus we see that, although we did not implement the economic aspect by adding a money need, the possession of money does influence indirectly how needs can be satisfied and the satisfaction of needs influences where money is spend and whether money can be earned.

9.6.2 Investigate the Effect of a Pandemic on Our Economy

We discussed most of the effects of the pandemic in Sect. 9.6 with our overall analysis of the second scenario. Here a pandemic causes people to get sick and possibly die. People who are sick (or dead) will not perform work. This causes a clear shortage of labour for workplaces, shops, hospitals, schools, etc. This can cause a supply chain problem, such as closure of shops or lack of stock to sell.

In our model, shops continue to operate if their employees are sick and at home, which could be compared to hiring temps in the real world. But we do expect that the effects of the large amount of sick people will be more severe than our simulations show due to these complicating factors in the supply chain. We could also expect this to happen with the healthcare system. If no capacity is available, e.g. due to sick doctors, we will expect to have a higher mortality rate and, thus, a bigger effect on the economy.

Another effect that we see is that sick people only shop for the bare necessities and spend less money on other things. We see this represented in the decrease of the velocity of money, as seen in Fig. 9.15, and the decreased profit of non-essential shops. Also the dynamic of people getting richer and spending more once the pandemic if over is reflected in this graph.

After the pandemic, the system recovers in our simulations. But this is also where our uncertainty of the economic effects increases. We assume, for instance, that people who do not work during the pandemic can instantly resume working at their old job. We also assume that shops and working places don't close down indefinitely after running a deficit for a while, they don't really go bankrupt. Although this seems a severe limitation of the model it is also a very simple way to model that some companies might go bankrupt (seize to operate), but new ones will be founded as soon as the opportunity arises. In our model these new companies are simply the old companies that resume operation. For longer term simulations it is an interesting avenue for further expanding the economic model: how to deal with closing establishments, letting them go bankrupt and fire all employees and once the pandemic is over founding new businesses and finding new employees (a job market?) and possibly having some shift in types of economic activities. As we do not consider these macro-economic effects in our models we leave this for future work.

9.6.3 Investigate the Effects of Measures Taken to Stop the Pandemic on Our Economy

The main measure investigated in the scenario is the 10-5 lockdown. As said before, this means that a strong lockdown is enforced once 10% of the population is infected and lasts until the numbers are down to 5% infected. This results in less people getting sick at the same time and spreading the infection of the total population over a longer period of time. The resulting dynamic is a wave pattern that slowly dies out, like the waves resulting from dropping a stone in a still pond.

This leads to a very different economic dynamic as we have abrupt changes instead of gradual effect without the lockdowns. Goods production goes down instantly, non-essential shops close and can no longer sell things, the velocity of money goes down and the government loses money by not receiving taxes. People with pensions or other government financing get richer as they can't spend their money. Shops affected by the lockdown run out of capital and can no longer pay their workers. Essential shops on the other hand make a lot more profit by being the only open shops and having customers with more disposable money. Once the lockdown is lifted we see that people go on a shopping spree due to their depleted needs, which has a small reinvigorating effect on the economy. But this also leads to a massive amount of contacts at the non-essential shops and could undo part of the epidemiological effect of a lockdown by creating a potential super-spreader event. This is illustrated in Fig. 9.26 showing the amount of contacts people have in the lockdown with wages scenario.

An interesting connection between the economic model and the epidemiological model is the duration of a lockdown. In our simulations the first lockdown lasts roughly 5 weeks and the next lockdown starts roughly 3 weeks after the first is lifted. During the beginning of first lockdown non-essentials shops have a reserve to pay wages, despite being closed. Most of them run out of money to pay the wages during the lockdown, which means that the workers of those shops need to use their own financial reserves to buy food at the essential shops. We can clearly see this in the increase of workers that end up in poverty (see Fig. 9.13) and the spike when the lockdown is lifted. Unfortunately the three weeks of no-lockdown isn't enough to replenish the reserves of the people or shops and we see that the effects of the second lockdown on the amount of people in poverty is stronger than the first.

An other metric that we track is the quality of life. During the lockdown period the average quality of life goes down at the start of the lockdown, but seems to stays stable at that level during the rest of the lockdown. This level of quality of life is higher than the quality of life level during a no-lockdown infection. That the quality of life stays stable also shows that the increased number of people in poverty only have a small to no effect on the average quality of life. It seems that the other effects of the lockdown have a bigger impact on the quality of life.

This gives us interesting things to consider when talking about lockdowns and the economy. Longer lockdowns have a bigger effect on the financial position of shops, workplaces and people, which is directly related to their reserves in capital

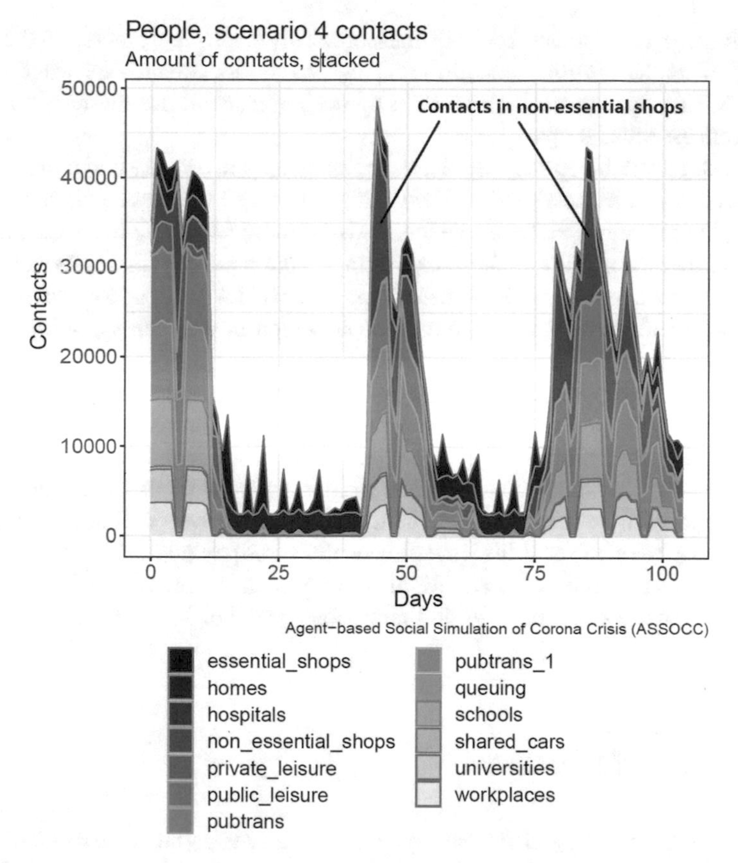

Fig. 9.26 Contacts in scenario 4 showing a big amount of contacts in non-essential shops after a lockdown is lifted

and goods. It also seems that repeated lockdowns have a big effect on the economy when the time between lockdowns is not long enough to fully recover. So on one hand we would prefer longer lockdowns to prevent multiple lockdowns (wave effect) with a lot of potential damage to the economy, but on the other hand one would like to have shorter lockdowns to prevent reserves from running out causing a lot direct harm to the economy.

9.6.4 *Test Various Economic Measures that Could Mitigate the Negative Effects Stemming from the Pandemic*

We investigated the effect of paying wages by the government of closed shops once the shops run out of capital to pay wages. The effect of this is that workers have more money available, which they in turn spend at the essential shops who seem to be the main other benefactors from this measure. This is also represented in the higher velocity of money in this scenario compared to the lockdown—no wage subsidy scenario. This does result in more debt by the government despite receiving more taxes from essential shops.

The extra available money also leads to less people in poverty and a bigger shopping spree once the lockdown is lifted. This could lead to an even bigger negation of the epidemiological effect of the lockdown as more people go to shops and cause super spreader events. Which is an interesting finding: paying (more) money to people who can't work will enable them to do more activities once the lockdown is lifted and brings them in contact with more people which in turn can spread the disease more effectively.

Other measures have not been investigated due to time constraints. But one could think about implementing measures to prevent shops from using all their reserves such that they have more leeway once the measures are lifted. We also realise we have effectively implemented a job security measure as workers are not being let go from their jobs. We expect that the quality of life of workers will decline further if they can lose their job and the government does not pay a subsidy or their wages to prevent them from being fired. We also expect that the recovery of the economy will be slower as people need to find new jobs instead of instantly resuming being productive.

9.6.5 General Discussion

Taking in all our results and discussions so far it seems to come down to a trade-off of health and lives versus economic impact. In our simulations, the repeated lockdowns (wave effect) are more harmful to the economy than just letting a virus spread in a natural way. Before we translate this result to the real world, we have to stress that there are many things that we haven't taken into that can influence our findings. For example:

Long term effects on health During the course of 2020, it became clear that quite some people who have been cured from COVID-19 have long lasting side-effects. These impacts their productivity and might cause additional expenses for a long time to come.

Reinfection We assume, that reinfections are not possible. But by now we have seen instances where this has happened. If this becomes a bigger issue the idea of "just letting everyone get infected and we're done" won't hold and the economic difference between no-lockdown and lockdown could be very different.

Productivity at home We assume, that people are 50% productive at home. We know, some people are more productive and some are less, heavily depends on their needs, family situation and type of job. Having empirical data on this would improve the realism of the simulations.

Public sector and the economy The only part of the public sector included in our economy model is the government as a financial institution. Elements such as public transport, healthcare & schools (other than paying wages) or many other public elements that are critical for the economy are not included for economic analysis.

Long term effect of loss of life on the economy In our model, people who die only seize to be consumers and are easily replaceable at their workplaces. In reality, they also take a lot of knowledge and know-how with them and are not as easily replaceable. How would this affect the economy?

Vaccines or other cures We have not implemented a cure for the disease, yet this could be compared to the increased number of immune people in the simulation. But the way one would vaccinate a population could affect the economy. Would you first want to vaccinate certain sectors of the economy? Or certain age groups? Or even certain geographic locations? All these choices will most likely have different economic effects.

9.7 Conclusion

In this chapter, we have shown how one can add another relevant aspect for society of a pandemic to the ASSOCC model. The homeostatic needs model as foundation gives us great tools to explore the problems and creates the main drivers for our agents to partake in an economy, By implementing a "simple" economic model such as the cycle of income, we can create a decent economic system to experiment with in the context of a pandemic.

These experiments have show the relationships between the epidemiological effects of a pandemic and the economy. We have spoken about the wave effect that we see due to lockdowns, which that has by time of writing this chapter also seen in real life by Statistics Netherlands (CBS) [7]:

> In 2020, the Dutch economy showed strong undulations: from severe contraction during Q2 to extreme growth in Q3; however, the latter trend was too weak to compensate for the contraction in the first two quarters.

The Netherlands has been in lockdown during Q2 of 2020 and lifted the lockdown in Q3 of 2020. In the end of Q4, a new lockdown has been set in place and it will be very interesting to see how this "lift" of the economy looks like once that lockdown is lifted.

In our model, we measure the quality of life to gain insight how society is doing. In the economic sense this you could see this see this as an alternative for GDP, one of the key metrics that is used to see "how well we are doing economoically". Despite not having measured the GDP and only the velocity of money we assume the GDP to have plummeted during the lockdowns and during the height of infections in the no-lockdown situation. But we found that these economic aspects only have a minor effect on the quality of life where other things, like not being able to meet other people, have more effect on the well being of people. The mental effects on well being, for example through needing fulfilment of self-esteem or belonging, is a serious aspect of the economic measures being taken. Of course if people have no money to buy food that will be a problem, but it seems in our analysis that the lack of consumption is not a very big influence on the well being.

In the end, it all comes down to how much value we give to a human life and physical health. Trying to save everyone by having a lockdown until the virus has disappeared is economically very costly. How we should value the infection of someone in an economic way is hard to say as we only have limited knowledge about the consequences. What are the long term effects of a virus? How immune are people after being infected or vaccinated? Without answers to these type of questions it will most likely remain educated guesswork on what the best measures will be if we want to optimize for both the economy and the well being of our population.

But what we can say is that choices have to be made. Choosing for better physical health by reducing contacts, e.g. with lockdowns, will mean more negative effects on the economy. Choosing to not limit economic activity will cost us more lives and will cost us a lot less money in the short run, but holds more uncertainty for the future.

Acknowledgements We would like to acknowledge the members of the ASSOCC team for their valuable contributions to this chapter. We also wish to thank Loïs Vanhée and Luis Gustavo Ludescher for their development work on the code for this chapter and their contributions to the conceptual model.

References

1. H. Dawid, DD. Gatti, Agent-based macroeconomics. Handbook Comput. Econ. **4**, 63–156 (2018)
2. G. Fagiolo et al., Validation of agent-based models in economics and finance, in *Computer Simulation Validation: Fundamental Concepts, Methodological Frameworks, and Philosophical Perspectives*. ed. by C. Beisbart, N.J. Saam (Springer International Publishing, Cham, 2019), pp. 763–787. ISBN: 978-3-319-70766-2. https://doi.org/10.1007/978-3-319-70766-2_31
3. U.S. Department of Commerce. Measuring the Economy: A Primer on GDP and the National Income and Product Accounts (2015). https://www.bea.gov/sites/default/files/methodologies/nipa_primer.pdf
4. A.E. Murphy, John law and richard cantillon on the circular flow of income. Eur. J. Hist. Econ. Thought **1.1**, 47–62 (1993). https://doi.org/10.1080/10427719300000062
5. S. Heidari, M. Jensen, F. Dignum, Simulations with values, in *Advances in Social Simulation*, ed. by H. Verhagen, et al. (Springer International Publishing, Cham, 2020), pp. 201–215
6. M. Bassetto, A game-theoretic view of the fiscal theory of the price level. Econometrica **70**(6), 2167–2195 (2002)
7. The Year of Corona (2020). https://www.cbs.nl/en-gb/news/2020/53/the-year-ofcoronavirus. Accessed 10 Jan 2021

Chapter 10
Effects of Exit Strategies for the COVID-19 Crisis

René Mellema and Amineh Ghorbani

Abstract Most focus during the COVID-19 crisis has been on which measures of governments are the most effective to curb the spread of the virus. Far less attention has been given to the best way to release the restrictions. In this chapter we discuss the various elements of the so-called *exit strategies* from different governments. We will group them in several types of strategies and compare the effectiveness of them. From our simulations we do not observe big differences in the outcomes from the exit strategies. Because the simulations are run on a small scale more research would be needed to see how the results could be translated to the real world. However, in practice we have seen that the effects of the different strategies in countries around the world have indeed had little effect on preventing a second wave.

10.1 Introduction

As the COVID-19 pandemic spread around the world, most governments around the world started to implement measures to limit the spread. A lot of these measures were quite invasive into peoples lives, such as the lockdowns that could be seen in France, Spain and Italy. While much of these were quite effective in limiting the spread, they mostly did not help in completely eradicating the virus. This meant that after a few months in lockdown, people wanted to go back to a more "normal" way of life. However, since the virus was still around, simply ending the lockdown was expected to lead to a so called *second wave* of the disease, which could still lead to exactly the same problems that the lockdown was meant to circumvent.

Therefore, many places have set up so called *exit strategies*, which dictate when which restrictions are lifted, and under which conditions. However, this of course

---

R. Mellema (✉)
Department of Computing Science, Umeå University, SE-901 87, Umeå, Sweden
e-mail: rene.mellema@cs.umu.se

A. Ghorbani
Faculty of Technology, Policy and Management, TU Delft, Jaffalaan 5, 2628 BX Delft, The Netherlands
e-mail: a.ghorbani@tudelft.nl

F. Dignum (ed.), *Social Simulation for a Crisis*, Computational Social Sciences,
https://doi.org/10.1007/978-3-030-76397-8_10

raises the question Which measures can safely be lifted, and which new ones need to be introduced in order to protect the people from the disease?

In order to investigate this question, we extended the ASSOCC model with groups of exit strategies, which allowed us to build coherent sets of measures that were activated at the start of the crisis in the simulation, and were lifted after certain conditions were met. Both the set of measures, as the conditions, were based on real world exit strategies as described in the ACAPS dataset.[1] Since certain measures such as testing and tracking and tracing were already investigated in other scenarios (Sect. 6 and Sect. 7, respectively), we focused on the opening and closing of certain types of locations and other ways of controlling movement.

This chapter is structured as follows. It starts out with some background and context, in particular a description of the ACAPS dataset, and the trends we saw in the real world. After that, we discuss the extensions made to the model, the exit strategies we designed, and the settings used to run the experiment. This is followed by a discussion of the results we got from running the simulation, after which we discuss the results and the overarching conclusions we can draw from it. We will end this chapter with some concluding remarks.

10.2 Measures and Exit Strategies

In order to answer the research question in a way that is useful to policy makers, it is important that our exit strategies and measures connect to what is happening in the real world. In order to base our exit strategies as much as possible on real world strategies, we have based our exit strategies on the ACAPS dataset, and its classifications. We start this section with a short description of this dataset. From this dataset some general classes or groups of exit strategies can be distinguished, which we discuss afterwards.

10.2.1 The ACAPS Dataset

The dataset is a very comprehensive set of government strategies of 193 countries around the world that have been implemented to combat the COVID crisis. For each country it lists the measures implemented in that country and indicates when it was implemented. Furthermore, it indicates whether it was an introduction of a new measure, or the relaxation of another measure. Furthermore, the measures are categorised into five topics:

1. Movement Restrictions, which covers things such as border closure or flight suspension

[1] https://www.acaps.org/projects/covid-19.

2. Public Health measures, including measures such as testing or isolating, but also awareness campaigns and mass isolation policies
3. Governance and socio-economic measures, which includes measures such as limitation of product imports/exports, but also things such as the subsidising of salaries and military deployment
4. Social Distancing Governance, which covers measures such as limitation of public gatherings and the closure of certain types of locations
5. Lockdown, which are the strict restrictions allowing people to leave their houses only on very specific conditions

The above five categories provided the basis for the implementation for the strategy scenarios in this chapter. We checked which measures were often simultaneously implemented or lifted by certain countries. Additionally we used some other sources of secondary data (such as the RIVM coronavirus recommendations[2]), to lead us to the following two important conclusions:

1. Countries have different overall *orientations* that determine how COVID measures are implemented and lifted
2. Countries use different preconditions that lead to the relaxation of measures

With respect to 1. we could distinguish the following three orientations:

- Business-orientation: countries with this orientation seem to give priority to relaxation of measures that are related to the economic well-being of the country. Therefore, it appears that businesses are the first to open up in the case of situation improvement. This was the case for example in Switzerland.
- Leisure-orientation: these countries seem to give extra attention to the social well-being of the society and their leisure time which at the same time affects the tourism activities of that country. Therefore, opening up of restaurants, beaches and museums seems to be the priority in case of situation improvement. Spain can be considered as an example within this category.
- Public service-orientation: Countries with this orientation seem to give priority to the availability of public services to support the functioning of the society. Therefore, lifting limitations on public transport or opening of schools seem to receive top priority, North European countries such as the Netherlands belong to this category.

We will use these orientations to create exit strategies for the simulations.

Countries also differed on the triggers or preconditions for the exit strategies. These could be based on:

1. Number of infections
2. Number of deaths
3. Number of available hospital beds/IC units
4. Effective reproductive number, R_t, normally written as R_0 in informal discussions

[2]https://www.rivm.nl/coronavirus-covid-19/actueel.

Finally, it became obvious that exit strategies consist of different phases in which restrictions were gradually lifted. Each phase having its own preconditions. In each phase several restrictions are lifted or relaxed. For example, the number of home guests allowed is increased and the cafes and bars are allowed to be open. Furthermore, countries normally keep monitoring the effects of various relaxations of measures, such that they could go back a phase (putting back restrictions again) if there were more new infections than anticipated. We will call this *snapback*.

Based on all these features, we designed exit scenarios for the ASSOCC model that combines the three orientations, the distinguished preconditions, and the lifting of restrictions spread over phases.

10.3 Scenario Description

For this scenario, the ASSOCC implementation was extended with a mechanism where, instead of triggering measures by hand or on some percentage of the population infected, measures could be enabled and disabled based on the relative number of infected, people in hospitals, or simply a certain amount of days that had passed. As said before each combination of criteria can lead to the enactment or lifting of a set of measurements at the same time. This starts a new *phase* in the simulation. The simulation goes through these phases linearly based on the criteria.

10.3.1 Model Extensions

In order to implement the various exit strategies, the model was extended with a system that automatically activates and de-activates the various measures. The phases that are implemented are:

- a `crisis-not-acknowledged` phase, during which the infections start;
- a `ongoing-crisis` phase, when restrictions are put in place;
- three exit strategy phases (called `phase-1`, `phase-2`, and `phase-3`), during which the restrictions are lifted in stages.

We start with a phase where there might be some infections, but no measures are deemed to be necessary yet. The second phase, `ongoing-crisis`, is triggered once a set percentage of the population (set by the parameter `acknowledgement-ratio`) get infected. Once this phase is triggered, the simulation also activates all the measures from a certain strategy at the same time. Then, based on one of several phasing out conditions (see Table 10.1), it will move onto the next phase, `phase-1`, which relaxes a certain set of measures. We distinguish three phases as that seems to correlate the best to the phases observed in the real world. The same triggers are used to move between each of the phases. I.e. if an exit strategy is triggered by having less than a certain amount of deaths per day, the different exit phases are all triggered based on number of deaths.

In the standard ASSOCC framework every measure that is used in a scenario is triggered by the governments `acknowledgement-ratio`. Usually this is the number of infections that is accepted before a crisis is acknowledged. We use the same trigger for all the measures in the exit scenarios that we explore. However, using these standard triggers means that sometimes a measure is deactivated due to the number of infected persons getting very low before the trigger for the end of a phase is reached. E.g. phase 3 might have a minimum length of 30 days or is ended based on the number of deaths being lower than 5. It is well possible that the number of newly infected people is below the threshold while there are still people dying. As will be shown later in the results, this situation does not happen often, and can be seen as a relaxation of all measures at the end of a run, a sort of implicit phase 4. Thus we have not changed the triggering mechanism of the measurements based on these exceptions.

Furthermore, in the real world data we see that governments sometimes go back to an earlier phase if the number of new infections is too high. In order to mimic this, we use a simple snapback procedure, which can be turned on or of by setting `phase-snap-back?` to true or false, respectively. In this case, if the number of infections goes above `acknowledgement-ratio`, the government phase will be reset to the `ongoing-crisis` phase.

In order to experiment with different conditions and phase lengths we use a number of parameters that allow to flexibly schedule the phases of the exit strategies. These parameters are as follows:

- the `minimum-days-between-phases`, that determines how many days a phase needs to be in progress before the condition to progress to the next stage is checked;
- the `day-gap-for-phasing-out-condition`, that is used to set the days that are taken into consideration for the phasing out condition (if needed);
- the `next-phase-condition-percentage`, that is used if the condition is dependent on a percentage;
- the `next-phase-new-infection-limit`, that is used if an absolute number of new infections was needed.

In reality the number of deaths is relatively small. Thus if an exit phase depends on deaths often the absolute number is used. The number of newly infected persons is much bigger and also fluctuates more, thus usually an average percentage of the total population over a number of days is taken to describe a trigger for an exit phase based on infected people. All of these parameters can be set by the user, but besides the `minimum-days-between-phases`, not all of them are relevant for all phasing out conditions. The descriptions in Table 10.1 include the parameters that are relevant for each condition.

The conditions are based on the preconditions that were reported in the ACAPS data set. The number of deaths precondition was ignored, since we only had a small number of deaths, so we judged that to be too volatile to be included. (Note that in

Table 10.1 The various phasing out conditions that could be used by the simulation

Name	Description
"35 days of quarantine"	This condition is true at the end of the global (35 days) quarantine as used in Chap. 7
"Only look at days since last phase"	Move on to the next phase after `minimum-days-between-phases` has passed since the last phase shift. Is used as a control condition
"#infected has decreased in day gap"	This condition is triggered if `#infected` is now lower than it was `day-gap-for-phasing-out-condition` days ago. Allows for settings such as: "The number of infections is lower than 3 days ago"
"Hospital not overrun & #hospitalizations has decreased in day gap"	The amount of hospitalizations has decreased compared to `day-gap-for-phasing-out-condition` days ago, and the hospital is currently not at maximum capacity. Allows for settings such as: "There are less people hospitalized than 3 days ago"
"Percentage immune"	When `next-phase-condition-percentage` of the population is immune, this condition is triggered. Allows for settings such as: "When 60% of the population is immune"
"New infections percentage of average over day gap"	When the number of new infections over the past day is lower by a certain margin than the average infections over the last few days. Allows for settings such as: "When the number of new infections is half of the average number of new infections over the past 10 days"
"New infections under limit"	When the total number of new infections over the last `day-gap-for-phasing-out-condition` days is below `next-phase-new-infection-limit`. Allows for settings such as "Only 10 new infections in the last 5 days"

the real world the total amount of deaths per country due to COVID-19 is below 0.1%. In the ASSOCC simulations this amounts to between 0 and 1 individuals dying during the simulation! We have compensated for this statistics, but still there are hardly ever 2 or more individuals dying per day). The "35 days of quarantine" condition is implemented to calibrate the system, but not used for the scenarios. The variable "only look at days since last phase" is implemented as a control measure, to see if the investigated condition for an exit strategy is better than just waiting a set number of days. The condition "#infected has decreased in day gap" is implemented as an early measure in the testing phase based on the number of infections. "hospital not overrun & #hospitalizations has decreased in day gap" is implemented to mimic the preconditions based on the number of available hospital beds. The "percentage immune" condition mimics the often not publicly mentioned goal of herd immunity, and the "new infections percentage of average over day gap" and "new infections under limit" conditions are added to mimic the preconditions based on the reproduction number R_t.

As mentioned before, several restrictions can be enacted or lifted in each phase. The mapping from phases to sets of lifted restrictions is what we call an *exit strategy*. An overview of the restrictions that are used can be found in Table 10.2. These restrictions are based on the exit strategies in the real world but adapted to the limitations of the ASSOCC simulation model. I.e. not all variations and refinements of restrictions can at present be represented in the ASSOCC model without a major restructuring of basic elements of the simulation and consequent changes of the agent deliberation cycle. We will get back to these limitations in the conclusion of this chapter.

The banning of travel is incorporated for a few reasons. First of all, a very common measure is the limitation of travel between towns, which is what this measure does in the simulation as well. However, there are also measures introduced related to tourism, that are not possible to easily represent explicitly in the simulation. Therefore this measure also incorporates those aspects.

The recommendations to work from home and the measures under the heading **Social Distancing** occur frequently in the dataset, and included for that reason. Similarly for limiting the capacity for public transport. The isolation of retirees was not something that showed up very often, but was often implicitly assumed and thus included.

As can be seen from the table, we have not implemented any governance measures such as states of emergency. This is due to the fact that the governance agencies such as police and army are not explicitly included in our simulation. Thus these governance measures would not have any effect on the results. They would be interesting for larger scale and more governance focused simulations.
We also do not include any socio-economic measures. Mainly these will be dealt with in Sect. 9, while in this chapter we focus on the health implications of the exit strategies. The lockdown is not explicitly taken into account, but can be mimicked by turning on most of the **Social Distancing** measures.

10.3.2 Design of the Exit Strategies

As mentioned before, the ACAPS dataset includes roughly three groups of measures, those with a business orientation, those with a leisure orientation, and those with a public service orientation. In the simulation, we represent these three broad categories and compare their effectiveness. In order to do this, for each strategy, we divide each of the restrictions up into one of the three exit phases, where it is most appropriate to lift the restriction. For example, for the strategy focused on the economy, we allow people to go back to work as soon as possible. Restrictions on retirees will be lifted

Table 10.2 The restrictions that can be included in an exit strategy. The rows with no name refer back to the categories of measures as used in the ACAPS dataset

Name	Description
Movement restrictions	
Banning travel to other places	This stops agents from leaving the town. See Sect. 3.4.10 for more information on the travelling mechanics
Public health measures	
Isolating retirees in their homes	Retirees requested to stay in their homes at all times. Note that retirees can still leave their homes, and even have to do so to go to essential shops
Recommendation to practice social distancing	Requires that the agents perform social distancing, and thus limit the spread of the disease that way
Requires to work from home	This measure asks workers to work from home if possible, but did not force them to do so
Social distancing	
Closing of schools	Closes the schools, forcing youth to stay at home during the day
Closing of workplaces	This measure closes workplaces, forcing workers to work from home, if possible
Closing of non-essential shops	This closes non-essential shops (no possibility to violate!)
Closing of universities	This measure closes the universities, forcing students and university staff to stay at home during the day (no possibility to violate!)
Closing of private leisure	This measure closes private leisure, so agents can not come together with agents from other households in their free time
Partial close of private leisure	Limits the capacity of private leisure, so agents can go to private leisure, but might be send away if there are already too many agents there
Closing of public leisure	This measure closes public leisure
Having busses work under half capacity	This measure limits the amount of agents that are allowed into a bus, and thus having to have more agents queue to get somewhere, but limiting the amount of contacts in a bus

later as these will not impact the economy. Based on the focus and priorities of the strategies, they are named: *business*, *leisure*, and *public services*. In Table 10.3 we indicate for each strategy which restrictions are involved and the order in which restrictions are lifted in the exit phases.

Per restriction in a row the columns indicate the phases in which that restriction is in place for the different strategies by an "X". The `ongoing-crisis` phase is

Table 10.3 The exit strategies used in the scenarios

	Business strategy				Leisure strategy				Public services strategy			
	C	1	2	3	C	1	2	3	C	1	2	3
Closing of schools	x	x			x	x	x	x	x	x		
Closing of workplaces	x				x	x	x	x	x	x		
Recommend working from home	x	x			x	x	x	x	x	x	x	x
Closing non-essential shops	x				x				x			
Closing universities	x	x	x	x	x	x	x	x	x	x	x	
Full closing of private leisure	x	x			x				x	x		
Partial closing of private leisure	x	x	x	x	x	x			x	x	x	x
Closing public leisure	x	x	x		x				x	x		
Practice social distancing	x	x	x	x	x	x	x	x	x	x	x	x
Busses at half capacity	x	x	x	x	x	x	x	x	x			
Isolating retirees	x	x	x		x	x	x		x			
Banning travel out of town	x				x	x			x	x	x	

indicated with a 'C', and the numbered phases are indicated with their number. We will compare these exit strategies on the amount of people that are infected over time. However, the scenarios also serve as a starting point to investigate general guidelines on how to design exit strategies to have maximal effect. Thus we will not only look at the number of infected people, but also where they get infected and how this might change over the different phases of the exit strategies.

Per strategy there are a number of expectations of their effect that we will use as the hypotheses to be tested in the scenarios.

The business exit strategy starts opening up workplaces, non-essential shops, and travel out of town in phase 1. Because of this, we expect workers to get more infected in phase 1. When they go home they might infect their kids. In the second phase, schools and parts of private leisure open up. We would expect this to lead to a second increase among all groups, except retirees, but mostly youths, who can now infect each other more in schools. In the third phase, retirees are allowed to come out of isolation, and public leisure will be opened up again. This might lead to another small uptick, depending on how many people have already been infected.

The leisure exit strategy in phase 1 starts with opening up non-essential shops and public leisure, and partially opening up private leisure locations. This should lead to an increase in infections in all groups except for retirees, who stay in isolation. In the second phase, private leisure opens in full, and travel out of town opens up again. The only measure opened up in the last phase is that retirees are allowed to come out of isolation. Because the phases do not differ all that much, the expectation for this scenario would be that it would not be too dependent on the phasing out condition.

Table 10.4 General settings for the exit strategy scenarios

Parameter	Value
`acknowledgement-ratio`	0.02 0.05
`condition-for-acknowledging-the-crisis`	"ratio infected>acknowledgement-ratio"
`minimum-days-between-phases`	15 30
`day-gap-for-phasing-out-condition`	15
`phase-snap-back?`	false
`migration?`	true
`probability-going-abroad`	0.03
`probability-infection-when-abroad`	0.02
`probability-getting-back-when-abroad`	0.12
`food-delivered-to-isolators?`	false
`only-setup-loosened-measures?`	false
`productivity-at-home`	0.75

In the public services exit strategy, the focus in phase 1 is on getting public services up and running again. Like the other two scenarios, it opens up non-essential shops, but unlike the others it directly gets retirees out of isolation, and it opens up busses at full capacity. Like in the other scenarios, this should lead to an increase in infections, but unlike the others here we expect it to be across all age groups. In the second phase, schools, workplaces, and public leisure open up, and private leisure opens up partially. Since this is more than in other strategies for phase two, we would expect a possible third wave here. In the third phase, this strategy opens up universities and out of town travel. Depending on how fast the disease spreads again, this could potentially lead to a fourth wave, but if enough of the population has been infected by this point, this will not happen.

10.3.3 Settings

The general settings for the exit strategy scenarios are listed in Table 10.4.

We use two different values for the `acknowledgement-ratio`, which determines how large a group in the population has to be infected before the government goes to the `ongoing-crisis` phase. These values are 0.02 (2%) and 0.05 (5%). The 0.02 was chosen in line with the experiments described in Sect. 7, and the 0.05 was chosen to get a higher number of initially infected reflecting a later government response as was observed in several countries. The `minimum-days-between-phases` was set to either 15 or 30, based on the two weeks and 1 month time intervals we saw in the real world. The `day-gap-for-phasing-out-condition` was set to the shortest of these two, so the simulation would not use data from the previous phase to determine if it could leave the current phase.

During the scenarios we did not use the possibility to snap back to the crisis phase again. Although it is realistic it distorts the comparison between the different exit strategies too much. Thus `phase-snap-back?` is set to false.

Since the model uses the travel feature for one of the restrictions, this feature needs to be turned on for the simulation. We set the `probability-going-abroad` to 0.003, so around 3 agents each tick will go abroad (if it is allowed), and they will come back according to `probability-getting-back-when-abroad` which is set to 0.12, so a trip will take around 12 ticks.[3] The probability that they will get infected during a tick while abroad is 0.02.

The setting for `food-delivered-to-isolators?` is turned off, because its effect on the number of infections during the `ongoing-crisis` is small and most real world strategies did not specifically account for food delivery for people that were in isolation. Since we want all the measures to be active during the `ongoing-crisis` phase, even if they are not relaxed in an exit strategy, we set `only-setup-loosened-measures?` to false.

We set the `productivity-at-home` to 0.75. This value is increased from the default value to make sure that the stores will not run out of goods. The goods of the stores are coming from workers in workplaces (partially representing factories or warehouses). We assume goods are still produced when the workers work at home, but with less efficiency. Therefore it might happen that stores receive too little goods and frequently run out of products to sell. Because the ASSOCC agent deliberation does not take into account that a store might run out of goods, this would cause all the agents to go to stores often in attempts to get the products they need and spreading the virus in the shops that get more busy than normal.

As triggers to move from one phase to the next, we used the "only look at days since last phase", "hospital not overrun & #hospitalizations has decreased in day gap", "percentage immune", "new infections percentage of average over day gap", and "new infections under limit" conditions. We will discuss the latter three of these in more detail, since they have parameters of their own. The percentage immune for the condition was set to 60%, which means that 60% of the population should have had corona and be cured. This percentage is higher than people at first thought would be needed in practice. However, in later research[4] it was claimed that the 60% figure is more realistic. So, we keep to this higher percentage, especially, because we have set the contagiousness in the simulation higher than in practice as well.

We run the "new infections percentage of average over day gap" with values for `next-phase-condition-percentage` of both 1 and 0.5. With the value of 1, this means that there are less infected persons today than the average over the last 15 days before. When the value is 0.5 the new infections today have to be less than

[3]The probability that an agent will come back after x ticks is $(1 - 0.12)^x$, so the chance of them being away for at least 12 ticks is 0.216.

[4]https://www.news-medical.net/news/20201204/Herd-immunity-threshold-far-higher-than-previously-thought-say-researchers.aspx.

half of the average new infections over the last 15 days. Thus with 0.5 we require a substantial decrease of infections.

The "new infections under limit" condition has two parameters: `day-gap-for-phasing-out-condition` and `next-phase-new-infection-limit`. We choose to set the day gap to either 3 or 5 to closely mimic real world scenarios, where the absolute number of new infections is normally quite low. Likewise we set the `next-phase-new-infection-limit` to either 10 or 5. This leads to a total of 4 settings with different levels of strictness.

The values for the other parameters are set to their default values and were not changed for the simulations in this chapter. Useful to note is that, under these settings, there were 1162 agents in total at the start of the simulation.

10.4 Results

Given the many parameters we can generate many results from this scenario. We will start with analysing some of the differences between the conditions and acknowledgement ratios we set up, and then we will continue with an analysis of the various strategies. All the plots in this section were created by aggregating multiple runs together and taking the average. For the plots where the exit strategy determines the colour of the line, there were 12 repetitions. For the plots where the acknowledgement rate determines the line colour, the number of repetitions was 18.

In all of these scenarios, the condition "only look at days since last phase" is used as a control for the conditions. If a condition performs similarly to this condition, then there is little to no benefit to using that condition at all, since it performs the same if we just waited a set number of days. This makes is possible to not just judge the length of the exit strategies under different conditions, but also if those conditions are useful to implement.

Because there is a lot of data and the plots can be a bit hard to read, we will start with going into detail on one set of parameters, before we discuss all the results. After that, in Sect. 10.4.2, we will discuss how the exit phases can be controlled and what their effects are. The questions asked here concern the interaction between the conditions, and their minimum length (`minimum-days-between-phases`). In Sect. 10.4.3, we discuss the effect of the different acknowledgement rates. The question asked here is: "What is the effect of a quicker government response on the spread of the disease?" Finally, we will discuss the strategies themselves in Sect. 10.4.4. Here, the questions asked concern the effects of the strategies on the spread of the disease, locations where the spread happens most, hospitalization and mortality, and the results on the economy.

10.4.1 Basic Settings

For this section, we will focus on the situation where the phasing out condition is set to “new infections under 10 over last 5 days”, the minimum days between phases to 30, and the acknowledgement rate to 0.05. Furthermore, we only focus on the Public Services strategy. The graph showing the number of infections for this case is shown in Fig. 10.1.

There are a few things to notice about this graph. First of all, it is clear that in this example, there are two waves of the disease, one during the lockdown, and one during the first phase. The wave during the lockdown shows that it takes some time before the lockdown takes effect. First people that were already infected before the lockdown become aware of this fact later and these people might also infect their housemates during the lockdown. During the first exit phase, places for people to get infected open up, so a new wave will start after that. The reason why there is not a third wave in this example, is because already a lot of agents have gotten infected, as can be seen in Fig. 10.2. In this graph, it can be see that as we enter `phase-2`, already 598 agents have been infected with COVID-19, which limits the spread of the disease, since there is now a more than 50% chance that if you meet someone outside, that they will be immune.

Something that the astute observer might also have noticed, is that in both Figs. 10.1 and 10.2 the number of infections start to rise before we enter the new phase. This is due to the fact that we are averaging over multiple runs, for both the infections and the time of the phase shift. Some runs, those where the phase shift

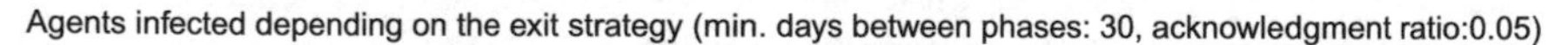

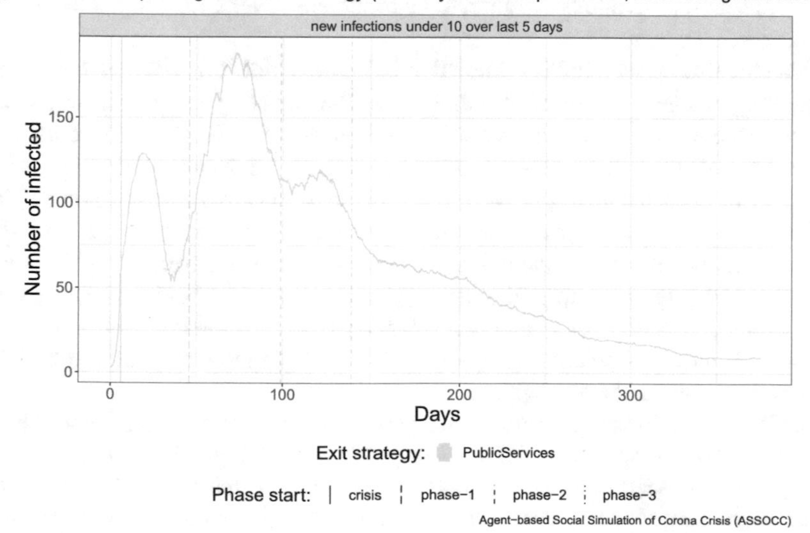

Fig. 10.1 Infection peak for the base case. The first vertical line indicates the government going into the crisis phase, consecutive dotted lines indicate transfers to the phases1-3

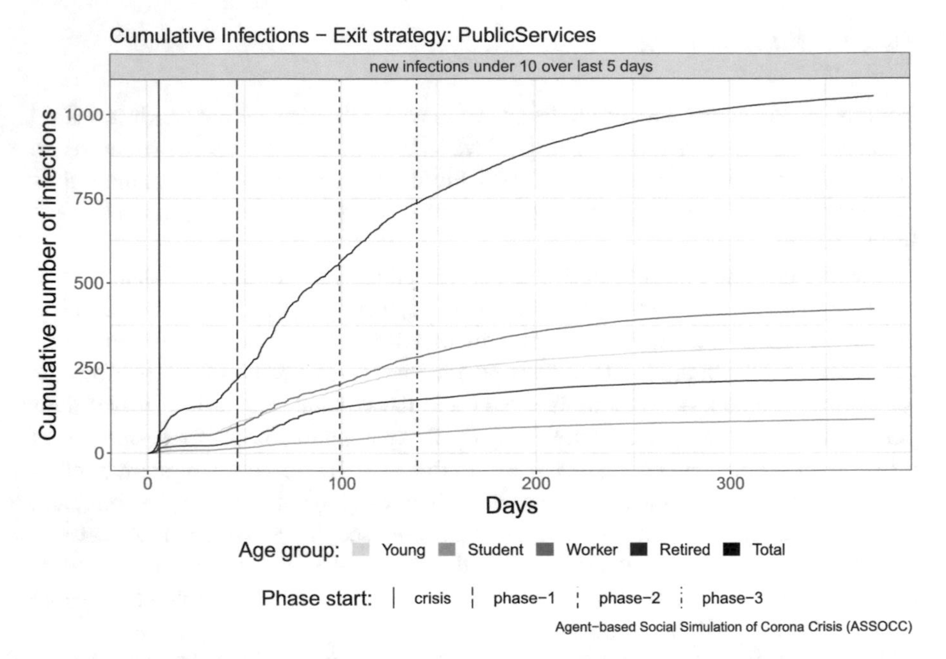

Fig. 10.2 Cumulative infections for the different age groups in the example case. The vertical lines in the figure indicate when, on average, a phase shift happened

happened earlier than average, start contributing to the new infections before the phase shift is displayed in the figure, which is where the average phase shift happened. This causes the number of infections to increase ever so slightly at the end of `ongoing-crisis`.

What can also be seen in the cumulative infections plot (Fig. 10.2) is that not all groups of agents get infected at the same time. We can see this more clearly when we look at the ratio of infected agents for each group (Fig. 10.3). Here we can see that the ratio of infected retirees increases quicker than that of the other groups in `phase-1`, but that the young get way more infected in `ongoing-crisis`. This latter one can be explained by the household composition. The young agents live together with one other young, and two workers as parents (on average). This means that if one young agent is infected when `ongoing-crisis` begins, they are very likely to infect one other young and two workers. The other age groups live together in pairs, so if they go into `ongoing-crisis` while healthy, they are more likely to stay healthy during the crisis phase.

In order to explain why the retirees are getting infected more often, we will have to look at where the agents get infected (Fig. 10.4). In this graph, for each type of location it can be seen how many agents get infected in that location. In this plot we can see that most agents get infected at home. The location type with the next highest number of infections occurring is the non essential shops. This leads us to

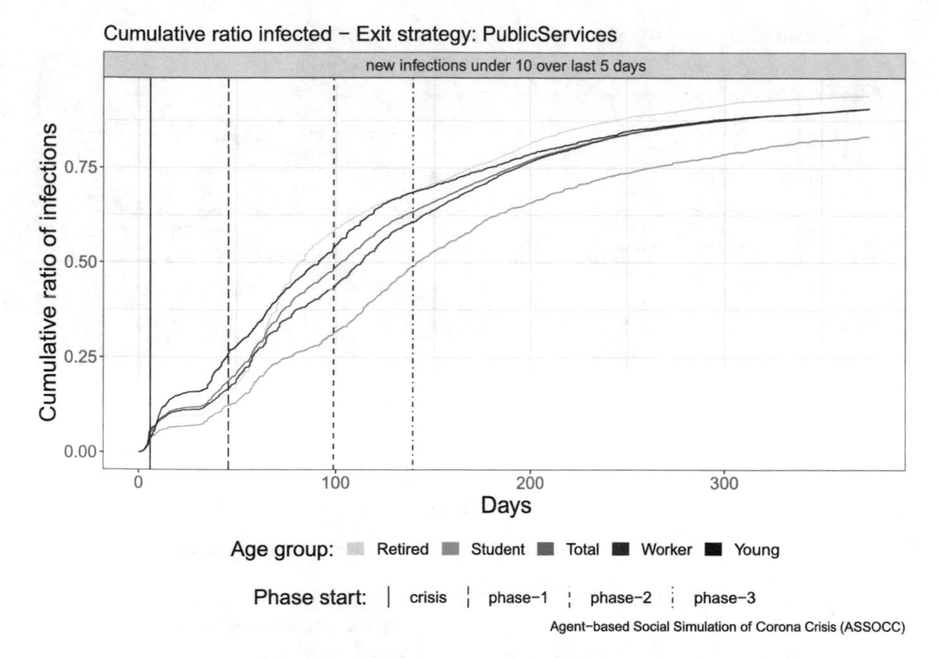

Fig. 10.3 Ratio of cumulative infections for the different age groups in the example case. The vertical lines in the figure indicate when, on average, a phase shift happened

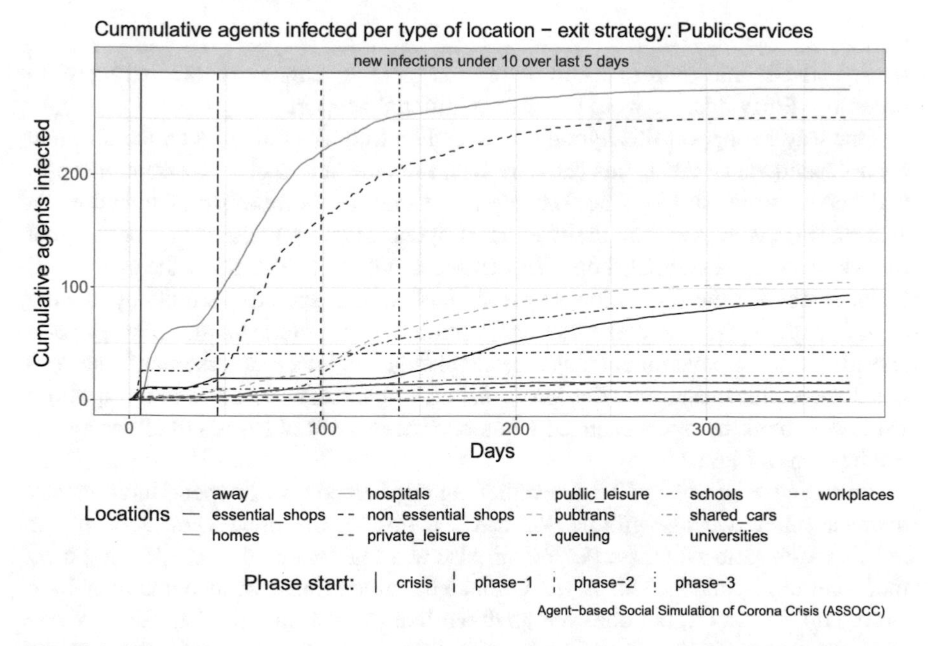

Fig. 10.4 The number of people infected in each type of location per day for the example case. The vertical lines in the figure indicate when, on average, a phase shift happened

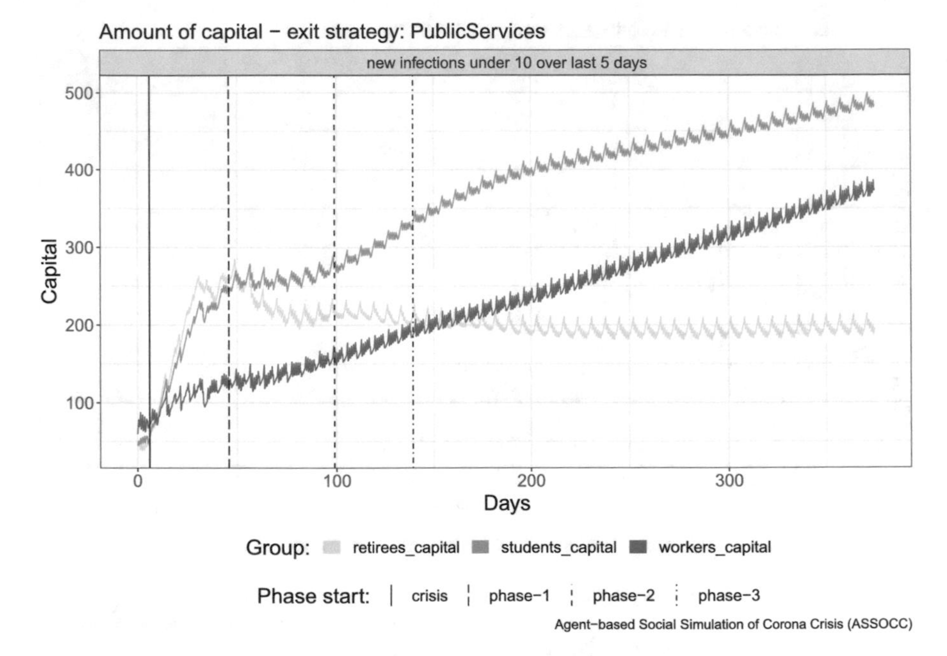

Fig. 10.5 The capital of the different agent groups over the course of the runs. The vertical lines in the figure indicate when, on average, a phase shift happened

the hypothesis that retirees spend more time there, and thus get infected more. In order to solidify this, we would need an additional support.

One way to support this hypothesis would be to look at how much the different agents spend, since the agents can only spend money at two places, essential shops, and non-essential shops. Since we see no change in the number of infections at essential shops, it is safe to assume that if agents start spending more money, that they do so at non-essential shops. We can assume that the income of the agents will be the same over the simulation, so we can look at their spending simply by looking at their capital. We can make that assumption because students and retirees get a subsidy from the government that does not change, and the workers salary also does not change. This means only the agents that no longer can go to work, or those whose work went bankrupt see a change in income. The amount of capital of all the agents can be seen in Fig. 10.5.

As can be seen in Fig. 10.5, the retired agents indeed start to spend more money than the other two age groups. We can see this by the downward slope of the red line after `phase-1` starts. We can also see that students start spending a bit more, but not by much. This is visible since the green line flattens out compared to `ongoing-crisis`, but does not go down like the red line. So therefore, we can conclude that retirees spend more time at non essential stores, since they are the group that spends the most money in `phase-1`.

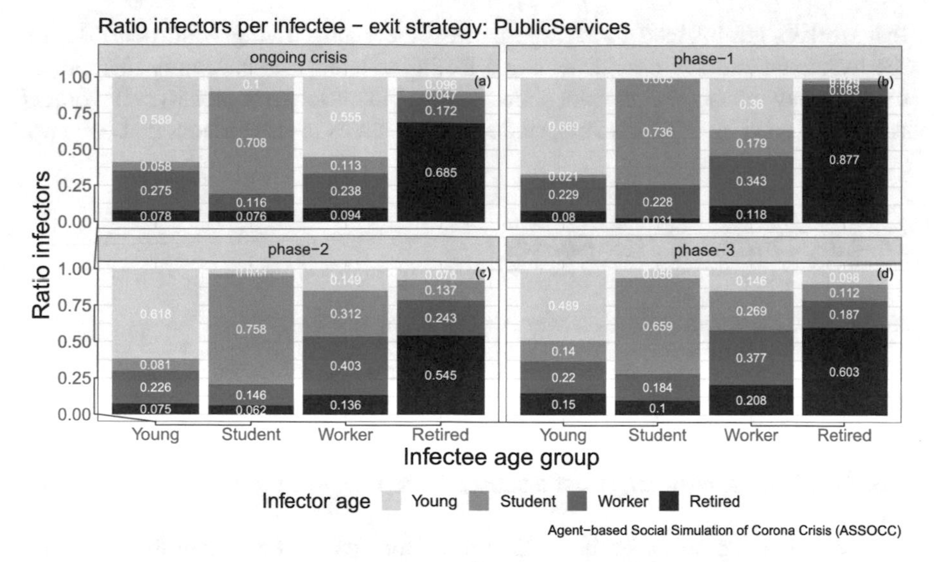

Fig. 10.6 The ratios of how many people of an age group infect another age group, split out by current phase, for the Public services exit strategy

Why retirees are the group that spends more is also easily explained when looking at an agents daily schedule, which is described in Sect. 3.7.1. Retirees have more free time and opportunity to go to shops, whereas students and workers have to spend time studying/working. Therefore, we would also expect that the people that retirees mostly infect each other at non-essential shops, so we would see a high level of retiree to retiree infections in `phase-1`.

In Fig. 10.4 we see that besides homes and non-essential shops, many people get infected in public transport, in schools, and in away travel. There are not many infections occurring during queuing. Thus it seems people do not have to wait for buses a lot. Therefore it seems that the infections in public transport are not mainly due to buses being overly full.

Besides the places where people get infected it is also important to investigate which type of people infect each other most. The plot for this can be found in Fig. 10.6. Since the places where people can meet are different for all the phases, it is split out by phase. The `crisis-not-acknowledged` phase is left out because it is very short (on average less than a week). For this plot it is most interesting to compare the different phases. For example, it can be seen that in `phase-1`, most retirees get infected by other retirees, but in the other phases this seems not to be the case. As mentioned before, this is expected because retirees will mostly meet other retirees in the shops.

In `phase-2` and `phase-3`, public and private leisure open up again, as do the schools and workplaces. This gets more types of agents out of their houses, who then all meet and can infect each other at work, leisure places and public transport. This contributes to a more spread out ratio of other infectors. It is interesting to see that workers get infected for a large part by the young (living with them) during the lockdown and this ratio decreases a lot in the later phases. Mainly this can be explained by the fact that the plot shows a ratio (not absolute numbers) and workers can get infected by other workers, students and retirees in other places that open up.

10.4.2 Control of Exit Phases

In this section, we will try to answer the following questions:

- Which conditions have an effect on the disease when compared to the control?
- Do longer phases help in flattening the curve?
- Do longer phases help in lowering the mortality?

And based on the answers, select a subset of the data to be used for the rest of the section.

The infection plots under the various conditions for both values of the minimum number of days between phases can be found in Fig. 10.7. The red graphs indicate the case where each phase lasts at least 15 days after which a next phase will start if a trigger condition is met. The blue graphs show the case when each phase lasts at least 30 days before the trigger conditions for a next phase are checked. When comparing the conditions to the control, "only look at days since last phase", we can see that some of the conditions we set have little effect. Taking the immunity of the population as criteria (Fig. 10.7b) has a big effect, but also has as a consequence that the lockdown lasts for almost 10 months! The criteria that seem most effective compared to just having fixed times for the phases are those that trigger a next phase when the average number of new infected people is below a threshold for at least 5 or 10 days. In Fig. 10.7f–i) we can see a significant reduction of the peaks while the phases are not terribly long yet.

The first thing that can be noted, is that in almost all conditions the minimum days between phases settings has an effect. This can be seen by looking at the difference between the blue and red graphs in the same condition (box). The blue graphs generally start the next phase later and have lower peaks than the red graphs. However, the exact effect it has differs between the conditions. For the "only look at days since last phase" condition, the next phases start immediately after the minimum number of days have passed. And it seems to have a similar effect on "hospital not overrun & #hospitalization has decreased in day gap" and the "new infections x percent of average over day gap" conditions. This means that the trigger conditions are met before the minimum number of days of a phase are passed and thus a next phase can always start when that minimum number of days has passed. We will explain why these conditions have so little effect later on. For the "new infections under x over last y days" conditions it seems to have little effect on the length of the exit phases, but it has a large effect on the length of the crisis phase. It seems that after 15 days the infection has reached its peak for the crisis phase, so the infections start to go down again, and under the minimum 15 days setting the exit phases start.

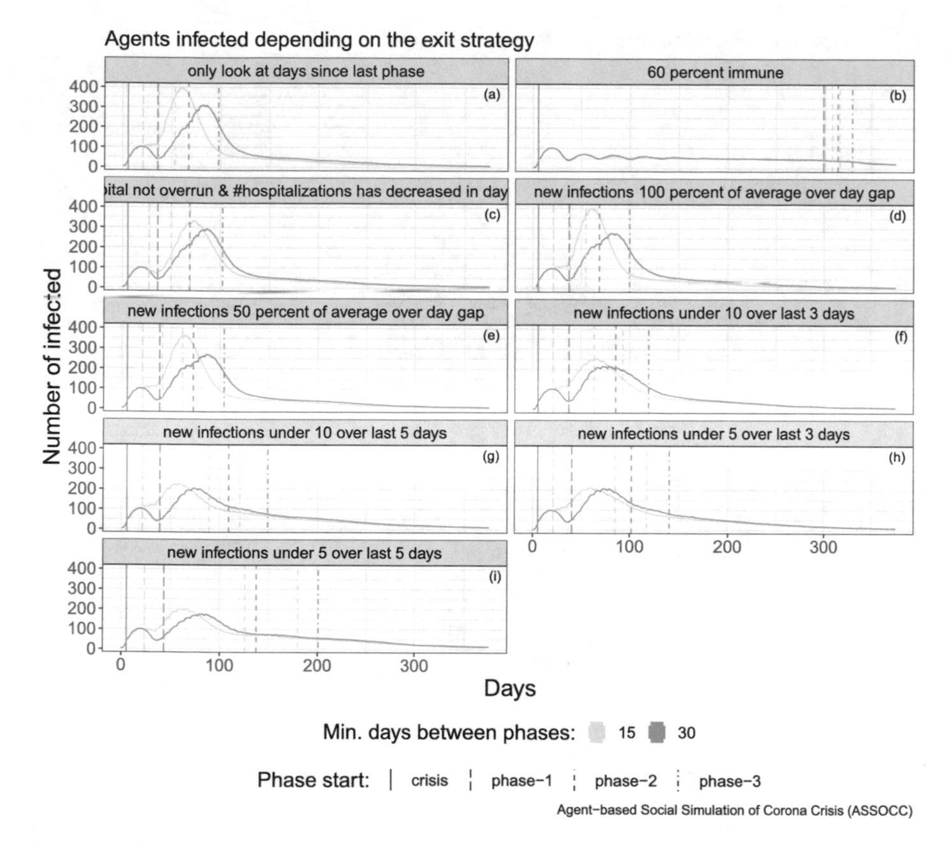

Fig. 10.7 Infection peaks under different conditions. The vertical lines in the figure indicate when, on average, a phase shift happened

However, in the minimum 30 days the crisis phase holds on a bit longer, shifting the phase shifts under those setting by around 15 days.

Due to this shifting of the start of the exit phase, there are also less infections when the exit phase starts. This by itself already lowers the infection peak, but as can be seen in Fig. 10.7a, the amount of infected agents at the start of `phase-1` only differs by 57.82, and the difference between the peaks is 86.79. The difference in the peaks is due to the fact that if one starts with less infected people when `phase-1` starts there are less people to infect each other in places that open up. This causes the disease to spread slower, which flattens the curve. Note that the accumulated amount of infected people does not differ between the two settings at the end of the simulation (see Fig. 10.9a).

As can be seen in Fig. 10.8, the settings have little effect on the total mortality, with all settings reaching a similar level of overall deaths. This can be explained by investigating the cumulative infections and hospitalisations (Fig. 10.9 and Fig. 10.10, respectively). Despite the flattening of the curves, a similar amount of agents get infected overall, and because the hospital is never overrun we can expect a similar

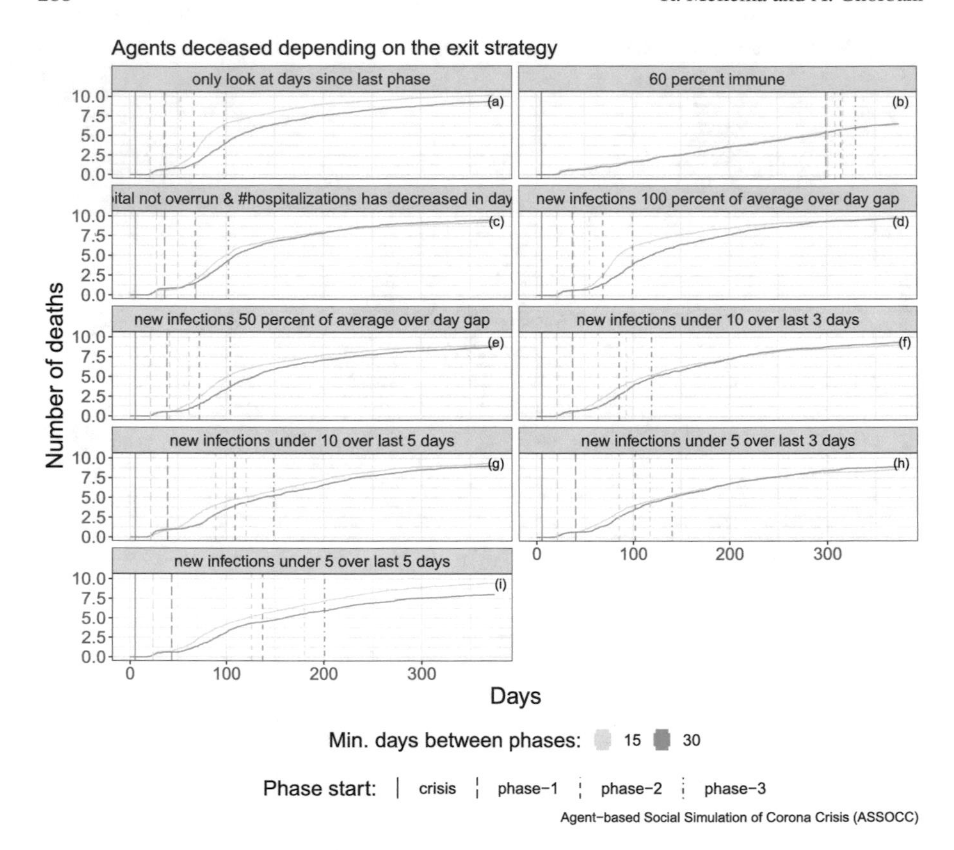

Fig. 10.8 Mortality under different conditions. The vertical lines in the figure indicate when, on average, a phase shift happened

amount of deaths in all settings. Note that this might be different in the real world, where hospitals have limited capacity and flattening the curve can have an effect to prevent people dying before being able to get to an hospital.

In order to see *why* some of the conditions have so little effect, we will investigate them in some more detail. First we will look at "hospital not overrun & #hospitalization has decreased in day gap". In order to go to the next phase under this condition the number of hospitalizations should decrease while the hospital is not full. In Fig. 10.10c it can be seen that the hospital is never overrun, since there are 13 beds in total, and only around 6 are taken at the peak of hospitalizations. This means we only need to focus on the number of hospitalizations going down.

From the same graphs one can see that the amount of people in hospital varies slightly over time which correspond to people being released and people being taken back in into the hospital. As can be seen from these variations, there are enough moments where there is someone released, but no-one is taken back in into the

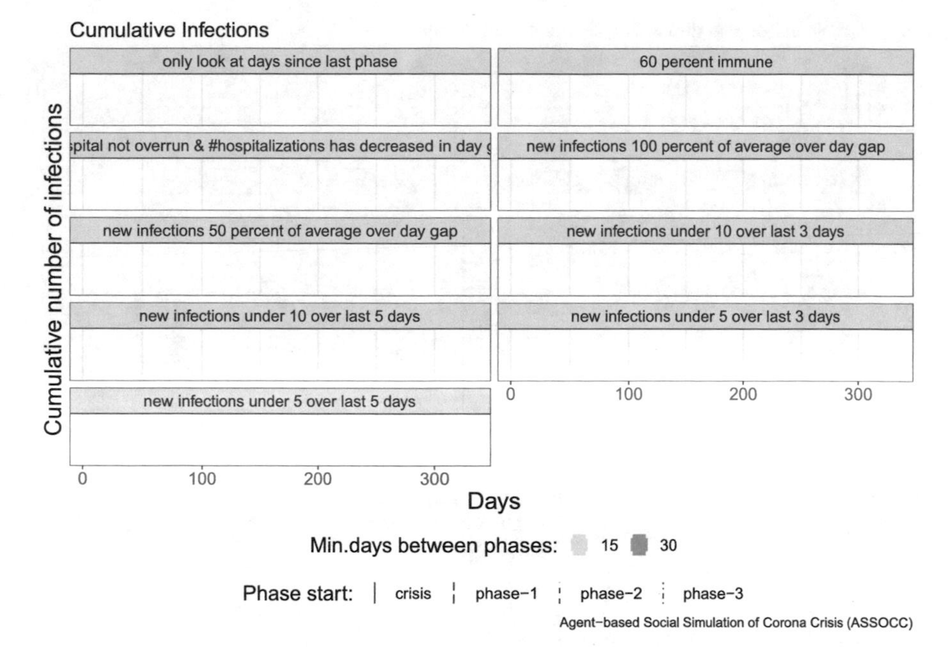

Fig. 10.9 Cumulative infections under different conditions. The vertical lines in the figure indicate when, on average, a phase shift happened

hospital at the same day. At these moments, the simulation will advance to the next phase, and since there are many of these moments, it will do this soon after the minimum days between phases has passed.

For the "new infections *x* percent of average over day gap" conditions, it is most informative to look at a plot with the new infections, and not the actual infections. This plot can be found in Fig. 10.11. In this plot, it can be seen that the amount of new infections on a day can be really volatile. This makes checking that the infections on one day are below a certain threshold as a criteria to move to the next phases is also very volatile. For example, if more people stay at home on a Monday, but went out the Saturday before and infected their housemates on the Sunday in between, the number of new infections will be lower than the average, but it can still be quite high. Checking the new infections under a set limit do not have this problem, since they sum the new infections over multiple days, and compare this amount against a pre-set, strict limit. This causes them to be less influenced by the changes in the new infections, and thus have less of a chance to move to the next phase prematurely, as the earlier two do.

Because the minimum days between phases of 30 days leads to better results (the flatter curve) with clearer waves, we will focus the rest of this analysis using that setting, using only the data where the minimum days between phases is 30.

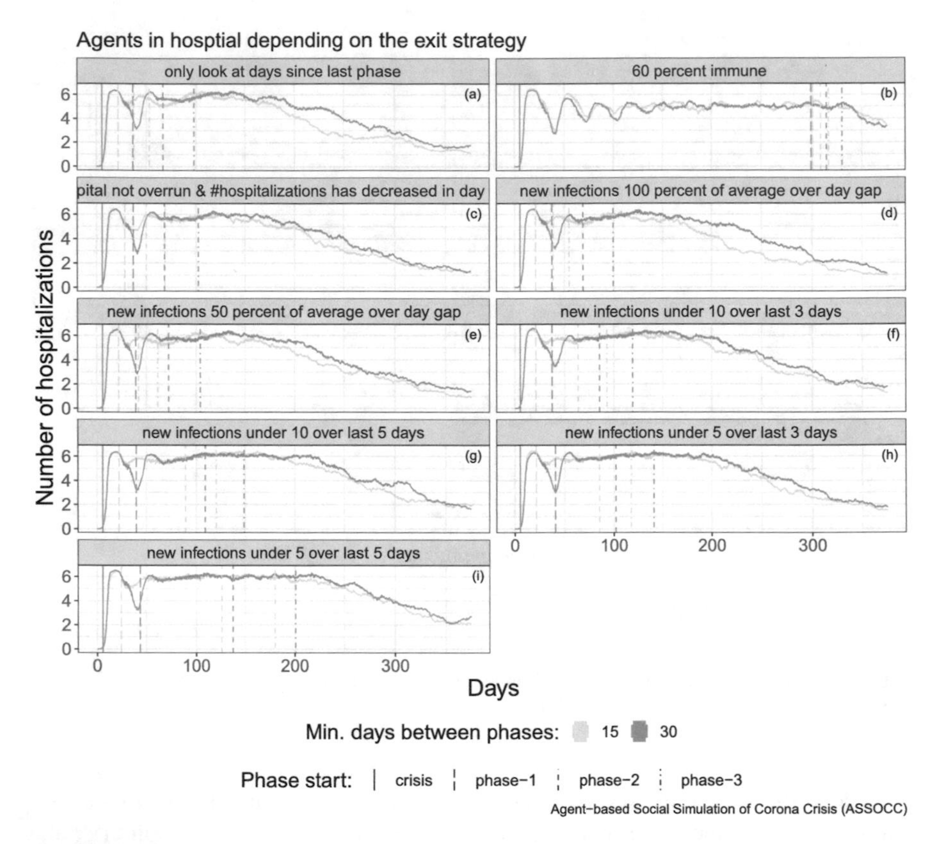

Fig. 10.10 Hospitalisations under different conditions. The vertical lines in the figure indicate when, on average, a phase shift happened

10.4.3 Acknowledgement Ratio

For this section of the analysis, we will only look at the effect of the acknowledgement ratio, which determines how fast the government initially responds to a certain amount of infections being known. Using this, we want to answer the following questions:

- Does a quicker government response help in flattening the curve?
- Does a quicker government response shorten or lengthen the phases in an exit strategy?
- Does a quicker government response decrease the rate of hospitalization?
- Does a quicker government response decrease the mortality rate?

We will do this by comparing the effect of responding after 2 or 5% of the population is infected. Note that these percentages are relatively high for complete countries but are realistic for the simulation with around 1000 agents.

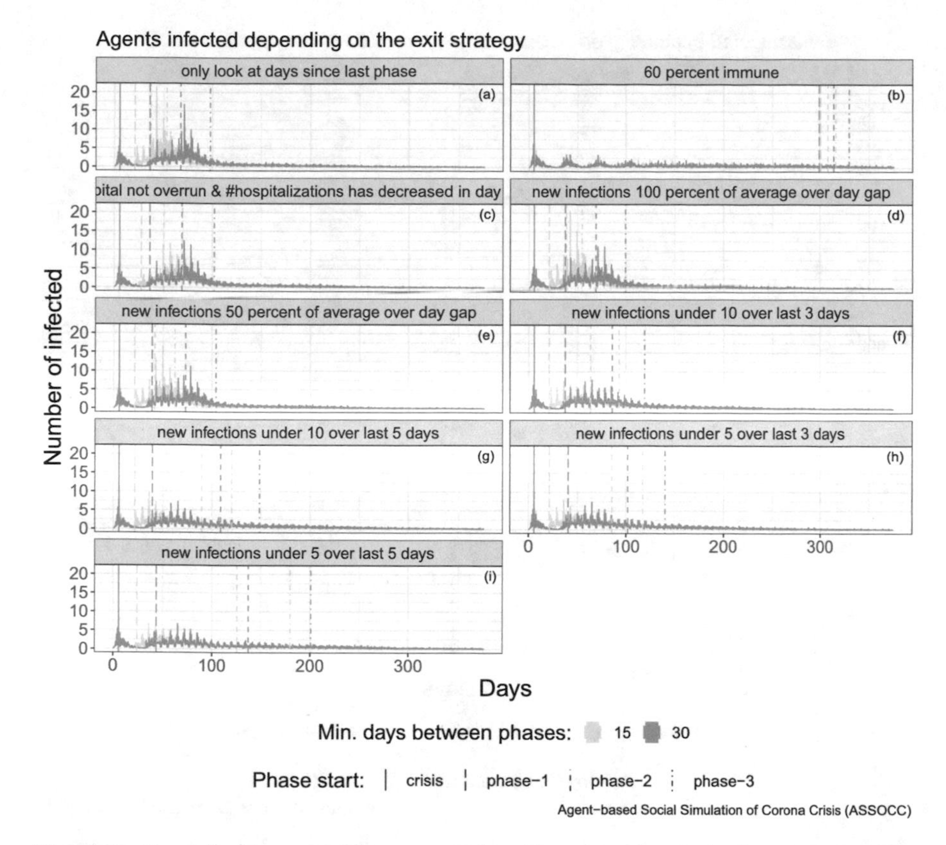

Fig. 10.11 New infections under different conditions. The vertical lines in the figure indicate when, on average, a phase shift happened

The infection peaks for the various trigger conditions can be found in Fig. 10.12. The red lines in the graphs now denote responding after 2% of the people is infected. The blue graph shows what happens if the government only reacts after 5% of the population is infected. As can be seen from this figure, a lower acknowledgement rate (and thus a faster government response) leads to a lower initial peak in the number of infections. This makes sense, because with a lower acknowledgement rate, the `ongoing-crisis` phase begins earlier, limiting the spread of the disease through a lockdown. However, maybe less intuitively, this also leads to a higher second wave.

The explanation for this higher second wave can be found in the cumulative infections plot Fig. 10.13. As can be seen in this plot, the number of people that have been sick at the start of `phase-1` is higher in the 0.05 case than the 0.02 case. This means that there are also more people immune in this case, and thus that they cannot get the disease a second time. This causes the second peak in the 0.02 case to be higher than the one in the 0.05 case, since the disease can spread easier through the not immune population.

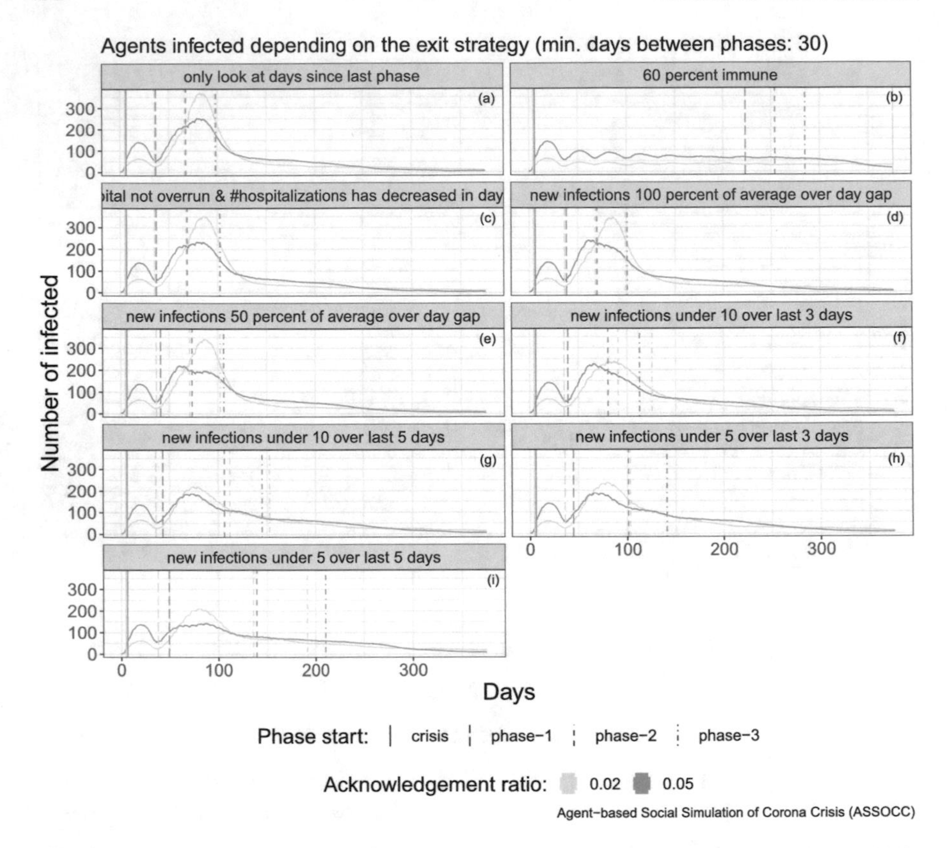

Fig. 10.12 Infection peaks under different conditions for the various acknowledgement rates. The vertical lines in the figure indicate when, on average, a phase shift happened

Similarly, a lower acknowledgement rate also sometimes leads to longer phases. For the herd immunity condition, this can be explained by the fact that the disease spreads less quickly at the start, so there are more people that still need to get the disease to get the required level of herd immunity. A similar explanation holds for the new infections conditions, where there are simply more people left to infect because the initial peak was lower, so the disease has more people to spread to, leading to a longer time with high infections.

When looking at the amount of deceased agents (Fig. 10.14), we can see that in some cases the acknowledgement ratio of 0.05 seems to have a higher number of deceased agents. However, as this difference seems to be at most 2, from the graph we cannot conclude that this setting leads to a higher mortality. In order to investigate this further, we have also created boxplots of the distributions of deceased agents for the different settings, which can be found in Fig. 10.15. From the boxplots it can be seen that there might be a difference under certain conditions, but these differences are actually not that big and because the boxes mostly overlap, they are mostly not

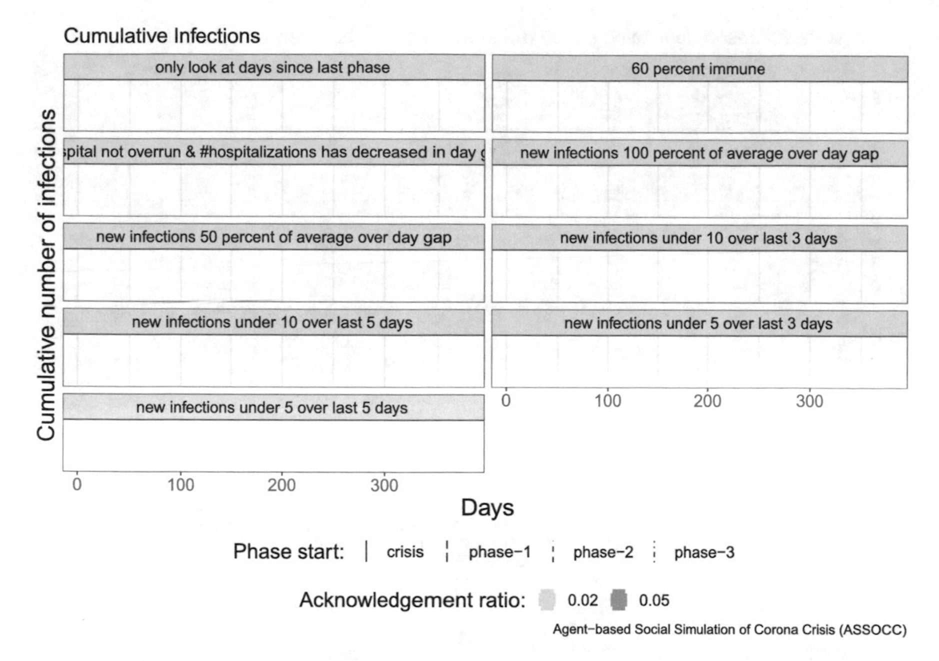

Fig. 10.13 Cumulative infections under different conditions for the various acknowledgement rates. The vertical lines in the figure indicate when, on average, a phase shift happened

significant. The exception to this is the 60% condition, but that one also had a higher number of cumulative infections, so that is in line with what is expected.

The acknowledgement ratios also have little influence on the number of hospitalisations (Fig. 10.16). While there are some differences, they are not consistent over conditions or days between phases, and they are very small. The only general conclusion that one could draw from these graphs is that when the government responds quicker, the number of hospitalizations is lower in the first phases and a bit higher in later phases. This might help if the hospitals have to run at full capacity at the beginning of the pandemic.

For the rest of the analysis we will focus on the cases with an acknowledgement rate of 0.02, so with a quick government response at the start.Thus assuming the most positive case.

10.4.4 Strategy Effectiveness

In this section, we will answer the following questions:

- What is the effect of the different exit strategies on the spread of the disease?
- Where do people get infected in the various strategies?
- Do the different exit strategies differ on the rate of hospitalization?

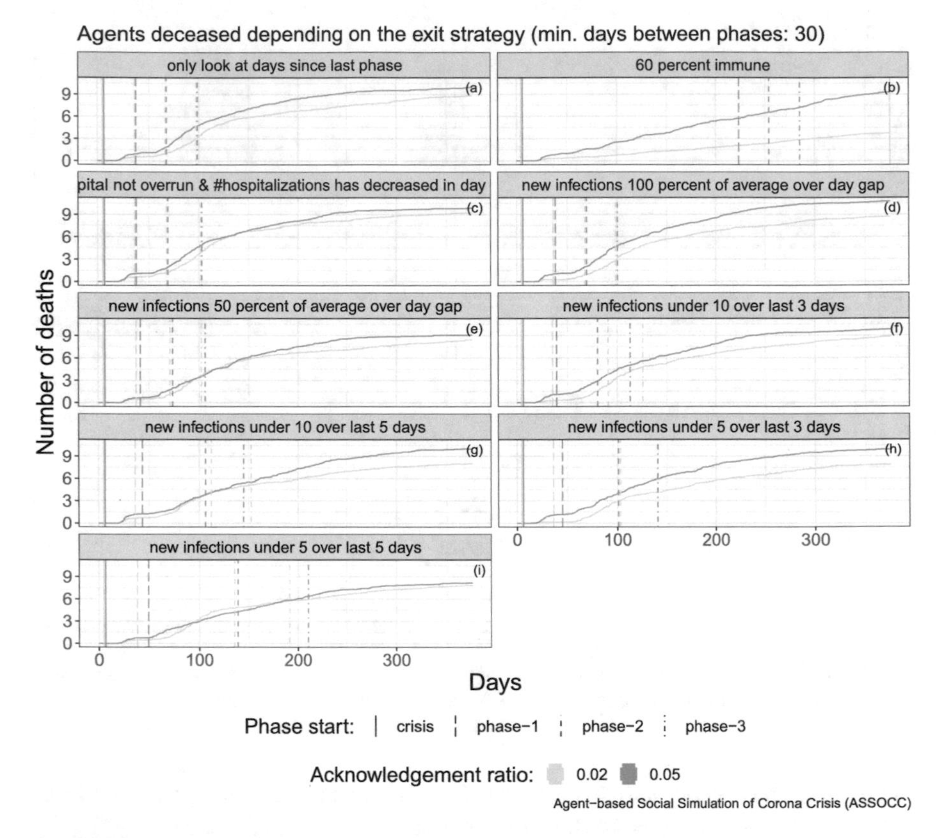

Fig. 10.14 Mortality peaks under different conditions for the various acknowledgement rates. The vertical lines in the figure indicate when, on average, a phase shift happened

- Do the different exit strategies differ on the rate of mortality?
- What is the effect of the different exit strategies on the economic situation of agents?
- What is the effect of the different exit strategies on the economic situation of companies?
- What is the effect of the different exit strategies on the economic situation of the government?

We will discuss the first question in the next subsection (Sect. 10.4.4.1). The second question will be addressed in Sect. 10.4.4.2. The third and fourth questions are discussed in Sect. 10.4.4.3. The last questions will be covered in Sect. 10.4.4.4.

Given the results of the previous sections we will now focus on the cases where the acknowledgement ratio was 0.02, and the minimum number of days between phases was 30, because these settings have shown to show the best effects. Therefore we compare the different exit strategies we devised and their effectiveness using

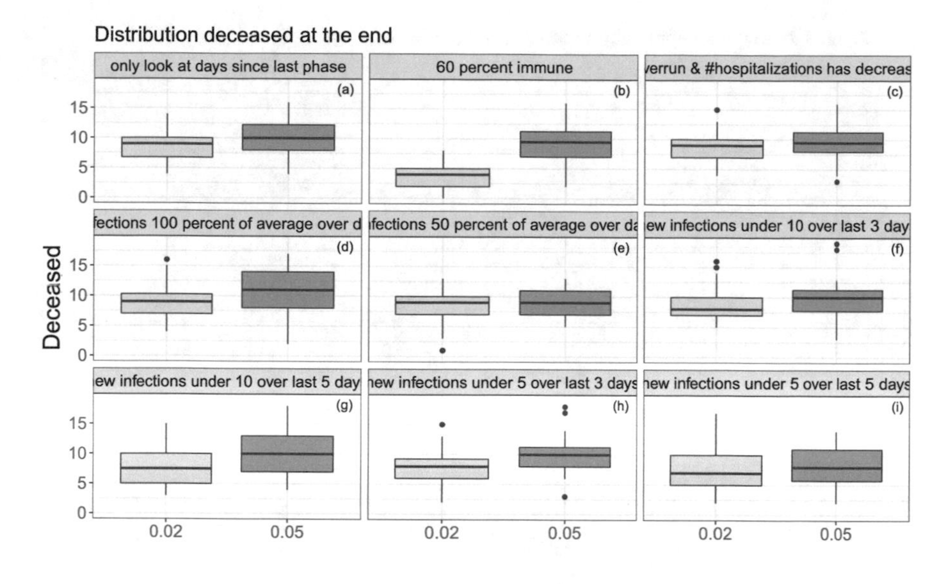

Fig. 10.15 Boxplots of the number of deceased agents at the end of the simulation for the two values of `acknowledgement-ratio`

these settings. Please refer to Table 10.3, which describes the strategies based on their restrictions and differences in the phases of the exit strategy.

10.4.4.1 Resulting Infections

We will start this section by making some general observations about the number of infected agents for the different strategies, the graphs of which can be found in Fig. 10.17. In these graphs the red line denotes the results from the business exit strategy, the blue line denotes the results of the leisure exit strategy and the green line the results of the public service exit strategy.

The first thing that can be seen in the plots is that there is always a second wave of infected agents. In some cases, notably the Leisure strategy in (d), (e), and (f), there is even a small third wave. These waves follow directly from lifting restrictions. This will allow people to go out more, giving the disease more of a chance to spread again (if one starts the exit too soon). The second wave is almost always the highest, with the exception of the 60% immune condition. However, in this condition we never leave the crisis phase, so the second wave we see here thus does differ from what is normally meant by a second wave. Given the fact that there is a second wave in most cases we could already conclude that the triggering conditions for `phase-1` are too lenient and should be even stricter than assumed here (based on real world criteria).

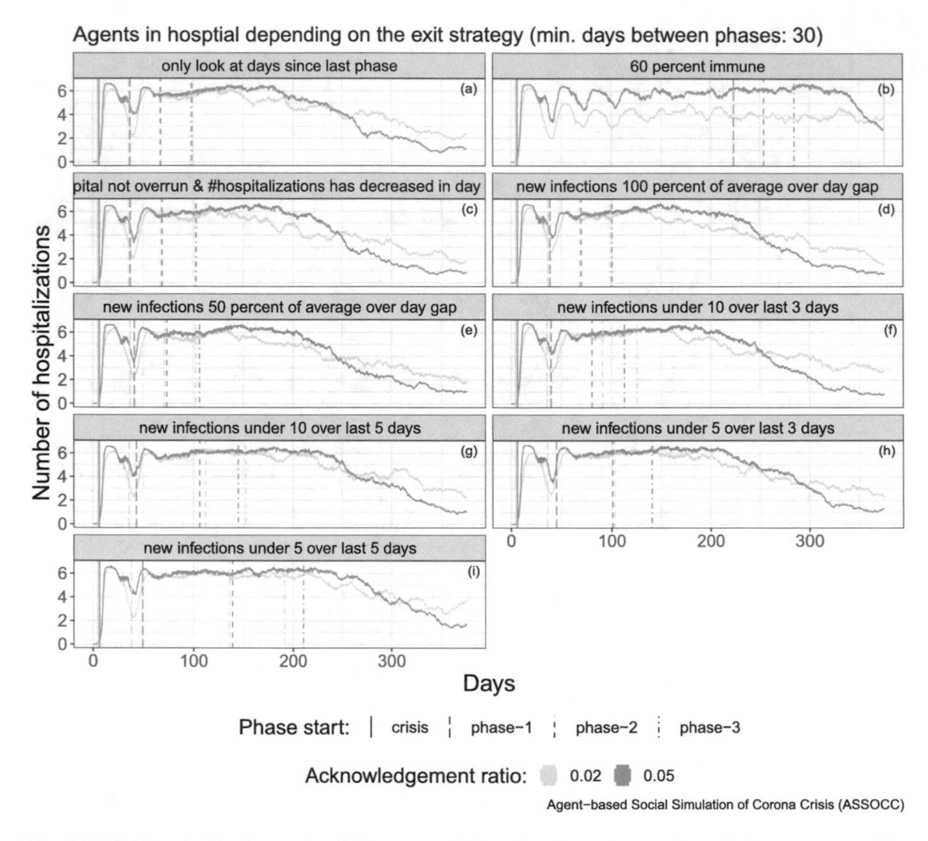

Fig. 10.16 Hospitalisation under different conditions for the various acknowledgement rates. The vertical lines in the figure indicate when, on average, a phase shift happened

The lack of a third wave is sometimes also caused by the fact that the simulation switches to the next phase while the second peak is still ongoing. This can be seen in (c), where the Public Services strategy is still at the top of its infection peak when the government moves into the next phase, making the peak wider instead of introducing a new one. So while there is not an actual third wave, the mechanism behind it is still there.

Looking at the differences between the strategies, we can see that the leisure strategy is performs the same under all conditions, and it always seems to create a similar peak with a maximum height between 200 and 300 people, even in plot (a), where the only indicator is time. The success of this strategy is due to this strategy keeping a lot of measures active, such as the closing of schools and workplaces, and keeping busses at half capacity. This means that there are less locations where the virus can spread thus preventing a big second wave.

The accumulated infected agents for this strategy are given in (Fig. 10.18). We can see the total accumulated number of infected agents in the green line. This line increases quickly in `phase-1`, and after that steadily continues going up until

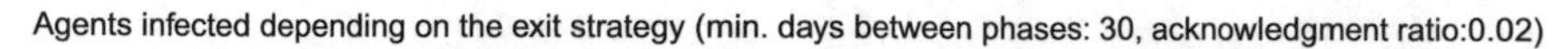

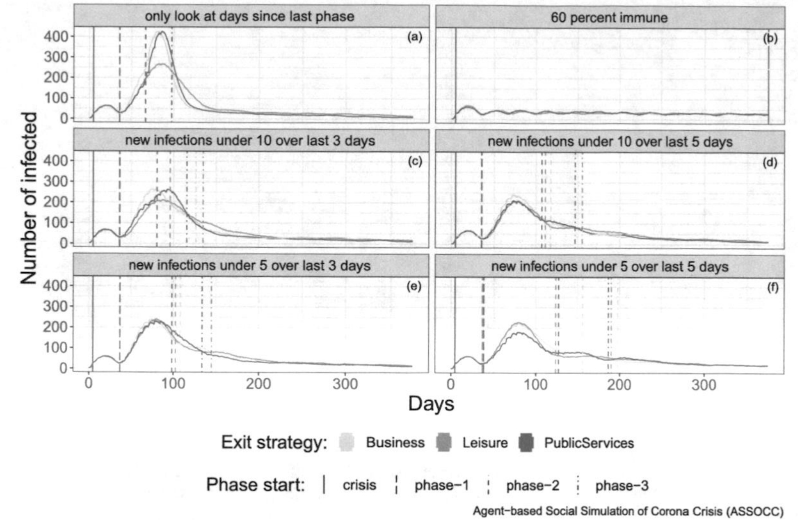

Fig. 10.17 Infection peaks under different conditions where the acknowledgement ratio was 0.02. The vertical lines in the figure indicate when, on average, a phase shift happened

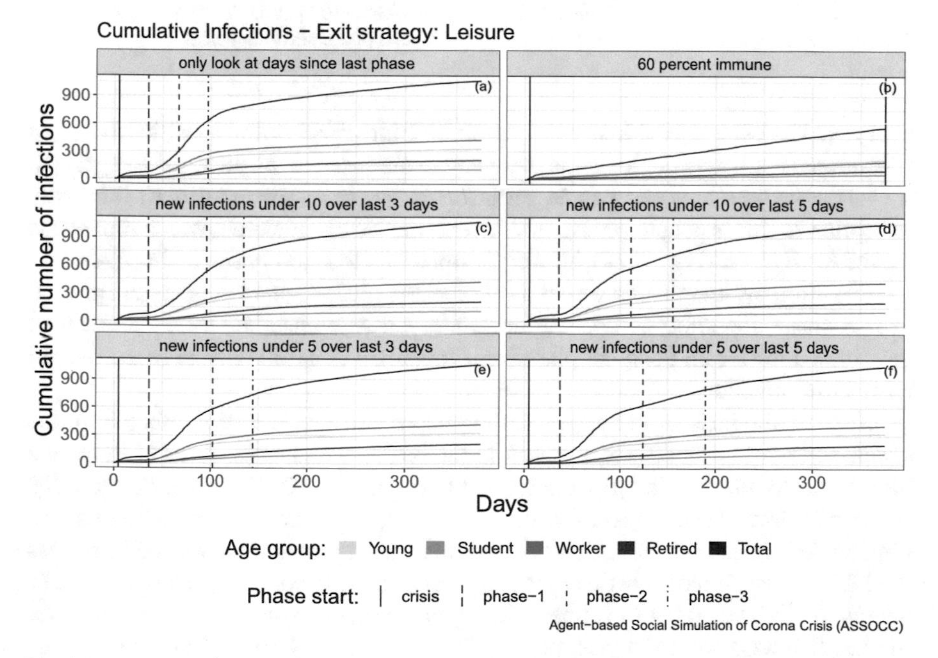

Fig. 10.18 Cumulative infections for the different age groups in the Leisure strategy under different conditions with the acknowledgement ratio at 0.02. The vertical lines in the figure indicate when, on average, a phase shift happened

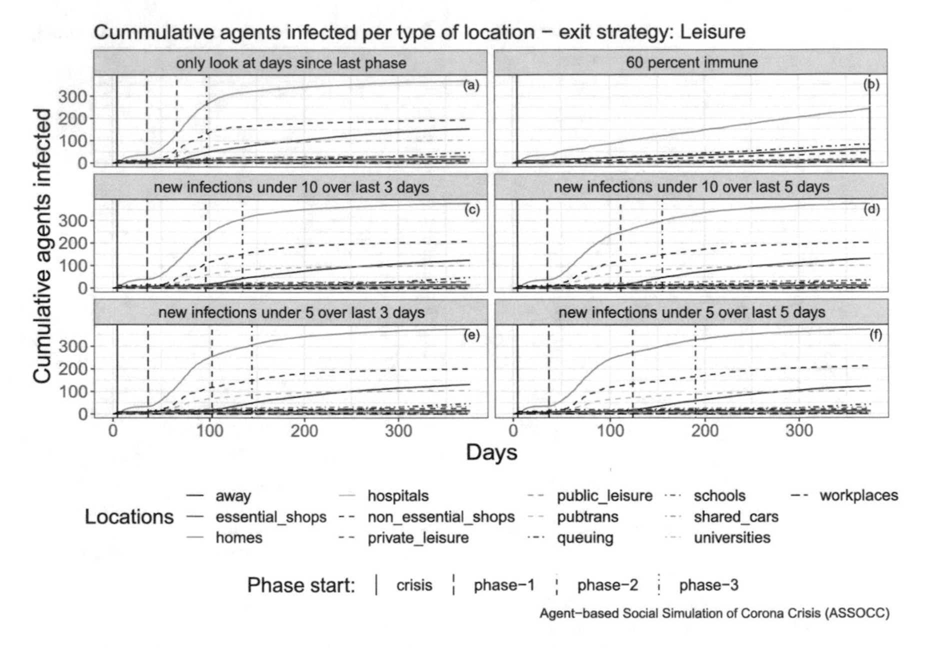

Fig. 10.19 The number of people infected in each type of location per day for the leisure strategy. The vertical lines in the figure indicate when, on average, a phase shift happened

everyone has been infected. There seem to be little effect of the specific triggering conditions that are used. Neither is there any difference for the different age groups.

When we look at where people get infected, we see a similar pattern. The only exception to this is people getting infected at the “away” location, which represents out-of-town travel (Fig. 10.19). However, since this location opens up in `phase-2` for this strategy, and there are also a lot of other places where people get infected in that phase, the effect of opening up this location is absorbed by the amount of new infections for all the other places and the effect is hardly visible in the overall number of infections.

Whereas the peak of the waves in the leisure strategy is hardly affected by the specific triggering conditions, the peak in the business strategy is influenced quite a bit by the strictness of the triggering conditions. In Fig. 10.17a, it can be seen that if we only take time as trigger for the next phase, the peak of the infection will be high, at 433 agents infected. However, adding the trigger that there need to be less than 3.33 agents infected per day over the last 3 days, lowers the peak down to 273 agents. If we take that even further to on average one person infected per day over the last five days, we get a peak of 245, which is 56.6% of the original number of infected agents at the peak.

Thus, as expected, if the amount of infected agents is lowered before we go into a new phase of the strategy, then we will no longer be at the top of an infection peak,

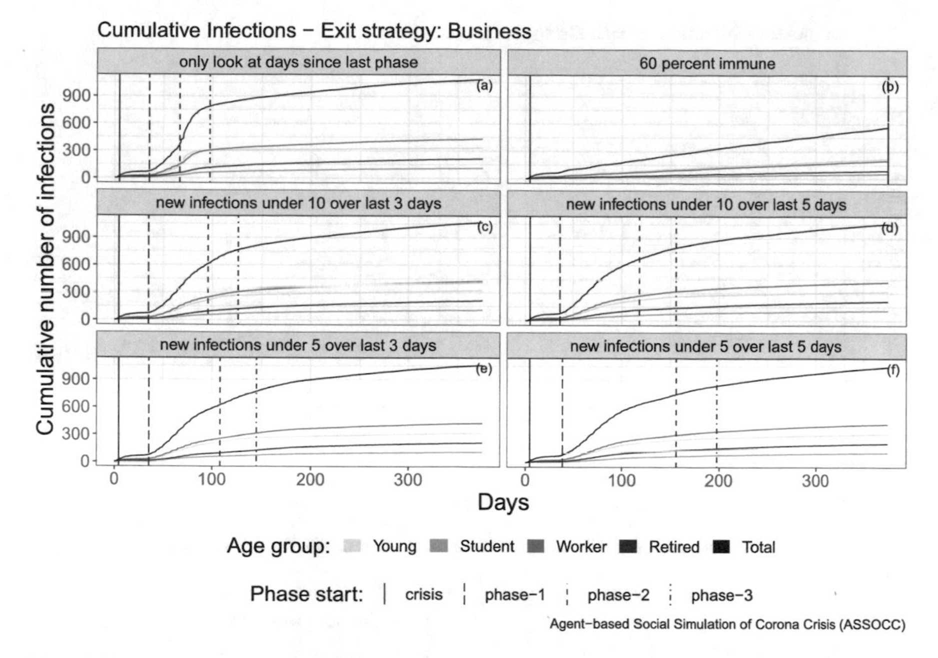

Fig. 10.20 Cumulative infections for the different age groups in the Business strategy under different conditions with the acknowledgement ratio at 0.02. The vertical lines in the figure indicate when, on average, a phase shift happened

and the highest infection peak will be lower. However, this in turn also means that it will take longer for the phase to end. We can observe this result exactly in the figures.

In the public services strategy, we see similar effects, but here we also get the additional effect that the stricter the phasing out condition, the more prominent its third wave is. We can explain this difference between the Public Services strategy and the Business strategy by looking at the accumulations of infected agents for the different strategies. The plots for comparison can be found in Figs. 10.20 and 10.21.

When looking at the green line (representing the total infected agents) in Fig. 10.20, there is a big increase in infected agents at the start, but a steady increase in later phases. The lines for all the age groups show a similar trend, with the exception of the youths (orange), which sometimes has a small bump. This bump can be explained by the schools reopening. This bump can also be seen in the infected per location, confirming that this is indeed due to the reopening of schools (Fig. 10.22c shows this the clearest).

In the public services strategy, on the other hand, we can see these kinds of bumps at multiple locations. Take for example Fig. 10.21f around the start of `phase-2`, where it is visible most clearly. Here we can see the number of infected workers increase, have the increase slow down, and then increase again for a bit, and there is a similar result for the youths. This latter effect is easily explained by the observation that there is an increase in the number of infections at schools at the same time

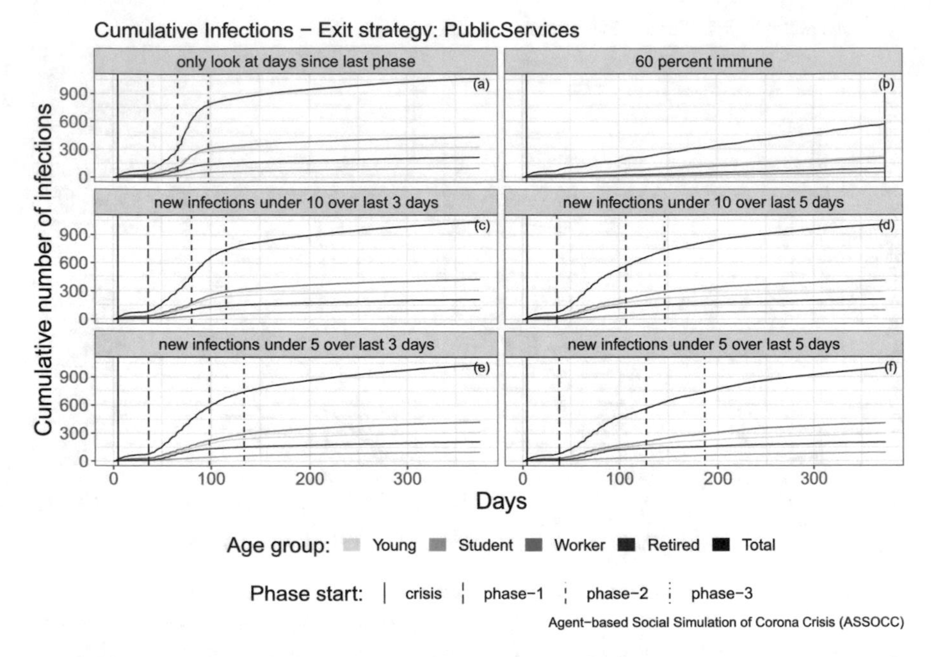

Fig. 10.21 Cumulative infections for the different age groups in the Public Services strategy under different conditions with the acknowledgement ratio at 0.02. The vertical lines in the figure indicate when, on average, a phase shift happened

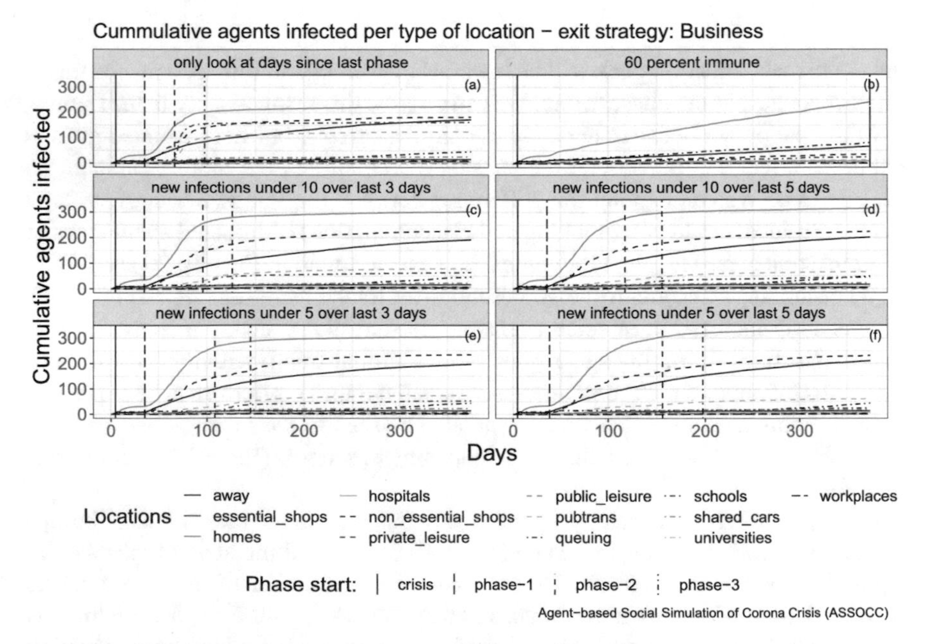

Fig. 10.22 The number of people infected in each type of location per day for the business strategy. The vertical lines in the figure indicate when, on average, a phase shift happened

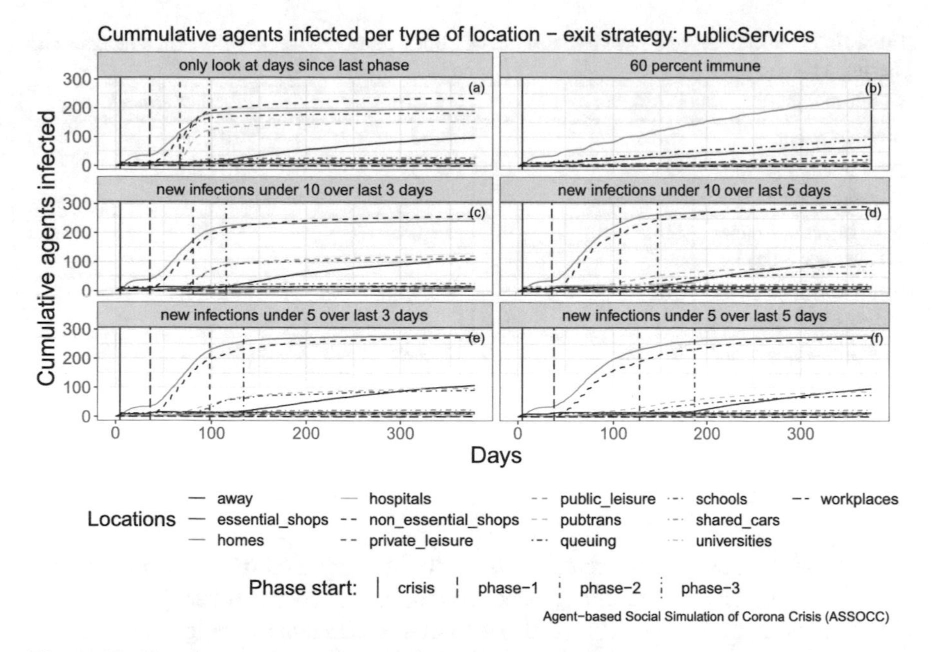

Fig. 10.23 The number of people infected in each type of location per day for the public service strategy. The vertical lines in the figure indicate when, on average, a phase shift happened

(Fig. 10.23f). However, the bump noticed for the workers does not have a similar bump in Fig. 10.23 and is thus less easily explained.

What we can see in Fig. 10.23f is that there are small bumps for multiple locations (notably homes, non essential shops, and private leisure) around this time, which together would explain the bump for the workers. This means that the opening of the schools increase infections at home, which then in turn also increases the infections between workers (and others) as an effect of having more infected people around.

10.4.4.2 Locations Where Infections Arise

In the above discussions we have only compared the strategies based on the number of infected agents. However, we also should compare them based on the locations where infections take place. It might be that just lifting a restriction on a certain location is the key to most of the development of the spread of the virus afterwards. Thus whenever this particular location is opened in whatever strategy it will determine the outcome. And if two different strategies overlap on this location then they will also have similar results. In order to make a comparison based on locations, it is easiest to look at cumulative plots of infected agents, since then we will not have to look for spikes in infections, but can just see in what place agents got infected the most. For the business strategy that plot can be found in 10.22, for the leisure strategy in Fig. 10.19, and the public services in Fig. 10.23.

Table 10.5 The number of people infected taking public transport over the different strategies and phasing out conditions

Condition_phasing_out	Business	Leisure	PublicServices
60% immune	95.25	95.50	107.83
New infections under 10 over last 3 days	125.33	149.08	137.00
New infections under 10 over last 5 days	116.50	137.08	99.33
New infections under 5 over last 3 days	118.17	149.92	110.92
New infections under 5 over last 5 days	104.58	143.08	102.00
Only look at days since last phase	170.50	150.00	165.08

In the plots, it can be seen that, no matter the strategy, a lot of agents get infected in their homes. This makes sense, since most strategies focus on keeping people in their houses as much as possible, and is also in line with other findings (Sects. 5.5.4, Fig. 6.15, and 7.5.3). We can also see that, no matter the strategy, a lot of infections happen in non-essential shops. Since the non-essential shops are one of the first things to open—in `phase-1` in all strategies—this also makes sense since that is then one of the only places where the agents can be not at home, which might cause them to leave their house and go there. A third place where a lot of agents get infected is the "away" location, which represents out of town travel. However, this last one is a lot higher in the business strategy than in the other two strategies, due to it opening up earlier.

Finally, public transport also seems to be a location where the spreading of the virus is easy and leads to many new infected agents. In all scenarios, public transport shows up in the top 5 of most infectious places. However, if we compare the number of infections for the different strategies, we can see that there are a lot more infections in public transport in the public services scenarios compared to the other scenarios. Since in the public services scenario the bus capacity has increased, more agents have the opportunity to take a bus together with other agents and thus have a chance to get infected. However, in the cases when the bus capacity is not increased (in both the Business and Leisure strategies), agents will spend more time waiting for a bus if they want to go to work, school or other place. Thus, these scenarios will lead to more infections occurring during the queuing. This is indeed what we can see in the results. If we combine the number of people infected in queuing at the end with the people infected in public transport, we can see that there are only small differences between the different strategies (Table 10.5). So, the total amount of infected agents caused by taking public transport does not differ all that much over the exit strategies.

Schools show up as locations with a lot of infections. This is particularly noteworthy, since they are not open for the entire duration of the simulation. However,

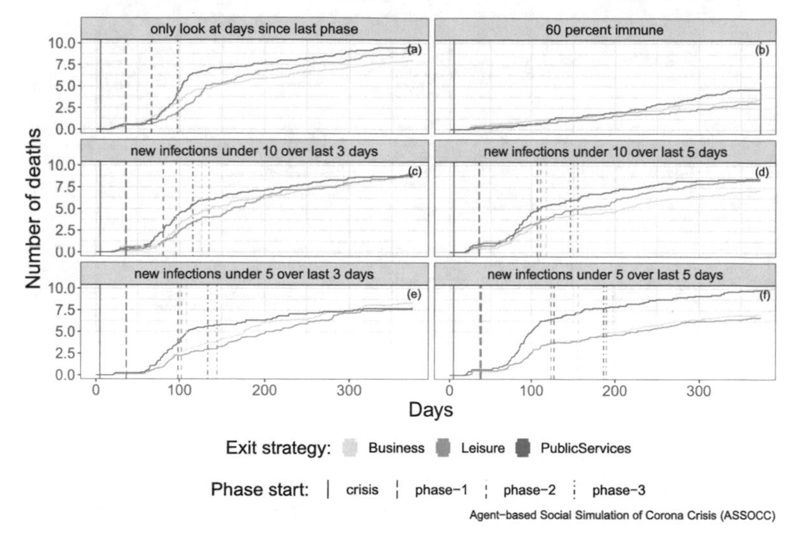

Fig. 10.24 Mortality under different conditions where the acknowledgement ratio was 0.02. The vertical lines in the figure indicate when, on average, a phase shift happened

we can also see that opening schools up a bit later can counteract this effect by quite a bit. Take for example plots (c) and (d) in Fig. 10.23. In (c) there are on average 139 agents infected in schools, but in (d), this number dropped down to 82. The difference in time between the closing of schools and their reopening in (c) and (d) is only 24 days (58 days versus 83 days). This difference on its own probably leads to the occurrence of a third wave in Fig. 10.17d for the public services strategy, while (c) in that same figure just had a wide second wave. So, we can conclude that opening schools in this scenario might lead to many infections.

10.4.4.3 Hospitalizations and Mortality Rate

Besides infections, there are also two other common performance indicators that are used to evaluate strategies, which are the number of deaths, and the amount of people that are hospitalised. As can be seen in Fig. 10.24, the differences between the number of deaths are not big, but there seems to be a pattern there, where the business strategy in general has the lowest number of deaths in total, with the public services strategies usually having most deaths and the leisure strategy ending in between. We can explain this general result by the fact that the public services strategy allows retirees to go out first. Because this is the most vulnerable group it makes sense that there would be a higher number of deaths than in the other strategies.

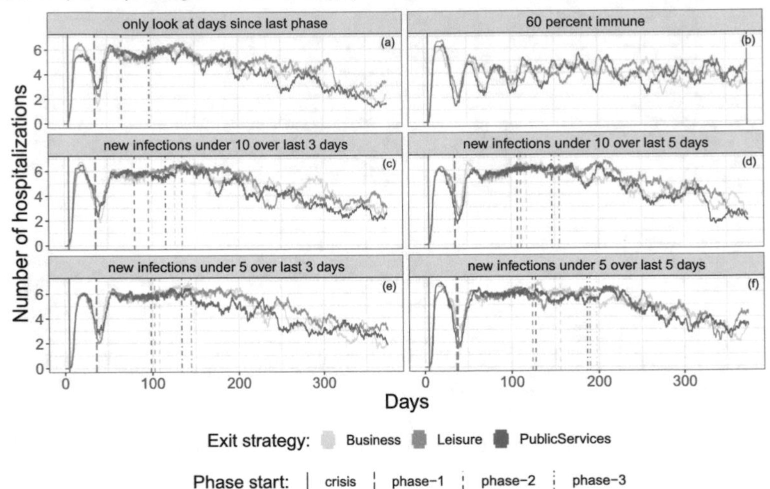

Fig. 10.25 Hospitalisations under different conditions where the acknowledgement ratio was 0.02. The vertical lines in the figure indicate when, on average, a phase shift happened

From Fig. 10.25, we can see that there is little effect of the strategies on hospitalizations. The overall differences at the end between the strategies tend to be not more than one agent. What is of note though, is that the maximum number of agents in the hospital in all scenarios is 6, while there are 13 hospital beds available. This means that in no case the hospital is overrun with too many infected people. This means that in the simulation, the exit strategies provide enough spread to stop the hospital from being overrun. This also explains why the "hospital not overrun & #hospitalizations has decreased in day gap" condition is not very effective in curtailing the spread, since the first condition is always true. This means that for the entire condition to be true, only one person needs to leave the hospital without a new one coming in before going to the next phase.

10.4.4.4 Results in the Economic Situation

Exit strategies also have an impact on the economic situation. Both on different groups of people as well as on the overall government deficit. In this section we evaluate the different strategies on how well they perform for the economy. First of all, we check the total effect on the society by looking at the effect of the exit strategy on the total amount of capital in the system. The plot for that can be found in Fig. 10.26. Here it can be seen that the amount of capital in the system is going up, which means that there is enough production to sell goods outside the system. It can also clearly be seen that the business strategy has the most goods being sold.

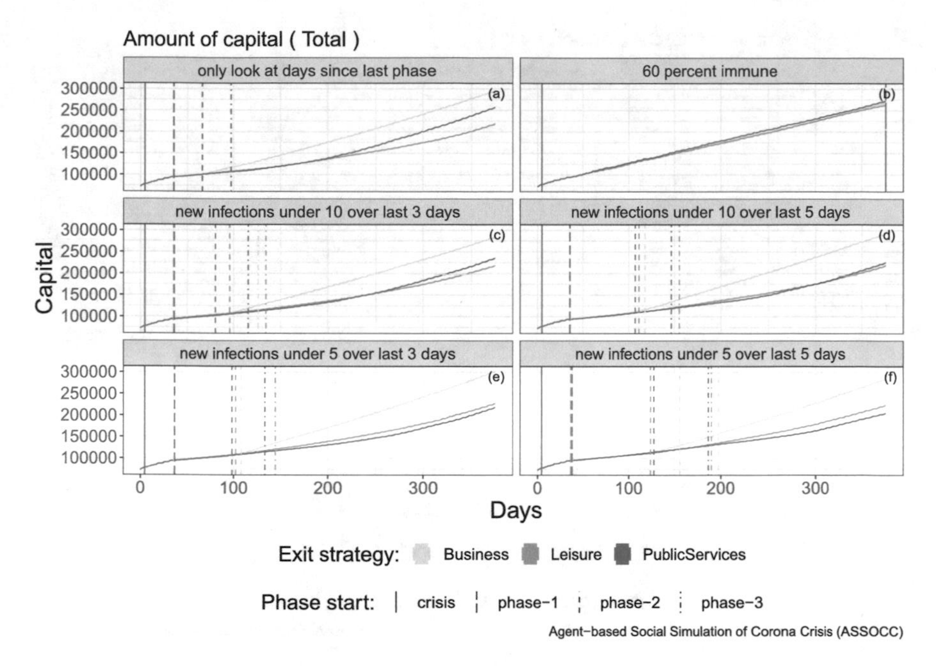

Fig. 10.26 The total capital in the simulation for the different exit strategies. The vertical lines indicate when, on average, a phase shift happened

The point where the business strategy gets this big increase the first time is after the start of `phase-2`, which lifts the recommendation to work from home. Lifting this restriction will increase productivity as people are assumed to be more productive (in general) at work.

We can also check the amount of deficit the government incurs under the different strategies. The graph for this can be seen in Fig. 10.27. Here it can be seen that under all exit strategies, the government does not make enough money in taxes to fix their deficit. As expected, the shorter the exit strategy, the smaller the government deficit. However, what is interesting to see is that the public services, and not the business strategy, seems to keep the government deficit smallest. This might be because of spending by retirees, which in the business and leisure strategies are asked to stay in isolation, but in the public services strategy are allowed to leave their house and thus buy things at stores. the taxes on these sales will help to decrease the government deficit.

The capital owned by the agent groups in the different scenarios can be found in Figs. 10.28, 10.29, and 10.30. Here we can see that the capital of retired agents keeps building up in the business and leisure strategies (because they have no way of spending their money), but that they do not have a similar buildup in the public services strategy. Furthermore, in these figures we can see that this buildup of capital is smaller in the cases where the difference between the various government capitals is also smaller. Therefore, this seems to support our theory.

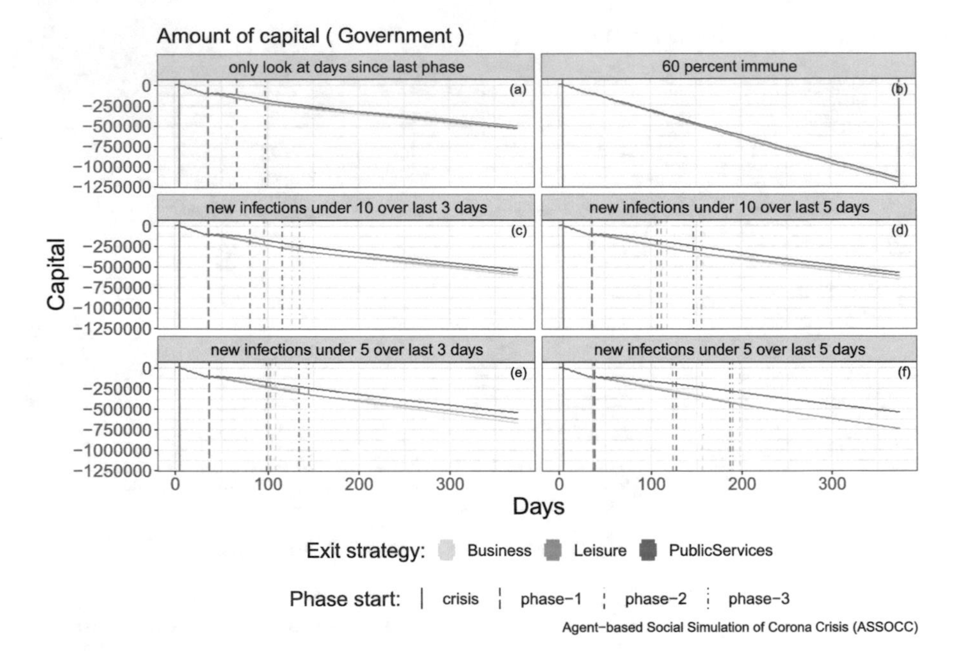

Fig. 10.27 The government capital in the simulation for the different exit strategies. The vertical lines indicate when, on average, a phase shift happened

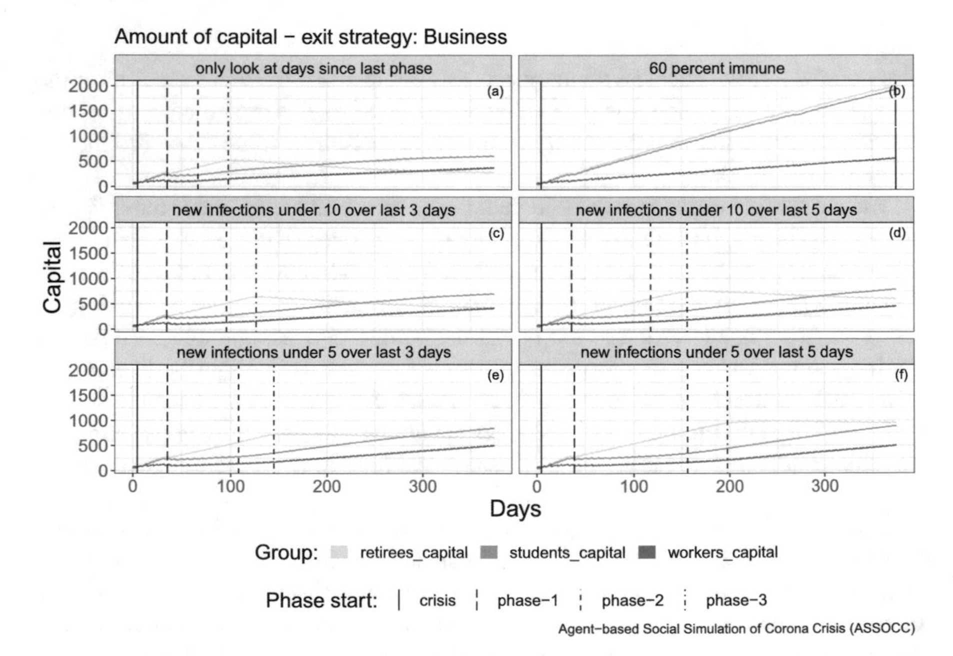

Fig. 10.28 The capital of the various agent groups for the business exit strategy. The vertical lines indicate when, on average, a phase shift happened

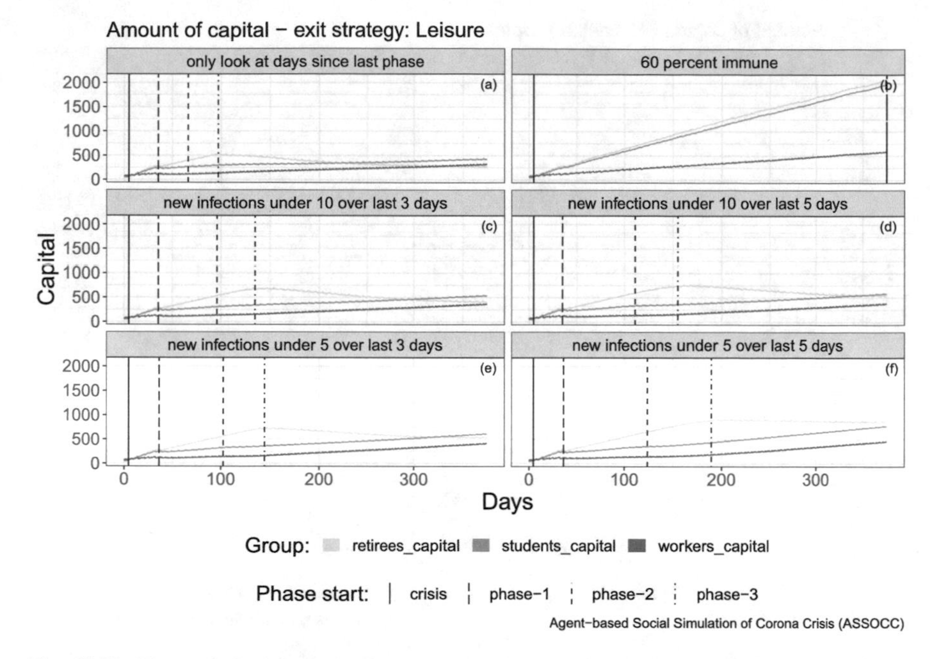

Fig. 10.29 The capital of the various agent groups for the leisure exit strategy. The vertical lines indicate when, on average, a phase shift happened

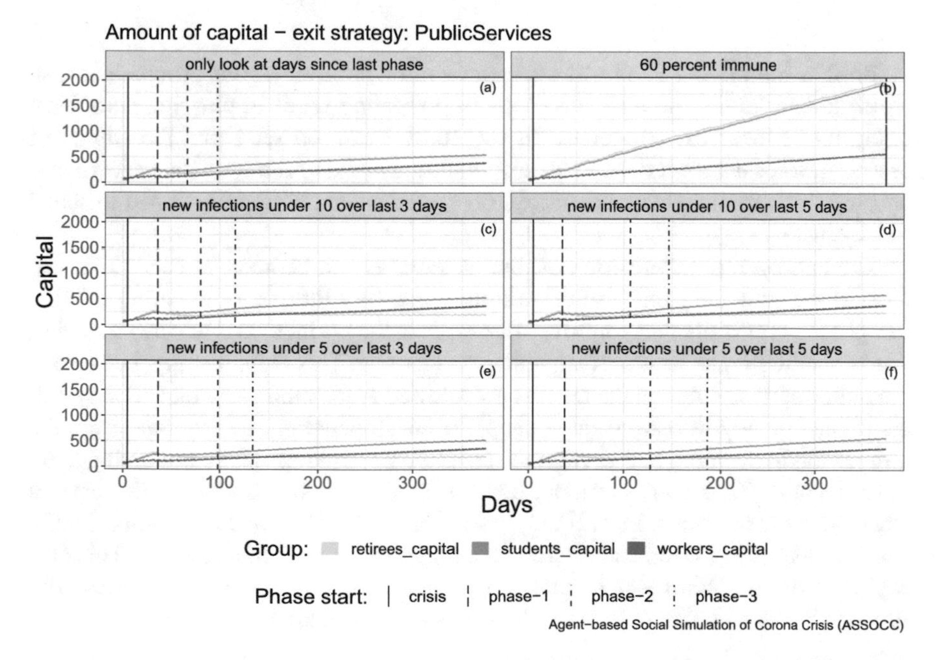

Fig. 10.30 The capital of the various agent groups for the public services exit strategy. The vertical lines indicate when, on average, a phase shift happened

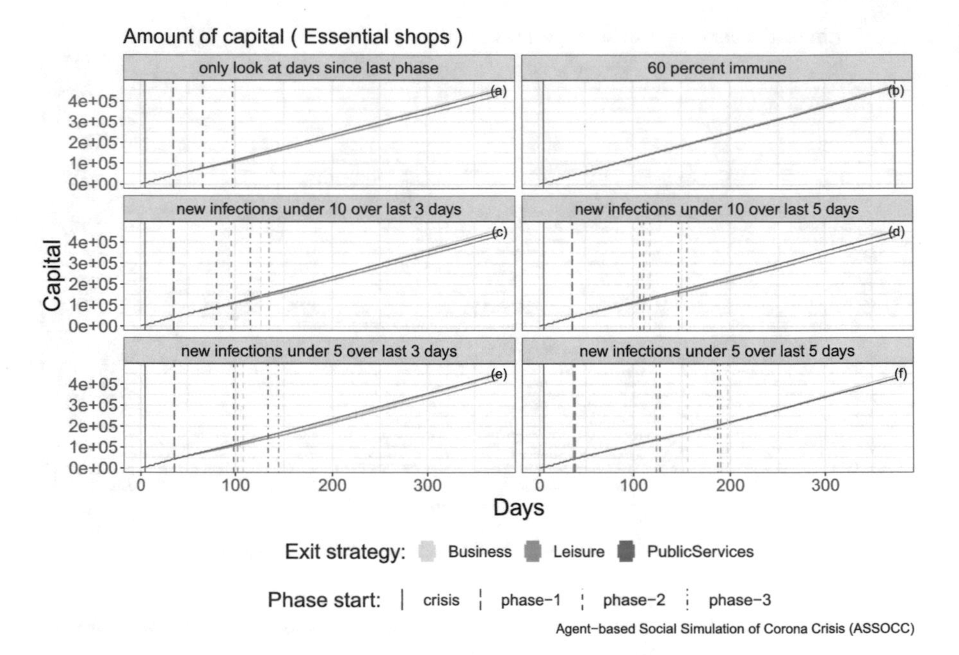

Fig. 10.31 The capital of all the essential stores combined. The vertical lines indicate when, on average, a phase shift happened

Besides the effect of the exit strategies on the different agent groups it is also worth to check how the different shops (representing retail in our simulation) are doing under these exit strategies. In Fig. 10.31 it can be seen that the capital of essential stores does not depend on any condition or exit strategy. This makes sense because the essential stores are always open, and agents will always need to spend money there.

The non essential stores do not fare as well, as can be seen in Fig. 10.32. For most of the strategies, many of these stores have very little income during the first few phases, and while they improve afterwards in the business and leisure strategies, this is not always the case for the public services one. This is interesting, because in the public service strategy the retirees are allowed to go shop early on and they will also spend money in non-essential shops. Ironically enough, this income from the retirees might be enough to prevent the shops to go bankrupt while not giving them much income. Thus there are more non essential stores left in the public services strategy, as can be seen in Figs. 10.33, 10.34, and 10.35. This causes the non essential stores to have more overhead relative to their income under this strategy. This also explains why the non essential stores have less capital under the conditions with shorter phases, since in those cases the stores are less likely to go out of business.[5]

[5]This can only happen if the stores run out of both capital and goods to sell.

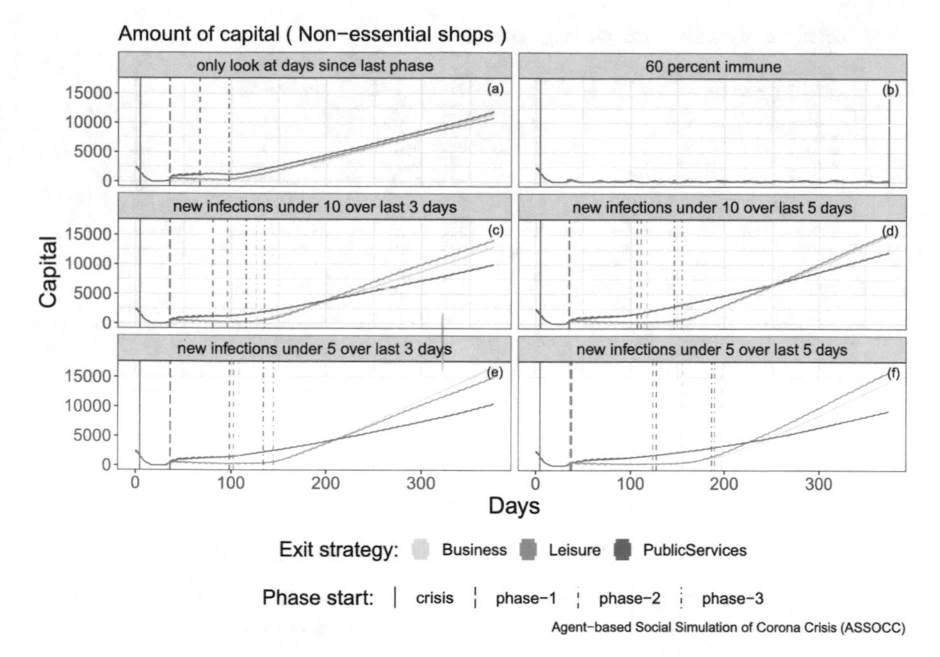

Fig. 10.32 The capital of all the non essential stores combined. The vertical lines indicate when, on average, a phase shift happened

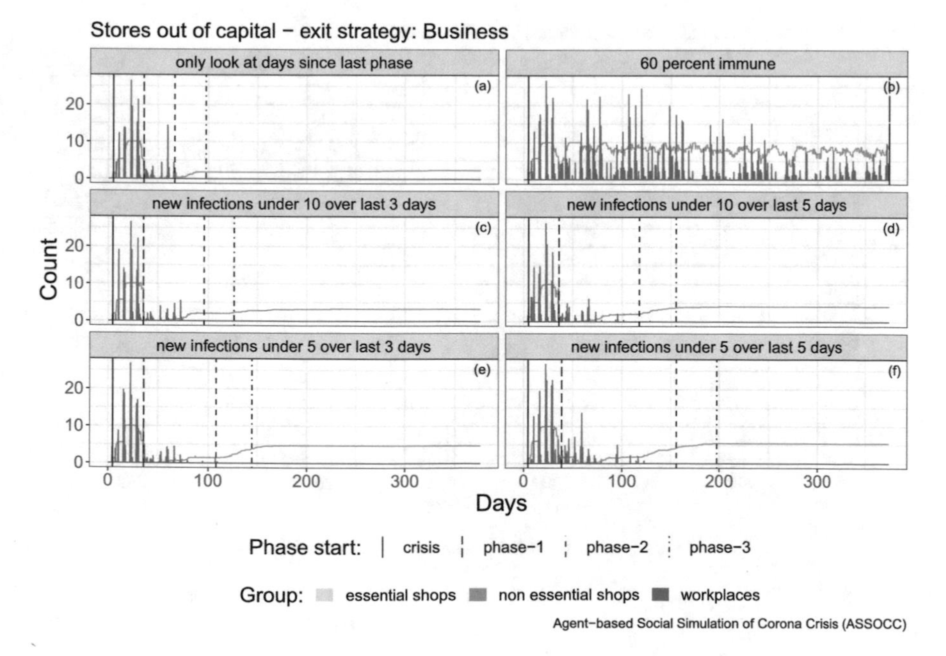

Fig. 10.33 The number of locations out of capital for the business exit strategy. The vertical lines indicate when, on average, a phase shift happened

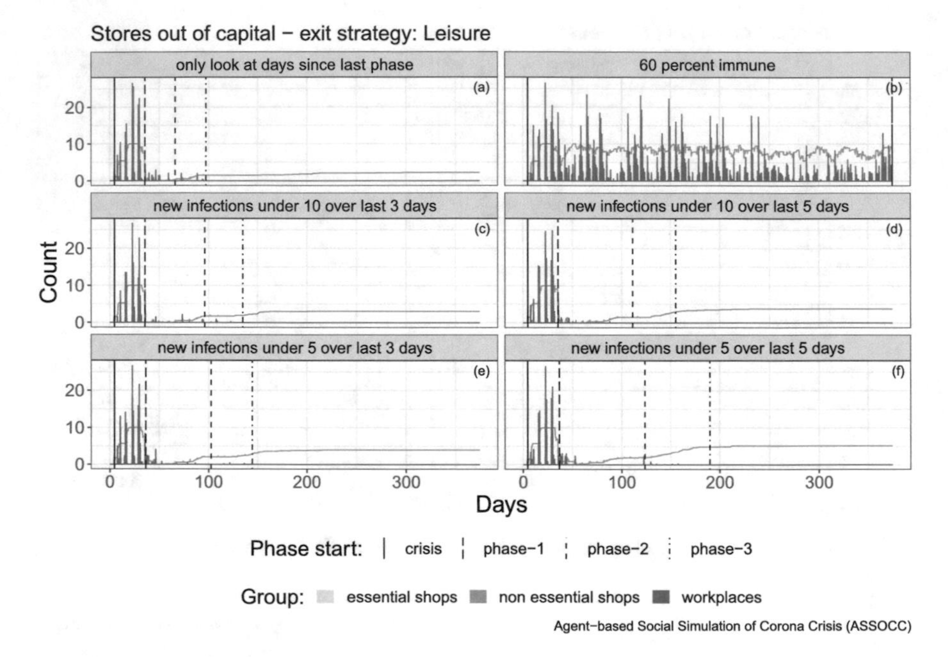

Fig. 10.34 The number of locations out of capital for the leisure exit strategy. The vertical lines indicate when, on average, a phase shift happened

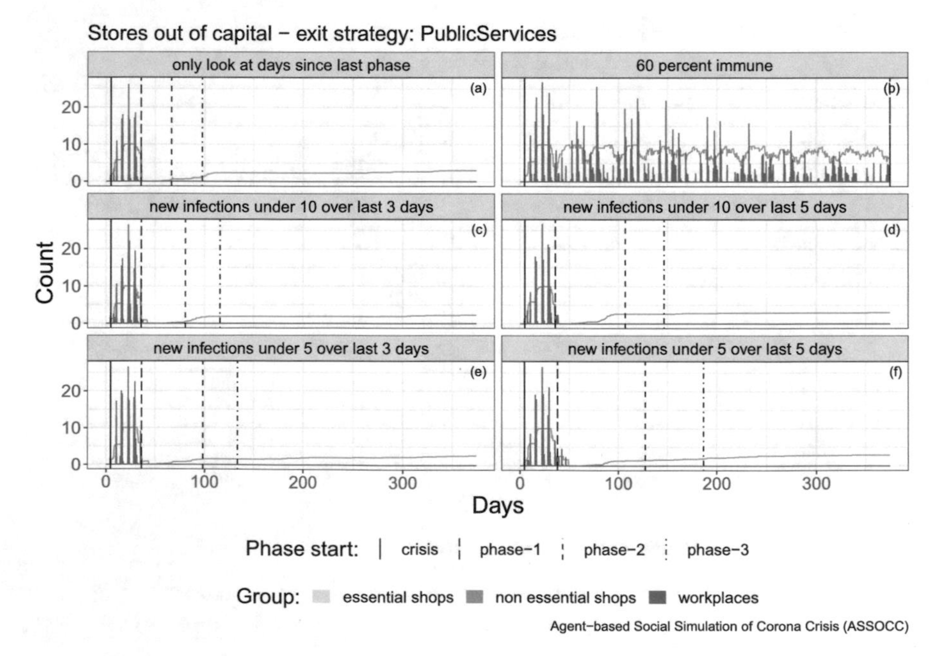

Fig. 10.35 The number of locations out of capital for the public services exit strategy. The vertical lines indicate when, on average, a phase shift happened

Finally we can also see in this figure that during the `ongoing-crisis` phase and sometimes the `phase-1`, the workplaces will run out of capital as well. They recover quickly afterwards, because we do not have the notion of bankruptcy for workplaces (which are representing factories). The reason for the decrease of capital of workplaces comes from the fact that they keep paying personnel, but if they work at home they are less productive. At the same time the non-essential shops are closed and do not need products. Finally, the essential shops only need products every now and then and thus in between these orders the workplaces can run out of capital. For our simulation this has no effect as we do not let the workplaces incur debts or go bankrupt. However, in the real world this might have more effect.

10.5 Discussion

Before we will give some final conclusions in the next section, in this section we will discuss some of the findings and interpret them as well in the context of the realism of the simulation and events in the real world.

The main reason why it is difficult to use the results of this chapter to make predictions for the real world the problem of the law of small numbers. Since the amount of agents in the simulation was limited, and the mortality in most cases was low, small changes in the random setup of households could have "big" consequences in the mortality or hospitalisation, since those number increased by one or two. Therefore, these results all need to be seen as trends rather then predictions. In general the difference between the strategies is more important than the results of any strategy separately.

Besides the law of small numbers, there is also the problem that due to a small amount of agents that are all relatively interconnected, the virus spreads fast when it gets the chance to spread. This leads to situations where after a second wave almost all agents have been infected already and there is no third wave due to herd-immunity. It is a big incentive to scale up the simulation in order to see these effects of exit strategies on the longer term as well.

Besides the scale of the simulation, the fact that it was build on top of the already existing ASSOCC simulation also meant that we inherited some of its technical debt. This is most visible in the way the exit phases are build on top of the old measurements system. In this system, measurements are activated once a certain percent of the population got infected, and de-activated once the number of infections dropped below that percentage. In order to trigger all the measures once the crisis started, we set this percentage for all measures equal to our `acknowledgement-rate`. However, this meant that in some cases (in particular the "new infections 60 percent of average over day gap" condition and at the end of `ongoing-crisis`) measures were dropped while this was not mentioned in the exit strategy. Since this did not come up often, was the same for all situations compared, and would have been hard to fix, we left it as is.

Another issue that we had to solve was the translation of real world measures to the elements available in our simulation framework. E.g. opening pubs and restaurants but only allowing people to sit and not more than a certain amount per table. Or restricting restaurants to take away services. It is unclear what the impact of those measures is on the number of customers or what they order, etc. Thus we have not implemented these more detailed measures, but tried to approximate them through more clear measures. In top of this the real world exit phases were sometimes well defined, but not followed in practice at all. Changes were continuously made and restrictions lifted whenever possible (even if criteria were not met yet). This made it impossible to validate the simulation results against specific exit strategies used by countries, or give specific advice to policy makers. Despite this, we could see some patterns that could inform decision makers. E.g. unexpected economic effects and little difference between some hot-disputed strategies on the number of infections.

What is also quite different from real life, is the amount of control of the behaviour of the agents. In the simulation it is easy to simply stop all agents from going to, for example, private leisure. However, in the real world this process is harder. While it might be possible to close down all restaurants, stopping people from hanging out at a friends house is a lot harder, but this is very important for containing the spread of the virus. We do have some measure of these effects, because when our agents cannot satisfy their socialising needs by going to private leisure places they will try to find other actions to satisfy their needs.

In this simulation we always had enough hospital beds to cope with all patients, namely 13 in total. On 1152 agents, that makes for almost 1 bed per 89 agents. A more realistic number would be in the range of 6 beds. However, this would mainly lead to a few more deaths and have little influence on the outcomes otherwise. So, we decided to leave the amount of beds at this maximum, which is in most countries the maximum that could be mustered if really needed.

Despite these caveats, there are some general conclusions that can be drawn from this scenario. The first of these is that increasing the length of the exit strategies helps in spreading the infection curve, since this allows the number of infections to lower again, which is a big factor in the spread of the virus. While maybe not exactly surprising, the simulation has shown that a waiting time of only 15 days, which was not uncommon in the real world, is not enough to properly spread out the curve. The same effect can be achieved by setting strong enough conditions for going to the next phase.

Similarly, it has been shown that a quick response can lead to a smaller initial peak of infections, which can then lead to a higher second wave, so in those cases extra care needs to be taken to not lift restrictions too quick! However, this might also be an effect of the small amount of agents. If the same criteria would have been used on country level the virus would have been almost extinct before restrictions are lifted. But having only around 1000 agents, it means that having 10 agents still infected is 1% of the population. Which is far more than targeted in the real world.

When it comes to opening up places, the simulation shows that opening up many different places might be the best solution. If there is only one (type of) place opened up where people can go, this place will attract a lot of people to it. This could be seen in the Public Services strategy. If we compare this with the Leisure strategy, where many different places opened up at once, we see less effect of the condition used, but also a lower spread.

On the other hand, this can also be used in a strategy explicitly. If you know a lot of people will get infected at a certain place, then this will also allow you to instate track and tracing measures or testing for those places specifically. This will allow you to track the virus in a targeted manner and protect the population that way. Furthermore, one could also instate certain conditions those places need to meet before they can open, such as ventilation or face masks.

Another finding is that schools also have a large effect on the spread, and thus their opening needs to be done in a safe manner as well. One approach (that we were not able to implement in this model) that could be a good model was that with limited capacity schools, such as happened in Germany. In this system, half of the class went to school in the one week, and the other half went the next week. This seemed to have led to less spread of the virus. However, it should also be stated that in some countries (like Sweden) where the basic school never closed, the spread of the virus seemd not to be affected at all.

Opening workplaces, on the other hand, seemed to have not much of an effect on the spread of the virus. This is probably due to the fact that many agents were working from home, only going back to the workplace after the pandemic had mostly subsided. This indicates that making sure people have a good working environment from home can also help in flattening the curve.

From the results we can also see that travel has a major effect. Both public transport, and out of town travel are big influences on the accumulated number of infected agents. For that, it did not matter if the bus capacity was halved, because if the capacity was halved, people had to wait in the queues more, where they could still spread the virus. However, it should be noted that by default, the bus capacity is 620, and the average number of people in public transport in a normal situation would be around 564.[6] So in the standard case, there is enough, but not extremely so, space in busses. Thus in that case more infection takes place in the bus.

Also it should be noted, that the agents do not take into account that the bus capacity is halved when deciding to take the bus or another form of transport. Thus they do not avoid the waiting in queues. Real people might react quite differently in this respect than the agents in our simulation did, which would lead to less people queuing for the bus, and thus less infections there. Also of note is that in the simulation, halving the bus capacity only made sure there are less agents in the bus, lowering the number of contacts there. In the real world also other approaches that limit the spread of the disease are possible.

Furthermore, we have seen that, while going back to work can have a positive effect on the economic situation for a factory or services company while keeping a

[6]This last number if based on the amount of agents of each age group, multiplied with the chance of that age group taking the bus if they needed to travel.

part of the population safe. For stores it might be better to allow all people to visit them, because the elderly have the time and money to actually buy things during the crisis phase, a luxury that company workers might not have.

10.6 Conclusion

In this chapter, we have shown how the ASSOCC framework can be used to implement a mechanism for representing various exit strategies. We tested three different exit strategies that were based on exit strategies that are applied in the real world. We compared the exit strategies on the number of agents that were infected, both the peaks as well as the accumulated numbers. We also investigated whether certain groups of agents would be more affected than other ones and where infections took place. Finally we also looked at the economic effects of the different exit strategies. For each of the exit strategies we used different triggering conditions to check which of these conditions would be the best to mitigate any second wave after the lock down phase.

The main, somehow surprising, observation is that the exit strategies differ very little with respect to the number of infections and the peaks of infections over time. They also differ little on the number of hospitalizations and deaths. The exit strategies do differ on *where* the infections take place, although for all of them most infections occur at home. The differences in places where more infections occur for the different strategies can further be explained by the order in which locations open up under the strategies. We have also concluded that it is possible to safely open up all kinds of locations for people to go to, as long as they do not attract too large a crowd.Based on the combinations of triggering conditions and exit strategies, we concluded that in order to have an effect, a condition for going to the next phase needs to be strict, otherwise they will trigger to soon and it will be no better than simply setting a time deadline.

An interesting result is the good economic results of the public service strategy relative to the business strategy. One would expect the business strategy to be better for business. However, it appears that opening businesses as quick as possible after lockdown does not have the long term positive economic effect.

Overall, we have shown that the ASSOCC model can be a valuable tool for policy makers and other interested parties to inform them on the possible effects that certain policies will have and especially the effects of different triggering conditions in combination with overall exit strategies.

Given the results, some future work becomes apparent that will make the framework even more valuable. A first of these is that the current system only allows for the relaxation of measures. As we see in practice that exit strategies are becoming cycles of lifting and reintroducing some restrictions again, it would be good to extend the framework with a gradual introduction of single measures instead of activating everything at once. In general this would allow for a more flexible reaction of instating and lifting restrictions based on the circumstances.

We have seen that countries use many different triggering conditions for the exit phases. In order to accommodate them easily it would be good to have some interface in which these conditions can easily be specified.This will allow for using flexible parameters that can easily be updated when circumstances change rapidly.

Finally, more measures could be included in the system. Currently most of the measures available in the simulation were controllable from the exit strategy system. However, all of these had to be added by hand in a quite cumbersome way. Besides this, the measures that were included in the simulation where also not all as detailed as the measures that were implemented in the real world. For example, the simulation could not deal with partial school classes, which was an approach that was used in some European countries. It should be kept in mind though that using more fine grained and detailed measures also requires that the deliberation of the agents is extended to take the new detailed aspects into account. E.g. whether or not to wear a face mask in some locations, or go there if not everyone else is wearing a face mask. These additional features will have a negative impact on the efficiency of the deliberation of the agents and thus indirect on the scalability of the system.

Acknowledgements We would like to acknowledge the members of the ASSOCC team for their valuable contributions to this chapter. In particular, we would like to thank Loïs Vanhée for contributing to the first conceptualization and implementation of the exit strategy scenario. Furthermore, we would also like to thank Mijke van den Hurk for proofreading and her valuable feedback. The simulations were enabled by resources provided by the Swedish National Infrastructure for Computing (SNIC) at Umeå partially funded by the Swedish Research Council through grant agreement no. 2018-05973. This research was conducted using the resources of High Performance Computing Center North (HPC2N).

Part III
Results and Lessons Learned

In the last part of the book we discuss the lessons we have learned from the ASSOCC project and also look into important future directions of research. Especially we discuss some challenges that need to be addressed by the research community if we want to make real-world impact with social simulations and have them used as standard support tools in times of crisis.

Chapter 11
The Real Impact of Social Simulations During the COVID-19 Crisis

Frank Dignum

Abstract Creating simulations for a crisis is not new. However, there have been no agent based social simulations that have been used during a crisis. In this chapter we describe the impact one can have with a simulation result, but also the requirements this poses on the simulation process. Basically it requires that one can quickly and flexibly generate many variations of scenarios, explain results in terms of the decision makers and extend the scenarios with new aspects that become important in a very short time.

11.1 Introduction

Although the social simulation community has made several social simulations for decision makers that have also been actually used in practice, most of the academic work has been limited to creating prototypes, pilot cases and proofs of concept. An excellent book discussing simulations for complex situations is [1]. It discusses both some experiences of people having created social simulations in policy making environments as well as some of the problems that arise. In [2] some pitfalls are given to be aware of when creating simulations that might be used by policy makers. Some important ones are the modelling assumptions that are made and loosing sight of the limitations of a model. In [3] it is argued that one has to be careful to determine where a simulation is used for. Prediction is only one possible (and very difficult) use of social simulations.

All of the issues that are mentioned in [1] are also relevant for simulations for crisis situations. However, in crisis situations some aspects become even more important and at the same time are not easy to tackle during a crisis. Based on our experiences during the COVID-19 crisis I would argue that the most important issue is the acceptance of the underlying assumptions of the agent based model. The ASSOCC framework is based on an abstract agent model as described in Chap. 2. Although *we* believe in the overall validity of this model (i.e. the components that are part of the

F. Dignum (✉)
Department of Computing Science, Umeå University, SE-901 87, Umeå, Sweden
e-mail: dignum@cs.umu.se

F. Dignum (ed.), *Social Simulation for a Crisis*, Computational Social Sciences,
https://doi.org/10.1007/978-3-030-76397-8_11

model and how they interact) it is by no means a generally accepted model of human behaviour. However, there is also not a more generally accepted abstract model of human behaviour. A model that is widely used in the software agents community is the BDI model that bases behaviour on beliefs, desires (goals) and intentions (plans). However, this model fits well with some type of rational, strategic behaviour, but is not suited to model any realistic behaviour of people in day to day situations.

In contrast, research communities in e.g. epidemiology and economy do have a number of commonly agreed upon models. In epidemiology the SEIR model ([4]) is the basis of most models in some way. In economics it is the rational or utility based reasoning model that is the basis of all theories. Differences arise by adding more elements or refining stages in this model. This has several consequences. First of all it allows for a very quick update of this model with virus specific parameters and running simulations very quick, because the software that implements the basic model is already there. A second, and possibly even more important, consequence is that all experts agree on this model, the decision makers also have heard of the model and everyone readily trusts the outcomes of the simulation. If results can also be explained with an intuitive story conforming expectations it becomes hard to challenge these outcomes.

A related issue that is important in a crisis is the trust and reputation decision makers put on the research institutes and their models. In the case of the COVID-19 crisis the situation is foremost seen as a health issue. As a consequence the decision makers will look for epidemiologists to provide answers. In case of natural disasters they will look at scientists that specialise in the particular area of nature that causes the disaster. Social scientists and computer scientists are not established parties for these crisis situations and thus their models will not be consulted by default. This is the case, even when the scientists from the other fields admit that their models lack important behavioural components! Components that are inherently part of the social simulations.

In the rest of this chapter we will discuss our experiences during the COVID-19 crisis with decision makers and official advisory committees more in depth. In the next section we will discuss why social simulation plays no role in advisory committees that are consulted by governments during a crisis (yet). We will discuss both the contributions but also requirements for being part of these committees. In the section after that we will discuss the processes during the crisis and what is required in order for social simulations to be used properly. Finally we discuss ways forward if we want to be part of the advisory infrastructure for decision makers.

11.2 Social Simulation as Tool for Advisory Committees

In times of crisis the government makes use of national committees to support their decision making. These committees are usually not formed ad-hoc for a crisis situation but are based on institutes that advice the government on certain areas of expertise. For the COVID-19 crisis the national health institutes are the most natural and also most used contributors for such a committee. During the economic crisis usually governments use some national economics or economic policy institute. As a rule the cause of the crises is not primarily social or behavioural. Therefore social science and social simulation experts will not by definition be involved in the committees advising the governments.

Taking the Dutch situation as example that we have closest experience with during the COVID-19 crisis, the government turned to the National Institute for the Public Health and Environment (RIVM) for guidance during the pandemic. The institute forms a so-called "Outbreak Management Team" (OMT) to actually advise the government during the crisis. In this OMT a number of epidemiologists, virologists and medical doctors representing the most connected expertise fields (such as IC care, virology, elderly care, etc.) are gathered. The standard tools of this committee are based on epidemiological theories. Those tools will indicate the main risk areas for infections and which measures should be taken to cut those out. These tools are built based on many years of experience and used in a variety of situations. In the COVID-19 crisis though these tools were lacking in the behavioural components. Due to the specific characteristics of the corona virus, human behaviour and reactions to restrictions form a large influence on the spread of the virus.

Even though this limitation of the models was recognised in March/April 2020 we have not been able to connect with the relevant persons within the RIVM to see whether our behaviour based models could assist in this respect. Of course, one can expect that in times of crisis the responsible persons are not apt to switch to an alternative model that has not proven itself over a longer period yet. However, also our offers to provide extensions to existing models or run simulations that could provide better estimations of parameters were not accepted. It is clear that a new player on the market is not granted a position at the table just like that. This is not unwarranted, of course! In crisis time there is no time to experiment with new tools. However, what is not good in this situation is that a committee admits the failure of its used models in some respects, but keeps using the same model for decisions on instating and lifting restrictions for a long time still. This has led to the following negative spiral:

1. instate a restriction of which it cannot be explained very well why it is necessary and will work
2. due to the unclarity people do not all obey the restriction
3. the restriction does not have the predicted effect either because it was not effective in the first place or people have not obeyed it sufficiently
4. new restrictions are instated as reaction to the situation, but it is now even more unclear whether these will work
5. people react to the increased negative consequences of the restrictions, reinforcing the negative cycle

In this case the pitfalls that were listed in [2] for social simulation models also applied to the epidemiological models. The fact that people are entrenched in their own models and stick with them despite their failure leads to an aggravation of the crisis and prolongment of lock-downs!

One of the biggest failures of the used models is that restrictions of behaviour do not only lead to a reduction of contacts in some places, but also to alternative behaviours that might negate the results of the restrictions! These types of consequences are difficult to capture in models where the human behaviour is represented as stochastic parameters. These connections are not present in the model and are easily overlooked or alternative behaviour cannot be estimated beforehand.

In the Dutch situation the above failures have resulted in the promise of incorporating behavioural scientists in the OMT. However, I have not seen any of that actually take place!

Ideally social simulation would be an integral part of the tool box of the national institutes that advise the government in crisis situations, because in most of these situations the human behaviour is an important ingredient in the outcome of any measure the government takes to solve the crisis.

The above discussion describes some of the practical reasons why social simulation is not taken up in crisis situations by advisory committees of the government. We also indicate some of the negative consequences of this. However, we should also be realistic and see what would be requirements for the social simulation models in order to be ready to be used in these crisis situations.

In the introduction of this chapter we already mentioned that an important requirement on the social simulation model would be that it is built based on some agreed upon set of concepts. This will facilitate the trust in the models created as there is a kind of general validation of the models components over all applications that they have been used in. Given the current state and discussions in the social simulation community it is unlikely that the community will agree upon such a universal set of concepts and models. There are discussions between proponents of keeping things as simple as possible and proponents of using descriptive models. See e.g. [5] for this discussion. Without wanting to take sides in this discussion it is difficult to convince other disciplines of the value of your models if within your own discipline you do not agree upon them yet. An important obstacle in the uptake of our results was the lack of scientific publications on the model that we have used as the basis for the ASSOCC simulations. There simply had been no time to write papers on this specific implementation yet and have them be accepted and published.

Fortunately we realised from the start that in the time of Internet and electronic publications it would be important to create a website for the project (simassocc.org) on which we could communicate our results. On this website we have published summaries of all results, but also overviews of the models used, interfaces to scenarios and the papers that were published in the last year as part of the project. Moreover, the website also had a link to all the software and its documentation. This website helped to get trust in our approach. The fact that we were transparent in all that we did and other groups could experiment with our models gave confidence that we did some solid work based on principles that could be checked.

Based on our experience we would argue that one should not expect to have one set of concepts that are to be used to model all human behaviour. The set of concepts that are most relevant depend on the context in which the behaviour is taking place and that is being modelled. E.g. long term behaviour (spanning several months or years) involves different deliberations than short term behaviour (spanning a few days at most). Thus maybe one way to move forward from this quandary would be to develop methods that indicate the set of concepts that are most relevant for a situation and having some standard way to combine these concepts into models for the simulation.

Having an accepted model of human behaviour is not the only thing required. This model should also be combined with the models used in the domain that causes the crisis. In the case of the COVID-19 crisis it means that this behavioural model has to be combined with an epidemiological model. In this book we show how we managed to do this with the ASSOCC framework and the SEIR model. Note that this combination is more than just specifying some API between the two models! In Chap. 12 we have extensively discussed how parameters can be mapped or should be generated by one or the other model.

Suppose we have a framework that would accepted and is also widely used for crisis situations and can be easily combined with the models from the domain area. The next requirement concerns the way the social simulations can be used. Keeping to the example of the COVID-19 crisis, the simulations should support the decision makers on the measures that are most efficient to end the crisis. This means that each of the potential measures should be easily implemented and added to the simulation and its effects visible and explainable(!). As we argued before, the crux in these scenarios is to not only restrict some behaviour, but also model the possible alternative behaviours that might arise from the restrictions. E.g. in a complete lock-down, when people can only go out to the supermarket and other essential shops, people will more often go to the supermarket than in normal times. In order to react quickly to each of the potential measures, these changes in the simulation scenarios should be easily implemented and run such that results can be shown within at most a couple of days. Moreover, the results should also be explainable in terms of analyses of all the changes that are a consequence of the new measure. Unfortunately, we do not have frameworks that have this kind of maturity!

Finally, the simulation framework should also be extendable with more domain aspects along the time. E.g. in the first months of the COVID-19 crisis, tourism was not a big item, but around June when the summer holidays were approaching this became a huge issue, both economically as well as from the health perspective. Thus the simulations should be extendable with this aspect, including the transport and the deviating behaviour of people during a holiday. Again, this is not feasible with the current social simulation frameworks! Adding new behaviour and interactions usually implies re-implementation of the code to incorporate these actions in the environment and deliberation cycles of the agents.

11.3 The Crisis

In the previous section we described some issues that need to be taken care of in order for social simulations to be useful during a crisis. In this section we will mention some of the requirements on the organisation that develops and supports the use of the social simulation tools.

As was made clear in the previous sections the advisory committees that will give advice to the government and other decision makers during a crisis are formed out of existing national institutions covering a field of expertise. Here we have the first hurdle to take in that social simulation tools are not used or developed by any of these national institutions. Thus there is no natural champion for the use of social simulation by the crisis committee. In the Netherlands there would be two possible candidates to host the development of social simulation platforms for the use in crisis situations (and more general for policy making). The first would be the Netherlands Institute for Social Research (SCP). The second would be the behavioural insights team within the Dutch government.

Both of these organisations are not very well suited for developing social simulations for crises. The SCP is doing research and gives advise on social and cultural aspects of society. Although social simulation might play a role in the methodology their aim is not primarily to give insights in potential (social) consequences of policies. Rather, they investigate the current consequences of policies and other changes in society.

The behavioural insight team within government might be the other natural place to develop social simulations. However, this unit is relatively small and their main aim is to see how policies might be reinforced through behavioural nudges. Thus they could use the social simulation tools, but are not an independent institute that could develop the social simulation tools.

The social simulations and the platforms and tools are developed by academics at different institutes around the world. Although many people devote a lot of time and effort on them there is no concerted effort to develop social simulation tools for this specific purpose. As a consequence bits and pieces of possible solutions are developed in a fragmented way, usually funded through a very specific project with limited funds and time. Thus there is no consolidation, neither a joint effort to keep developing and maintaining the tools.

We found ourselves in a similar situation in the COVID-19 crisis when developing the ASSOCC framework. We were one of the many parties that stood up in academics and tried to support the decision makers with the simulations that we built. There was no time to create an effort based on e.g. the European Social Simulation Association. These academic associations are meant to foster scientific exchanges of knowledge, but are not equipped to set up teams of researchers for the purpose of setting up social simulations for a crisis. There is no funding nor other resources to do this.

When we started with the creation of the ASSOCC framework in March 2020 the main motivation was that we as social simulation researchers had something to contribute to the world in the COVID-19 crisis. Based on the type of research that we had been done over the last decade or more we believed that we could build a flexible platform on which we could quickly develop scenarios to test all kinds of measures and see what their effects might be in reality. There was no funding, but a lot of motivation and thus the ASSOCC framework is mainly developed in people's own time!

Because we were not directly connected to any national institute that got concrete requests from government we relied on public discussions and connections we had to see which would be the most important scenarios to develop. Hardly an ideal situation, but it worked to some extent as can be seen from the results reported in this book. The only way that we, as academics, could get our results in the national debate and possibly considered by politicians was by going through the media and individual contacts with politicians.

Because these parties are most interested in results that would be contradicting the government policies we got most media attention with our results on the effects of the tracking and tracing app. Our simulations showed far less effect of this app than what was predicted by the "official" models that were used by government. This was widely cited in the media and also got scientific interest. However, when we were asked to explain the differences we had no readily available answer!

We had set up the scenario and run the simulations with the sole goal to produce the results as quick as we could in order to be in time to influence the public debate in The Netherlands. While we did manage that it was difficult to explain why we got different results than established models. Since the time of publication of the results we have spent some months to actually assemble those explanations. A complicating factor was that there was no single factor where we differed from the other models that could explain the different results. In the end there were a number of assumptions and dependencies that each contributed to the difference and in combination all made sense. So, we can conclude that we have a satisfactory story, which in the mean time is also validated by reality, but it came a few months too late! We simply did not have the resources to do this any quicker.

During the spring of 2020 we would have to balance our time between developing a new scenario, adjusting some existing scenario, presenting and explaining results to media and doing some consultancy and advising to people from the national institutes and committees. As a group of (mainly young) academics we were not very well equipped for all these roles, neither did we have the time to actually do the activities properly. Especially, considering that we did not receive any funding for the whole project and were managing this in our spare time.

From the above one might get the impression that the whole ASSOCC project was a huge failure. However, this is certainly not the case. Due to the publicity around our results and subsequent contacts with politicians and journalists we have indirectly still influenced the debate around some of the measures. That is already a huge victory for the social simulation community. With the above discussion we try to indicate what are the requirements for the development and use of social simulation to play a more structural role in the crisis management of countries. We have shown that such a role can only be taken up with social simulations tools that are better equipped for crisis situations. But it is not just a matter of developing better tools, in order to play a substantial role in the crisis management teams we also should be embedded in the organisational structures on national level and the expertise that we bring to the table should be accepted by the other parties.

In the last section of this chapter we will discuss which steps could be taken by the academic social simulation community in order to be able to step up to this challenges.

11.4 Real Impact

Before discussing possible ways forward to create more impact with social simulation in crisis situations it is worth to check whether other groups from the social simulation community have had more success to have real impact on governmental measures. A very short answer to this is:*no*. There has been one publication [6] of a simulation that was used to advice the Australian government on school closures, air traffic restrictions and the effectiveness of social distancing. It is interesting to see that one of the authors has an affiliation with the Marie Bashir Institute for Infectious Diseases and Biosecurity at the University of Sydney, which lends this work extra credibility. Although the simulation was used by government and also widely cited in the media there were no subsequent simulations or more permanent membership of advisory committees resulting from this work. The only other work I know of that has actual impact in epidemiology is that of Epstein [7], which pertains to epidemics in general. Epstein is professor in epidemiology and thus in a position to be consulting with epidemiologists. But also Epstein has not been member of advisory committees for the government of the USA during the COVID-19 crisis. So, this confirms that the position from which one communicates with the domain experts is very important for the acceptance of the social simulation tools and results that are used. But even in these cases there is no structural place for bringing in the results of social simulations.

So, what can we do to ensure that the social simulations have the impact they deserve in crisis situations? Basically, this will require a long term view and persistence.

First of all we have to make sure that the social simulation models that are used are widely accepted and supported. This will allow trust in the behavioural aspects of the simulations. As we have argued before, there is not one universally applicable set of concepts that can be used for the agent deliberation architecture. Therefore it would be most important to develop a methodology that supports choosing those concepts that are most important for modelling a particular situation. Having such a methodology would facilitate any discussion on the use of a particular model for a simulation. In some sense we argue for a circumvention of the KISS versus KIDS debate ([5]) by stating that the complexity of the models is dependent on the application of the simulation. In some sense we follow the classical heuristic given by Schelling in [8]: the model should be simple enough to understand the results of the simulation, but not too simple as to render only trivial results that could have been obtained analytically or follow directly from the assumptions. What we have shown in the results from the ASSOCC framework is that explanation sometimes may be complex but possible through a thorough analysis. Thus we should not abandon a slightly complex model on the basis of this paradigm too quickly! Moreover, we have also shown that in some cases the results stemming from a complex model seem very simple and could be obtained through a much simpler model. However, analysis can show that the simple result is sometimes achieved through a combination of more complex interactions between phenomena that lead to a simple result in "normal" situations. However, the advantage of the complex model might only come to light in changing conditions such as a crisis situation where the simple model fails to give the right results while the complex model still can represent realistically what is happening.

In this context we also reiterate the importance of taking into account the purpose of the simulation when creating the model [3]. This purpose also will determine the scope of the concepts that are important for the simulation. In Chap. 2 we already argued that given the dependencies between different aspects of life and their importance during a crisis situation it is advisable to use models that contain some more abstract concepts, such as values, to be able to connect the different aspects of life. Although values also play a role in every day life they usually do so implicit and do not have to be modeled explicitly. From this experience we can see that the usefulness of a model depends on the purpose and situation for which a model is used.

Following this methodology we need, of course, an implementation framework that facilitates the creation of the simulations that are needed to support the decision making for the crisis. This framework should contain at least the structures with which all the concepts can be represented and standard ways to connect these. Thus it needs to be of a higher level than e.g. Netlogo which allows to program agent deliberation cycles and actions, but does not support specific social concepts. We have shown for the case of *norms* that adding a social concept to a model has many repercussions [9]. Norms do not only constrain certain behaviour, but also have a motivational component and thus lead to different, alternative behaviour. When a norm such as "social distancing" is added it is not just restricting the contacts between people, but also influences the places they might visit. Moreover, people also selectively violate the norm. E.g. two people that fell in love recently do not keep social distance.

We use the above example just to illustrate that the requirement of creating a proper social simulation platform that can be used readily for crisis situations is non-trivial. Building such a framework is a long term effort and it also needs to be maintained and updated based on experiences and new insights over a long period. Just like platforms such as Repast and Netlogo this requires funding and resources over a long period of time. Several groups have been doing research on models for agent deliberation in social simulations and some implementations have been made based on these. However, we require a concerted effort to lift these efforts onto a level that is also readily usable and adjustable by people outside the academic social simulation community.

The above issues are primarily of concern for the academic community. They involve academic research in developing social simulation frameworks that are better usable for crisis situations. We will elaborate more on these aspects in the next two chapters. However, in order to have real impact with a social simulation framework it should also be adopted by a notional institute that is involved in the crisis management process. We believe that we have made a very first step in this process with the ASSOCC framework during the COVID-19 crisis. Some national institutes in some countries have become aware of the existence of social simulation tools and their potential use during the crisis. A way forward would be to connect to the people within these institutes on a more permanent basis and keep them abreast of the developments in social simulation research. Potentially this could lead to some social simulation experts becoming part of these institutes or even setting up social simulation groups within these institutes.

In several countries there exist national organisations for (applied) research that have natural connections to government and the national institutes involved in the crisis management. Ideally the development of the social simulation framework would be hosted at such an institute in cooperation with academic groups. This could assure long term financing of the effort, while creating trust in the framework and the right connections to the committees that would use the framework in times of crisis.

11.5 Conclusions

We can conclude that the results from the simulations of the ASSOCC framework had some impact during the COVID-19 crisis. Mainly they played a role in the discussions around some measures the government was contemplating. Basically the results had impact when they were contradicting popular beliefs. In these cases they were picked up by the media and were used in the debates. During these interactions it became clear that the ASSOCC tools were not mature enough to provide quick and thorough analyses of the results. These analyses are needed to back up and explain the results. We were able to create these explanations, but they came available only after decisions had been made already. In order to have more structural impact we need to have more mature tools.

In the other hand we cannot expect that academic groups are equal discussion partners in a crisis situation if these groups come out of the blue with some new results. In order to be incorporated in the crisis management process we have to connect to the national institutes involved in the crisis management on a more permanent base. This means that we have to make the people from these institutes aware of the value of social simulation and provide them with the tools and evidence of this value. This involves a long term concerted effort that exceeds that of having an occasional joint research project. It should consist of a co-development of the tools and possibly having permanent positions of social simulation researchers at these institutions.

Acknowledgements The authors would like to acknowledge the members of the ASSOCC team for their valuable contributions to this chapter. This research was partially supported by the Wallenberg AI, Autonomous Systems and Software Program (WASP) funded by the Knut and Alice Wallenberg Foundation.

References

1. B. Edmonds, R. Meyer, *Simulating Social Complexity* (Springer, 2015)
2. L. ni Aodha, B. Edmonds, Some pitfalls to beware when applying models to issues of policy relevance, in *Simulating Social Complexity* (Springer, 2017), pp. 801–822
3. B. Edmonds, Different modelling purposes, in *Simulating Social Complexity* (Springer, 2017), pp. 39–58
4. R.C. Cope et al., Characterising seasonal influenza epidemiology using primary care surveillance data. PLoS Comput. Biol. **14**(8) (2018)
5. B. Edmonds, S. Moss, From KISS to KIDS-an 'anti-simplistic' modelling approach, in *Multi-Agent and Multi-Agent-Based Simulation*, ed. by P. Davidsson, B. Logan, K. Takadama. (Berlin, Heidelberg: Springer Berlin Heidelberg, 2005), pp. 130–144
6. S.L. Chang, N. Harding, C. Zachreson et al., Modelling transmission and control of the COVID-19 pandemic in Australia. Nat. Commun. 11.5710 (2020)
7. J.M. Epstein, Modelling to contain pandemics. Nature 460.7256 (2009), pp. 687–687
8. T.C. Schelling. *Micromotives and Macrobehavior*. (WW Norton & Company, 2006)
9. C. Pastrav, F. Dignum, Norms in social simulation: balancing between realism and scalability, in *Proceedings of the Fourteenth International Social Simulation Conference* (2018)

Chapter 12
Comparative Validation of Simulation Models for the COVID-19 Crisis

Fabian Lorig, Maarten Jensen, Christian Kammler, Paul Davidsson, and Harko Verhagen

12.1 Introduction

Modelling and simulation approaches are applied in a variety of scientific disciplines for analysing, planning, and optimising complex systems or phenomena. Simulation is not limited to applications related to computer science or information systems research [2] and has also become an accepted method, for example, in social sciences and medicine. But also for solving practical problems, e.g., in manufacturing, logistics, or engineering, the use of simulation is increasingly common. Due to its popularity and versatility, simulation is even referred to as a third pillar of science between theory and experiment [1].

The broad application of simulation puts high requirements on the credibility and validity of the generated results. Especially when used for decision support, but also when used to better understand phenomena, it is important to validate the behaviour of the simulation model and to assess how accurately it represents the real world [3].

F. Lorig (✉) · P. Davidsson
Department of Computer Science and Media Technology,
Internet of Things and People Research Center, Malmö University, 205 06 Malmö, Sweden
e-mail: fabian.lorig@mau.se

P. Davidsson
e-mail: paul.davidsson@mau.se

M. Jensen · C. Kammler
Department of Computing Science, Umeå University, 901 87 Umeå, Sweden
e-mail: maarten.jensen@umu.se

C. Kammler
e-mail: christian.kammler@umu.se

H. Verhagen
Department of Computer and Systems Sciences, Stockholm University, PO Box 7003, 164 07 Kista, Sweden
e-mail: verhagen@dsv.su.se

F. Dignum (ed.), *Social Simulation for a Crisis*, Computational Social Sciences,
https://doi.org/10.1007/978-3-030-76397-8_12

A common approach for validating a simulation model is to compare the results of experiments to reference data that has been collected from real-world systems [4, 5]. In the case of a simulation of the spread of COVID-19 or other societal phenomena that are related to it, the validation of the models by means of real-world data is challenging. Especially in the early phase of the pandemic, there has been a lack of empirical data on the dynamics of the spread but also a lack of understanding of the underlying infection mechanisms. This makes it difficult to apply conventional validation approaches and methods.

In this chapter, we discuss and apply an approach for the validation of simulation models for the COVID-19 pandemic by comparing their results among each other. To this end, a formal comparison between the equation-based epidemiological model developed by Ferretti et al. [6] and the ASSOCC model is carried out. This includes the calibration of the different models in terms of assumptions that are made. The comparison is based on a simulation study of potential effects of Tracking and Tracing Apps (TTAs) as well as the required acceptance rate for the app use to be reasonably effective. This corresponds to the concept of "model alignment", with the goal of determining if two models can produce corresponding results and whether they can substitute one another [7].

The chapter is structured as follows: First, an overview of challenges and methods for the validation of agent-based social simulations is provided. In Sect. 12.3, the mathematical model by Ferretti et al. [6] is presented, which is used for the comparison-based validation. Section 12.4 presents the reference scenario of introducing a TTA that was simulated with both models to generate comparable output data. Comparability is also affected by different assumptions the models make. Thus, in Sect. 12.5, differences between the models are outlined and necessary adaptions are described. Finally, in Sect. 12.6, the simulation results from both models are presented and evaluated.

12.2 Background: Validation of Agent-Based Social Simulation

This section outlines the challenges associated with validation of agent-based social simulation models and provides an overview of suitable methods for verification. In particular, comparative validation is introduced as method that can be used when there is a lack of real-world data.

12.2.1 Methods for the Validation of ABSS Models

To increase the trustworthiness of simulation models and to ensure the quality of the generated results, verification and validation (V&V) approaches are applied. Gilbert and Troitzsch [8], distinguish between verification, which is concerned with the model working according to the modeller's expectations, and validation, which assesses whether the model adequately represents the reality.

To this end, verification comprises testing approaches that evaluate the model's consistency with the underlying specifications. More specifically, it consist of the assessment whether the implemented model corresponds to its theoretical specification. Verification approaches and techniques analyse if the model's implementation in program code is correct. In other words, verification answers the questions of whether a model was built *right* [9].

Validation, in contrast, assesses a model's correspondence to its requirements. This is related to both the model's behaviour as well as its intended purpose and a model is considered *valid* in case it corresponds to the behaviour of the target system. Hence, validation can be considered as "the process of determining the degree to which a model is an accurate representation of the real world from the perspective of the intended uses of the model" [10, p. 719].

As modelling requires abstractions of the real-world system, it is challenging to build a truly valid model that adequately represents every relevant aspect of the underlying system. Thus, it is not feasible to evaluate models based on their general truth or accuracy. Instead, Parker [11] recommends the evaluation of a model's adequacy depending on its purpose. In this regard, the purpose of the model might for instance be prediction or exploration and a simulation providing different candidate explanations that can be used to conclude the ultimate explanation might be sufficient. Beisbart [12] concludes that even results that are not highly credible might be useful.

According to Sargent [5], the accuracy that is required to consider a model to be valid must be in an acceptable range that is sufficient for the respective purpose of the simulation. A model, that reflects the real-world system's behaviour in a satisfying way can, thus, be considered as a *valid model*. Established techniques include, for instance, extreme condition tests, validation against historical data, sensitivity analysis, and structural validation. Validation, in this regard, investigates if the *right* model was built [9].

Murray-Smith [13, p. 102] describes the relation between verification and validation as follows: "compared with verification, validation is a more open-ended task in which comparisons are made between model behaviour and behaviour of the real system for the same conditions". He outlines that validation should be considered as a process of building up trust in a model and its generated results. Moreover, he underlines that the acquisition of data for evaluations a model's results is challenging. For many applications of simulation, it is not possible to perform experiments with target system, e.g., due to inaccessibility, non-existence, or to not jeopardise the system. Thus, only historical data is available, which needs to be used both for the model development and calibration as well as for its validation. He distinguished between two approaches to validation: quantitative approaches that use such data from real-world target systems and qualitative methods, that are more subjective and rather rely on expertise and experience.

Ultimately, the credibility of a simulation depends on its underlying conceptual model and "a simulation can only ever be as good as the conceptual model on which it is based" [14, p. 250]. In contrast to the implemented simulation model, the conceptual model (or scientific model) consists of objectives, assumptions, simplifications, inputs, and outputs. It described what will be modelled and can be considered as a simplified representation of the target system that is independent of a simulation framework or a specific implementation [15]. The assessment of a model's credibility, however, is more comprehensive and besides validation also includes the evaluation of the model's design, data that has been used, reporting, and interpretations.

Another technique that is closely related to validation is calibration. Here, the goal is to find a parametrization that results in a desired behaviour of the model [16]. It is an iterative process in which the values of the model parameters are altered with the goal of minimising the deviation between the outputs of the model and the observations from the real-world system under similar circumstances. Methods that can be used for this include optimisation algorithms, for instance, simulated annealing [17].

For the validation of agent-based models in social sciences, Ormerod and Rosewell [18] suggest two stages: the description of the phenomenon that is investigated and the testing for realistic agent behaviour. In particular, the authors recommend the definition of criteria according to which the model's output can be judged and assessed followed by the comparison and implementation of different approaches for achieving the desired behaviour.

12.2.2 Comparative Validation of Simulation Models

The validation of the ASSOCC model by means of traditional methods is challenging as real-world data is missing which the model's behaviour can be compared against. In retrospect, data on the daily number of infected individuals can, for example, be used to assess whether the model is capable of adequately reproducing the dynamics of disease transmission under specific circumstances. Yet, it is the model's goal to simulate different interventions for prospectively containing the spread of the virus. For such ex ante analyses, specific data is usually not available and other approaches need to be applied to assess the model's validity. This is a challenge for both the ASSOCC model but also for other models that address phenomena that have not yet been thoroughly studied.

Kleijnen [19] discusses different approaches for the validation of simulation models in case data from the real system is not available or only limited. In case no data is available, Kleijnen recommends the systematic experimentation with the model to obtain data that can be used for assessing the model's quality [19]. One example of an experimental design that can be pursued here is the *one factor at a time* design that allows for a *what-if* analysis of the model.

Among different validation techniques that rely on historical data, Sargent [5] outlines the comparison to other models as suitable for analytical or mathematical models. This can also help to identify different threats to model validity such as hidden underlying assumptions that might bias the conclusions or invalid assumptions made during the development of the model [20]. Since the beginning of the COVID-19 pandemic, a great number of simulation models has been developed and published, which can be used for the validation of the models among each other.

12.3 The Reference Model: Network Model of Social Interactions

To investigate the validity of the ASSOCC model, we pursued a comparative approach. The model that we used to compare our model against was developed by Ferretti et al. [6] and in the following, we refer to this model as the *Oxford model*.

12.3.1 Mathematical Modelling Versus Behavior-Based Modelling of Pandemics

In contrast to the ASSOCC model, that pursues a behaviour-based modelling of the pandemic, the Oxford model follows an equation-based mathematical approach to modelling transmission dynamics. The advantage of behaviour-based models lies in the sophisticated representation of human-like behaviour and decision-making by means of Artificial Intelligence. For a given situation, each individual will deliberate on an appropriate action based on environmental circumstances but also personal factors. To this end, moving between locations or complying to interventions is not predetermined by a contact network but a result of the prevailing circumstances at that point of the simulation.

This, however, requires extensive computational resources as the current state of mind and the individual deliberation process need to be calculated for each individual agent. As a result of this, the scalability of such models can be limited and a trade-off needs to be made between the number of simulated agents and the level of detail of the decision-making. In Chap. 13, a discussion on the scalability of simulation models as well as on resulting challenges is provided.

As described by Bonabeau [21], it is challenging to model non-linear behaviour that results from certain thresholds or rules by means of differential equations. Moreover, mathematical equations tend to smoothen the effects that can be observed in a system. Considering transmission processes, where local deviations can amplify and lead to global phenomena, this might not be desirable.

Mathematical models, in contrast, are well suited when the number of states of the individual is small [22]. For simulating the transmission of diseases on the large scale, SIR compartment models are commonly used. In such deterministic non-linear differential equation models, the disease state of groups of individuals of a population is described by assigning them to a distinct compartment that corresponds to their state, e.g., susceptible, infected, or recovered [23]. There are numerous extensions of the classical SIR model to consider further states or courses of the disease. These include, for instance, an exposed state representing the incubation period or an immunity state after the infection. By means of global transition probabilities, the change of states can be efficiently simulated for a large population.

12.3.2 Introduction of the Network Model Approach

In order to compare the models we selected a specific scenario to generate results that can be compared. Hinch et al. [24] conducted a study in which they made use of the Oxford model to simulate the effects of digital contact tracing apps. Based on their report, we provide a description of the Oxford model in this section.

The Oxford model is a mathematical model, where the individuals have demographic profiles that are based on UK census data, belong to households that are defined based on survey data, and are part of different small-world interaction networks that also represent work places or schools. The allocation of individuals to networks depends mostly on the age of the individual. Based on these parameters, the spreading of the virus is simulated by calculating the probability that an infected individual infects another individual when interacting.

The probability of individuals infecting each other upon a contact in the small-world interaction networks depends on personal factors such as the state of infection as well as the infectiousness of the transmitting individual, the susceptibility of the infectee (which depends on the age of the individual), or the type of network in which the contact occurred. The compartment model used for modelling disease progress is an extension of the SIR model [23] consisting of 11 different states. There are multiple infected states to distinguish between the existence and severity of symptoms and the need for treatment in hospital or ICU. Moreover, individuals can be susceptible, recovered (immune), or dead. The transition between the states depends on age-dependent transition probabilities. The values of the remaining parameters of the model are shown in Table 12.1.

Table 12.1 Disease parameters of the Oxford model

Incubation time	6 days (sd: 2.5 days)
Infectiousness	Gamma-distributed (mean: 6 days)
% asymptomatic individuals	18% of all groups
Contagiousness	0.29 (asymptomatic), 0.48 (mildly symptomatic), 1.0 (severely symptomatic)
Generation time	6 days
Doubling time	3 days (R0 = 3.4) to 3.5 days (R0 = 3.0)

In their study, Hinch et al. [24] simulate a population of one million individuals, which represents the typical size of an organisational unit (NHS trust) within the National Health Service in England. In addition to digital contact tracing, the model includes other interventions such as social distancing and lockdown to limit the spread of the virus. Here, a lockdown is modelled such that people are required to stay at home, which in turn reduces the number of interactions they have on their occupation and other random networks by approximately 71%. The authors assume that 80% of individuals with symptoms will self-quarantine with the members of their household. Moreover, those older than 70 years will continue their quarantine when the lockdown is relaxed. Individuals with symptoms quarantine for 7 days whereas household members without symptoms quarantine for 14 days. Approximately 2% of the individuals will violate lockdown restrictions each day.

In the Oxford model, a 35-day lockdown is initiated once a 2% infection rate is reached. This results in a 20% decrease of contacts outside households and a 150% increase within households. Digital contact tracing is implemented with a 7-day memory and a 80% registration rate of social interactions is assumed to quarantine those that shared networks with individuals having self-diagnosed symptoms. These symptoms are not necessarily a result of COVID-19 and even without the use of the app most individuals will quarantine once experiencing symptoms.

As can be seen from the above, the Oxford model does in fact include some behavioural aspects of individuals. Yet, they are implicitly modelled as parameters of the interaction probabilities and are assumed to be constant over the time of the simulation.

12.3.3 Adapting the ABSS Model

To be able to compare the generated results, the ASSOCC model was adjusted to reflect the assumptions the Oxford model makes as well as the investigated scenarios. Especially the disease and the contagion models had to be adapted.

With respect to the comparability of the disease models, we extended the initial event-based disease model. In our initial disease model there was a distinct sequence of disease stages, starting at the incubation period, going to more severe stages and eventually resulting in death. At each stage there was a probability of becoming healthy. The adaptation of the disease model includes both the introduction of transition probabilities to distinguish between different disease paths (either asymptomatic, mildly symptomatic and severely symptomatic) as well as a tick-based progress of the disease in accordance with the disease states defined by the Oxford model. The disease state transition probabilities are shown in Table 12.2, the disease state transition times are shown in Table 12.3. For a more extensive explanation of the variables and a figure of the disease transition model see Sect. 3.5.

Similarly, the contagion model needed to be adapted to the Oxford model. Rather than adaption of the occasions when contagion can occur, this involved adjusting the probability a susceptible individual will be infected when being at the same

Table 12.2 The probabilities for transferring to a certain disease state

Name	Young	Student	Worker	Elderly
Probability-become-asymptomatic	0.18	0.18	0.18	0.18
Probability-become-mildly-symptomatic	0.79	0.73	0.6125	0.34
Probability-become-severely-symptomatic	0.03	0.09	0.2075	0.48
Probability-severely-symptomatic-hospitalised	0.02	0.04	0.1025	0.15
Probability-severely-symptomatic-recover	0.98	0.96	0.8975	0.85
Probability-survival-in-hospital	0.01425	0.025	0.074	0.5499
Probability-dying-in-hospital	0.98575	0.975	0.926	0.4501

Table 12.3 The time parameters for the adapted disease model of the ASSOCC model to correspond to the Oxford model. The parameters represent number of days

Name	Mean	Standard deviation
Time-asymptomatic-recovery	15	5
Time-no-symptoms-to-symptoms	6	2.5
Time-symptoms-to-recovery	12	5
Time-symptoms-to-hospital	5	–
Time-hospital-to-recovery	3.25	3.60555
Time-hospital-to-death	4.25	2

location as an infectious individual. Since there are different types of locations where individuals presumably have different proximity to each other and interactions we implemented a density factor (Table 12.4). The density factor is higher at homes and schools where people often have many close interactions, while in contrast the density factor is lower at workplaces and public leisure as at those places people usually have less direct interactions or are more spread out.

The ASSOCC model allows for the adaption of household compositions to represent different countries or regions. To represent Great Britain, as simulated by the Oxford model we set the ratio of different types of households according to Table 12.5. In accordance with these ratios we simulated 391 households with more than 1100 individuals. The simulation of 1 million individuals, as is the case for the Oxford model, is not possible with the ASSOCC model due to limitations of the NetLogo simulation framework and the sophisticated decision making modelled for each individual. To further adapt our model to the Great Britain scenario the number of places where individuals can meet was set according to Table 12.5.

Table 12.4 The density factor for locations and transport in the ASSOCC model

Name	Value
Density-factor-essential-shops	0.30
Density-factor-homes	1.00
Density-factor-hospital	0.80
Density-factor-non-essential-shops	0.60
Density-factor-private-leisure	0.30
Density-factor-public-leisure	0.10
Density-factor-public-transport	0.50
Density-factor-queuing	0.60
Density-factor-schools	1.00
Density-factor-shared-cars	0.80
Density-factor-walking-outside	0.05
Density-factor-workplaces	0.20
Density-factor-university-faculties	0.20

Table 12.5 Parametrization of the ASSOCC model to represent Great Britain

Name	Value
Ratio of households with adults only	29.2%
Ratio of households with retired couples	31.2%
Ratio of households with families	0.36%
Ratio of households with multi-generation	0.036%
Hospitals	4
Essential shops	10
Non-essential shops	10
Places for private leisure	60
Places for public leisure	20
School classes	12
University faculties	4
Workplaces	25
Beds in hospitals	unlimited

12.4 Reference Scenario: Tracking and Tracking Apps

The scenario which is used to compare the ASSOCC model with the Oxford model is the scenario presented in Chap. 7, i.e., the introduction of an app for tracking and tracing of contacts.

12.4.1 Simulating the Effect of Tracking and Tracing Apps in Pandemics

Tracking and tracing of the contacts of an infected person is a standard procedure to control the spread of a virus, as pointed out in Chap. 7. However, doing this manually is not easy and might delay the process of informing potential contacts. Furthermore, tracking and tracing should be done from the most likely time and place when the person became infected, however since the COVID-19 virus has on average seven days until symptoms show, it is very difficult to determine where a person go infected. This makes effective and precise manual tracing even more difficult. This can increase the risk of inaccurate tracking and tracing of an individual's contacts. TTAs have been suggested as be a valuable tool to support this process and meet these challenges. Some of these apps use the Bluetooth function present in every smartphone to keep a list of contacts a person has, given that Bluetooth is on and open. Once a person is in Bluetooth range of another person, such that the phones can connect, this encounter is automatically added to the contact list of the person. Given a potential positive test results, information can be send out to everyone in the contact list. This has the advantage to move away from laborious manual track and tracing and can increase the accuracy of the contacts, as it is not fully reliant on peoples' memory.

12.4.2 The Scenario and the Model

Given our simulation model, abstractions had to be made for implementing a TTA. Usually, only contacts longer than 15 minutes are registered by the app. Due to our tick based days, every contact at a specific location is captured. However, the locations have been configured in a way to compensate the loss of this time dimension. This also has the consequence that we take an ideal approach on the TTA and assume that the TTA is working perfectly, all contacts are tracked correctly, tests are done immediately, results are known, and infected people go into quarantine.

The TTA users are distributed randomly over the population, based on the ratio of app users. Furthermore, priority is given to anxious users (ratio can be set by a slider), as we assume that those people are more likely to use the TTA compared to other people. The TTA starts tracking seven days before the end of the global quarantine until the end of the simulation run. Every contact also using the TTA will be registered at every location and stored for the defined amount of time. When a user exhibits symptoms, every contact in the list of that user will get tested. In case of a positive test result for the user or a contact, every contact of that specific person will also get tested. This testing chain expands until everyone indirectly connected to the infected person is tested, given that recursive testing is enabled. More implementation details and a discussion of the abstractions made can be found in Chap. 7.

The scenario investigates the effect of using a TTA on the spread of the virus. To do so, a variety of different app usage ratios have been investigated. The results focused on the infections over time, the amount of cumulative infections over time, and infections between age groups. Furthermore, the amount of contacts and infections for different location types, e.g. home or shops, have been investigated. In addition, the effect on hospitalisations and tests needed was explored. Readers who are interested in an in-depth analysis and discussion of our results are referred to Chap. 7.

12.5 Differences Between the Models

The first step in the comparison between the two models is to assess all assumptions made in both models and translate those into constructs that can be compared between the models. Thus, the simpler disease and contagion model of ASSOCC is adapted to reflect the more sophisticated disease dynamics of the Oxford model. One of the most important distinction that was added to the ASSOCC model was the distinction of symptomatic and asymptomatic infected individuals, the symptomatic individuals were also split between mildly and severely symptomatic. The symptomatic individuals are more infectious than asymptomatic individuals. On the other hand, the Oxford model assumed that 18% of the people are asymptomatic while this number was age dependent in the ASSOCC model. We changed this in the ASSOCC model as well to create equal conditions. The new disease model also included more specific age based parameters such as the probability for transferring to a certain disease state and an updated contagiousness model. With the new (more sophisticated) disease model, the ASSOCC simulations were rerun. This did not lead to changes in the results as our initial disease model worked well. However using the more elaborate disease model prevents later critique on having a too simplistic model.

The other main aspect where the models differ are behavioural aspects. In the Oxford model, the behavioural aspects are fixed and values for parameters are based on general literature. The number of interactions is, for instance, based on [25]. The Oxford model considers interactions within households, workplaces, or random, using some networks connecting agents. For each of these three networks, a standard number of interactions per day is assumed that will drive the number of infections arising in each of these networks if one of the agents in such a network is infectious. For adults, this results in approximately 11 interactions plus household interactions and children around 12 plus household interactions, while elderly have around 6 interactions plus household interactions per day with some distribution around those numbers.

In the ASSOCC model, we do not consider a fixed parameter for the number of interactions. Rather the interactions follow from the activities performed by the agents, e.g. when many agents have the need to go to the shop the amount of contacts there will be higher then when only a few agents go to the shop. Dependent on the agents age and needs the agents go to work but also to shops, leisure places, school, hospital, socialise with friends and take transport. The age distribution of the agents is dependent on the general demographics of the UK. This household distribution generates the agents of different ages. This is explained in more detail in Chap. 3.

All of these aspects generate a number of interactions per day for each of the agents in the population. These numbers are on average much higher than the ones assumed in the Oxford model. Part of this comes from the fact that everyone in the same space is counted as interacting with each other, while in the Oxford model the number of interactions is based on survey results concerning the number of people participants remembered to have met in a day. However, one does not remember meeting all the people one meets in the bus, while they can still infect you if they are close enough. In the ASSOCC model, children have about 60 interactions per day on average, students around 85, workers around 65 and retired people around 40 interactions, that is with the Great Britain household profile and about 1100 agents. The average number of contacts for the agents is 60 contacts a day. Chapter 3, Table 3.2 shows how these numbers of expected contacts are calculated in more detail.

A second point where the Oxford model used other research was the amount of people that will violate quarantine regulations. In their model, they assume every day 2% of the people will violate quarantine regulations. In our model, the violations will dynamically arise from the need for autonomy and belonging. These needs are normally partly satisfied for adults by going to work where they meet with colleagues. However, when working at home the needs are less satisfied, but people also have a conformance need and will in general stay home and work during the quarantine. This leads to a high violation of the quarantine regulations during the weekend when people usually do not have to work and the need for belonging is more urgent compared to the need for conformance. Thus we see the violation not as uniformly 2% each day, but lower during the week and higher during the weekend. The higher violation during the weekend leads to more contacts and thus more spread of the virus during the weekend.

The models also differ in the total number of agents that are simulated. The Oxford model simulates 1 million agents while the ASSOCC model has around 1100 agents. This of course has implications for the absolute amount of infected, hospitalised and deceased agents after a simulation run, given the factor of almost 1000 of difference between the models. However, whether there is a relative difference is something we determine from the results. There could be a difference as having more agents could create more pockets for the virus, however this is mostly dependent on the network size settings. The ASSOCC model has a higher contagiousness setting in order for the virus to not die out early on, which can happen due to the lower amount of agents. One thing to note is that the Oxford model scales the y-axis of the output graphs and also their data to a population of 65 million [24]. This has been done for readability purposes, however the simulations were still performed with 1 million agents.

In terms of policy, there are some differences between the models. In the Oxford model, all of the elderly are kept in quarantine (shielding the elderly) after the global lockdown ends. The ASSOCC model does not apply this type of shielding.

The Oxford model consists of six different scenarios that represent different implementations of the TTA. We are mainly interested in scenario 1 and scenario 6 as these are closest to our implementation. Scenario 1 of the Oxford model has no tracking app active. This is a baseline scenario and can be compared with our 0% app usage setting. Scenario 2 is not relevant for us as there is no recursion in this scenario. The other scenarios release quarantine or do not have testing, therefore they are less comparable to our implementation than scenario 6. Scenario 6 has the TTA active and this app uses recursion. When an individual gets symptoms and self reports his household goes into quarantine. The individual is tested for the virus and with a positive result the other contacts of the individual are notified and placed in quarantine. In the ASSOCC model, an individual who gets symptoms and is user of the TTA is tested directly. When the individual is positive all of its contacts get tested. When one of those contacts is positive, they also have their contacts tested since recursive testing is enabled.

The Oxford model has 56% of the population using a TTA. In our model, we have five different ratios of app users 0.2, 0.4, 0.6, 0.8 and 1, which respectively generates 14.1%, 28.2%, 42.5%, 56.6% and 70.2% of the population using the TTA. Therefore the most comparable result for the TTA scenario will be between the Oxford model scenario 6 results and the 0.8 app usage ratio results of the ASSOCC model. Table 12.6 shows an overview of the relevant components in both models.

Table 12.6 An overview of the relevant components in the Oxford model and the ASSOCC model

Setting	Oxford model implementation	ASSOCC model implementation
Contacts per day	About 10 + household contacts	Average of 60, average of 10 during lockdown
Agents in quarantine	Uniform, 2% breaks quarantine every day	The agents break quarantine based on their needs, breaking quarantine is more frequent in the weekends
Lockdown	Workplace contacts reduced to 20% and household contacts increased to 150%	Agents stay home
Shield group	Elderly keep in (100%) quarantine after lockdown	No shielding of specific age groups after lockdown
TTA effect	Scenario 6 uses recursion and only quarantines contacts when the index case is tested positive	We test all the contacts and do recursive testing
TTA failure	80% of modelled contacts are saved by the app	All modelled contacts are saved by the app
TTA usage	56% of population uses the app, different variations	Different Test limits
Testing	No testing limits	No testing limits

12.6 Comparison of Results

The results we compare are the experiment results from Chap. 7 with the Oxford results for scenario 1 and scenario 6, given in [24]. The ASSOCC graphs have adjusted axis and are placed next to Oxford model graphs for easy comparison. We will look at the shape and trend rather than give a precise scientific comparison. We are not interested in showing whether the models are exactly the same or quantify how different they are, but rather we want to see where and why the models are similar or different and link this back to their implementation and conceptual choices.

12.6.1 *Daily Incidence*

One of the most important graphs is the daily incidence or daily new infected graph (see Fig. 12.1). These types of graphs are often shown in the news and used to evaluate whether interventions should be loosened or tightened.

The ASSOCC model has slightly more than 30 (2.7%) newly infected per day for the base setting while the Oxford model has over a million (1.9%) per day, this absolute difference is explained by the number of agents which is much higher in the Oxford model. However the graphs follow similar trends. Both graphs start with a slowly increasing daily incidence that becomes faster until the quarantine is started. Here, the line in the ASSOCC model becomes less steep until it starts to go down,

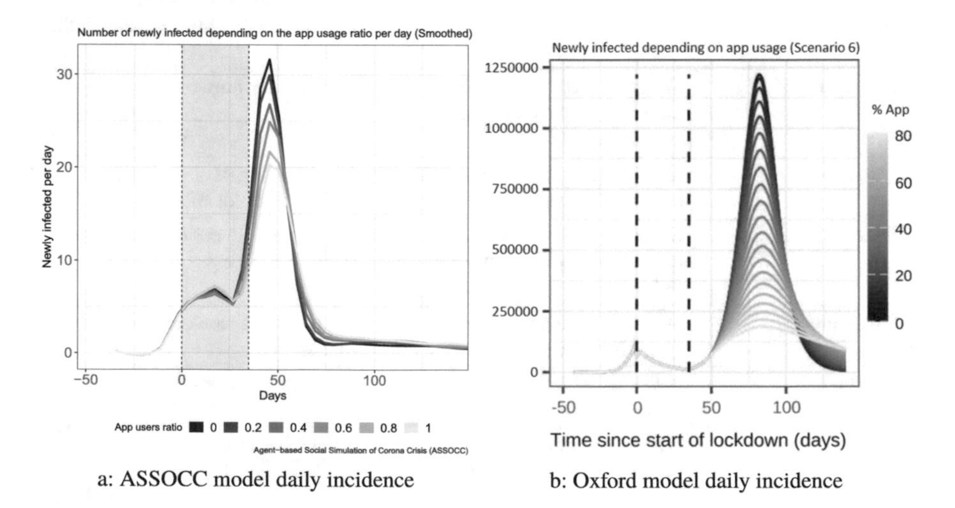

a: ASSOCC model daily incidence b: Oxford model daily incidence

Fig. 12.1 Comparison of daily incidence. The Oxford model has a doubling time (T2) of 3 days. This was the settings where the virus is most contagious, which makes it more comparable to our model. The Oxford model plot (right) has a scaled y-axis that is scaled such that it represents the daily incidence with a population of 65 million while in the Oxford model the daily incidence directly goes down. An explanation for this could be the higher number of daily contacts in the ASSOCC model combined with quarantine happening less uniform. This makes a lockdown less effective since as long as there are still a relatively high amount of contacts the spread of the virus will not go down quite as much as in the Oxford model. According to the average contacts graph in Chap. 7, the average number of contacts drops to around 10 during this lockdown period which is similar to the non-lockdown state normal amount of contacts in the Oxford model. Therefore, the average number of contacts in the ASSOCC model are always higher when comparing the models in the same state (lockdown or non-lockdown).

In the Oxford model, agent contacts are calibrated based on a study on actual social contacts [25]. In this study, participants were asked how many contacts they could remember talking to. However, this does not correspond to the number of contacts that were potentially infected as COVID-19 does not only spread by talking to another person but can also spread when in close proximity to an infected person or through toughing infected surfaces [26]. Therefore, to more realistically model the spread of the virus, we have to consider more contacts as, for example, people in the same office or classroom can indirectly infect each other.

After the end of the lockdown, we see for both models that the daily incidence increases quickly. The ASSOCC model has a steeper slope which is probably caused by the daily incidence still being relatively high before the end of the global lockdown. With more individuals infected, the virus can easily and more quickly spread when the restrictions are lifted. Not shielding the elderly in the ASSOCC model may contribute to this effect. In the Oxford model graph, we can see that a higher uptake in app usage drastically decreases the peak of the second wave, while this is not the case in the ASSOCC model which shows a milder decrease. This difference may also be explained by the higher number of contacts in the ASSOCC model and the much higher number of daily incidence at the end of the global lockdown. Both graphs have a fading out line after the second wave.

Another difference we can see is that our model has a less smooth graph while the Oxford model has a smooth graph, for example the peak has its sides going down with almost the same slope and at the start of the lockdown the daily incidence goes down directly. The Oxford model is more smooth due to it being a more mathematically and probability theory based model and having more agents. Our model has less agents but the agents also have more variety in their behaviour which generates a graph that shows more variety and is less smoothed out.

12.6.2 Cumulative Infections

Figure 12.2 shows the cumulative infections in both the models. The ASSOCC model (on the left) shows that there is only a slight decrease between no app users and 0.8 app usage in the cumulative number of infected. Since there are 1126 individuals in the ASSOCC model, about 75% of the total population gets infected. The Oxford model

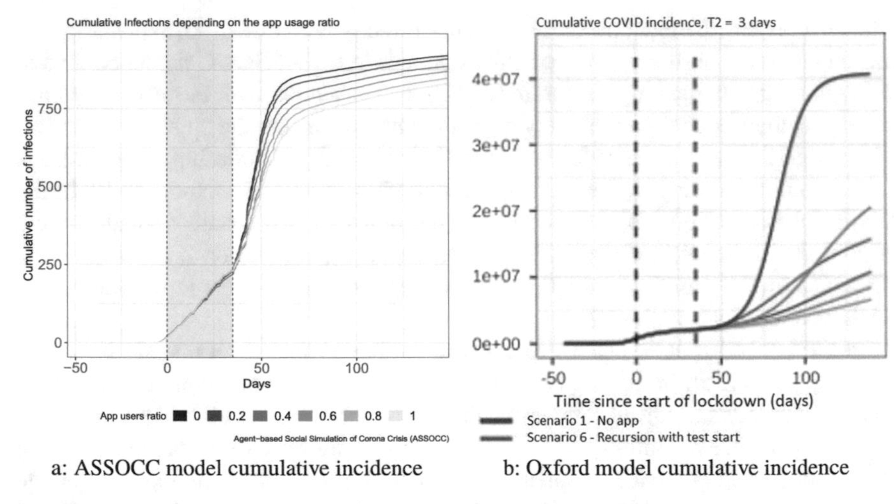

a: ASSOCC model cumulative incidence b: Oxford model cumulative incidence

Fig. 12.2 Comparison of cumulative incidence, T2 of Oxford model is 3 days

shows there is more than a 50% decrease of cumulative infected when we compare the red line (scenario 1) with the brown line (scenario 6). The Oxford model graph is scaled to a population of 65 million, which means in scenario 1 about 62% and in scenario 6 about 23% of the population gets infected.

The results differ probably due to the differences in daily contacts, higher number of infected after quarantine, and individuals breaking quarantine.

The results are comparable in the sense that they follow the same trend, especially for the baseline. At the beginning of the simulation, only a low number of infections can be observed until the virus starts spreading faster and faster (just before the start of the global lockdown). During the lockdown, the curve increases steadily and gradually becomes less steep. This effect is more prominent in the Oxford model. However, at the end of the lockdown the lines increase drastically until after 100 days it becomes flatter and fades out. These trends can be explained in the same way as explaining the daily incidence in Fig. 12.1.

12.6.3 Required Quarantined Individuals

Figure 12.3 shows the individuals that are supposed to be in quarantine. We can see that during lockdown in both graphs all the individuals should be in quarantine. Directly after lockdown all the lines drop down to about 20–30%. The baseline scenario with no app usage increases later than when there is a TTA active. Both these lines form a curve by going up and down after a couple of weeks. The ASSOCC model curve starts almost right away and is less stretched out which is probably caused by the quicker spread of the virus mentioned in earlier subsections. The Oxford model

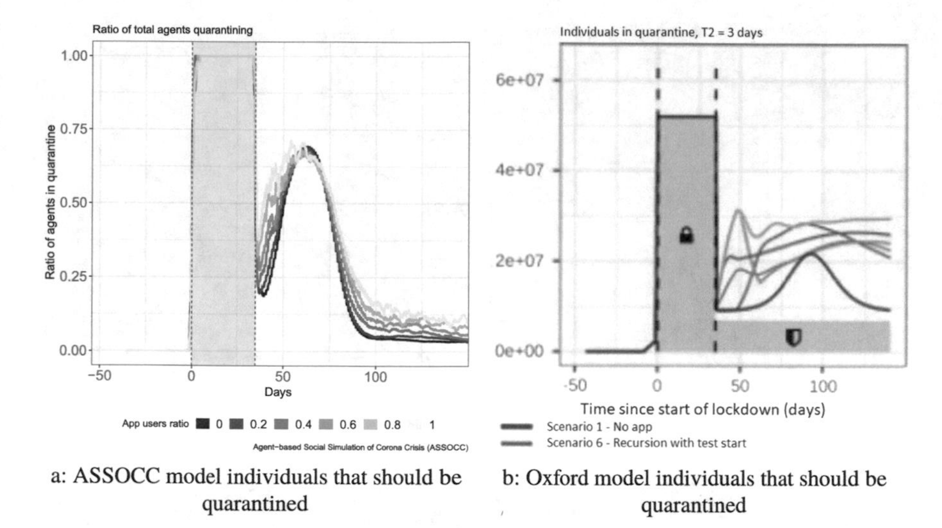

a: ASSOCC model individuals that should be quarantined

b: Oxford model individuals that should be quarantined

Fig. 12.3 Comparison of individuals that should be quarantined, T2 of Oxford model is 3 days. This number is different from the individuals that are actually quarantining

line starts to slowly increase after lockdown and hitting a peak at about 90 days after the start of the lockdown, after which is slowly decreases.

The app usage lines follow the same trend in the sense that they both increase quicker after quarantine compared to the red lines. However, while the ASSOCC lines does not show much difference when comparing it to other app usage ratios. The Oxford model with the TTA shows a quick increase at day 50 that stays high for some weeks until it slowly starts dropping again at day 90, which differs from the baseline scenario.

The differences are probably caused by the difference in rate of infection. The ASSOCC model has in all cases a quick spread of the virus, whether there is a TTA or not. The Oxford model in contrast has a slower spread and less spread when the TTA is introduced (Chap. 7).

12.6.4 Daily Tests

Figure 12.4 shows the daily tests *performed* of the ASSOCC model and shows the tests *required* in the Oxford model. In the ASSOCC model, it can be clearly seen that without TTA there are no tests performed, while with a TTA there are only a few tests performed until a week before the end of the quarantine. Then the TTA gets initiated and the number of daily tests increases very steeply. It then decreases with almost the same speed as it increased until it hits a platform where it decreases more gradually. The Oxford model shows hardly any testing before the end of the global lockdown. When there is no TTA the number of tests increases only after day

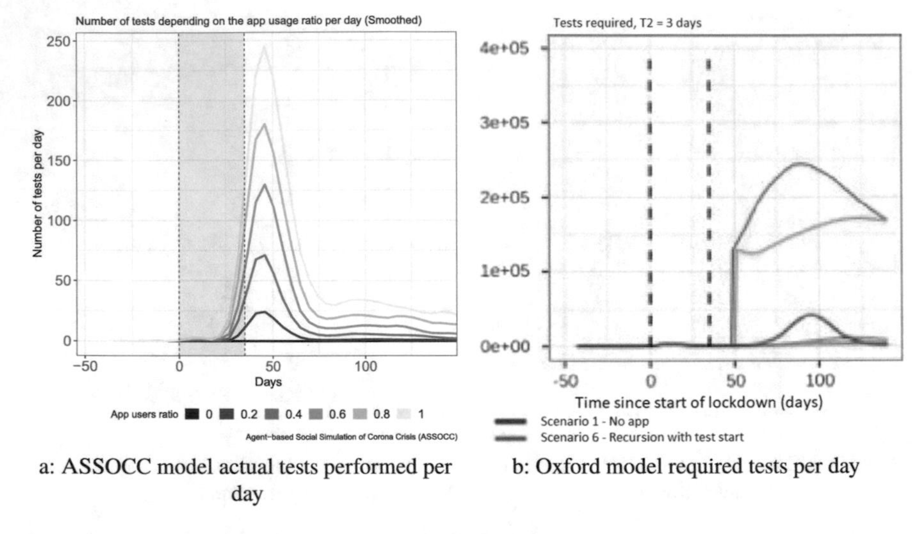

a: ASSOCC model actual tests performed per day

b: Oxford model required tests per day

Fig. 12.4 Comparison of daily tests, T2 of Oxford model is 3 days

60, forms a bump, and then goes down. In scenario 6 the number of tests raises very steeply and also form a bump after which it slowly decreases after its peak at 80 days after quarantine. Both the ASSOCC and the Oxford model show that a high number of tests is needed when using the TTA with recursion.

When we compare the app usage curves, they both increase very steeply at the beginning in both models. A difference is that the ASSOCC curve reaches its peak earlier and goes down earlier than the curve for the Oxford model. Also in these graphs, we can attribute the difference in width and timing of the curves to the difference in spread of the virus.

12.6.5 Discussion

The validation of simulation models usually requires the availability of real-world data. In case of the COVID-19 pandemic, such data was not available as the exact medical effects of the virus were not discovered. Furthermore we performed the analysis during the crisis rather than afterwards meaning the data of the effect of interventions could not be known. To this end, we validate our model by comparing it against the results of another simulation model that aims at investigating similar questions. After adapting the models to each other, we simulated the introduction and effects of TTAs on the dynamics of virus spread in different scenarios. Here, we analysed and compared four different performance indicators: daily incidence, cumulative infections, required quarantined individuals, and daily tests. It must be kept in mind that purpose and scale of the two compared simulation models differs.

While the ASSOCC model simulated approximately 1100 agents with a high level of detail, the Oxford model simulated 1 million individuals with more homogeneous characteristics and less diverse behaviour.

The maximum daily incidence without TTA that can be observed in the two models is 1.9% (Oxford) and 2.7% (ASSOCC). After the end of the lockdown, a rapid increase in daily incidence can be discovered in both models. The ascent that can be observed in the ASSOCC model is steeper compared to the one of the Oxford model. This can, for instance, be due to the relatively higher number of contacts each agent has in the ASSOCC model as well as due to the generally higher incidence that is present in the ASSOCC model after the lockdown ends.

The cumulative number of infections over the simulated period of time shows a similar shape for both models. After a shallow warm-up phase, there is a slight increase of cumulative infections during the lockdown, followed by a drastic increase shortly after the lockdown ends. While the increase in the ASSOCC model occurs rather promptly, there is a small delay in the Oxford model. This might be due to the assumption that elderly will continue to quarantine themselves after the lockdown. However it is most likely caused by the lower amount of current infected at the end of the quarantine, with a lower amount the virus will spread more slowly. Overall, the cumulative incidence in the ASSOCC model is 75% and between 23 and 62% in the Oxford model. At the same time, there are more infections occurring during the lockdown in the ASSOCC model which leads to other slightly different conditions once the lockdown is ended.

When comparing the number of individuals that should be quarantined during the simulation, this difference in the assumptions the models make on the behaviour of specific age-groups can also be observed. While the ratio of individuals that requires to be in quarantine drops considerably right after the lockdown ends. Yet, in the ASSOCC model, this initial drop to approximately 30% is directly followed by a steep increase up to 70%. In the Oxford model, this increase can also be observed, however, depending on the simulated scenario, it occurs with a delay of up to 90 days.

Finally, the number of daily tests can be seen as an indicator for the sensitivity of the models. For both simulations, the number of daily tests is low before the start of the lockdown and during its initial phase. Then, once the TTA is launched, the number of daily tests increases clearly. This is because of tests that are performed as a result of the contact tracing of an infected individual.

In summary, the behaviour of the two models is similar for the simulated scenarios. Smaller differences can be observed, yet, they can be related to differences in the assumptions the models make. Moreover, the differences in both scale and level of detail of the individuals is another factor which influences the results and might lead to variations.

12.7 Conclusions and Future Work

To analyse the effects different interventions have on the spread of COVID-19, a variety of simulations models has been developed. For the use in decision-making processes, the data generated by these models must be trustworthy. Traditionally, the validity of the model and its results is assessed by comparing the observed behaviour with data from the target system. In the ongoing pandemic, there is a lack of suitable data and accordingly, many validation cannot be applied. In this chapter, we discussed the suitability of different validation approaches for assessing the quality of simulation models for the COVID-19 pandemic. To this end, we conducted a comparison-based validation of the ASSOCC model against the Oxford model using the scenario of introducing TTAs.

A major challenge of the comparison-based validation was the adaption of the models in terms of their parametrisations and the assumptions they make. This included the adaption of both the contagion and the disease model. Moreover, the inputs and parameters of the models needed to be compared to identify corresponding variables. Differences in the modelling paradigms as well as in the scale of the models further affected the comparability of the results. While the ASSOCC model pursues a behaviour-based macro-approach, in which a smaller number of individuals is modelled with a great level of detail, the Oxford model is a mathematical model with a great number of relatively homogeneous and simple individuals. As a result of this, different questions and theories that can be investigated using each of the models. Comparing different epistemic outputs of the models revealed a similar trend, yet, smaller differences can be observed. However, these can most likely be attributed to the models' differences in scale and level of detail.

As the COVID-19 pandemic progresses, more data is getting available on the spread of the disease and how different measures can affect the spread. In the future, these data can be used for the application of further validation techniques as well as for the assessing the quality of the comparison-based validation. Additionally, comparisons with other available simulation models can be performed for a more comprehensive investigation of how results vary between different modelling approaches. Finally, a promising step towards more trustworthy and accurate simulations might be combination of different simulations models. By this means, minor inaccuracies or shortcomings of different models can be compensated by other models. Similar approaches, so called *multi-model ensembles* are, for instance, pursued in forecasting and prediction, e.g., in weather simulations.

Acknowledgements We would like to acknowledge the members of the ASSOCC team for their valuable contributions to this chapter. The authors wish to thank Loïs Vanhée for having developed the simulation and part of analysis based on which the results of this chapter are established. This research was partially supported by the Wallenberg AI, Autonomous Systems and Software Program (WASP) and WASP—Humanities and Society (WASP-HS) funded by the Knut and Alice Wallenberg Foundation. The simulations were enabled by resources provided by the Swedish National Infrastructure for Computing (SNIC) at Umeå partially funded by the Swedish Research Council through grant agreement no. 2018-05973. This research was conducted using the resources of High Performance Computing Center North (HPC2N).

References

1. R. Axelrod, Advancing the art of simulation in the social sciences, in *Simulating Social Phenomena* (Springer, 1997), pp. 21–40
2. S. Hudert, C. Niemann, T. Eymann, On computer simulation as a component in information systems research, in *International Conference on Design Science Research in Information Systems* (Springer, 2010), pp. 167–179
3. A. Tolk, "Epistemology of modeling and simulation, in *Winter Simulation Conference (WSC)* (IEEE, 2013), pp. 1152–1166
4. J.P.C. Kleijnen, Verification and validation of simulation models. Eur. J. Oper. Res. **82**(1), 145–162 (1995)
5. R.G. Sargent, Verification and validation of simulation models. J. Simul. **7.1**, 12–24 (2013)
6. L. Ferretti et al., Quantifying SARS-CoV-2 transmission suggests epidemic control with digital contact tracing. Science **368.6491** (2020)
7. R. Axtell et al., Aligning simulation models: a case study and results. Comput. Math. Organ. Theory **1**(2), 123–41 (1996)
8. N. Gilbert, K. Troitzsch, *Simulation for the Social Scientist* (McGraw-Hill Education (UK), 2005)
9. O. Balci, Verification, validation, and testing. Handbook Simul. **10**(8), 335–393 (1998)
10. W.L. Oberkampf, T.G. Trucano, Verification and validation benchmarks. Nucl. Eng. Des. **238.3**, 716–743 (2008)
11. W.S. Parker, II-Confirmation and adequacy-for-purpose in climate modelling, in *Aristotelian Society Supplementary* Vol. 83.1 (Wiley Online Library, 2009), pp. 233–249
12. C. Beisbart, What is validation of computer simulations? Toward a clarification of the concept of validation and of related notions, in *Computer Simulation Validation* (Springer, 2019), pp. 35–67
13. D.J. Murray-Smith, Verification and validation principles from a systems perspective, in *Computer Simulation Validation* (Springer, 2019), pp. 99–118
14. A. Gelfert, Assessing the credibility of conceptual models, in *Computer Simulation Validation* (Springer, 2019), pp. 249–269
15. S. Robinson, Conceptual modelling for simulation Part I: definition and requirements. J. Oper. Res. Soc. **59**(3), 278–290 (2008)
16. J. Banks et al., *Discrete Event System Simulation* (Pearson, 2014)
17. P. Kasaie, W. David, Kelton. "Guidelines for design and analysis in agent-based simulation studies, in *Winter Simulation Conference* (IEEE, 2015)
18. P. Ormerod, B. Rosewell, Validation and verification of agent based models in the social sciences, in *International Workshop on Epistemological Aspects of Computer Simulation in the Social Sciences* (Springer, 2006), pp. 130–140
19. J.P.C. Kleijnen, Validation of simulation, with and without real data, in 1998-22 (1998). https://ssrn.com/abstract=138010
20. B.B.N. de França, G.H. Travassos, Simulation based studies in software engineering: a matter of validity. CLEI Electron J **18.1**, 5–5 (2015)
21. E. Bonabeau, Agent-based modeling: methods and techniques for simulating human systems. Proc. Nat. Acad. Sci. **99**(suppl 3), 7280–7287 (2002)
22. H. Rahmandad, J. Sterman, Heterogeneity and network structure in the dynamics of diffusion: comparing agent-based and differential equation models. Manag. Sci. **54**(5), 998–1014 (2008)
23. W.O. Kermack, A.G. McKendrick, A contribution to the mathematical theory of epidemics, in *Proceedings of the Royal Society of London*. Series A, Containing papers of a mathematical and physical character, vol. 115.772 (1927), pp. 700–721

24. R. Hinch et al., Effective configurations of a digital contact tracing app: a report to NHSX, in En., Apr. 2020. url: https://github.com/BDI-pathogens/covid-19_instant_tracing/
25. J. Mossong et al., Social contacts and mixing patterns relevant to the spread of infectious diseases, in *PLoS Medicine*, vol. 5.3 (2008), p. e74
26. T. Galbadage, B.M. Peterson, R.S. Gunasekera, Does COVID-19 spread through droplets alone?, in *Frontiers in Public Health*, vol. 8 (2020), p. 163

Chapter 13
Engineering Social Simulations for Crises

Loïs Vanhée

Abstract Building social simulations during crises for decision-support largely expands the concerns and partly differs from building classic academic simulations: stakeholders are under high pressure and need fast and reliable answers, whereas public concerns and theoretical foundations for modelling the core aspects of the crisis regularly evolve as the crisis unfolds. Building upon the experience gathered by engineering the ASSOCC simulation, this chapter is dedicated to study how to frame the engineering activity for simulating during crises. This study includes an extended list of quality criteria of simulating in crises; an adaptation of software engineering methods to simulating during crises; and a guide and analysis for scaling up social simulation.

13.1 Introduction

The elaboration of a model is the result of the combination of many people conducting many activities. The success of a social simulation project, particularly during a crisis, is deeply tied to the many decisions ruling the complex interplay between people, activities, goals, resources, and techniques: selecting a wrong foundation for the simulation and major regression occurs every sudden turn taken by the crisis; seeking a perfect model and results will come too late for being of use for stakeholders. This chapter focuses design and engineering considerations: how, as a team striving to complete tasks towards producing prospective simulations, is it possible to build good models? By building upon classic literature on building software systems and our experience building the ASSOCC simulation, this chapter is dedicated to studying how classic methods are twisted by the crisis and how to these methods can be adapted for simulating in crises.

L. Vanhée (✉)
Department of Computing Science, Umeå University, SE-901 87 Umeå, Sweden

GREYC, Université de Caen, 14000 Caen, France
e-mail: lois.vanhee@umu.se

F. Dignum (ed.), *Social Simulation for a Crisis*, Computational Social Sciences,
https://doi.org/10.1007/978-3-030-76397-8_13

Classic engineering of social simulations is performed in contexts with soft time constraints. For these simulations, designers usually afford to perform linearly the classic sequence of social simulation activities: (1) develop a conceptual map of the case, usually through an extensive literature study of existing theories and related social simulations; (2) define the key salient aspects of this case, based on the concerns of the user or scientific study; (3) list the key components for modelling these aspects and their relations; (4) implement these components; (5) test, evaluate, and validate; iterate for fixing the bugs and modelling errors (or show discrepancies in the theories); (6) present the results.

The classic process for building social simulation is severely altered when simulating in crises. In crises, this six-step process fails from the very first steps, as phenomena of interest, theories, and key aspects are originally undefined and will likely change multiple times over time. In the COVID- 19 crisis, the scientific knowledge about crisis-dependent aspects was originally very limited and changed multiple times over the course of the crisis: what are the symptoms? How is the virus transmitted? How fast? Who is more sensitive to the virus? One day, children were risk groups to be protected at all costs, another they were immune, then high-risk spreaders, and later asymptomatic low-risk spreaders. In the beginning of the crisis, discussion were about closing basic schools and/or high schools. Later, workplaces were closed and in many countries complete lockdowns were imposed. Sometimes with curfews and sometimes without. The matter of face masks only became an issue a few months into the crisis. *Each of these new issues either involves additional aspects of the world that should be included in an (ever-expanded) simulation or revisions of former aspects, which may jeopardise some fundamental assumption of the model and thus entail to having to redo a large part of the simulation*, in a weekly basis. The chapters in Part II of this book reflect the many scenarios that had to be checked, each involving its own set of aspects. Besides this volatility, social simulations for crises are non-standard for the following reasons:

1. Simulations for crises serve several purposes at the same time (from Chap. 1)
2. Simulations for crises have to connect many aspects of reality and, thus, have inherent complex models (from Chap. 2)
3. Simulations for crises have to be extendable and adaptable due to rapidly changing situations in the crisis for which they are built
4. Simulations for crises are built under high time pressure of the crisis
5. Simulations for crises should be able to present results on an almost anytime basis
6. Simulations for crises should be scalable if they are to produce statistically relevant results for the real world crisis.

As a consequence, significantly adapted methods are required for simulating in crises. Studying related work, epidemiological models offer inspirational examples as these models succeed in being quick to produce results both early and over the course of the crisis. Typically, epidemiological models rely on the SEIR model [1], which is very straightforward to calibrate based on constant data acquired early on (e.g. propagation rate, incubation time). However, there is no existing SEIR-like approaches ready to be exploited at the moment and proposing one would take significantly longer than available during a crisis.[1] Moreover, such a SEIR-like approach would still need extensive adaptation and calibration work, as social simulation covers many more aspects than the health-centred SEIR model (e.g. economy, effectiveness of lockdown, mental exhaustion).

This chapter is dedicated to studying the question of how to build social simulations in the context of a crisis. As a method, we first define the quality criteria and issues to be addressed while developing simulations for a crisis, in Sect. 13.2. Then, the next sections are structured by adapting the classic software engineering methods from [2] to the case of simulating in crises, notably engineering activities in Sect. 13.4, software life cycle in Sect. 13.5, team organisation in Sect. 13.6, and architectures and platforms in Sect. 13.3. Moreover, a practical method for optimising and assessing the limits of scalability and optimisation of the software is introduced in Sect. 13.7. For these sections, we introduce the state of the art in engineering, describe how it is altered and to be adjusted for simulating crises, and specific examples from ASSOCC. By doing so, we hope to help the engineers of future crises in the many uneasy decisions and tradeoffs that are to be made in a hurry; proposing alternative ways of considering the design process of simulating in crises; and for be best preparing new tools for crisis simulations, discussed in Chap. 14.

13.2 Quality Criteria of Engineering Simulations in Crisis

As indicated in the introduction, simulation-building in crisis introduces concerns that expand and sometime differ from classic simulation-building. As a matter of enabling for arguing for the relevance of an approach, it is important to introduce specific quality criteria, which describe good properties of building simulation in crises (e.g. models without regression). These criteria can then be used for driving discussions towards which methods are appropriate in such a context. Whereas classic simulations focus on the qualities of the model (e.g. results accuracy), criteria should also cover the process and non-model related outcomes: the most accurate model has little value if stakeholders reject it due to its (rough) interface.

As overall quality variables, classic simulation quality criteria remain of course very important. However, beyond classic simulation quality criteria, connecting with stakeholders is very important: the model should be related to practical and user-friendly interfaces, for stakeholders to understand and trust the model and to effectively use it for supporting their decisions. Moreover, the model should also support trust-building: beyond a sound scientific validation, the model features should keep up over time, propose relevant plausible trajectories, and avoid making visible mistakes (which are usual in the early iterations if simulations). Good models should also be relatable to stakeholders, which can be achieved by making the model visually appealing, easy to explain, with has visible common-sense properties, with the closest correspondence to what the stakeholder actually manages (e.g. 1:1 population,

[1]This question of what can be done in this regards is further discussed in Chap. 14.

geographic layout matching the domain of the stakeholder). Good models should also be usable for many stakeholders, with minimal "one-shot" uses. Stakeholder concerns are interestingly distributed: every stakeholder tends to be associated to a very specific domain that differs from the next stakeholder (e.g. a specific region, a specific business) while sharing many aspects with other stakeholders, usually visible "terminal" aspects (e.g. R0, compliance rate of the quarantine). Therefore, there is an opportunity for extensive reusability of the model, while being also bound to stakeholder-specific effort. Good models should also be easy to adapt to unforeseen changes in the driving theories (e.g. degree of infectiousness of children) and stakeholder concerns.

As a matter of producing a (non-exhaustive) comprehensive list of criteria, we expanded the criteria of decision-support for pandemic from [3] with the criteria that proved to be central during the elaboration of the ASSOCC simulation (items marked with a * originate from [3]):

- Relevance for decision-support
  - Stakeholder concerns: does the model cover specific stakeholder concerns?
  - Prospective analysis*, counterfactual ability: does the model allow stakeholders to consider multiple courses of actions (e.g. switchable measures)?
  - Retrospective analysis*, explication-making: can traces and run be inspected for explaining the behaviour of the simulation? Can these explanations be made understandable?
  - Risk management support*: does the model provide relevant data for risk analyses?
- User-friendliness
  - Setup ease and ability*: how much effort is required for setting up the simulation? Can the software easily be deployed on any computer?
  - Usability*: how easy is it for the user to interact with the simulation (e.g. setting the right scenario, processing the output)? Can multiple format be given for setting the simulation parameters?
  - Relatability: does the stakeholder easily relate to the model and output of the system? How difficult is it for the stakeholder to assess that the components of the system are behaving as expected (e.g. visible agents doing concrete activity)
- Extendability and Specifiability
  - Responsiveness: how long does it take from deciding to add additional features or a given specification and first stabilised results?
  - Sustainable efficiency: what is the cumulative cost for implementing additional extensions?
  - Speed/quality alternatives: is it possible to implement multiple compromises between time, quality, and technical debt (e.g. rush a prototype and sustainable extension-building)?
  - Monotonicity: what is the expected regression or code to be re-written when later changes or extensions will be made? Can inaccuracies or visible erroneous

behaviour occur without being noticed regarding a feature due to later changes in another feature?
  - Re-Adjustability: how much effort is required for adjusting to changes in the theory? Can the system be automatically adapted/calibrated using data?

- Scientific features
  - Theory grounding: how direct is the translation of the theoretical foundations into the model and the implementation? Has the implementation technical intricacies?
  - Evaluability: are there facilities for testing that the system internally behaves correctly?
  - Validability: can the model output be validated against theories easily? (e.g. facilities for generating output, interactive generation of graphs)
- Model features
  - Resolution*: what is the resolution of the units of the simulation? Notably, population, time and geographic resolution (how many people, hours, m^2 are represented by every agent, time, and space units)
  - Scalability*: how many individuals and how much space can be represented by the simulation within reasonable computation time?
  - Genericity: how much is the model independent from context-specific assumptions?
  - Descriptiveness: how detailed and concrete (vs. abstract) are the various aspects?
  - Breadth: how many aspects are covered by the model? (e.g. transport model, demographic model)
  - Decision precision: how detailed is the decision process? Are these processes easy to adapt to new situations?
- Computation effectiveness
  - Computational complexity*: what is the theoretical complexity of the model? How much changing the value of the various variables (e.g. number of agents, resolution) impacts performance? Is the model well-optimised?
  - HPC support*: can the model be deployed on HPC for increasing the size of the precision of the simulation?
  - Real time capabilities*: how long does it take to get results?

A core design challenge lies in integrating the multiplicity of these criteria, which are sometimes mutually-opposing, within the high time pressure set by the crisis (e.g. resolution vs. computation effectiveness; model genericity vs. model descriptiveness). Failing on some of these aspects entails a severe degradation of system relevance, either by providing limited insights or by losing the trust of stakeholders. The ASSOCC model discounted on the number of represented individuals due to the inherent scalability limitations of the platform, which was a recurrent questioning from stakeholders: stakeholders can hardly reasonably justify their decisions based on models that are significantly smaller than the target community.

13.3 Software Architecture and Platform-Oriented Design

Software architectures seek to define the central concepts and structures on which the system is built [2]. Software architecture is relatively fuzzy concept that covers multiple meanings and uses, often simultaneously. Software architectures describe *overarching structures*, as would be a blueprint for a building (e.g. the simulation has five types of agents, each having specific tasks). Software architectures describe *core concepts for designers*, similarly to icons, working principles, and values (e.g. we seek to achieve high precision and large scale). Software architectures describe *core components* of the system (e.g. the model has an economy, the system has excellent plotting abilities). Software architectures are also often *tools for communicating with stakeholders* about what the system can and cannot do (e.g. our system can model the activity of any business up to 200 employees).

Defining the architecture of the system entails defining the general approach for solving the problem, usually involving committing significant initial effort for longer-term benefits. Thus, defining the architecture is highly critical for the design of the system as a whole, as it impacts what the system can (and cannot) become. Architectures are central for design effectiveness (e.g. by identifying recurrent structures that can be factored in one software component, by developing code that serves many users), for expandability (e.g. by making the system coherent and simpler to understand), for system quality (e.g. by dedicating more emphasis on the core aspects of the model), for teamwork (e.g. as an artefact for driving decisions) and for prospective users (e.g. for users to identify the qualities of the system). Committing to no architecture is committing to low effectiveness in the long run: the system will be composed of redundant extensions for solving immediate problems, which will stack up causing incremental engineering costs. Committing to a wrong architecture involves losing significant amounts of work in the future, either through re-design or ineffectiveness (e.g. hard to build, expand, maintain).

13.3.1 Background

Reference [2] focuses on architecture as overarching structures, as it is the most classic approach in computer science and the most direct one to capture with specific methods and explicit representations (e.g. diagrams). This approach introduces classic means for structuring software: programs with subroutines (i.e. the system is seen as a combination of parallel procedures that depend on each other), abstract data type (i.e. the system is seen as a composite object that is built and manipulated for representing an evolution over time), implicit invocation (i.e. the system is defined as loosely-coupled components, each that carry some specific operations), pipelines and filters (i.e. the system is seen as a sequence of transformation of a flow of data over time), repository (i.e. the system is seen as an entity that stores and searches through data), layered (i.e. the system is seen as a composition of services, which can call lower-order services). The task of the architect consists in matching the structure that best fits the task and instantiating this structure to the specific task at hand.

Architectures are tightly tied to *platforms*. Platforms, a loaded concept,[2] are considered here as artefacts (theories, ABM-models, software) designed for serving many stakeholders (e.g. multiple clients, an ecosystem of possible stakeholders, certain user types). For example, NetLogo is a wide platform for social simulation and game engines are platforms that often include specific simulation components and structures (e.g. gravity, geopolitics maps) that can be customised in a variety of specific games (e.g. Clausvitz[3] provides the facilities for modelling geopolitics, economics, and military warfare from roman empires to World War 2). Platform-centred design fundamentally impacts software design, as the software is built for a (prospective) community. Numerous additional tradeoffs are to be dealt with, notably how many people are covered by the platform, how well are they served by the platform, and how much effort is required by the user for exploiting the platform? For instance, the NetLogo platform covers relatively well the whole social simulation community, but its users have to involve significant effort for implementing the specific simulations they are interested in. A simulation platform dedicated to traffic management will better serve simulators for road planning who may just plug in the drawing of a route, but will be out of scope for a pandemic. Platforms can be built on top of each other (e.g. the ASSOCC platform is built upon the NetLogo platform).

13.3.2 Architecture, Platforms and Simulating in Crisis

Defining a strong appropriate architecture is an essential aspect for simulating in crisis. In particular, the architecture is critical for achieving the required functional needs (acquiring sufficient effectiveness for keeping up with the crisis, making the system extensible, and specifiable for many stakeholders) and communication needs (developing a platform with a strong validation and identity).

The difficult question consists in determining what to include in this architecture. The overarching computational structures introduced in the previous section are relatively uninsightful for describing social simulations, beyond describing aspects such as event-based or ticks-based. As for many simulation, the overarching computational structures of ASSOCC are layered structure for the overall simulation loop and abstract data type structure for representing the many components of the model.

Whereas seeking definite answers goes beyond the scope of this chapter, we believe that certain architectural core components are to be integrated in architecture for simulating in crisis. Given the high volatility of theories and concerns, the architecture needs to be founded in and structured around fundamental, immutable, and universal core components that are of relevance in such a crisis. Other, more volatile components, are to be included as extensions, so they can be better replaced

[2] We leave out the use of platforms as media for interactions.

[3] https://en.wikipedia.org/wiki/Paradox_Development_Studio.

as the crisis unfold. Typically, models founded in statistical decisions, epidemiology, governmental measures, and economy demonstrated regular over-specificity as new events occurred and thus were to be regularly revised (e.g. later statistics highlighting new patterns for people to violate the quarantine). The challenge lies in setting the appropriate core of for such a simulation.

As a concrete approach, we built the architectural core of ASSOCC on the fundamental cognitive elements that appear to be at play during the crises, as described in Chap. 2, e.g. norms, values, social networks, and needs. Unless encountering a behaviour-changing virus, psychological dynamics are fundamental, immutable, and universal in the context of a crisis: the state of the art in psychology is highly unlikely to be fundamentally altered by a pandemic. Other aspects, such as epidemiological model, and governmental measures are extensions to be added to this core, with the (unsurprisingly correct) assumption that they will change over time.

Committing to this core involves dedicating more emphasis to the psychological model and consider the other aspects as modular components. Such components should keep their inner workings private and be minimally entangled with the rest of the code. This approach is highly relevant for long-term sustainable effectiveness and extensibility of the model, as well as by having the simulation primarily relying on components that have been solidly tested and validated by being used multiple times (rather than all components depending on updates performed the week before). In ASSOCC, despite the high variability of the aspects of interests for stakeholders, the cognitive components remained highly stable throughout the project, except for minor adaptations of the existing structures for including the effect of new aspects. Moreover, relatively early in the project, we could have a high confidence that these models were working properly and providing sound insights for stakeholders.

Then, based on these components, the designer can build issue-specific models for answering specific questions and keep building a larger and larger model (with more and more aspects) while minimising the (re)design effort and regression risks. As a specific example, the disease model is accessible through a limited interface: the agent only gets a notification when believing to feel symptoms, the action module can test whether an agent is sick (when getting tested) and the contagion model can access the stage of the infection as it matters for contagiousness. In turn, this limited interface allowed to change the disease module three times with zero regression for the rest of the system.

The curfew example provides a direct example of extensibility. In January 2021, many countries were considering including curfews in their containment portfolio. It took us 48h for implementing this feature, from first discussions to sharing stabilised conclusions with stakeholders. The formalisation step consisted in studying how are curfews intended to be fulfilled in the real-world, which we could model as: “curfews are governmental restrictions, similar to lockdowns. They restrict access of certain facilities (leisure places, shops) during certain periods of the day (evenings). Restricted public facilities will be closed.”. As a model, it simply extends the range of prohibitions that can be set by the government, for which pre-existing decision components have already been introduced by the lockdown. The curfew affects two decision components: which locations are forbidden and which locations are open (forbidden locations related to business activities will be closed, others will still be reachable). By having built our model on advanced decision components, the pre-existing decision machinery takes care of adapting the rest of the decision process (e.g. planning model, needs model) *for zero additional design costs*. As example of side-effects that occurred without our intervention, agents would not try to get to closed places during curfews and would gain in e.g. compliance when respecting the curfew by avoiding forbidden locations. Implementing curfew within the agent process is (literally) a matter of 15 lines of NetLogo code, 5 within the agent deliberation and 10 for defining four modes of activation of the curfew (depending on when the lockdown ends). The final iteration is a piece of code encapsulated in an independent file, entailing virtually no design complexity for later steps of the process and no more management complexity than an extra chooser in the interface for selecting under which conditions the curfew is activated. The rest of the process is a classic generation and interpretation of results, using the now well-streamlined scripts for running them on the local HPC.

Looking at the ASSOCC project as a whole, we built (one of) the largest NetLogo simulation ever created, up to reaching the technical limits of the editor, within six months of labor in suboptimal conditions (lockdowns, all online). Despite this extremely high level of complexity, the ASSOCC model remained constantly up to date with the latest scientific findings and public concerns, providing adaptive decision-support and participating in the public debate for the ongoing concerns and decisions of stakeholders.

Moreover, this architectural core was very effective for defining clear a communication on what our simulation can and cannot do for stakeholders. Due to relying on psychological theories, which do not offer models as precise as e.g. physics, and orienting towards abstract models, it was very clear that the ASSOCC model should focus on *plausibility studies* rather than e.g. precise numeric predictions. Plausibility studies seek to provide the simulation user with reasonably admissible trajectories that the world can get into and sensible explanations for why and how such trajectories can occur. These projections are relevant for decision-support in crises by helping them to challenge and broaden the internal assumptions and expectations from stakeholders with aspects and dynamics that are difficult to foresee mentally a priori. Crises involve unexpected situations that can deeply challenge internal models: obvious factors a posteriori can be blurred by urgency and pressure. For instance, many people were convinced of the success of track and tracing, whereas the trajectories of ASSOCC were providing concrete arguments of why these apps can reasonably fail, founded in sensitive aspects that were not considered by stakeholders statistical decision models at this point time. By committing to the fringe of feasible contributions offered by our model, rather than seeking to match unfeasible objectives (e.g. perfect prediction, often expected from the general public), we can be clear in what we offer to our collaborators.

Likewise, this architecture could be applied for defining a solid conceptual orientation: by admitting cognitive precision and plausibility as a core value of our solution, we could easily align our decisions to a common goal. As an example of a difficult decision, we had to chose whether to lower the resolution of our model for larger scalability, as larger numbers of agents became a prime interest for stakeholders. How far are we willing to trade off the cognitive integrity of our agents for the sake of displaying larger numbers? How would it impact the plausibility studies we propose? Whereas technically possible, reducing the resolution would have brought us far away from the very foundations of the project and the scientific roots it is built on, so we decided to opt out. Chapter 14 will discuss how both can be integrated.

Regarding the quality criteria, the cognitive-centred architecture we propose is very suitable for achieving high expandability, through high responsiveness (by having a solid basis ready), a sustainable efficiency and high monotonicity (progressive consolidation and integration of features). The architectural design, and notably the platform-centred approach we selected is sound for satisfying the specific concerns of many stakeholders for moderate additional design costs. Moreover, by capitalising over the needs of multiple stakeholders, this platform-centred design allows for intensive factoring of the produced tools, thus covering general engineering concerns: decision-support support, model features, computation features, user friendliness; as well as scientific features by performing platform-level validation of its components. As drawbacks, platforms can involve more complicated setup costs for user and software complexity. They also require a precise and insightful anticipation, as wrong early decisions can entail significant re-fitting costs later in the crisis.

13.4 Software Building Activities

Six classic activities are usually deployed for producing software according to [2]: formalisation, modelling, implementation, testing, validating, and maintenance. For each of these activities, a set of approaches and tools are put forward. This section studies the relevance of these activities and tools for simulating in crisis. Then, Sect. 13.5 studies general strategies for combining these activities within the whole life cycle of the software. Many of these terms software engineering terms have also been borrowed by social simulation, with slightly different meaning. In case of ambiguity, a "C-" prefix will be added before the computer-science engineering version and (e.g. C-modelling refers to C-models used when building a software).

13.4.1 Background

Formalisation is concerned about making the needs of stakeholders (usually client or users) as clear, explicit, and formal as possible. Formalising is related to four concrete sub-activities. *Requirement elicitation* is the activity of informally understanding the problem, usually by eliciting the conceptual models implicitly used by stakeholders. Requirement elicitation is usually a negotiation step, in which the software analyst critically analyses the contradictions in demands and expected efforts and gains, and then constructively engages in with stakeholders with recommendations and counter-proposals. *Requirement specification* is the activity of describing the product to be delivered. *Requirements validation and verification* is the activity of reaching a consensus between stakeholders and software designers over the specification. Last, *requirement negotiation* is the reiterates a negotiation activity, based on the formalised requirements and more advanced plans.

Modelling focuses on translating the output of the formalisation step into formalised components to be implemented (i.e. turning goals to implementation plans). This activity is often dedicated to producing numerous computer-science formal representations, usually using the Universal Modelling Language. These representations include use-case diagrams (identify the actors and how they interact with and through the system), class diagrams (identifying entities of the software and their relations), state machine diagrams (identifying how entities can evolve), sequence diagrams (identifying the order of operations of the components of the system), communication diagram (identifying the ways the various components exchange data), and component diagrams (identifying combination structures of the various components of the system).

Implementing (or Software Design) is the activity of turning the formalised model into running code. Implementation methods can be grouped in three categories. First, methods for turning algorithmically complex problems into simpler problems, using for instance functional decomposition (e.g. decomposing the software as functions calling each other), dataflow design (e.g. decomposing the software as processing flux of information), object oriented design (e.g. decomposing the software as objects themselves composed of objects). Second, methods for finding for patterns that increases the efficiency of writing, using, and extending code. These patterns include abstraction, modularity, information hiding, complexity, and system structures. Third, methods for removing coding tasks by identifying and using pre-existing code for covering the most usual structures (design patterns).

Testing and **Validating** activities revolve around checking that the code is working as it should, according to the model (testing) and according to the stakeholder formalised needs (validating). Related methods detail how a software can be shown correct, including what to test (e.g. test criteria, flows of data, controlling constructs), why to test (e.g. checking faults, building confidence), and how to test (e.g. edge cases, base cases).

Maintenance studies what are the usual temporal dynamics of maintenance (e.g. an Poisson-like distribution of problems), causes for a software to require maintenance, and how to best maintain a system (e.g. local corrections, refactoring).

13.4.2 Software Building Activities and Simulating in Crisis

For simulating in crises, we advise to keep the formalisation and modelling activities as informal as possible. The prime aim of these activities consists in maintaining a strategic oversight over the features that are important for the general public, over the stakeholders the simulation is dedicated to (formalisation), and over the core aspects to be included for the features to be covered (modelling). Requirement elicitation has been found to be particularly effective for constructively negotiating what to integrate (or not) in the model, in line with the architectural orientations depicted in Sect. 13.3. For implementation, the greatest emphasis should be given on keeping the code as modular as possible (i.e. minimal interdependence between modules, notably secondary modules), pairing aspects to module as much as possible. The testing and validation steps should be kept relatively simple initially: a direct unit testing for verifying that the last increment is expected to be correct and a few checks "with the eye" that the main dynamics of the simulation are preserved. In testing activities, automated testing facilities should be added when the growth of the model makes verification "with the eye" too ineffective, as described in Chap. 4. Then, in a later stage of the process, another take should be given on these activities, notably for consolidating and exploiting the implemented contents, i.e. critically assessing and documenting the model (e.g. ODD, documentation of the key features) and the outputs (e.g. observed emerging tendencies). More details are given about this two-phase organisation in Sect. 13.5.

These shortcuts are motivated by the specifics of simulating in crises. Classic software-building activities are adjusted for (1) allowing to set a contractual relation with a third-party, (2) achieving high efficiency (i.e. factoring for reducing the amount of code to be written for code-intensive activities), (3) overcoming algorithmically difficult problems (e.g. chess-playing system, Google-like search engine). These activities also assume an high volumes of work (making pre-documentations worth the benefit) in relatively stable environments with stakeholders with explicit requests. None of these points apply for simulating in crises: no contract is to be signed on what the simulation should do; the overall volume is relatively low (the 8000 lines of the ASSOCC project is a huge NetLogo simulation but a low-volume software development project); and an overall low complexity (all algorithms used by ASSOCC can be produced by a good BSc computer-science student); and the environment is uncertain, with stakeholders who seek rather than give explanations. Overall, these activities involve extensive red-taping either with stakeholders (formalisation, validation) or with developers (modelling, verification), which has little benefit when simulating in crises.

The hard challenge of implementing social simulation in a crisis lies in achieving *sustainable effectiveness*, i.e. (1) turning conceptual descriptions of general phenomena into working implemented software; and (2) retaining this ability over time. The first point relies on relatively specialised "hand-eye coordination" skills for interactively fixing relatively low-level issues (e.g. proficiency at mapping real-world structures to computer-science structures and in turning computer-science structures into code) rather than on a deliberate slow-pace explicit elaboration of all details on a white board. The second point is mostly a matter of engineering and code-writing discipline. Overall, beyond the architectural core that should be developed with great care, if the designer starts to produce complicated diagrams before implementing any feature, then likely it will be infeasible to keep up with the crisis. Due to being skill-based, it is difficult to describing simply these abilities in a book, beyond enumerating dogmatic engineering platitudes (e.g. the code of each feature should be stored in different file). As a consequence, simulating in crises seems to necessarily require such specialised skills in the development team. However, it seems that only one or two specialists in the team are sufficient for achieving reasonable results. More details on team organisation are described in Sect. 13.6.

The proposed adaptation of the classic activities offers a solid compromise for meeting the extendability criteria (notably responsiveness, sustainable efficiency, speed/quality alternatives, and re-adjustability). Satisfactory scientific quality is achieved by performing extensive but delayed C-validation. However, algorithmically challenging model features, notably scalability, can involve greater costs due to cutting short formalization and C-modelling. Moreover, delayed C-testing and C-validation introduce moderate risks of regression, though major errors are unlikely to happen due to being generally very visible. This approach is relatively neutral with regards to other aspects when considered directly, but the efficiency gains indirectly benefits other features as more attention can be directed to them.

13.5 Software Life Cycle

Software life cycles captures how to combine the activities presented in Sect. 13.4 for elaborating the software as a whole. Various methods and techniques a proposed, such as fully completing every step before moving to the next, or working by small increments, which are to be triggered depending on the context (e.g. problem, development team, constraints).

13.5.1 Background

Multiple methods have been proposed for conducting software life cycle activities.

The Waterfall life cycle progresses linearly along the software building activities, with possible backtracks if errors are found: first formalise detailed requirements, then turn these requirements into a fully-fledged model, fully implement this model, thoroughly test this model and have it validated by the stakeholders, and last, perform some maintenance if stakeholders find later bugs or desire additional features. This approach has the benefit of being highly efficient for well-defined stable problems, at the expense of being less adaptive to later changes and having greater risks of fully missing the target.

The Agile life cycle, a more recent approach, takes the opposite direction by relying on the repetition of the shortest possible loops (one day to one month), during which all the activities are performed for achieving smaller-scale goals. The overarching paradigm focuses on maintaining a steady visible progress through growing a workable solution over time. Some extreme applications of the method seek a constant maximisation of the short-term value provided to stakeholders. Multiple variants have been proposed: *incremental development* focuses on adding one feature at a time; *rapid application development* focuses on maximising the benefit for fixed time-blocks (e.g. two weeks); *prototyping* seeks to drive showcase possible long-term directions for the development; *extreme programming* focuses on high-quality coding practices with extensive testing. Agile methods are relevant for uncertain contexts, for ensuring immediate and regular gain while using the context for further clarifying what is to be built next (e.g. when clients refine their needs as they use a first working version of their system). However, unless a sound architecture is devised early, agile methods often achieve medium efficiency, as they fail to capitalise on factorable elements, re-write and refactor code over time, and create technical debt (e.g. fast-written code ending up raising more costs due to extension, testing, and maintenance). More details on architecture are provided in Sect. 13.3.

Classic strategies for building social simulations follow globally a waterfall-like approach, with a recent advocacy for agile methods [4]. Usually, the modeller performs an extensive study of the theory and of the phenomena to be replicated (formalisation) before reaching the next step. Then, the modeller extensively seeks for the mechanisms that cause the emergence of these phenomena (first modelling) before continuing. Usually, this step is heavily documented and this documentation is included in a preliminary paper. Only then, extensive implementation effort is developed in the simulation, with usually a testing phase and generating graphs for pre-validating the model. Last, extensive validation through detailed graph production verifies that the simulation behaves according to theories. As in the classic waterfall model, every step can involve some regression to former steps, typically when results do not match the expectations, then the modeller revises the implementation, then the model, then the theory. This approach is sound, as the phenomena to model and its definition remains relatively stable over the process, except when theories are partial.

13.5.2 Software Life Cycle for Simulating in Crisis

As a general approach, we argue for a heavy parallelised incremental development approach, which combines one loop for the identification activity, many parallel loops for the development of the simulation, and multiple parallel loops for the development of surrounding modules.

The identification activity is dedicated to maintaining a strategic oversight, over both general public interest and prospective stakeholders to engage with. The purpose of this activities consist in sketching scenarios that covers congruent interests and features and that fits the structures of the architecture (more details about architecture in Sect. 13.3).

The development activity is split in four phases: the preparatory phase, the implementation phase, the consolidation phase, and the communication phase. Each development activity is related to a scenario (or a set of related features, including core features of the architecture). The phases of the development activity are completed sequentially. The *preparatory* phase is dedicated, at the conceptual level, to relate this scenario to a collection of phenomena to be replicated and aspects to be included, including a survey of related social-science literature and related news (formalisation, modelling activities). The *implementation* phase is dedicated to perform the implementation activity described in Sect. 13.4, i.e. turn the conceptualisation from the preparatory phase into workable code and complete first tests and a general validation. The *consolidation* phase is dedicated to further test and validate and then to develop scientific-grade documentation (introduce in text theories and phenomena to replicate, present the model in the ODD, explain the behaviour of the model through extensive plotting), performing formalisation, modelling, testing, and validation activities. The communication phase is dedicated to providing results to stakeholders and news producers. Multiple development loops can be handled in parallel, one activity per scenario or set of features, allowing for large-scale optimised teamwork (more details about teamwork in Sect. 13.6).

Besides the simulation-building activity, the visualisation and plotting modules can be built using (non-parallel) incremental development (more on these modules in Chap. 4). These modules, which can be developed on independent loops, are not further discussed here as they are already extensively covered by classic software development literature.

In a crisis, many aspects and concerns arise at once, and many features that can be packaged in relatively independent scenarios. Due to the high uncertainty inherently raised by crises and the high volatility of theories and interest over time, relying on parallelised incremental approaches offers a strategic oversight, a high reactivity towards the most urgent needs, and reduces the risk of committing to features that end up becoming of secondary interest for stakeholders. This high parallelisation also is also fitting for loosely bound (academic) groups (more on this in Sect. 13.6). In addition, completing these activities and phases offer high grounding and stability for the results while remaining in line with the current state of the crisis. As a drawback, such an approach requires a very solid architecture, implementation skills, and engineering discipline for sustaining its development. Sustainable development of social simulations is deeply rooted in maintaining a solid conceptual integrity, which is severely harmed by the chaotic environment intrinsically raised by pursuing in many scenario/iterations in parallel. More details about the architecture are provided in Sect. 13.3.

Regarding the quality criteria introduced in Sect. 13.2, this approach allows covering stakeholder interests by achieving high reactivity on the highest-impact questions and keeping up with the latest concerns raised by the crisis. This approach also allows matching high scientific standards and offers a good stability of the results before engaging with stakeholders. This approach, when combined with a good architecture and implementation ability and discipline, is also highly extendable and sustainable and offers monotonic progression through incremental additions. The early development of the architecture allows for early genericity, while the parallel development of the scenarios allows for developing a broad model relatively fast. To our knowledge, this approach has little drawbacks, besides its requirements and possibly engaging a moderate amount of effort in studying secondary tracks that fade out of interest of stakeholders before being implemented.

The ASSOCC project started by a first startup phase of a month, where the core of the model was built jointly with a minimal viable simulation and the collaboration was tested. At this point, the main essential features and core structure of the architecture were in place, thus, allowing for long-term stability and extensibility. Then, we started to develop independent scenarios in parallel (as described in Part II), based on the key aspects raised by the general public and the needs of various stakeholders. Most of the intended scenarios were brought to completion in the final version, except some features that faded away by being overshadowed by other features (e.g. public transport scenario did not reach its final stage as the track-and-tracing apps scenario became prevalent).

13.6 Team Organisation

Software engineering often involves the collaboration multiple individuals, both for delivering the product within reasonable deadlines and for combining specialised knowledge and skills. The success of software development activities is deeply tied tied to how well the team, its coordination and management are adjusted to the software to be built, the context at hand, and objectives (e.g. effectiveness, efficiency). Teamwork when modelling in crises is particularly central, due to the high pressure for remaining up to speed and updated with fast evolving theoretical findings and stakeholder concerns.

13.6.1 Background

The literature can be split in three parts [2]: management styles, coordination mechanisms, and team organisation.

Management styles describe the overarching values and concepts that individuals rely on for driving their actions towards a coherent coordination outcomes. Two complementary axes are brought forward: relations and tasks. The relation axis defines the prevalence of being cognitively close to the other individuals (e.g. understanding the other, agreeing). The task axis defines the prevalence of effective task-completion (i.e. effectiveness). These complementary axes lead to four archetypal management styles:

- *Separation* (low relation, low task): this approach seeks to maximise efficiency, i.e. the produced volume per hour. Individuals focus in maximising the overall production and progression, keeping the completion of tasks and updating others as a side effect of this activity (e.g. engaging in many projects at once, for capitalising over factorable code). This approach is fit when contributors can operate in independent silos.

- *Relation* (high relation, low task): this approach seeks to maintain a high degree of cohesion and proximity between the individuals. This approach is fit for environments that make difficult to introduce short-term goals but that requires participants to understand what the others are doing for working in a common direction.
- *Commitment* (low relation, high task): this approach focuses on achieving individual goals with deadlines. Individuals work relatively independently from each other.
- *Integration* (high relation, high task): this approach focuses on maintaining a cohesive team to work together towards specific self-given goals. This approach is fit for goal-driven exploratory activities that require complementary competences or with tight deadlines.

Coordination mechanisms describe structures for coordinating individuals and that individuals can use for coordinating. Reference [5] describe five archetypal coordination structures:

- *Simple structure* is based on direct supervision: a coordinator directly allocates tasks to subordinates. This coordination mechanism fits simple and dynamic environments, where the required skills can be anticipated while adjustments are required on the spot (e.g. shifting workforce from development to maintenance due to irregular requests from users).
- *Machine bureaucracy* is based around laying stable procedures for performing the activity. This coordination mechanism fits simple and predictable environments, where the process for solving tasks can be completely specified and combined with other steps. This coordination mechanism fits stable and simple environments that can be heavily standardised (e.g. standardised deployment).
- *Divisionalized form* is based on dividing the organisation in relatively independent silos that only depend on the standardisation of each other's output (e.g. multi-step transformation activities).
- *Professional bureaucracy* is based on giving professionals a high degree of freedom for performing their activities (i.e. highly-trained responsible individuals with specialised skills). This coordination mechanism suits environments that require the relatively independent actions of specialists that rely on each other ability (e.g. hospitals).
- *Adhocracy* is based on leaving specialists dividing and managing the completion of tasks on their own. This form of organisation is suited for complex and unpredictable environments that require advanced skills (e.g. research projects).

Team organisation focuses on describing which forms of social relationships are suited for the various types of activities. These patterns include:

- *Hierarchical organisations* revolve around leaders-subordinates relationships, where leaders are responsible for giving tasks to subordinates. This form of organisation can be expanded to multiple layers for larger groups.
- *Matrix organisations* revolve around spreading tasks to individuals across teams. This structure suits particularly groups with specialised individuals that are required in multiple parts of the project.

- *Surgeon teams (or Chief programmer teams)* revolve around allocating critical activities (usually the actual implementation, as it directly impacts the whole project) to a expert developer, who is supported by the rest of the team, which manages important but less-critical activities, such as preparing the specifications and tests, checking the code, validating, and documenting. This approach fits cases with significant disparities between the expertise of the members and for critical development activities (i.e. activities which outcome can impact the rest of the software).
- *Agile and SWAT (Skilled With Advanced Tools) teams* revolve around a strong informal cohesion of polyvalent members, usually with similar skills. The Agile team usually assumes a greater adaptability from its members for changing approach over time. This approach fits stable teams that work together on the long run.
- *Open Source Software Development teams* revolve around a community of loosely coupled individuals. Usually, these individuals are not bound to the project by formal obligations and, unlike most approaches, their commitment cannot be assumed nor "coerced" but negotiated (e.g. personal commitment to the project or its outcome, indirect personal interest, such as recognition). Usually, these teams self-organise along four levels of implication: core team (developing new features), co-developers (review code and fix bugs), active users (submit bug reports, request features), and passive users. The core team usually performs most of the implementation work.

13.6.2 Team Organisation and Simulating in Crisis

General patterns can be devised based on the general characteristics of simulating in crises, though decisions are to be adapted to every team. As core characteristics, simulating in crises involves a high emphasis over task-completion, as the value of delayed tasks (e.g. delayed consolidation and communication as depicted in Sect. 13.5) quickly decreases over time. Simulating in crises requires highly trained experts. Part of the required expert knowledge is predictable and belong to standard management, modelling, and computer-science skillsets (e.g. modelling, implementation, maintenance, reporting, communication) and part of the expert knowledge is situational to the crisis and thus requires to train or to connect to experts (e.g. economy, psychology, public measures). The inherent difficulty hinders centralised decision-making. Implementation-related activities are project-critical (most implementation actions directly impact the value of all other activities, errors are costly, they are unsafe to perform in parallel).

As an overarching analysis, the best fitting management styles are commitment followed by integration; the best fitting coordination mechanisms divisionalised forms and professional bureaucracies followed by adhocracies; the best fitting team organisations are surgeon teams, followed by matrix organisations, and agile and SWAT teams. Commitment and integration are task-heavy management styles, which is required for swift completion of tasks and thus maximising the value of the project.

Divisionalized forms (when possible) and professional bureaucracies offer the independence and room for expressing the required expertise with maximal parallelisation. Matrix organisations, surgeon teams and agile and SWAT teams further support the expression of expertise, either transversely across aspects (e.g. matrix organisation, surgeon in the surgeon teams) or centred per aspect (e.g. agile and SWAT teams, support in surgeon teams). The surgeon teams is particularly relevant because implementation tasks are critical and difficult to parallelise and require expert skills if applying the recommended compression of formalisation and modelling activities (described in Sect. 13.4). The integration management style and adhocracy coordination mechanism can apply and are usual in classic interdisciplinary social simulation activities, however they entail significant synchronisation and management costs that may harm responsiveness and sustainable development. Such a form of organisation corresponds to an ideal that can be applied for dedicated development teams.

For academic actors, the open source software development team organisation is the most sensitive option available, as they have little means for ensuring the involvement of contributors and cannot enjoy dedicated development teams. This approach allows notably to open the activity to a wide range of contributors, while giving more space for the most involved contributors to have greater impact. For such teams, it is very important to agree early on how (academic) rewards are to be shared over time (e.g. planned publications, authorship), as it can otherwise raise conflicts that can be very detrimental for the project as a whole. However, this team organisation is inherently risky as being tied to uncontrollable and sudden disengagement of core actors, due to factors such as paper submission deadlines, starting semesters, and ending funding.

As a side note, the academic system actually introduces some fundamental systemic issues when it comes to developing simulation in crises. First, because crisis-response requires a full-time focus on the crisis, whereas the academic organisation is rather tied to continuous numerous medium-intensity long-term engagements and obligations that cannot be interrupted for the time of a crisis without severe repercussions (e.g. teaching obligations, being excluded of collaborations). Second, because the systems does not incentivise contributors for the development and notably implementation of the simulation: this is activity is a heavy duty activity, to be performed under difficult conditions (constant time pressure), and risky (the project can fail or stop from one day to the next) while there is a significant risk from the developer to be left out of the author lists when papers are written about the project. Quite the opposite, the publication systems incentivises individuals who wait for the heavy implementation work to be done by someone else and engage in last-minute paper-writing activities. Perversely, the more successful is a project, the higher is the incentive for external actors to try stepping in and thus drain development resources; and for contributors to stop developing the model for writing about it before someone else does.

The ASSOCC team was lucky to have a very favourable composition. One contributor, the instigator, is a professor with has a solid experience on leadership, communication and has an extensive background on the conceptual side of social agent design; around 10 contributors, the majority of the team, are pre-doctorate researchers without marked expertise besides an interest for science and are all relatively directly related to the contributor; an associate professor, the lead developer, who is a long-term collaborator of instigator and has an extensive experience is engineering social simulations and implementing social agents; two late/post-doctorates, one with a general expertise in social simulation and the other on Unity for social simulation are also relatively related to the instigator. By a contingency of fortunate factors, most of the members of the ASSOCC team had little external constraints during the heat of the crisis and could (and did) fully invest in the project. This composition is very fortunate for solving leadership questions, as the instigator is also the sole professor of the group, has significantly more experience and has direct (sometimes hierarchical) relations with all the members. Last, the most advanced members of the group were used to work with each other and shared the complementary range of skills that covers most of the critical skills for developing such a simulation: from connecting to the stakeholders to the development of effective models, fully-fledged high-quality architecture, and interpretation and communication of the results. The presence of numerous early stage researchers, all related to the instigator, was very beneficial for effective scenario-centred parallelisation and high commitment to the project.

Overall, the instigator performed autonomously the identification activity and the communication phases per scenario, as presented in Sect. 13.5, while offering an overall direction for the group. For every scenario, a surgeon team organisation was set up, involving one or two early stage researchers as "owners" of the scenario and the lead developer for conducting the implementation. This organisation allowed to achieve heavy parallelisation of the activity while centralising the most sensitive operations around more specialised contributors. In addition to this organisation, the early split between the modelling and the GUI modules allowed to rely on a divisionalized form between these two modules. This split allowed to cut down the coordination costs between the two teams without harming global goals. ASSOCC relied on a commitment-heavy management style (high focus on tasks and more moderate on relation).

As a routine, every day, a meeting allowed every owner to present their findings and challenges, followed by strategic decision-making for choosing features to fix or implement within the next day(s). The strategic decisions usually involved a bi-partied debate: on the one hand, the instigator presented stakeholder-driven ideal features to include in the model whereas on the other hand the lead developer assessed the feasibility and sustainability of including these aspects in the model. This constructive confrontation of concerns, under the drive of the architectural choices, has been a determining factor for converging to outcomes that are both feasible and useful. The schedules of the members got tilted over time, with a part of the team working during the day (08:00–20:00) and another during the night (12:00–04:00). This adaptation increased the effectiveness of the team by having the key features implemented overnight and then tested in the feature owner before the daily meeting (13:00).

This organisation was very beneficial for allowing every member of the core team to exploit their skillset while ensuring steady progress over time. Early researchers owning features allowed the team to develop and maintain very early a solid overview on possible aspects, develop a general expertise, and develop concrete results, while keeping a manageable challenge for their level of experience. This organisation combined very effectively with the chief programmer organisation, leading to a maximum parallelisation of the effort with minimal coordination overhead.

The proposed approach offers high relevance for decision-support, user-friendliness, and scientific features, due to leadership activities given to the member turned towards stakeholders and by involving early researchers as support, with some concrete benefit of turning their feature into a publication. Model features, computation effectiveness, responsiveness, and re-adjustability are well supported by the chief-programmer organisation, as it allows to preserve the highest coherence of the model and ensures that the highest development standards are in place. The extendability of the model (responsiveness, sustainable efficiency, speed/quality alternatives) is well supported by the daily negotiation between stakeholder concerns and technical possibilities.

13.7 Computational Optimisation and Scalability

As the crisis unfolded, scalability became an increasingly central aspect of interest for stakeholders. The ASSOCC platform achieves mitigated results regarding this scale, as the platform only simulates up to 4000 individuals for one-hour-long simulation time. This result is due to a contingency of factors, including architectural core orientations, which promotes a strong emphasis on the psychological model, thus involving a "1 agent per individual" degree of resolution, advanced psychological models, a strong orientation towards extensible implementation that hinders optimisation quirks, and the computational inefficiencies that are inherent to the NetLogo platform.

Despite this orientation over quality rather than quantity, optimisation efforts have been developed to rule out any significant complexity issues. We wanted to optimise up to the point any further significant optimisation would also entail significant compromises in terms of model quality according to our architectural orientation. As such an optimisation is both useful and rarely documented, we describe here the motivation and processes we applied for determining when to stop optimising, notably by identifying hard upper bounds on the number of agents.

13.7.1 Reaching the Limits of the Model

The first step when starting to optimise consists in deciding what can be compromised for computational efficiency and subsequent scalability. In our case, we gave high precedence to *model integrity*, i.e. preserving the closest proximity between model and implementation. This integrity involves a strong avoidance against lowering the resolution (e.g. one agent representing a thousand of individuals) and avoiding computer-science quirks (e.g. representing boolean agent variables using integers). As a principle, moderate computer-science optimisations were tolerated if they allow lowering the complexity (e.g. caching resource-heavy computations when repeated by all agents), maintain readable code and do not significantly alter the model.

Computational complexity theory is a standard tool for setting concrete limits of the scalability of a program. Computational complexity relates how changing the input of the program impacts the expected computation time. The complexity of a well-optimised ABMs usually scales relatively linearly to its number of agents, the complexity of the decision of every agent, the number of ticks, and the degree of locality of agent interactions. More formally, the complexity of a well-optimised ABMs is usually $O(n \times (i + d) \times t)$, where n is the number of agents, i is the number of interactions an agent has with other agents, d is the complexity for the decision process (usually in $O(1)$ or $O(i)$) and t is the number of ticks. i is dependent on the locality-related factors (e.g. number of other agents sharing the same gathering points) and assumes a number of interactions linear to the size of this locality (e.g. shaking hands with the other agent presents). d is context-dependent but is often constant time or linear to size of the locality (e.g. size of the family, set of considered actions or located activities). *In less technical terms, if a piece of code (1) needs to be there, (2) cannot be further optimised, and (3) needs to be run by every agent, then the time taken by running this piece of code unavoidably linearly increases with the number of agents (i.e. twice more agents $\approx$ twice more time).* Practical heuristic optimisation methods can be based on the previous principle: run your simulation at full capacity (i.e. with the largest number of agents possible before which the system is considered too slow); check which parts of the program took the longest run-time; check if these parts are essential to the program; if they cannot be further optimised, then a concrete bound on the scale can be set.

As a specific example of the analysis on the current version of ASSOCC, we wanted to run on a desktop computer a standard 1500-ticks simulation within 25 min, which equates one tick per second. The system is run with $n = 2864$ agents (1000 homes) for $t = 10$ ticks, and took around 10s to complete. The NetLogo profiler indicates how long time the program exclusively spent running every function. The most time-consuming function is "infection-status". Over 10 ticks, this function tool a total active time of 1.7 s, being called 8061393 times with less than 0.001s per call. Time can either be saved by optimising this function or reducing the number of times this function is called. This function is a constant-time ($O(1)$) function, as it is a nearly-direct reporter. Thus, no significant performance gain can be expected. Whereas the number of calls is high, a rough estimation suggests that this number is not significantly higher than it should be (the infection status is required many times every tick, for every agent, multiple times for every decision, and for every interaction) and a cross-comparison with lower value of n indicates that the number of calls scales relatively linearly with n. Overall, assuming this function or its use can be optimised, only a moderate gain is to be expected.

As a matter of bringing strong formal evidence, we can search for the most expensive function for which we can explicitly demonstrate that the number of call is minimal and that no significant optimisation can be expected. "global-prevalence-of" has a total execution time of 0.551 s, is called 228696 times with every call costing 0.002 s. This function, which computes the need satisfaction of a given located activity, is called a reasonable amount of time: each of the $n = 2864$ agents assesses $d = 10$ plans per tick for every of the $t = 10$ ticks. The content of the function itself is hardly optimisable, as it is a constant-time $O(1)$ function that simply computes a weighted sum of 12 values. As a matter of illustrating the limit, even in (irrealistically optimistically) assuming that the rest of the code executes instantaneously, the cost of the "global-prevalence-of" function entails that the simulation cannot host more than $n = 57280$ agents before taking longer than one round per tick. In other words, due to the computational costs inherently brought by this specific function (which is an essential part of the needs model), the simulation cannot scale up beyond $n = 57280$ agents within given constraints.

13.7.2 Scaling Up

As indicated in the introduction, this bound to 4000 agents is a consequence of our architectural decisions, for preserving theoretical, conceptual, model integrity while preserving a clean, stable, and expandable the code. However, different tricks and compromises can be setup for raising this number.

Assuming the preservation of the integrity of the model, one can reasonably expect a multiplication factor between 100x and 3000x through a variety of techniques. A 5–10x performance gain on NetLogo appears to be reachable with moderate cost and sacrifice of other aspects, notably on readability and minor model alterations. A 5x–20x performance gain can be expected through investing significant effort in advanced computer-science quirks and the use of a lower-level language (Java or even C), given the state of the art in videogaming and hardcore optimisation techniques. Another 30x/60x improvement can be expected for a relatively affordable design cost, through parallelisation on a single computer, which is relatively easy for multi-agent systems. Assuming a Google-like computational power and proficient engineers, a 1:1 simulation of the human population seems to be within reach (high parallelization power offered by the agent-oriented paradigm, ability for storing locally data). Though, the soundness of the investment of resources with regards to the gain is questionable at this point, as the low-level engineering effort is likely to be severely detrimental regarding other aspects of the model, notably reactivity. As a side note, as parallelization is tightly related to scalability, it is often very easy to parallelize the execution of independent experiments with more CPUs. High agent scalability for high cost is to be considered with limited interest: whereas it offers a great technical demonstrations that can help convincing stakeholders, it will be very expensive to run the many simulation required for exploration and validation.

Another approach for scaling up consists in altering the model, notably by reducing the resolution as done for instance by [6], which allocated one thousand of individuals to every agent. We did not take this approach in ASSOCC, as our models relied on very deep individual psychological dynamics and personal interactions. Whereas it is possible to stretch statistics-oriented decision processes from one to a thousands individuals with lose relations, the same stretch is more questionable when variables are very personalised: how does a 1000-individuals agent with 3% of which being sick and 2% being aware of it decides on whether to meet friends? The extension is feasible, but at the expense of severely revising the model and making explanations significantly more convoluted.

13.8 Conclusion

This chapter studied how to engineer social simulations in crises through the lenses of general software engineering. Notably, this chapter introduced core criteria for assessing the quality of the process of building simulations in crises, coupled with an adaptation of the classic software engineering methods and approaches to the context of simulating in crises. The obtained experience by designing ASSOCC shows that engineering social simulation in crises introduces important extended concerns and differences with regards to general software engineering. These changes alter general aims of the approaches to specific methods, yet retaining similar central concepts, structures, and activities.

As a matter of introducing integrative concepts, engineering simulations in crises is about **sustained effectiveness**, i.e. achieving specific goals (e.g. implementing core features, solving scenarios) and retaining the ability of achieving specific goals later on. The challenges of simulating in crises lies, within a *moderate time-pressure*, in achieving *many goals*, each of which being relatively *simple* to solve, which can be solved *relatively independently except at the code level*, where interdependencies can be required and are easy to introduce unnecessarily. Therefore, the problem of engineering simulations in crises differs from most of classic engineering purposes, which is mostly dedicated to maximise efficiency (i.e. volume or value produced per unit hour) or solve hard algorithmic problems.

Setting sustained effectiveness as a general goal then helps setting concrete decision canvas for solving engineering matters, from high-level organisation to code-level implementation choices. From a holistic perspective, sustained effectiveness implies to develop and rely on a solid *extensible architecture*, for driving design and collaboration decisions, communication, and achieving the sufficient efficiency that is required for sustainability. This architecture should strive to define the *core*, i.e. the immutable elements of the development activity (e.g. values, orientations, concepts, implemented structures), defining what the simulation is good at. This core then defines which elements should be given the priority regarding development resources: how much extra complexity, engineering time and computational costs are we willing to invest in a feature? For a secondary feature, a simpler model that reduces the interdependencies with other features is a good tradeoff. Sustained effectiveness also frames the requirement on the activities to be conducted. In particular, most of "red-taping" documentation, which is a core element of software engineering, can be left after the development process, once a working solution is found. In terms of activity organisation, most of the activity can be conducted in parallel, except for the implementation activity due to the tendencies in social simulations for most aspects to indirectly influence each other. A highly parallel incremental approach can be put in place. This approach is actually effective as the only operation that cannot be parallelised (fast modelling and implementation) involves a workload that is manageable by a small team (though with the constraint of significant prior experience). In turn, these constraints drive a coherent team organisation, by having many feature experts developing in parallel and relatively independently the various scenarios and a small developing team for implementing them all. The feature experts prepare the ground and consolidate upon the developing team, thus allowing optimal efficiency by minimising interdependencies and capitalising on the expertise grown by all contributors.

Altogether, engineering simulations crisis is an exciting scientific venue, as it involves a complete retake on engineering activities. Whereas the final purpose is to develop useful code, the means by which this code is to be delivered over time and subsequent development organisation widely differ. As a final positive note for the community, this engineering experience highlights that developing social simulations in crisis is feasible and can be achieved for moderate costs. A few engineers dedicated to and trained for crisis-response, who can grow validated and trusted crisis-response solutions in off-crisis periods, coupled with a network and resources for pulling scientists in case of a crisis appears to be cost-effective workable solution for achieving significant impact for future crises.

Acknowledgements The author would like to acknowledge the members of the ASSOCC team for their valuable contributions to this chapter, with a particular thank for Frank Dignum, Fabian Lorig, and Harko Verhagen for their support writing this chapter. Moreover, the author wishes to thank Nicolas Payette for offering his expertise as a NetLogo platform developer for confirming our scalability and optimisation analysis of the ASSOCC platform. The author also wishes to thank the NetLogo team, which was a necessary component of the success of the project.

References

1. R.C. Cope et al., Characterising seasonal influenza epidemiology using primary care surveillance data. PLoS Comput. Biol. **14**(8) (2018)
2. H. Van Vliet, H. Van Vliet, J.C. Van Vliet, *Software engineering: principles and practice* vol. 13. (John Wiley & Sons Hoboken, NJ, 2008)
3. D.J. Heslop et al., Publicly available software tools for decisionmakers during an emergent epidemic-Systematic evaluation of utility and usability. Epidemics **21**, 1–12 (2017)

4. A. Schmolke et al., Ecological models supporting environmental decision making: a strategy for the future. Trends Ecol. Evol. 25.8, 479–486 (2010). http://linkinghub.elsevier.com/retrieve/pii/S016953471000100X
5. H. Mintzberg, Structure in 5's: a synthesis of the research on organization design. Manag. Sci. 26.3, 322–341 (1980). issn: 00251909. 10.1287/mnsc.26.3.322. https://doi.org/10.1287/mnsc.26.3.322
6. T. Blakely et al., Determining the optimal COVID-19 policy response using agent-based modelling linked to health and cost modelling: Case study for Victoria, Australia. medRxiv (2021)

Chapter 14
Agile Social Simulations for Resilience

Maarten Jensen, Frank Dignum, Loïs Vanhée, Cezara Păstrăv, and Harko Verhagen

Abstract In previous chapters we have described the results and analysis of social simulations for crisis situations based on the experiences of the ASSOCC framework. Whereas we managed to build an implementation in a very short time, based on many years of fundamental research, such an implementation is inherently limited due to the many tasks and challenges that are to be dealt within high time pressure (e.g. keeping up with the many emerging public concerns and specific stakeholder issues with relatively unstable theoretical background while dealing with scaling up technical challenges). This chapter proposes and discusses what can be prepared for future crises, from fundamental conceptual building blocks, to decision components, and technical implementations. Building upon the experience of what we missed and had to create or deal with(out) during the crisis, this chapter is dedicated to point how, as a community, we can be ready to adaptively respond to future crises.

M. Jensen (✉) · F. Dignum · L. Vanhée · C. Păstrăv
Department of Computing Science, Umeå University, SE-901 87 Umeå, Sweden
e-mail: maartenj@cs.umu.se

F. Dignum
e-mail: dignum@cs.umu.se

L. Vanhée
e-mail: lois.vanhee@umu.se

C. Păstrăv
e-mail: cezarap@cs.umu.se

L. Vanhée
GREYC, Université de Caen, 14000 Caen, France

H. Verhagen
Department of Computer and Systems Sciences, Stockholm University,
PO Box 70003, 16407 Kista, Sweden
e-mail: verhagen@dsv.su.se

F. Dignum (ed.), *Social Simulation for a Crisis*, Computational Social Sciences,
https://doi.org/10.1007/978-3-030-76397-8_14

14.1 Introduction

By 2017, dozens of decision-support models dedicated to pandemic management were available, yet no major success of social simulation for supporting the crisis can be reported today [1]. There has been one social simulation model that has actually been used to base decisions of the Victorian government in Australia on [2]. This simulation is more based on individual statistical behaviour than incorporating social influences. As pointed out in Chap. 13, the constraints raised by simulating during a crisis are many and they are challenging existing methodologies and tools for building social simulations. Either the simulation system is built *a priori*, building upon assumptions and modelled features that are not likely to hold or correspond to stakeholder concerns during the crisis. Or the simulation system is built *during* the crisis, in which case the required activities for achieving a fully-fledged development are hindered by time-constraints. In both cases, these simulation systems fail in their tasks of supporting decision-makers by failing to build and maintain a high level of trustworthiness, itself caused by a failure to sufficiently cover at once the (too) numerous simulation challenges. These include the simulation being too simple, small scale, imprecise, out of context, difficult to adapt, etc. ASSOCC, which falls within the category of simulations built during crisis, is no exception to this. Our model is stitched together and its success almost exclusively results from a solid theoretical model, our extensible approach, and day and night commitment, which allowed us to iteratively add the required components within the very tight time constraints. This fast-tracked development was necessary to "keep up with the crisis", enabling is to analyse and give advice on relevant and ever shifting concerns of policy makers. However, as the model grew, design costs incrementally increased, as the structures became more numerous and intertwined. Design choices that were necessary for being relevant early and throughout the crisis (e.g. NetLogo prototyping) ended up introducing inherent limitations regarding other challenges of simulating in crisis (e.g. scaling up the number of agents, spatial models).

Is simulating in crises inherently unfeasible, as models made a priori are likely to fail and models made during the crisis are likely to be too limited or to arrive too late? The experience accumulated during the ASSOCC project encourages us to follow a third path: *models able of fast deployment and adaptability* (or sustainable effectiveness, as described in Chap. 13). We propose to build and consolidate further on the solid theoretical conceptual foundations of ASSOCC that offer a good trade-off between realism and scalability. Furthermore, the core components of the ASSOCC platform software can be redesigned in order to lay better foundations for scalable simulations that can be extended and instantiated to concrete stakeholder concerns. Chapter 13 described our engineering stance for keeping the design as extensible as possible once a crisis started. The current chapter is taking the experiences from ASSOCC and see how they can be applied to prepare better for the next crisis. As we went through the process of building simulations for a crisis, we now better understand what is needed for building such extendable models. We can move one step forward and consider what theoretical, modelling, and implementation elements could have been produced beforehand for better handling the challenges of simulating in crises, including saving development time.

We take our lessons learned to create the foundation for a simulation platform that can be used to support the resilience of society in the face of crises. A more generic platform that allows to quickly build specific scenarios for all kinds of crises enables simulation of all kinds of crises before they appear and check their effect on society. This supports experimenting with measures that can be taken to better cope with such a crisis. Thus, rather than waiting for the next crisis to happen and create simulations for the specific crisis at hand, we can build a more resilient society by developing a set of modules that can be adapted to potential crises and test the value of the set by continuously simulating possible crises and measures to prevent them or possible responses to them. Such a model would allow to perform counterfactual studies, such as assessing the consequences of suspending all traffic between countries worldwide within a week after the outbreak of a pandemic. No doubt this would curb the pandemic, but also for immense costs on the economy. And imagine that some virus is detected every half year and it is not clear how dangerous a virus is right away. Would traffic be halted every time? What alternatives would there be? And at what costs and risks? But besides these obvious pandemic-related applications we should also be able to see for instance how we could best respond to refugees flooding neighbouring countries when wars erupt. How to support refugees best (also for the long term), while respecting and supporting the local population as well. Or how to cope with natural disasters on a large scale such as floodings, draught, etc.

In the rest of this chapter, we describe some of the key theoretical, modelling, and implementation tools that are necessary to support the general requirements described in the next section. We argue that building the foundations for a new framework for crises simulations should start from the basic principles of the agent decision-making model and incorporate all possibilities for making these models scalable right from the start. In the next section we will describe a number of generic components that are common for simulations for crisis contexts. In the section after that we will describe solutions for modularizing the software in order to support the explicit modeling and of components such as actions, plans and needs. In Sect. 14.5 we will discuss the concept of context as a way to achieve scalability in the context of complex and abstract agent models. In the following section, we will discuss some tools for covering expectedly recurrent needs of developers and users of the social simulation platforms in a crisis. These tools include user interfaces for the public, for showing results and give interaction possibilities and solutions for specifying complex scenarios and analysis of the data. Then, we will briefly discuss how social simulation platforms that are built on the principles of the previous sections can be used to test the resilience of society as well. We do not have to wait for the next crisis to create a simulation for it. In the last section we will draw some preliminary conclusions and sketch necessary steps to achieve simulations for resilient societies.

14.2 General Requirements for Agile Simulation System Development

Aiming for one simulation with all the aspects for preparing to any kind of crises is obviously intractable for complexity reasons. For example, geographical components are far more important for certain crises than for others. However, it should be possible to incorporate standard models for these components in a way to construct a decent simulation for each crisis situation. We keep in mind that the platform should support the requirements for simulations for crisis situations:

1. Simulations for crises have to connect many aspects of reality and Thus, have inherent complex models (from Chap. 2).
2. Simulations for crises should be scalable if they are to produce relevant and valid results for the real-world crisis.
3. Simulations for crises have to be extendable and adaptable due to rapidly changing situations in the crisis for which they are built.
4. Simulations for crises should be able to present results on an almost anytime basis, including in a format that is understandable for stakeholders and with an important ability for interaction (e.g. define scenario, counterfactual).
5. Simulations for crises are built under high time pressure and Thus, should be easily composable, maintained and used.

All these requirements cannot be fulfilled simultaneously. Thus, trade-offs have to be found and also new ways to create building blocks for a simulation platform that supports all requirements in the best way possible. We propose to do this by creating components and support for model components that are inherently important in simulations for crises. E.g. the fact that the simulation should be easily extendable with new aspects means that it should be possible to specify new types of actions and their effects and integrate them in the rest of the model. Thus, the platform needs some support in specifying actions on a more conceptual level (rather than having to program them directly into Netlogo or Repast).

Given the requirements for each simulation the following aspects are all linked and need to be modeled explicitly such that they fit together:

1. temporal resolution
2. spatial resolution
3. actions and their effects
4. deliberation process

Developers should be able to choose the resolution of the temporal dimension of the simulation depending on the time scale of the important actions and decisions that play a role in the simulation. In the ASSOCC simulations we chose to have each tick represent 6 h as that would give the opportunity to distinguish between work and evening activities and at night people would sleep. However, this granularity does not allow to record the time people stay in a shop and having choices to go out of a shop quickly when it is busy, etc. All these aspects would have to be modelled within the shopping action. Using a much higher resolution of the simulation would allow to make more fine grained decisions, but it would be very inefficient if most important decisions are made only a few times a day. In order to remedy the inefficiency of a high temporal resolution one could make an implementation where the tick is representing a very short time, but not all agents will have to make a decision each tick. Efficiency measures like this are quite common in other platforms such as video games. However, these techniques are not much used in social simulation platforms yet.

Similar arguments as for the temporal resolution can also be made for the spatial resolution. Having a very fine grained resolution makes it possible to simulate in-building situations (such as how persons move in a shop or work space). However, if we also want to model the difference of the spread of Covid between villages in the country side and big cities we need large scale spatial dimensions as well. Again, if we have the highest resolution available the decision making on how to move in the spatial environment will become very inefficient as routes need to go from moving in a room to moving between work and home. Like with the temporal dimension one would like the spatial granularity for decision making to be dependent on the action that is performed. Where a kind of default routes are assumed on the levels below the one that is decided on by the current action. I.e. if a person goes from work to home the action can use a default route covering every spatial point between work and home without having to explicitly plan those points.

In specifying the actions and their effects we include all the different models of the different aspects of life that have to be incorporated in the simulation. E.g., if the simulation should contain economic aspects, we should model the economic consequences of all actions as well. If the health aspect is important we have to model the health consequences of actions (such as infecting others or increasing a risk to get infected, etc.). Although most actions have a very obvious effect in one aspect of life, we should always consider all (social) side effects of an action as well. E.g. taking a bike to go to work has the effect that I get at the work place. However, taking a bike instead of a car can also influence neighbours, children, friends and family to take the bike more often. It should be possible to specify the actions all explicitly for the simulation in order to facilitate the extendability of the simulation. I.e. it should be possible to add more actions later on. Some of the work of specifying actions and their effects can be taken from the planning literature [3] and can be done in a way similar to how the effects of actions on needs were specified conceptually in Chap. 3.

Finally, all these aspects have to come together in the deliberation process of an agent. For example, if the economic aspect of life is included in the simulation there should be preferences over economic effects of actions and preferred economic states for the agent. So, the needs model that was used in ASSOCC requires needs that are affected by the economic effects of actions and that balance with other needs. Thus, it should be possible to directly specify new needs, combine needs, and check that they complement each other. In Chap. 2 we have shown how the current needs model is based on a more fundamental model of values, motives, affordances, and social constructs. This underlying model can serve as the basis for a tool that can be used to specify the needs model in a coherent and consistent way.

For example, the needs model as we have used for ASSOCC will become a bottleneck for more elaborate simulations. Every time an action is added to the simulation the whole needs system has to be recalibrated to account for the extra action. Thus, there arises a need to reason about actions and their effects on needs in *contexts*. This will isolate the consequences of actions to a limited number of needs and thus, avoid some of the large interdependence between all needs and actions. Besides the readjustment of the needs system, if agents have to create plans over high levels of granularity temporal and spatial dimensions there needs to be a kind of goal/plan generation mechanism that is very efficient. Resulting plans should be able to be used as basic actions on the other levels of the spatial and temporal dimensions again. Thus, ways to connect actions at different levels of granularity would be necessary. Note that some of this work can be adopted from Agent Oriented Programming Languages [4].

14.3 Bootstrap Simulations

Epidemiology models such as the SEIR model are very effective for responding early in a pandemic crisis by featuring *working abstract structures and dynamics* that can provide answers on the first day, and can then be refined depending on the situation. Whereas simulating societies is inherently more complex, the same principle can be applied for social simulations. A set of social simulation (pre-)models can similarly be implemented ahead, which would feature abstract representations of the highly regular structures and dynamics that are to be expected in virtually any crisis. Typically, cognitive processes such as those as described in Chap. 2 are likely to be present in most cases (e.g. norms, values, needs). Metaphorically, the idea is to develop a bootstrap model or a "Swiss army knife", a versatile starting solution that is available and can easily be refined for matching the specifics of the crisis.

A bootstrap simulation is a fully-working simulation that covers most of the classic structures and dynamics that are of influence in a crisis. These structures are abstract (in the sense of non-descriptive, not tied to a specific application). Similarly to a Swiss army knife, all core functions are covered, yet in a relatively rough manner. This concept differs from generic architectures, which offer programming directions but no working simulation. For example, the Belief Desire Intention (BDI) framework provides the building blocks, structures and execution facilities for executing practical reasoning. However, there is no BDI-based simulation model ready to be used at the start of the crisis.

Beyond working implementations, the concept expands to other levels of development of the simulation. For certain aspects, it may be difficult to converge on a final implementation due to having to commit to situational assumptions (e.g. a model of happiness). For such aspects, making progress on developing advanced conceptualisations and pre-models can help being more effective for reacting to a crisis (e.g. pre-models of happiness derived based on need satisfaction).

Abstract models, as described by [5], provide solid examples of the approach and components for building such a bootstrap model. Abstract models seek to introduce the fundamental structure to be replicated while minimising the specification to a given context (e.g. norms and social inequalities in general [6] versus descriptive models that replicate details of a specific population, in a specific period and specific area [7]).

Multiple examples of such key abstract aspects to be integrated in a bootstrap model can be derived from the ASSOCC experience. Based on the ASSOCC experience the bootstrap model would contain a spatial area large enough to contain a build-up area and possibly some water and country side landscape. It would be populated with agents that have the basic needs identified already in Chap. 2 and the actions available to each agent to fulfil a daily pattern of life fitting a number of age brackets and standard roles and functions. Thus, we would start from a standard population living a stable life in a geographical neutral area. In the ASSOCC model **needs** are the central concept to create an abstract state that is independent from specific actions and can be used to balance different behaviours in dynamic environments. We have already argued that therefore it is good starting point for any model for simulating in crises. As crises tend to deprive needs and trigger coping behaviours that are bound to specific dynamics (e.g. most people end up violating quarantines, individual context such as social situations can cause deprivations to be more severe and Thus, more frequent violations), the needs model also can be used to generate new types of behaviour and interactions. Giving different priorities to needs between individuals also allows easy variations of behaviour between individuals without giving them rigidly different behavioural rules. **Time and spatial** models have important impact on most crisis-response scenarios. They should be included at least in a primitive form (e.g. network-based space) for consequent expansion to specific requirements (e.g. geographic data). As stated above we can start with a standard daily time scale and a spatial area covering a (small) town. A **demographic** model is important (e.g. age, activity), as it can impact other aspects such as institutions, roles, activities and norms (and like in the COVID-19 case also the disease process). **Disease** models are obvious for modelling for pandemics. Disease models entail sub-models such as models for modelling the evolution of the diseases and models for representing contagions. **Institutions** such as **norms**, **social networks** and **organisations** are also critical entities that influence agent behaviour and are highly stressed by crises. In crises, agents often behave abnormally, which can yield important emerging phenomena and the setup of new norms (e.g. stockpiling). Moreover, norms and institutions can be used to model new restrictions and measures to cope with a crisis. Thus, having some standard regulations to start from will give a good idea about how they will likely influence behaviour in various ways. Finally, the crises can change the relation between individuals and organisations (e.g. trust influencing defiance or compliance). Thus, we also need to have a rudimentary form of organisations. This can then be used to show how organisations impose rules on their employees (like e.g. working at home) and the reactions of agents on these rules. But it can also be used to show how organisations may have different types of needs and priorities and Thus, try to influence both the public, employees as well as government.

Economy is also often an important element for stakeholders and the public, particularly in for modelling long-term pandemics such as the COVID-19. Moreover, economic status can have a significant impact on individual behaviour (e.g. closing workspaces, social welfare, poverty). Due to economy being very complex to model and to calibrate, economic principles and rules can be a sufficient starting point for the next crisis. Finally, the simulation should include the primary structures for tying these elements together. For instance, sickness can influence need deprivation, income, and occupation of space.

14.3.1 Implementation Prospects

From a structural standpoint, the selection of the right platform is critical for a simulation, as this model is meant to be a foundation that is to be extended later on. Among possible candidates, Repast appears to be the most relevant candidate for this task, as it is a very solid software package for building simulation functions, featuring interesting characteristics, notably for high scalability through native programming languages and parallelisation facilities. Repast alleviates some scalability issues since there are versions in Java and C++ which can be used to directly implement some parts of the simulation. Python-based libraries, such as AgentPy[1] and defSim (Discrete Event Framework for Social Influence Models)[2] are also relevant approaches as python is relatively effective for fast-prototyping, at the expense of moderate performance. Python also includes data processing and plotting libraries, for streamlining the plotting process. Using Repast instead of Netlogo Thus, supports:

1. more advanced complex modelling
2. extendability
3. greater scalability, more than 100.000 agents with Repast HPC [8]
4. representing results in more appropriate ways
5. simplify maintenance, through step-wise debugging

The NetLogo platform, while offering higher-level agent primitives, is not really sufficiently suited for large implementations and large scale simulations.

As the bootstrap model should be easily extendable but also quickly setup, a compromise should be found between extensibility and how far the model can be implemented a priori. More advanced implementations usually involves more assumptions and more rigid structuring, increasing the costs for adjusting the model, similarly to (overly) descriptive models. An intermediate solution consists in relying on a

[1] https://agentpy.readthedocs.io/en/latest/.

[2] https://github.com/defSim/defSim.

collection of modules whose effective connection has been extensively tested prior to the crisis. Moreover, different levels of pre-development can be considered for the various modules, from pre-models (conceptual maps but no specific models), pre-implementations (a general idea of the model dynamics, but no working and optimised implementation) and implementations (a fast software package that is ready to be expanded). For example, the economy model could be excluded in the bootstrap model, but rather be included at a later stage. Economic principles and rules can already be defined in advance.

Regarding concrete examples, the classic SIR model [9] is the most straightforward implementation of the epidemiological model of viruses. Regarding organisations and norms, extensive research has been dedicated to providing abstract structures, such as Opera+ [10], Operetta [11], Moise+ [12] and MAIA [13], which provide advanced facilities for implementing organisations, norms, rules and more, to build a structure for an agent-based model. There also exists a wide variety of geographical models, such as cellular automata for modelling for example forest fires! [14], flooding [15], oil spilling [16] and more. Cellular automata provide interesting compromises between generality and spatial representativeness. For the temporal component, a tick-based component appears to be a sufficient fit between precision, simplicity, and effectiveness in the case of a pandemic. However, other models can be considered, such as event-based models if agents make decisions on very different timescales.

14.3.2 Bootstrap Simulation and Crisis Requirements

Bootstrap abstract all-purpose simulations have the prime benefit of shortening start-up time for a team developing the simulation architecture and a workable simulation cycle. Therefore, such simulations allow for early response and a fast reaction to specific stakeholder concerns. Moreover, the developers can rely on early models that are relatively more stable, streamlined with output generation (both user-friendly GUI and plotting facilities) and have been scientifically validated before the crisis. This stability allows being granted early trust from stakeholders and lower risks for missing core components throughout the process of building the simulation.

Finally a higher level abstract structure is needed to connect these elements in the bootstrap simulation. This is more extensively explained in Chap. 2, however we want to mention it again due to its importance. Needs and values can play an essential role here to connect the different systems (Fig. 14.1) and to prevent the whole system becoming too rigid.

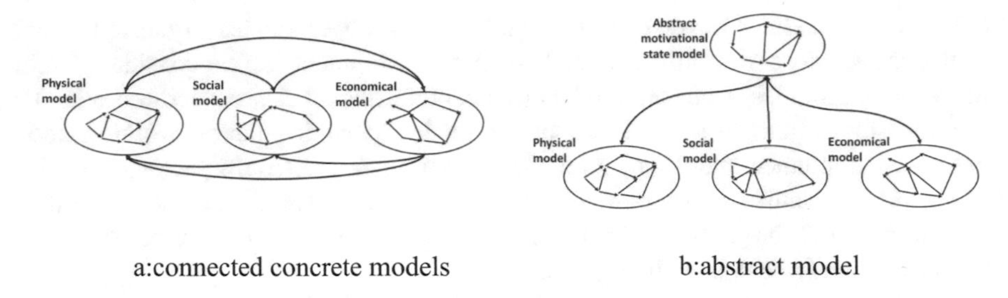

Fig. 14.1 Concrete models versus abstract model (also in Chap. 2)

14.4 Modularisation

Agent-based models can handle a higher degree of complexity than other types of models, but this ability can come at a cost. Large models which include many interacting aspects of a real world scenario can become difficult to manage surprisingly quickly, especially if they are developed iteratively and additively, without much advanced planning, as is the case when building a model while the crisis is still unfolding. In a complex model like the ASSOCC model, which allows strong interconnectedness between its components, any change to one part of the system can ripple through the whole system. This means that every time the model is updated to fit the new incoming information coming (which is quite often in a crisis), the whole model needs to be tested to check for undesirable new behaviour, the necessary changes need to be made to bring the model back into optimal operating parameters, and then the success of these changes needs to be evaluated through a new round of testing. If the updates affect the core of the model, this process may take many iterations of testing-adjustment-testing until we can be sure there are no undesirable behaviours or artefacts being produced during simulations. With a complex enough model, the adjustment process can quickly become the most time consuming activity for the modelling team, and soon afterwards, it becomes impossible to make new changes and guarantee the model behaves as designed at the same time.

One way to maintain control over a large and complex model is to break it down into modules. Modularisation introduces separation between different parts of the model such that changes to one part of the model do not ripple too far out and impact other parts of the model in unforseen ways, keeping the time costs of updating the model low, and ensuring that it is far more difficult to develop a model so complex that no one can guarantee it behaves correctly. In the context of simulation, the main advantage of modularisation is that it allows the addition, removal and modifications of modules in such a way that the functionality of the other modules is minimally impacted, allowing for quick development of simulation scenarios. In the context of crisis conditions, this is a particularly desirable characteristic of the system being developed since things change at a rapid pace and in unexpected ways.

For instance, during the development of the ASSOCC model, we used two different disease sub-models in our scenarios. We started out with our own, then later switched it out for the Oxford model since it was deemed to be the more realistic of the two. The process was relatively straightforward because this particular sub-model was not strongly intertwined with the rest of the model, especially not with the decision making framework. Had we decided to swap out a sub-model enmeshed with the decision framework, things would have been much more difficult.

14.4.1 Modular Social Simulations

When it comes to complex social simulations, modularisation is highly desirable, but difficult to attain at anything other than a high-level conceptual level. Being able to break down a large model into manageable pieces that can be developed relatively independently, can be modified without the changes rippling out to the rest of the model, and can be replaced with different versions as needed is a very good way to have a model that is easy to adapt and expand. A modular approach could mean building a library of modules that can be easily combined to make new models or new versions of the same model, which would make crisis development much faster and responsive to unexpected new developments that may arise as the crisis unfolds.

This difficulty arises from the fact that the aspects of a social simulation are designed to interact with one another in order to create a functional simulation world the agents can understand and act within and upon. For instance an economic module and a transport module are not monolithic pieces that connect with one another in clearly defined ways, except in the conceptual sense. In practice, these modules are collections of smaller pieces which connect across modules in various different ways. Often, these pieces vary between versions of a module, which makes it impossible to design standardised connections for other modules to plug into, and thus, swapping and combining modules is an easy conceptual exercise, but a problematic one when we get to the implementation stage.

This issue affects the decision framework of the agents as well. These internal mechanisms modules should function correctly regardless of which modules they have to work with. New modules may introduce the ability to perform new actions or make existing actions impossible. The agents have to be able to perceive and interpret new elements from the new modules. The needs framework requires balancing to account for the new elements.

For this reason, the conceptual modularisation of a model does not map cleanly onto its implemented counterpart and even conceptually modular models tend to devolve into the code equivalent of a hairball when they are implemented.

While the strongly interconnected nature of social simulation modules poses a significant hurdle in the way of developing strongly modular models, in particular when transitioning from the conceptual model to its implemented counterpart, this does not mean a partial modularisation would be without merit. While it may never be possible to develop fully modular models where new scenarios can be build simply by mixing and matching pre-built modules from a library, except maybe in a handful of very particular cases, a modular approach can still ensure that the model is much easier to expand, adapt and maintain. Conceptual modularisation, even if it never translates into the final implementation, is a useful development practice. It helps organise the simulation world in segments that can be worked on separately, which is of particular importance in the case of larger models where details might get lost and then be hard to find and correct later on.

Some aspects occur across all simulations, such as time scale, physical space and decision framework. Depending on one's particular focus, other aspects may also be shared across simulations, such as the presence of an infectious disease, as is the case in ASSOCC. A good modularisation effort can start by first defining these shared modules, identifying their scope, elements and behaviours, and then the points where they interact with other simulation modules. Some of these modules can remain fixed across all scenarios, providing a core other modules can be built around.

14.4.2 Related Work

This kind of partial modularisation, restricted to either model domain or model element is what we see in the literature. For instance, mobiTopp [17] is a modular agent-based travel demand modelling framework. It offers a number of modules, such as destination choice, travel mode choice or commuter ticket ownership, all very domain specific, and very targeted. Rather than approaching the whole transportation system as one module, the mobiTopp platform breaks it down into its distinct elements and behaviours. These, in turn, can be modelled with various degrees of complexity within the module, as required.

As an example of a modular shared model element, we can turn to FAtiMA [18], which is an agent architecture, consisting of a core reasoning algorithm which can be combined with a number of other modules which extend the agent's reasoning process in various ways. Such modules include reactive, deliberative, cultural or theory-of-mind modules. Modular decision architectures for agents are not uncommon in general, but they are mostly of interest within automation related fields, such as robotics [19] or self-driving cars [20] and have not been used within social simulations. In general the agent deliberation model in social simulations is relatively straightforward, as most social simualtion models focus on a single phenomenon and therefore modularisation of the deliberation is unnecessary.

14.4.3 Implementation Prospects

Since we do not intend to limit the platform to any one model domain (such as pandemics or city environments), there is no one approach to slice the world into definite modules, and therefore no one way of translating the conceptual modularisation into code. As such, the platform offers complete control over the choice of implemented modularisation to the modeller.

We do, however, offer a couple of supporting tools. First, a meta-model visualisation tool that can be used to keep track of the relation between the conceptual and implemented modularisation, and also visualise the state, structure and connectivity of the current implemented modularisation. We already have the ODD document and general code documentation to record all this information, but their creation, maintenance and use is a lengthy and laborious process. A visual, even interactive, meta-model can be read and understood much faster, letting the developers zero in on which conceptual modules are yet to be implemented fully, which code constructs are they actually mapped onto, and which are the points where different implemented modules interact, which is vital for keeping track of complexity as the model grows.

The second tool is the option to use a composition principle, rather than an inheritance principle, when developing the code for the modules. In complex systems which contain many different elements, inheritance hierarchies can become too large too be effectively managed anymore, slowing down development, and turning difficult to extend and maintain. Entity-component-system is a compositional alternative architectural pattern, in which every element (called an entity) is created by combining any number of pre-defined components. This allows for much more flexibility in the types of elements that can be built within an application because there is no underlying hierarchy to keep track of. These entities are strictly data objects. Their behaviour is controlled by systems which operate on an entity's components. Any system which can operate on a given component set, can operate on any entity which contains that component set, which makes it very easy to define behaviours for related, but not identical entities. For instance, a simulation may contain several types of vehicles, which can all be implemented as entities containing some shared components (such as speed, position, number of passengers, destination) and some type specific components (fuel type for powered vehicles, but not for bikes; or owner for privately owned vehicles, but not for those belonging to public transport, for instance). All of these vehicles can be moved through the world with a single system which operates on any entity with the components speed, position and destination (see Fig. 14.2). This completely eliminates any need to hierarchically organise the objects in a simulation, and ensures that any new objects can be added at any time without the effort of fitting them in an existing hierarchy or, in the worst possible case, redesigning of said hierarchy.

The downside of using such a compositional approach is that the number of components can quickly balloon beyond what can be comfortably managed. This mainly happens if the application contains many entity types which require specific systems. If this is the case, a hierarchical inheritance pattern, such as the well known OOP, is much more preferable. Ultimately, it is up to the modeller to decide which path to take towards implementation. It is clear that the above ways of modularisation can best be done in a more powerful language such as Java or C#. This reinforces the choice for Repast as implementation platform as it is easy to implement modules directly in Java or C# and then incorporate these in the simulation.

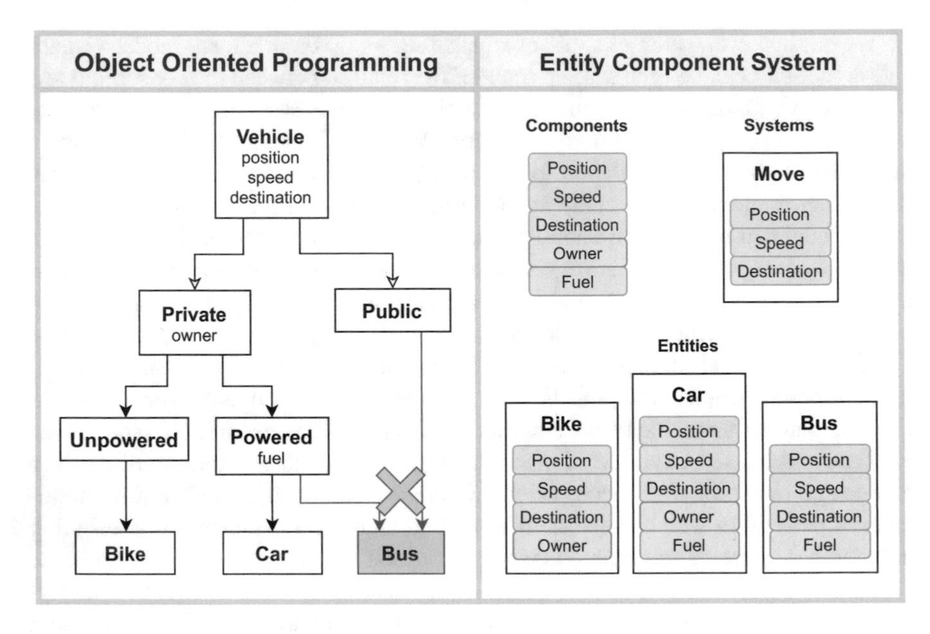

Fig. 14.2 Object Oriented Programming (OOP) versus Entity Component System (ECS)

14.5 Contexts

We have argued in several places that the agents should have abstract and complex models in order to cope with the diverse aspects of life in a crisis where the environment changes rapidly and repeatedly. However, the complexity of the decision model will become a bottleneck for the scalability of the simulation. At each tick the agent needs to make a decision on which next action to take. This decision is influenced by all different relevant aspects (e.g., economic, social, health in ASSOCC) and action selection balances all these aspects and possibly the consequences of a choice for future possibilities. Whenever the model is extended with new aspects and actions the decision making process has to be revised to incorporate this new choice. This will also require a careful calibration of all needs. One can easily see that connecting all aspects in one central decision mechanism is not viable if we want to scale the simulation to 50.000 or 100.000 agents.

As is usual in these cases it makes sense to investigate how humans cope with all this complexity in the real world and check whether we can use some of those mechanisms for the agent deliberation process. In general, humans do not always use the same decision mechanism for every decision. For some decisions many aspects are taken into account, while for other decisions the automatic pilot is used. E.g. when buying a car humans will take some time to assert their preferences, compare all kinds of available models and consider price, comfort, functionality, environmental impact, etc. But when buying milk in the supermarket they probably just walk to the shelf where they always get the milk from and pick the packet they always buy. In a similar vain they do not get up every morning and start wondering what they will do that day. Usually they go to work and know more or less what has to be done that day. When going to work by car they also will just stop at red traffic lights without considering whether to comply with the rules or to violate them.

So, for practical daily decision making there are two options:

1. take only a limited amount of aspects into account
2. use different decision mechanisms depending on the type of decision

Both of these options require placing the decision making in a context. For the first option the context is used to determine which aspects to take into consideration. In the second option the context determines which decision mechanism to use. Thus, reasoning in context seems a reasonable way to keep grip on the deliberation process in a complex and dynamic environment. Although the use of context seems intuitively correct it is not very straight forward to determine what is meant by *context*. The Merriam-Webster dictionary defines context as follows:

> The interrelated conditions in which something exists or occurs

Although the definition is clear, it gives little insight into determining whether a certain element (such as objects in the same space, persons present, time of day/year,...) is a relevant part of a context or not. We will not give some definite definition here, but in the next section we will discuss some related work on the use of context for decision making. That will give some direction to how context can be perceived and that will be used as the basis for some first steps on how to use context in the section after.

14.5.1 Related Work

Context in the sense of environment in which to act is of course common in principle in all social science theories and social simulation models in which agents are to act. However, here we do not use context in this sense but rather as a meta-level context, a characterisation of what the important element of the decision-making environment are and their influence on how decisions are made. The environment here includes even the social environment - which other agents are represented in the decision-making process, if any - apart from the physical environment. Moreover, most models of motivation-action causation focus on only one decision-making process. For instance, the (in)famous BDI models explicitly model all agents as goal-directed, where utility-maximisation is the mechanism to select an action. Thus, any and all motivation needs to be translated into utilities in order to be processed by this mechanism. In the case of for instance social concepts such as values this becomes cumbersome and easily leads to hard-coded solutions minimising the option to change agent model. An early comprehensive description of different action motivation mechanisms is the work of Weber [21]. He considers four different social action ideal types (meaning these are analytical categories that allow for a mixture of these to be applied in practice). A social action is an action containing subjective meaning attributed by the acting agent considering the behaviour of other agents. The four social action ideal types are:

1. Traditional Social Action. Means and ends are decided by the social customs of the society. Means and end for a certain action are already decided by social convention. These actions might need no deliberation.
2. Goal Rational Action. The goal is derived from the desires of the agent and means and ends are decided by the goal that needs to be achieved in an effective and rational way. As the purpose of the action is to fulfil some other goal and is treated as a means in itself i.e. the action is instrumental.
3. Value Rational Action. Here, the means and goals are defined by a person's value system. Rationality also includes a judgement of aesthetic, religious or constitutional values. Goals are subjective and do not necessarily result in material benefits. Means are chosen for their efficiency while the ends are justified by their value.
4. Affective Social Action. This action type transcends the sphere of rational decision-making by including the emotion of the individual. Thus, there is no means and ends maximisation but the action can be carried out in the heat of the moment for instance.

In psychology we can see the work of Kahneman [22] as an example of context-dependent decision-making mechanisms. Based on empirical data showing issues with the one size fits all utility-maximising view on human choice, Kahneman describes two different decision-making mechanisms:

- system 1 - fast, automatic, emotional, and unconscious.
- system 2 - slow, logical, calculating, and conscious.

The complexity of the information on which a decision is made is one of the key aspects deciding how humans will make their decision, with experience over time moving the decision-making from system 2 to system 1. Apart from this Nobel Prize in economics winning work, others in economics have tried to incorporate normative considerations into the preferences of the agent, e.g. altruism [23]. This transformation of preferences reduces the reasoning on the normative level to goal-directed decision-making, not unlike the value rational ideal type of Weber described above.

The same drive towards minimising cognitive processing as Kahneman describes is the driving force in one of the few frameworks for contextual decision-making developed in agent-based social simulation, the Consumat model [24]. It is a generic framework developed to simulate human behaviour in consumer decision situations based on different theories and models from (cognitive) psychology that are seen as applicable in different contexts. Originally the Consumat model consisted of four decision-making mechanisms: repetition, deliberation, imitation, and social comparison. The choice of the decision-making mechanism is mapped onto an agent"s needs and (un)certainty concerning the results of behaviour. High levels of satisfaction and certainty lead to repetition, low certainty and high satisfaction result in imitation, high certainty and low satisfaction lead to deliberation and low certainty and low satisfaction cause social comparison. A later version of the Consumat model ([25]) also includes satisfaction and uses social networks to inform the social comparison mode. This is an extension less well founded in social theory than the (in psychological theory) well-founded original model. Besides, not all decision-making is akin to consumer decision-making.

A recent development of context-sensitive decision-making in social simulations is the Computational Action Framework for Computational Agents (CAFCA) [26]. CAFCA is a framework based on the idea that decisions are always highly contextual, i.e. dependent on an agent's interpretation of a situation. In order to also ensure the needed generality, CAFCA is a two dimensional framework of contexts, corresponding to many of the frameworks described above. One dimension is the sociality mode - do other agents play a role in the decision-situation and if so,what role. The second dimension is the level of reasoning that is involved. Each dimension has three elements: the sociality dimension encompasses the individual, social, and collective level while the reasoning dimension consists of automatic, strategic, and normative reasoning respectively. Thus, the framework distinguishes nine contexts (see Fig. 14.3. In the individual mode the agent interprets the decision as independent of others. In the social model agents recognise other agents in the situation but see themselves as distinct from or in competition with them. In the collective mode the agent not only recognises others but perceives itself as belonging to the others, as a member of a collective or team. The sociality levels of social and collective correspond to a transformation of agency, from the individual to the collective or team [27] in the case of the collective level. This means that the other is not seen as an

	Individual	Social	Collective
Automatic	repetition	imitation	joining-in
Strategic	rational choice	game theory	team reasoning
Normative	(institutional) rules	(social) norms	(moral) values

Fig. 14.3 The CAFCA model

incumbent on the decision making but as a positive force for achieving a joint endeavour. On the social level, agents are independent and interdepent yet retain their own agency. The nine contexts can be used to analyse what to include in a model under development or to analyse an existing simulation system to investigate if and how the relevant decision-making mechanisms are implemented. From an agent perspective, the main issue when using the CAFCA framework is to determine which context an agent is in. This is no problem if CAFCA is used as an ontological framework that is used by the designer of the agents of a social simulation, but it becomes an issue if the agents have to decide themselves in which context they are in order to determine which kind of deliberation mechanism to use. If determining context is as computationally expensive as the deliberation itself we did not gain anything by introducing contexts. In the next section we will briefly discuss some steps to cope with this problem.

14.5.2 Deliberation in Context

As argued in the previous sections we can make the decision making process of an agent dependent on the context. However, the task of recognizing the context in which an agent is, is in itself a difficult task. Not in the least, because we have not defined what a context exactly is. I.e. in the previous section we have seen some properties that could be used from a current situation to determine a context, but it is not clear when these properties are salient. The only work that seems to address this point to some extend is that of Edmonds. Edmonds develops contextual social behaviour in such a way that decision-making is made dependent on the recognition of the context and context recognition is made dependent on context learning (cf. [28, 29]).

In this section we will discuss some steps to make contexts practical in social simulations. The big challenge is to make a good trade-off between using contexts as fixed, predetermined structures that classify situations and indirectly determine how the deliberation process is done in each class and using contexts as some structure that is built upon during every deliberation step of an agent and can adapt to all circumstances. If we make the use of contexts too elaborate it will not serve its purpose of being an instrument to achieve scalable simulations. If it is too simple it ends up as a very rigid structure that can hinder the realism of the simulation to an extend that it prevents getting any good insights from them. Thus, we aim for a general but simple mechanism to determine the right context for each deliberation that keeps the amount of aspects considered at the deliberation at the minimum required. We base the ideas on some work on social deliberation for software agents as described in [30].

Although we do not give a definition of all aspects of contexts, we will assume that some aspects are always part of the context and serve as the starting point of deliberation in context. These basic aspects are:

- space and time
- activity and purpose

Space and time are the physical elements that are always present in the environment at the start of a deliberation. Thus, they have a kind of special status in contexts as they can always and easily be ascertained. The activity or purpose of a deliberation gives a kind of internal context that is always present. It indicates the wider goals that an agent has for which the current deliberation tries to choose an appropriate action. We use *activity* as a term to denote a set of coherent actions. Thus, the activity is a context of the actions that are potentially part of it. E.g., the activity might be "shopping" and the deliberation which shop to go to first. The purpose gives the reason for the deliberated action, e.g.g we shop for clothes or food or a car.

14.5.2.1 Determining the Deliberation Mechanism

The elements given above can often already determine whether there is a default action to take. If the agent has daily cycles that have the same actions at a certain time of the day (e.g. getting up at 7am, having dinner at 18pm, going to work at 9am) the agent can assume it is in a repetition context and follow the default behaviour. This default behaviour can be checked against the current goals the agent has to see if it contributes to these goals. The same can also be done the other way around. The agent starts with a recurrent goal and chooses the habitual action for that goal. It then checks whether the place and time are fitting for that action.

The above describes the core starting point of deliberation in context. It is a process where internal deliberation context (in the form of the activity or purpose of the decision) and external deliberation context are in a constant feedback loop each determining the other. If the assessment of the context based on the above elements is that a habitual action is possible that action will be performed and the outcome assessed.

In case no default action is available, e.g. because the time and place and purpose are not combined in a way that there is a fitting action that has already been performed many times, the deliberation moves out of the "repetition" box from the CAFCA model. Which context to move to is determined by the properties of the purpose. If the purpose is purely functional, e.g. getting some food, we move to the "strategic choice" context. This involves deliberation about an optimal way to get food. In order for this deliberation to function more aspects of the context can be sought out and incorporated. E.g. is there a supermarket on the way I already planned to take. Is there a lot of traffic near one supermarket and not near another. Does one supermarket have some favourite products for sale.

If the purpose is more social, the agent can move to the "imitation" context and gather information about what other agents are doing in the same situation. E.g. everyone goes to the mall. In this way the deliberation determines which aspects of the context are being used, while the context also determines which type of deliberation is most appropriate. Important to notice is that this is not a two step process where first a context is determined and subsequently a deliberation mechanism selected. It is an iterative process where the deliberation steps are built from simple quick steps needing little contextual elements towards more and more elaborate deliberation mechanisms that require more input from the context and will seek out more and more elements to feed the deliberation.

In general this mechanism of constructing the context is starting from the "repetition" box in the top left of the CAFCA model towards the bottom right. Whenever it is needed to move to another box, the perspective on what constitutes the context also changes and in general grows larger. With repetitive actions the agent will look at the immediate physical environment to check whether the default action is possible. Are the right affordances available as resource for the action, is there enough time available to perform the action and can any obstructions be expected? When moving a column to the right other agents in the environment are perceived not just as objects, but as agents. Thus, in that case the context is enriched with the social and personal properties of those agents. It also means that in the effects of actions the agent considers the social effects and Thus, looks at the elements of the context that can influence those effects. E.g. helping someone to gain status within a group, requires the group notices the support action. Thus, the agent should see whether the group is in a position to observe what is happening. In the right most column of the CAFCA model we also include some models of other individuals in the context that can help to decide how to take their interests into account in the deliberation. Thus, now the context includes the goals and needs of the agents that are either directly or indirectly involved in the activity.

In general the context builds up from a very concrete temporal, spatial, and short concrete activity that is very physical oriented and short term oriented towards contexts in which longer time spans, larger spatial dimensions, and more abstract activities are incorporated.

14.5.2.2 Determine Which Needs to Consider

Possibly depending on time and place, the purpose is primarily connected to one or two needs that are fulfilled by that purpose. These are the needs that are considered in deliberating over the actions. Thus, although the action can have effect on more needs, the agent only considers this subset of needs that are the focus of the purpose. The needs fulfilled by a purpose might depend on place and time. E.g. shopping done over the weekend might fulfil a social need to perform a joint activity and also to get luxury items (fulfilling the need for comfort). During weekdays there is less time available and shopping is mostly related to the need of survival.

A second important feature that determines the needs considered during the deliberation over an action is the frequency with which the action is taken and (often) the consequence of the choice of action on future possibilities. E.g. buying a car or house is done only rarely. Therefore more elements of the long term context are taken into account and more different needs are considered. This also ties in with the fact that a more strategic or normative context is considered, which requires more elaborate reasoning and Thus, more contextual elements to fill the parameters for the reasoning process.

14.5.3 Conclusion on Contexts

The previous sections have given some ideas on the use of context for the deliberation of the agents in the social simulations. We have argued that these contexts should not be fixed beforehand but should be dynamically constructed. However, it is clear that this is not possible (yet) with the present day implementations of agents in social simulations. It requires a radically different approach to the agent deliberation cycle which should be far more situational. Some aspects that have to be taken care of before contexts can be used in the sense described above are:

- agents need a history of previous performed actions
- agents need to be able to compare situations to check whether they are compatible enough to perform a habitual action
- agents would need at least a mechanism of deliberation that can move between simple repetition all the way to value-based reasoning
- agents need a theory of mind for contexts including other agents
- aspects of the context that are used on different levels should somehow be accessible for the agent
- we need to have specifications of activities and purposes and their connection with needs

Given this list it seems contexts make the simulation more complex rather than more scalable. However, many of the elements above can be taken care of at design time and do not have to be generated on the fly. E.g. comparing a current situation with a standard situation to check whether a habitual action can be taken can be done very simple by checking a few available parameters. That can be the location and period of time and maybe the current activity. All of these are readily available. If these habitual actions take care of 70–80% of the actions most of the deliberation becomes very simple and effective. This means the remaining 20–30% can use some more computational power. But even for these situations one could e.g. specify on forehand which agents belong to the context in a certain place, time and activity. Those agents' parameters can be made visible and Thus, directly usable for the social context of an agent. In this way many deliberations can be kept simple and computationally cheap. Of course, it remains to be seen how well this set up works in practice. But that this is the way to go for social simulations for crises is certain.

14.6 Stakeholder and Scientific Models

A crisis like the Covid-19 pandemic naturally requires a coordinated response from many different people and institutions. Building a model of such a crisis with the goal of capturing as much of its inherent complexity would need to address all these different stakeholders, and they have vastly different communication styles and expectations. This becomes even more important when we consider the vastly different backgrounds and interests of these stakeholders and how much more complex an agent model of this size can be compared to the more commonly used mathematical and statistical models most stakeholders are likely to be familiar with. Accurately communicating all this complexity, simulation results and their meaning is acutely important if we want our work to be understood correctly and applied responsibly. Given this context, the classical academic approach of communicating though papers and the occasional presentation is likely to be insufficient and we will need to additionally build a graphical user interface, which offers far more tools and flexibility we can apply towards our communication goals.

14.6.1 Stakeholder Interface

When building an interface for our model, the first thing we want to know is who our stakeholders and what their communication strategies and needs are. We can then separate them into different groups based on how they want to interact with the model and what they'd like to learn from it. It is unlikely that all stakeholders involved would respond to the same type of interface, but it is very likely that the overlap would be significant. As such, we would design a number of targeted communication strategies which may include a GUI, charts, texts etc.

Based on our ASSOCC experience, we can start with a list of stakeholders requiring different conceptual GUIs for the communication and interaction styles we observed:

- the scenario user: this requires an interactive interface that allows stakeholders to dive into the simulations and experiment with different scenarios and parameterizations. Scenario users are more interested in how the model works and what's in it than in properly analysing results. On top of interactive graphics the interface should also include information about various components and interactions so that a user can easily learn what is going on in the model. This information should be delivered in short to-the-point bits, on demand, be unobtrusive, but easy to request as needed, and include minimal jargon.
- the scientist: scientists prefer a massive numbers of charts, well-organised into categories. They also prefer access to raw data and meta-data. The impatient analyst can be provided with the simulation-charting pipeline so they can setup their own experiments and/or with extra tools to they can make plots of the data in new ways

as needed. The scientist's goal is to make sure the model generates useful results that make sense

- the modeller: for modellers ODD documents for abstract models are essential as well as modules and components that can quickly be assembled and modified to keep documents up-to-date with the state of the model. The modeller wants to either make use of parts of the model in their own models (which may have variable commonalities to our model), fit the model to their own world context (strong commonality), or extend it further. In any case, they need to look under the hood, to understand how the model is actually built.
- the participant (ambitious as this a heterogeneous group): participants like to contribute to the building of the model from already existing "semantic bricks" and thus be involved in the modelling process. This would require the development of a participatory modelling protocol on top of the interfacing tools, but is a natural next step of the strongly modular approach we advocate for in this chapter.

Given the vast effort required to build GUIs in general, we would prefer to have all these styles as part of a single modular application, similar to the current ASSOCC setup. Since any stakeholder can move back and forth between styles, the application should easily transition between any of them on demand, a feature that is not consistently implemented across ASSOCC modules at the moment. Designing one GUI that would cover all the stakeholder groups is feasible, but we do not consider it the best choice. Too much information, especially unwanted and unneeded information, would make the use of the GUI difficult and tedious, rendering moot the time and effort invested in its development.

14.6.1.1 Related Work

Stakeholder engagement in relation to modelling efforts has become commonplace, and is well studied. There are many degrees and types of involvement, and different goals for bringing the model and stakeholders together. The process of how the stakeholders interact with one another, with the model, and with the researchers are of utmost importance. These aspects are discussed in detail in [31] and [32].

While many efforts have been made to involve and communicate with stakeholders, ranging from more traditional static visuals to interactive 3D environments and role playing games, our proposed interface has the potential to be the most comprehensive and extensive so far.

14.6.1.2 Implementation Prospects

In keeping with the modular approach described in the previous section, our GUI must also be modular. Each of the modules available for use in models should, therefore, be developed in tandem with a graphical counterpart which will be loaded into the GUI when the module is loaded into the model. It is very important that separation be maintained between a module and its graphical counterpart so that any module can be loaded without also loading its associated graphics. This plays a significant role for large scale simulations since they are generally run without graphics, which are extremely computationally costly.

In order to reduce the effort required to develop these graphical counterparts, we can identify shared elements and develop graphical templates for them, which can then be customised and used as needed. These templates require that their module element counterparts share the same data representations, which shouldn't be an issue given that they should be designed from the ground up to interface with other modules and with the decision framework and this cannot be achieved without shared representations.

Once the data representation is systematised, we can build our own basic plotting/graphics modules (such as info labels/boxes being displayed on ticks, premade line/bar/pie charts that need only to be connected to a data source to work, and standardised extensible/customisable representations of various elements of the simulation, i.e. red signifies the presence of infection, light/dark colour palettes for occupied/empty buildings, representations of density, presence and absence of certain elements etc.

While general user GUIs are not a particular focus of agent based simulation, they are extensively studied in other fields. Research from the human computer interaction field and information visualisation can point towards the proper methodologies and design patterns, we just need to follow them in order to assemble a good library of elements and use-patterns that we can mix and match as needed.

Given our existing experience, a good approach would be to keep R for charting, but extend it with interactive capabilities by wrapping it in a desktop app (written in Java or C#) to allow the user to interactively generate custom charts, then style and save them. We would also keep using Unity for the extended live interactive interface. We have already assembled a library of assets, pre-assembled scenes and scripts that connect to outside applications (such as Netlogo), which can be easily adapted and extended for various future scenarios. Unity also allows extensions to its editor, which means we can build editor extensions that would allow stakeholders to build their own scenarios in a participatory modelling setting with far less effort than it would take to provide the same functionality in a live build.

14.7 Integrating for Society's Resilience

Any complex system can be subject to shocks that force it to transition to an entirely new, and possible undesirable, state which would take a significant effort to reverse. For realistic society systems, which lay towards the high end of the complexity spectrum, these can range from pandemics to political unrest, sudden natural disasters to prolonged pressures brought about by the changing climate (for example sea level rise [33, 34] or droughts [35–37]) or by unplanned for demographic shifts [38, 39]. Any of these can cause significant damage to property and daily life, which is enough reason to strive towards more resilient societies. However, any of them can also set off a cascading effect, one improperly mitigated crisis giving rise to another and another, making it even more vital that there are measures in place to prevent, address and reverse these crises.

At the same time, the resources a system can allocate towards resilience are limited and choices must be made about where, when and how these can be used best to limit the damage a crisis can inflict. However, crises also involve a significant amount of uncertainty regarding their timing, severity, extent and effects, and this makes it extremely difficult to make allocating resources and setting up preventative and mitigating measures in advance. Mistakes at this stage can be quite costly, both in terms of wasted resources, and the damages done by a mismanaged crisis.

14.7.1 Social Resilience

Simulation is a method researchers and stakeholders can use in order to get a better understanding of the uncertainties involved concerning the development and effects of a crisis and the effectiveness of countermeasures. Building the models while the crisis is unfolding, as we have done with ASSOCC, is one way to use simulation. Even more so, in a virtual sandbox interested parties can instantiate any number of particular societies, inflict any number of crises, and try any number of countermeasures at any time, even before the crisis occurs. In keeping with the subject of the book, we can investigate the prevention and management of pandemics. Modellers do not need for the World Health Organization to declare the existence of a new pandemic. They can already check with stakeholders and experts what type of scenarios or crises should be modelled and experimented with. For example, reports are emerging of new viruses such as first cases of a new influenza virus,[3] which could be used as a basis for new experiments. Researchers and policy makers can gain a better understanding of the dynamics at play and be more prepared for when the crisis hits by running and experimenting with different scenarios. At the same time, having a collection of models and scenarios already implemented makes an excellent basis for quickly bootstrapping new ones as the crisis unfolds, making the simulation approach more responsive to real world developments. Furthermore, more stakeholders can be involved in the process with the proper model interface and participatory protocols. Thus, the model can even reach the communities that are likely to be affected by the potential crisis and who are likely to be overlooked by more traditional approached to resilience research.

There is one caveat that must be mentioned: the uncertainty inherent in these types of scenarios means that any model developed well in advance of a crisis is unlikely to match the real-world situation well enough to be useful. There already exist many models of resilience on various types of socio-ecologic systems and the crises which

[3]https://bnonews.com/index.php/2021/02/russia-first-human-cases-of-h5n8-bird-flu/.

may affect them [40–43], and while they aid preparedness they are usually not very adaptable as they are usually tailored to one specific scenario.

In order to make this resilience modelling approach more viable, any models developed before the actual crisis must be easily and quickly adaptable to fit the situation on the ground. In the case of pandemics, many modellers used the SEIR standard, which is easily parameterizable to any kind of infectious disease. What we aim for is a platform that can facilitate the development of models similarly standardised and adaptable to crises in general as the SEIR model is to pandemics in particular.

14.7.2 Integration and Resilience

A platform geared towards modelling for resilience and also towards fast adaptation of the models for crisis should first and foremost provide a way to quickly develop and adapt scenarios for running experiments. All the elements described in this chapter so far aim to help with this.

The platform aims to provide a number of abstract modelling elements that can be instantiated and customised to fit the desired models and scenarios. These elements can further be developed into modules. Once the core modules are defined and implemented (together with any variants that may be desirable), other modules can be developed and added to extend the simulation and create new scenarios. The context reasoning framework we provide for the agents is aimed at efficiently handling larger and more complex scenarios without significantly increased use of computational resources. The extended model interface will allow stakeholders to interact with the simulations, to provide feedback and to contribute to the development of new scenarios, Thus, extending the reach and applicability of the models.

Using this new iteration of the ASSOCC platform would begin with a deliberation framework for the agents based on the one described in Sect. 14.5. The needs of the agents are then customised to reflect those of the population being studied. Multiple versions of this core module can easily be created by varying the number, type and priority of needs. Then, we define the agent actions and tie them to the needs, which allows for the easy creation of variants again, by introducing different actions or by varying their connection to the needs.

Once the decision framework is calibrated to ensure the agents follow a normal daily life pattern of behaviour during what is termed the normal state of the simulation, the rest of the world can be built. First, the time scale and spatial modules should be added in, as they are foundational to any other world aspect that will be introduced later. From here, it is up to the modellers and their chosen subject. City or countryside modules can be added; economy, transportation, agriculture, industries and services, recreational facilities, geographical, geological and ecological systems, anything relevant to the model, or just a scenario in particular.

After the world is built and is functioning properly in a normal state, we can introduce the crisis. If we stay with the subject of pandemics, various infection modules can be created, reflecting the existing assumptions about how the studied disease spreads and manifests in the population. If the crisis is, for instance, a flooding event, various scenarios can be created to study different versions of such an event, depending on when it occurs (at night vs during daylight, during a work day vs during a holiday, spring vs autumn etc.), where it occurs (in a residential area, an agricultural area, an industrial area etc.), and the kinds of disruption it causes (damaged electrical grid, damaged transportation arteries, damaged residential areas etc.). Multiple crises can be combined by adding modules for these crises to the model. For instance, an infectious disease outbreak during an ongoing flooding event is a sadly common occurrence due to lost access to electricity and clean water, and often crowded conditions in areas set up for those displaced by the crisis.

As a final step, the countermeasures are introduced in the model. These can serve to prevent, slow down or mitigate the effects of a crisis. These modules can include several responses grouped together to form a coherent strategy, or can each be implemented as their own separate module to be added in any combination, or be studied one at a time.

This layered approach geared towards the creation of many related models and scenarios makes it possible to study a crisis before it occurs in many possible forms and from many possible angles, together with any number of countermeasures. It also ensures that when the crisis actually occurs, the modellers have already compiled a number of relevant simulation modules and ready-made scenarios, which can greatly accelerate the development of a crisis model as the crisis unfolds.

14.8 Conclusions

When we realised during the ASSOCC project that scalability would be a major issue while developing the social simulations in Netlogo, we decided already early on that we would port the simulation to Repast later on to gain some efficiency. However, we also realised that just re-implementing the current model in Repast would gain us some efficiency but not enough to really run simulations with 100.000 agents or more. Thus, we decided that before porting the simulation to Repast we should also take the time to investigate the foundations of the model and see where we would have to adjust things to keep an extendable architecture with an abstract model, while also allowing for maximal efficiency. From this investigation followed a number of strategies rather than a specific new architecture and implementation.

In this chapter we described the strategies and argued why these strategies are necessary for any social simulation for crisis situations. Basically these simulations have to balance a complex, abstract deliberation model (in order to cope with suddenly changing circumstances and having to combine many aspects of life) and scalability in order to create simulations with many agents that can show statistically relevant results. Moreover, within a crisis we need to be able to build the simulation quickly and Thus, need building blocks that can easily be assembled on a high, conceptual level. This also facilitates the addition of new aspects during a crisis.

All these requirements led to two main strategies for a new social simulation framework. This framework for social simulations for crises should be highly modular in order to be very flexible, have a library of predefined components that can be easily assembled for new simulations. The components for this framework can be based on the elements that were already used in the ASSOCC project. These have proven to be quite robust and easy to compose. However, the framework does not have to be limited to these components. E.g. a component for goal planning would be desirable for bridging different time scales, where the planning gives some consistency over the smaller time scale and the resulting plans can be seen as basic actions on the larger time scale.

It is also clear that any abstract model that connects all different aspects of life of an agent will become a bottleneck for the deliberation cycle. We therefore propagate contextual deliberation that will take into account as many aspects of the context as is needed to make a realistic decision in a particular situation. It means that not all aspects are always taken into account and that the deliberation mechanism is varying from very simple to very complex depending on the situation (based on the CAFCA model). This contextual flexibility also opens the way to move away from the uniform, synchronous deliberation cycles of all agents per every tick to a more flexible system where only agents that really need to change behaviour make a new decision in a tick.

Although the above strategies for building the new framework seem inevitable it is also clear that they will lead to a situation where the persons creating the simulations have to be more knowledgeable about the internal working of the model and consequences of design decisions (which modules to use and how to compose those) on efficiency and scalability. Thus, a very important accompanying issue for developing the above new framework will be the developing of a solid methodology to design and implement social simulations with this framework.

Acknowledgements We would like to acknowledge the members of the ASSOCC team for their valuable contributions to this chapter.

References

1. D.J. Heslop et al., Publicly available software tools for decisionmakers during an emergent epidemic-Systematic evaluation of utility and usability. Epidemics **21**, 1–12 (2017)
2. T. Blakely et al., Determining the optimal COVID-19 policy response using agent-based modelling linked to health and cost modelling: case study for Victoria, Australia. medRxiv (2021)
3. Malte Helmert, Concise finite-domain representations for PDDL planning tasks. Artif. Intell. **173**(5–6), 503–535 (2009)
4. C. Baroglio, J.F. Hubner, M. Winikoff, Engineering multi-agent systems. in *Proceedings of the 8th International Workshop on Engineering Multi-Agent Systems* (2020)
5. N. Gilbert, Computational social science: agent-based social simulation (2007)
6. C. Castelfranchi, R. Conte, M. Paolucci, et al., Normative reputation and the costs of compliance. J. Artif. Soc. Soc. Simul. 1.3, 3 (1998)

7. J.S. Dean et al., Understanding Anasazi culture change through agentbased modeling. in Dynamics in human and primate societies: agent-based modeling of social and spatial processes (2000), pp. 179–205
8. Nicholson Collier and Michael North, Parallel agent-based simulation with repast for high performance computing. Simulation **89**(10), 1215–1235 (2013)
9. W.O. Kermack, A.G. McKendrick, A contribution to the mathematical theory of epidemics, in *Proceedings of the Royal Society of London. Series A, Containing Papers of a Mathematical and Physical Character*, vol. 115.772 (1927), pp. 700–721
10. J. Jiang, V. Dignum, Y.H. Tan, An agent-based interorganizational collaboration framework: Opera+. in *International Workshop on Coordination, Organizations, Institutions, and Norms in Agent Systems* (Springer, 2011), pp. 58–74
11. D. Okouya, V. Dignum, OperettA: a prototype tool for the design, analysis and development of multi-agent organizations, In *AAMAS (Demos)*. (Citeseer, 2008), pp. 1677–1678
12. J.F. Hübner, J.S. Sichman, O. Boissier, MOISE+ towards a structural, functional, and deontic model for MAS organization, in *Proceedings of the first international joint conference on Autonomous agents and multiagent systems: part 1* (2002), pp. 501–502
13. A. Ghorbani et al., MAIA: a framework for developing agent-based social simulations. J. Artif. Soc. Soc. Simul. 16.2, 9 (2013)
14. Ioannis Karafyllidis and Adonios Thanailakis, A model for predicting forest fire spreading using cellular automata. Ecol. Modell. **99**(1), 87–97 (1997)
15. Yi. Li et al., Spatiotemporal simulation and risk analysis of dam-break flooding based on cellular automata. Int. J. Geogr. Inf. Sci. **27**(10), 2043–2059 (2013)
16. M. Gług, J. Wąs, Modeling of oil spill spreading disasters using combination of Langrangian discrete particle algorithm with Cellular Automata approach. Ocean Engineering **156**, 396-405 (2018)
17. Nicolai Mallig, Martin Kagerbauer, Peter Vortisch, mobiTopp-a modular agent-based travel demand modelling framework. Procedia Comput. Sci. **19**, 854–859 (2013)
18. J. Dias, S. Mascarenhas, A. Paiva, Fatima modular: towards an agent architecture with a generic appraisal framework. in *Emotion Modeling* (Springer, 2014), pp. 44–56
19. Wei Li, Notion of control-law module and modular framework of cooperative transportation using multiple nonholonomic robotic agents with physical rigid-formation-motion constraints. IEEE Trans. Cybern. **46**(5), 1242–1248 (2015)
20. S. Singh et al., A modular framework for adaptive agent-based steering (2011), pp. 141–150
21. Max Weber, *The Theory of Social and Economic Organization* (Free Press, New York, 1947)
22. D. Kahneman, *Thinking, Fast and Slow* (Farrar, Straus & Giroux, 2011)
23. M. Rabin, Incorporating fairness into game theory and economics. Amer. Econ. Rev. **83**(5), 1281–1302 (1993)
24. W. Jager, Modelling consumer behaviour. Ph.D. thesis. Universal Press, Veenendaal, 2000
25. W. Jager, M. Janssen, An updated conceptual framework for integrated modeling of human decision making: The Consumat II. in *Proceedings of the Complexity in the Real World Workshop, European Conference on Complex Systems* (2012)
26. C. Elsenbroich, H. Verhagen, The simplicity of complex agents: a Contextual Action Framework for Computational Agents. Mind. Soc. **15**(1), 131–143 (2016)
27. R. Hakli, K. Miller, R. Tuomela, Two kinds of we- reasoning. Econ. Philos. **26**, 291–320 (2010)
28. B. Edmonds, Complexity and context-dependency. Found. Sci. **18**, 745–755 (2013)
29. B. Edmonds, The sociality of context. in *Computational Social Science or Social Computer Science: Two Sides of the Same Coin*? ed. by T. Balke et al. (SocialPath 2014 Sintelnet, 2014)
30. F. Dignum et al., A conceptual architecture for social deliberation in multi-agent organizations. Multiagent Grid Syst. **11**(3), 147–166 (2015)
31. A. Voinov, F. Bousquet, Modelling with stakeholders. Environ. Model. Softw. **25**(11), 1268–1281 (2010)
32. A. Voinov et al., Modelling with stakeholders-next generation. Environ. Model. Softw. **77**, 196–220 (2016)

33. K. Thorne et al., US Pacific coastal wetland resilience and vulnerability to sea-level rise. Sci. Adv. 4.2, eaao3270 (2018)
34. P. Lu, D. Stead, Understanding the notion of resilience in spatial planning: a case study of Rotterdam, The Netherlands. Cities **35**, 200–212 (2013)
35. A. Gazol et al., Impacts of droughts on the growth resilience of Northern Hemisphere forests. Glob. Ecol. Biogeogr. **26**(2), 166–176 (2017)
36. G.G. Haile et al., (2019) Droughts in East Africa: causes, impacts and resilience. Earth Sci. Rev. **193**, 146–161 (2019)
37. X. Fu et al., An overview of US state drought plans: crisis or risk management? Nat. Hazards **69**(3), 1607–1627 (2013)
38. V. Guiraudon, Economic crisis and institutional resilience: the political economy of migrant incorporation. West Eur. Polit. **37**(6), 1297–1313 (2014)
39. R. Paul, C. Roos, Towards a new ontology of crisis? Resilience in EU migration governance. Eur. Secur. **28**(4), 393–412 (2019)
40. L. Shenk, C. Krejci, U. Passe, Agents of change- together: using agent-based models to inspire social capital building for resilient communities. Commun. Dev. **50**(2), 256–272 (2019)
41. Z. Kobti, R.G. Reynodls, T. Kohler, A multi-agent simulation using cultural algorithms: The effect of culture on the resilience of social systems. in *The, Congress on Evolutionary Computation*. CEC'03. Vol. 3. IEEE. 1988–1995 (2003)
42. J. Simmie, R. Martin, The economic resilience of regions: towards an evolutionary approach. Camb. J. Reg Econ. Soc. **3**(1), 27–43 (2010)
43. M. Schouten et al., Resilience-based governance in rural landscapes: experiments with agri-environment schemes using a spatially explicit agentbased model. Land Use Policy **30**(1), 934–943 (2013)

Chapter 15
Challenges and Issues for Social Simulations for Crises

Frank Dignum, Maarten Jensen, Christian Kammler, Alexander Melchior, and Mijke van den Hurk

Abstract In the previous chapters we have described our experiences in simulating for the COVID-19 crisis. We also described how we envision that we can get to a simulation platform that will be more supportive for simulating a next crisis. In this chapter we will recapture a number of the challenges that need to be faced in order to make social simulations grow to their real potential in crisis situations. The challenges range from theory and conceptual models to implementation and use of the simulations. Moreover, we argue that the development of the simulation platform should be accompanied with an effort of the community to connect to established advisory boards and committees in order to integrate the social simulations in the normal set of tools that are used by crisis teams.

F. Dignum (✉) · M. Jensen · C. Kammler
Department of Computing Science, Umeå University, SE-901 87 Umeå, Sweden
e-mail: dignum@cs.umu.se

M. Jensen
e-mail: maartenj@cs.umu.se

C. Kammler
e-mail: ckammler@cs.umu.se

A. Melchior · M. van den Hurk
Department of Information and Computing Sciences, Utrecht University, Princetonplein 5, 3584 Utrecht, The Netherlands
e-mail: a.t.melchior@uu.nl

M. van den Hurk
e-mail: m.vandenhurk@uu.nl

A. Melchior
Ministry of Economic Affairs and Climate Policy and Ministry of Agriculture, Nature and Food Quality, Bezuidenhoutseweg 73, 2594 AC Den Haag, The Netherlands

F. Dignum (ed.), *Social Simulation for a Crisis*, Computational Social Sciences,
https://doi.org/10.1007/978-3-030-76397-8_15

15.1 Introduction

Social simulation projects provide more knowledge than just the phenomenon to study. It teaches us the limitations of the theory, models, software tools and methodologies. Often these do not gain a place in papers reporting the results of a project. These papers focus on the successes and the insights gained in the phenomenon studied. In our case we reported these results in Chaps. 5–10. Although we certainly believe that our approach has been successful in the COVID-19 crisis, we also experienced many challenges along the way. These challenges are of utmost importance to learn and keep developing social simulations for crises. We have the hope that this chapter is read as a challenge to a wider community of social simulation researchers and practitioners to pick up one of the issues and help built a future generation of social simulation platforms that will support modellers to quickly react in times of a crisis in order to adequately support decision makers. But we also hope that these new social simulation platforms can be used to investigate how to create more resilient societies that can cope better with crisis situations.

In this chapter we will describe a number of major challenges in different areas. First we will discuss challenges for the conceptual model of the social simulations as this lays at the heart of the success of any social simulation. We will specifically pay attention to the relations between different levels of aggregation of agents into micro-, meso-, and macro-level. How can we model the relation between e.g. individual economic preferences, organisational economic behaviour and macro economic dynamics. These relations become important if the scales of time and/or space are shifting from small to large. Next we will describe the major challenges related to the implementation and especially the scalability issues. Then we investigate some challenges in the area of methodology and project management. We conclude the chapter with some recommendations and future research topics.

15.2 Conceptual Model

We will discuss the needs model first as this is the most central and difficult part of the conceptual model. After that we discuss some of the other aspects that face a similar challenge as the needs model, which is that we need to find the right trade-off between an efficient and scalable model and a realistic model.

15.2.1 Needs Model

As already argued before in other chapters (Chaps. 2, 13, 14) the needs model of the agents is the core of the ASSOCC framework and determines both its success in terms of flexibility and robustness as well as it determines its main limitations.

As described in Chap. 2 the homeostatic needs model is a concrete implementation of the more foundational model depicted in Fig. 2.7. The main advantage was that this needs model kept the deliberation model relatively simple. However, it also has some disadvantages that are challenges for its use:

1. **Balancing needs**: Having actions purely based on the homeostatic needs model may require a lot of balancing/tweaking as agents could for example go bankrupt because they spend all their money. Or they do not care about the laws and go out of their supposed quarantine very frequently.
2. **Lack of social rules**: things like habits, norms and social practices are supposed to be implicit in the needs model. This means that the deliberation over actions has to take these social aspects into account implicitly as well and they disappear in the code.

Balancing the needs becomes problematic because we only have an implicit connection with the value system. In a value system as depicted in Figs. 2.3 and 2.4 there is a dependency between the priority between different values. This means that balancing the concrete values is based on these implicit ordering and cannot be done in any possible way. This helps the calibration as shown in e.g. [1]. In Chap. 8 we included cultural value priorities to balance the needs which already gives some more stability, but is limited to value priorities of fixed cultures.

In the same way the needs model supersedes the norms, habits and social practices component. It means that habits cannot be modelled separately. Habits would make it easier to create patterns of life that calibrate needs. Social practices and norms could do similar things by creating standard interaction patterns. However, modelling the social rules separate and combining them with the needs model also makes the model more complex again.

In the other hand the homeostatic needs model facilitates an implementation where explicit planning for goals is not necessary and even is difficult to incorporate. In one hand this is a good thing as the homeostatic model is more robust in dynamic environments. However it also means that longer term goals can only be implemented through the needs model through careful calibration such that all actions in a potential plan to reach a goal are sometimes chosen and fictive goals will be reached.

The above illustrates the recurrent challenge for social simulations for crisis situations. In one hand they need a complex and abstract deliberation mechanism, while in the other hand efficiency and scalability require the model to as simple as possible. Making a model simpler (like was done by collapsing several aspects in the homeostatic needs model) means that other aspects have to taken care of implicitly. In the case of the needs model this means that the needs model has to be very carefully calibrated in order to create natural behaviour. Whenever new behaviour is added this calibration has to be repeated and with every addition becomes more complex. In Chap. 14 we have indicated some possible ways to attack this challenge, but it is far from solved yet.

15.2.2 Realism Versus Efficiency

The challenge discussed in the previous section on the needs model is an example of a more general challenge. In the ASSOCC framework we have chosen to model many aspects, but have a simple model for each aspect. This means that many aspects can be extended to incorporate some more realistic features:

1. **Economics**: Although we have incorporated economics in the ASSOCC framework, the agents only have an economic desire to be able to pay for things they really need. It means that there are no incentives to save for a car, a house or holidays, etc.
2. **Spatial model**: The spatial model is very simple. This made it impossible to implement some measures that were seen in the real world, such as only allowing restaurants to open with four people per table.
3. **Location types**: As a simplification, we grouped together a lot of locations under a single banner. This made it impossible to implement certain real world measures. For example, private leisure places were both bars, restaurants, people's homes, and sports places, so opening up outdoor sports places first can not be implemented.
4. **Ages**: We only have four age brackets, which are more or less connected to different daily life patterns. A more fine grained age difference would have given more realistic epidemiological properties. We could have distinguished more age groups if we had more agents in the simulation and could still have some statistical relevant results for each age group.
5. **Households**: We took the most common household types from demographic models. However, as a consequence we do not have households with singles. Partly this is due to the scale of the simulation.
6. **Behaviours**: There are limited amount of actions available. That makes for a limited amount of variation in behaviour. However, more available actions also means more complex deliberation and more difficult calibration.

The above list is far from complete. The recurring theme is that some details could be added to get a more realistic simulation. For each of the above aspects it is also intuitively true that some more details would give "better" results. And even worse, for most of these issues it is not very difficult to imagine what should be done to add the additional features. For some, such as additional behaviour, this might be quite a bit of work, but conceptually not very difficult. Thus there is a tendency to agree that this should be added especially if such a feature would make it better possible to model certain (possible) measures.

However, this also leads us on a slippery slope that many others have trodden as well. We can add more and more details and never be ready. See [2] for a discussion on this phenomenon. The question is whether adding more details will actually make a significant difference for the results that can increase the value of the simulation for its purpose. E.g. differentiating ten age groups instead of four would give us a more detailed view of the number of infected people per age bracket. However, it is doubtful that this would influence the main results in any of the chapters in this book in a way that based on this more realistic simulations a different policy decision should be taken. Thus the main challenge with social simulations for crisis situations is to find the right balance between simplicity and realism for each aspect that is modelled. The underlying idea being that we push the realism as far as we can without getting a model that is too complex to allow for a scalable and efficient simulation. And doing that while getting results that are good enough for the decision makers!

15.2.3 Social Networks

Social networks have a major influence on the behaviour of individuals. In particular in times of crises, when we are forced to deal with unique events and how to react upon them, we tend to look at how others that we are close to behave. During the COVID-19 pandemic for example, lots of measures where imposed, like social distancing and wearing face masks in public places. Adoption of new norms like these are affected by social networks, i.e. if we see all our friends or colleagues wearing a face mask we are more likely to wear one too. We have implemented the social networks in the ASSOCC framework such that the adaptation of agents to new imposed measures would be more realistic.

We were challenged by the fact that we had to connect the influence by the social networks on a meso-level with the individual needs on a micro-level and the cultural values on a macro-level. We integrated the social influence with the micro-level by some of the individual needs of the agents. The conformation need, for example, is modelled by setting a higher expected gain in satisfaction when an agent chooses an action that the majority of the social network performed previously. We connected the social relations with the macro-level by generating the networks based on value similarity between the agents.

The ASSOCC model could also be used to do more analyses regarding social networks. For example, we see a lot of civil unrest emerging during the COVID-19 pandemic, leading to protest and (violent) riots. Also, a crisis like this seems to be a breeding ground for the rise of polarisation of the population. We see this phenomenon happening in a lot of countries, with on one hand people that strive for more measures to stop the spreading of the virus, and on the other hand people who see the measures as a freedom restriction, imposed by the government.

From research on radical behaviour we know that social networks play an important role in this process [3]. It functions as a factor on meso-level, together with personal needs on a micro-level and ideologies, with norms and values, on a macro-level. We have already integrated these different levels in the ASSOCC model. However, we implemented the social networks in such a way that it was sufficient for the social related needs, as we were mainly interested in the effect of these networks on a micro level. In order to do research on the above mentioned phenomena, the model should be extended.

A first challenge is the integration of social groups. People not only have social connections, but also belong to social groups. A social group is more than a network of relations alone, as social groups consists of members sharing the same values and acting according to group norms. Deviation of those group norms can lead to some form of punishment. For example, if you are not wearing a face mask while all your friends do, this might lead to not being invited to social gatherings anymore. A second challenge for the model would be the salience of a social identity that comes with a social group. We all belong to multiple social groups and depending on the context a corresponding group identity will become salient. In times of a crisis the group that we identify with can be a different group than the ones we identify with in normal circumstances because of a change in priority of needs. Especially this change in salient identities in times of a crisis might contribute to polarisation within a community.

In order to let phenomena like radical behaviour and polarisation emerge in the ASSOCC model both concepts of social groups and social identities should be implemented. This means that multiple extensions need to be made. First of all, agents need to be able to reason about which action to choose where group norms and potential deviation of those norms should be taken into account. Secondly, an agent should have the ability to choose what social group he wants to be a member of and with which agents he wants to connect. This means that the relations between agents have to become dynamic. Also, an agent should be able to reason about his needs and which relations will help him satisfy those needs. Thirdly, the population of agents should be big enough such that multiple social groups emerge that differ significantly when looking at value importance. We only used around 1000 agents in our current simulations, which means that the number of agents within each subgroup is relatively small. If we want to use variation in values in order to have social groups with enough agents, the size of the population should be increased.

15.2.4 Macro Economics

In ASSOCC we have modelled micro economic rules and principles. The model is simplistic in that we do not create market mechanisms to determine prices etc. The main objective is to see the flow of money during the crisis. In order to balance the flow of money we more or less created a closed economy, where the government and banks are the buffers of money that can absorb and provide money where needed. Although Chap. 9 shows that we already can get some interesting results with just this model in connection with the other components and the needs of the agents, it is clear that the economics model is far from complete. One of the first questions that came up when interacting with decision makers from the government is whether it would be possible to see macro-economic effects of the crisis and also effects of measures on the macro-economy.

We assumed that we would not be the first modellers to try to connect a macro-economic model to a micro economic model, but have been unable to find any literature on such combinations. There is enough literature on (social) simulation for economics, but this is mainly focused on generating macro-economic effects from individual agents' behaviour (see e.g. [4] for a good example of this work). However, for our simulation we would need to have both the micro and macro economic model. The micro economic model for the purpose that we already indicated before. The macro economic model provides input on unemployment rates and ripple through effects of some sectors doing very bad during the crisis or actually have more business during the crisis. Thus the macro economic model should also influence incomes, costs and employment figures on the micro economic level.

The challenge comes from the fact that our simulation with around 1000–2000 agents does not represent all sectors of a country's economy. For instance, we do not have agents working in agriculture or tourism. Thus if we want to represent the macro economy of a country there will be sectors that can be generated based on the agents in the ASSOCC simulation and their work activities, but we have other sectors for which such a set of agents is not available. To remedy this we looked for a macro economic model that would compatible with an agent based approach. We ended up with the so-called balance-sheet model [5]. This model was already used by J. Musschoot to investigate macro economic effects of the corona crisis.[1] The balance sheet divides the economy into a number of sectors. In the balance sheet one can see the flow of capital from one sector to another sector. The following Table 15.1 is a simplified version with fictive numbers that are reasonably realistic for some year in a European country. Using the table one can create a "flow" model where capital flows from one sector to another.

This macro economic flow model can be operationalised by creating a macro economic agent per sector of the balance sheet. These agents can be relatively simple. They receive money from other sectors and spend the money again in other sectors again. Overall this should generate a flow that is more or less balanced. The sector agent can again be modelled using a homeostatic model (see Fig. 15.1). In this case the agent has one container that receives all money from the other sectors. If the money is below a certain threshold the agent will decide to spend less money in other sectors as well. This amount can vary per sector and thus we have a container per sector that the agent spends money in. The amounts given in the balance sheet table can be seen as the target amounts that a sector agent should receive and spend over a year. Thus an agent can distribute such an amount exactly evenly over all ticks in a year. However, some sectors will earn more or less in certain times of the year. E.g. tourism earns most money in summer (or in mid winter for skiing resorts). Using the containers as buffers the agents can balance these variations over a year.

While the macro economic model seems intuitive and can also be used to analyse the effects of the crisis by just draining some of the containers of sectors and/or stopping the flow between sectors. It is mainly useful to see ripple effects of measures like closures of shops, restricting transport etc. One can also see what measures a government could take to prevent some of the consequences. I.e. subsidise a sector

[1] http://blog.janmusschoot.be/2020/03/31/interlocking-balance-sheets-and-the-corona-induced-sudden-stop/.

Table 15.1 Sectorial balance-sheet. It is an abstraction of https://www.cbs.nl/nl-nl/publicatie/2019/29/nationale-rekeningen-2018

	Agriculture-Essential	Agriculture-Luxury	Manufacturing-Essential	Manufacturing-Luxury	Services-Essential and Finances	Services-Luxury and Hospitality	Education-Research	Households	Government	International
Agriculture-Essential	0	10	20	10	25	20	5	10	30	30
Agriculture-Luxury	10	0	10	10	20	30	5	10	25	50
Manufacturing-Essential	1	0	0	10	40	30	10	20	30	70
Manufacturing-Luxury	1	0	20	0	40	40	10	40	35	70
Services-Essential and Finances	10	5	30	30	0	40	5	110	50	80
Services-Luxury and Hospitality	10	50	20	30	40	0	5	100	25	80
Education-research	1	1	10	20	10	1	0	40	0	5
Households	20	50	30	50	60	80	10	0	45	50
Government	20	10	20	15	50	10	40	50	0	20
International	120	120	100	100	80	80	5	20	15	0

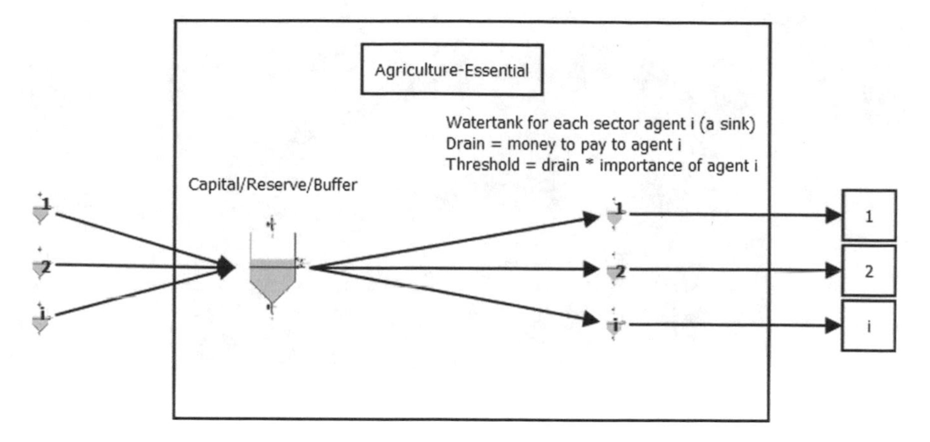

Fig. 15.1 Example sector agent architecture

or allow some limited trade or pay salaries etc. The difficulty is in connecting this macro economic model with the micro economic model we already had in ASSOCC. Remember that this micro economic model was also designed as a closed system. However, when connecting the macro economic model we have to recognise that there are sectors missing in the ASSOCC simulation and that money flows to and from these sectors. Some of these flows are completely outside the ASSOCC simulation, but some of them are connected to activities in the ASSOCC model. An impression of the connection between the models is given in Fig. 15.2. Some of the flows between sectors is present in the micro model as well. E.g. between education and government. These flows have to connected in the two models, such that as the flow changes in one model it will also influence the flow in the other model. One could opt for the influence in this case always to go from the micro model to the macro model and thus generating the corresponding sector flow.

There are also flows between a sector that is not represented in the micro model and a sector that is represented in the micro model as well. In these cases we have to *open* up the micro model and connect the flow from the non-represented sector to the agents in the micro model belonging to the represented sector. In the ASSOCC simulation this could be the agriculture sector delivering products to supermarkets (essential shops). However, this opens up the economic cycle of the micro model and thus we have to make sure that the resulting flow of money gets compensated elsewhere in the model. The influence in these connections can go both ways. If in the micro model e.g. the supermarkets sell more food they will need more products from the agricultural sector. This will lead to an increased flow at macro level. In the other hand if the agricultural sector spends less money in the industrial sector because many factories close during the crisis, this should be reflected in the factories in the micro model as well.

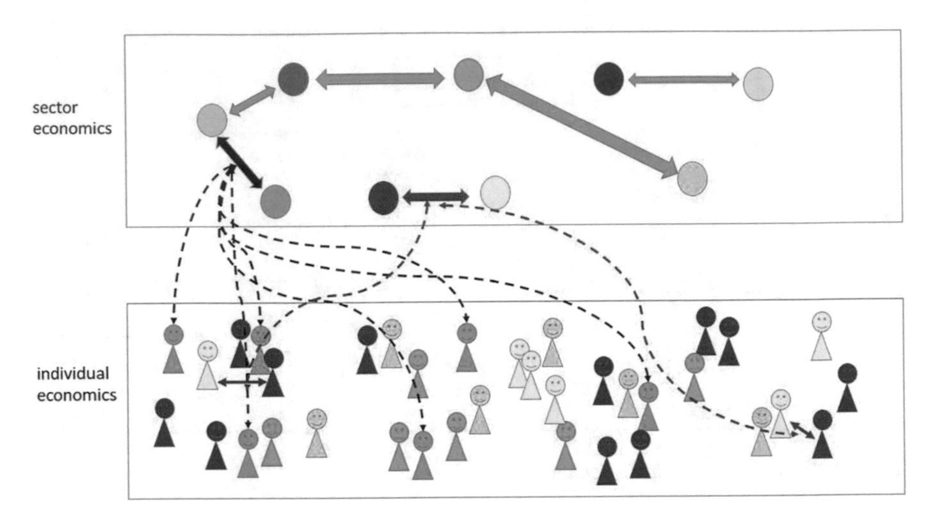

Fig. 15.2 Micro and macro economy

Finally there are flows between sectors in the macro model where both sectors are not present in the micro model. There the flow of money is not represented in the micro model.

In general we can see that the two models influence each other in both directions. In total we also have to have a balanced macro model and a balanced micro model. In the end we did not incorporate the macro economical model in ASSOCC because this last calibration step appeared to be very difficult. Some sectors are widely represented in the micro model, while others are present, but only marginal. Therefore one cannot use one influence factor that takes the total economic activity at micro level and translates that to a corresponding activity level at macro level. One has to calibrate these parameters with the relative importance of sectors, the relative amount of persons working there, etc.

Given the above difficulties the macro economic model was not included in the ASSOCC framework as it would make the economic component too sensitive to all these calibration factors and could easily lead to unwarranted results. One of the main problems, of course, is that not all sectors were represented in the micro level model. Basically this was a result of the decision to represent a (small) town in the simulation because this would lead to the most representative results for the pandemic. Due to the scalability issues it is difficult to extend the micro level model to include all sectors in a representative scale. However, this issue can be solved if the scalability can be extended and we are able to run simulations with around 100.000 agents. With this amount of agents it becomes possible to realistically represent all sectors.

15.3 Scalability and Productivity Tools

Scalability became of significance as stakeholders started to see the value of agent-based systems, but could not confidently rely on their results due to the number of simulated individuals being too far from the numbers of real individuals they deal with. Thus, achieving scalability is an important challenge for achieving impact. It is theoretically impossible to fit more than 100 million agents on a high-end computer and in practice, and more practically, simulations with more than 10.000 agents, particularly with advanced cognitive models are rare.

A challenge that we have not explicitly discussed yet is the use of high performance computers to run large scale simulations. Within the ASSOCC project we already made use of HPCs (The HPC2N in Umeå[2]). Basically the HPC was used to run many simulation runs in parallel. In many scenarios we wanted to compare different settings of measurements and for each setting we ran the simulation 30–40 times to average out some accidental occurrences of events. By using a HPC we could run most of all these runs in parallel and be done in an hour or two. Although this is already a valuable use of the HPC it becomes really interesting if we also can make use of the parallel computational power of a HPC within a run.

In general the emerging results of a social simulation comes from the interactions between the agents. Thus if sets of agents are separated we more or less take away their interaction opportunities and thus prevent the emergence of the results. However, not all agents interact with all other agents (all the time). Thus one might group agents on e.g. neighbourhood where they live or by age group, because those properties determine many of the daily interactions in the ASSOCC simulations. It is clear though that these are a kind of naive solutions and some more solid solutions should be found to make sure that the results of the simulations are at least comparable to the ones where agents are not split up. Thus one might do many test runs on a small scale simulation to see which splits into parallel parts of the simulation give results that are within acceptable boundaries to the original simulation. In this way the parallelization for the large scale simulation can be made giving a good balance between efficiency of the simulation and results that are minimally hampered by missing connections in the simulation.

When scaling up the simulations the need for good productivity tools is also increasingly important. Currently, developing, testing, validating, analysing, and comparing necessarily involves highly technical activities and abilities. This requirement causes high inefficiencies and ineffectiveness due to coordination efforts (the "feature owner" often differing from the implementer). An extensive array of productivity tools can be put in place and benefit the social simulation community: e.g. a visual interface for connecting the cognitive components with the new feature to be included and tweaking this connection later on, user-friendly facilities for facilitating analysis, such as interactive plotting devices (e.g. select the variables to be plotted in a list, tick a few boxes) and methodologies and tools for comparing models (e.g. argument structure, supported by plots).

[2]https://www.hpc2n.umu.se/.

We want to single out the challenge of facilitating analysis for large scale and complex simulations. Analysis of the results of a simulation is done for many purposes. It ranges from checking whether an emerging phenomenon is an interesting feature of the model or actually a bug in the code, to comparing two simulations made by different groups. Currently, this activity involves a combination of general insights on what the feature should be doing (both by itself and in relation with other features), the implementation in the simulation code of the adequate instrumentation, the implementation of aggregating features for presenting this instrumentation (e.g. plots, graphs), the execution of the model, and the interpretation of these features and the confirmation or not of the replication of the expected phenomenon. In a crisis, this whole process, often carried out by multiple people (e.g. the feature owner and the lead developer as introduced in Chap. 13), can become ineffective and inefficient as many iterations are required. Our comparison with the Oxford model and our search for specific details took multiple weeks and around 1.5 person-month.

15.4 Project Management and Methodology

One issue that we have discussed before in Chap. 13 is the formation of a team and the management of this team for building and maintaining the simulations, adding new aspects, improving the platform in terms of usability for users as well as designers, etc. As was remarked there, in the ASSOCC project we managed to get together a number of people that had some time, commitment and skills to work on the project goals. It should be reiterated that getting the right team of people together is of the utmost importance to make a project like ASSOCC work!

For a "normal" academic project we would write a proposal, get reviewed and, if lucky, get funded. Between the project proposal and the start of a project normally a year can pass. It is clear that we cannot take this approach when reacting to a crisis. However, unlike some disciplines that are used to deal with crisis situations, we do not have the possibility to free up people and start a project whenever a crisis is apparent. There is no funding or infrastructure to establish a team of people that can work on a project like ASSOCC full time for some months.

This is a major challenge! The ASSOCC team was asked several times during the crisis period why something (a specific element or a scenario) had not been done yet or whether we did investigate whether the results of our simulations were corroborated by reality. These were very legitimate questions, but we just did not have the resources to follow every part of the research. (If we had those resources, this book would probably be twice as thick).

The reason that the ASSOCC team worked is the lucky circumstance that around half the team consisted of people that worked directly or indirectly with Frank Dignum as PhD student or research engineer. The ASSOCC project fitted good enough in the general research goals of all these people to be able to commit themselves to the project. But even then the project work has been done for a large part outside office hours! This is only possible in special circumstances like the

COVID-19 crisis and only for a limited period of time. Thus this is not a sustainable model! We cannot depend on a coincidental pool of Ph.D. students being available to run a project like this. Thus we really need to find ways to make time and resources available in the long term for efforts on social simulations for crises if we want to make real impact.

The special circumstances of the team of the ASSOCC project had some advantages as well. Because most members of the team are working together or have worked together in the past there was no need for discussions on the best way to approach the project. Having a common ground on the modelling approach and way to implement the simulations is crucial to act quick and also react in agile ways to new developments. Thus having a team that *thinks alike* is very important for such a team to succeed in a crisis. Crisis response teams often train for years to establish a common ground that allows them to react quick and as a team during a crisis. The same is important for a team building social simulations in a crisis! Note that in these cases having a *common approach* is more important than having an *optimal approach*! This leads to the observation that building social simulations for crises requires a group of social simulation researchers and engineers that are trained in the same way of (agent based) modelling and can be made available at short notice by their organisations. Preferably this would consist of a core team that is working at an institution to develop the social simulation platform, theory and methodology permanently and which could be extended in times of crisis with other people (as was also already concluded in Chap. 13).

Having a team that participates on a purely voluntarily base and where most of the work has to be squeezed in between the normal work also has some consequences for the team management. In a more formal project structure one can also create more formal roles and responsibilities that can be enforced. In the ASSOCC project and in any other project that is non-funded and based on voluntary efforts the arrangements have to be established on common agreement. When this works it works very well, because everyone is also voluntarily committed to the team efforts. Thus rules are followed from internal motivation rather than an external enforcement. In a crisis situation where the work pressure is high and time is short this really makes a big difference! Thus the challenge is to have a team for which the social practices fit with the team rather than using some practices that are theoretically optimal but do not fit the team!

Without having a formal team structure there are also many issues that are common when working with a large group on a software project and that are exacerbated by the time pressure and availability of all members. We list some of the more common ones that are also of interest for future endeavours of building social simulations for crises.

1. **Documents and ideas are everywhere**: even with a neat start and ordered directories on google drive it suddenly ends up with a massive amount of documents that all describe different aspects in different ways. Many of them are not updated and incomplete. This is partly caused by a model growing quicker than the documentation and very pressing deadlines which puts proper documentation at the

lowest priority. Secondly, during a fast moving project there are many perspectives on the project with each its own documentation requirements. Thus it would make sense to have a pool of documentation that can be accessed in different ways. This issue is especially important when new people join the team and need to get up to speed on the project. In the ASSOCC project we actually created a starting document that served as a guide to find all the other documentation that a new member would need to see. In general the challenge is to find a good way of documenting the project while not overloading people with time spent on documentation or reading it.

2. **Documenting the complete conceptual model of the simulation**: The ODD [6] turned out not to be very suitable to document the simulation model for complex and more realistic simulations. There is a large emphasis on describing every part of the model, followed up by how these variables are interrelated in the scheduling. While this is good in order to understand how the implementation follows the conceptual model, the document explodes when a simulation is very complex and has many interrelated elements (and multiple people work on the same simulation). In the ASSOCC project we started working with a full ODD and an abbreviated version for communication purposes. There is a real need for an adjusted standard that gives better guidelines on especially the behaviour model.
3. **Communication**: with many people working at a distance on the same project the need for communication is high. We used a weekly meeting with the whole team and ad-hoc meetings with subsets of people whenever specific issues needed to be discussed in more detail. Although some minutes of meetings were kept, there is actually too little documentation left of most meetings. This made it difficult to keep track of all decisions made. The challenge here is to find a balance between time spend on documenting decisions and carrying out the decisions.
4. **Spaghetti code**: it is already very tricky to keep the code of a simulation model clean when working on it. This has to do with the iterative nature where especially with the KISS (Keep It Simple, Stupid) approach a model starts out relatively simple, and complexity grows when more elements are added. This approach causes problems in the implementation as some variables and functions may become deprecated or should have been implemented in a different way for the new conceptual changes. This issue is multiplied when more people work on the same model at the same time, as changes in one part may heavily influence changes in the other parts. The use of git[3] is a necessity here but cannot truly prevent the code from becoming messy. Thus there is a real software engineering challenge to keep the code clean while working in a cyclic methodology with many people. Having a simulation platform that allows more natural for a modular approach would already support keeping the code more structured.
5. **Many tweaks and quick fixes in the code**: this is part of working on a simulation model in crisis under high time pressure. Under time pressure quick fixes are used to at least get some feature working (properly). This happens especially

[3]https://git-scm.com/.

if a fundamental solution would need a major restructuring of the code. Such a restructuring often also involves code other persons are working on and thus halts the progress of the whole project for some time. The need to balance between keeping progress quick enough to reach deadlines and keeping the code well structured is a very difficult one. Deadlines we experienced were of the kind to produce results of a scenario before a debate would take place in parliament. Thus being late would possibly make a lot of the work useless as coming too late to influence a debate. However, many quick fixes can lead to a problem in the long term as it is very difficult to see whether the behaviour now actually stems from the behaviour system or a quick fix in that behaviour system.

6. **Parameters**: In most social simulation models there are around 10 or maybe 20 parameters that can be set for a simulation run. In ASSOCC there is a magnitude higher number of parameters (around 170). This makes the management of these parameters an issue on itself. Most of them can be found in the initialisation files however they are still initialised in different places, while some are also hard coded somewhere in the code and thus extremely hard to find. The Oxford model has a good example on how to approach this for the modeller. They use an external .csv file to load in the default variables and for different scenarios they have different .csv files. These can work well for a modeller that knows exactly which parameters have to be set to which value for a specific type of scenario. However, for a normal user of the simulation we would need a tool that supports setting up scenarios and the corresponding parameters. This tool could check for consistency and take care of dependencies between parameters.

Although all the above issues are not big scientific challenges to be investigated by social simulation researchers, they are all of the utmost importance for the success of the project. Having a methodology and support tools in place at the start of a simulation project in a crisis situation will have huge positive impact on the success of the project.

15.5 Real World Impact

In Chap. 11 we have discussed extensively about the issues that have to be solved in order to get real world impact with social simulations. The more important conclusions from that chapter are that if we want to have real world impact, we need at least:

1. Continuous development and publishing on social simulations for crises
2. Involvement in national and international crisis institutions
3. Co-development of social simulation tools for crises

Besides these points we need of course many good and useful simulations and results that actually show the added value of our approach. Most of this cannot be done by a single research group, but should be supported by a larger social simulation community that is ready to work towards these goals. Of course, they do not have to be the only goals of the social simulation community. But in pursuing these goals we will also tackle many interesting social simulation research questions and advance the field as a whole! And if we can do this while having real world impact that would be something very valuable for this community as well.

The previous issues have all to be addressed in order to get real world impact. However, we will face another challenge when the goal of real world impact is achieved. What happens when the government asks us to run a simulation to check the effect of a complete lockdown for a period that lasts until there are less than 10 infections in the country per day? How long do we think this lockdown will last and how will people react to it?

We could run a simulation with ASSOCC (maybe an improved version) and make some plots and even make it possible to interact with the simulation to try out several scenarios. However, would we dare to give a firm answer to the questions of the government? Scientifically we would probably have to say that we can only give an indication. There are always many uncertainties and parameters that were not included in the model that will make any prediction very uncertain as well. But could we give a rough estimation in e.g. number of months it would take? And a percentage of people that might violate the lockdown per month rounded on 10% estimates?

This (hypothetical) situation gets us back to the question when the simulation is good enough to dare make such a prediction [2]. Even with a rather simple model a prediction can already be made. However, how certain are we of the reliability of the results? And can we put a number on the certainty of the results? The latter is the biggest problem. We cannot guarantee that the results are correct within a certain error margin. Moreover, adding more details (or data for that matter) to the simulation does not necessarily make the error margin smaller. So, is social simulation as a scientific field not mature enough to do this type of work? It seems other disciplines can make predictions and indicate error margins. Epidemiologists and virologists have been giving predictions in the time of the pandemic, economists are giving predictions on the development of the economy all the time.

It should, however, be noted that the predictions given by those disciplines are often way off and in times of crises are not even near to what is really happening. The trust in these models is based on the regular prediction their models make in normal times when the environment and behaviour of people is rather stable. This allows for averaging out many uncertainties and using a simple behavioural model based on data. In times of crisis these assumptions on human behaviour are no longer valid! We have seen some of this in Chap. 12 where we compared the ASSOCC simulations with an epidemiological model. Thus the certainty that other disciplines have on the prediction of their models comes from pushing the inherent uncertain human behavioural factors out of the model and incorporating this only through some (stochastic) variable that is based on average behaviour in stable situations.

Thus we can conclude that the social simulation models are not inherently inferior to the models from other disciplines. The main difference is that we put the uncertainty of human behaviour squarely in the centre of our models. That makes the simulations and thus the predictions a bit more volatile. However, this vulnerability also gives us a huge advantage. First of all, if we have human behavioural models in the centre of the simulation we can also adapt these models to the observed reality during a crisis. We can add alternative behaviours, new motivations, changing preferences and based on that create alternative scenarios. This can be done without needing a massive amount of new data to calculate some new values for the parameters in a model that does not include the human behaviour explicitly.

Moreover, we can also explain the results of the simulated scenarios based on the human behavioural models underlying the simulation. Thus we can trace results and see which type of behaviour or interactions underlies those results. By tracing the results back to these human behavioural elements we can create powerful explanations that are readily understood by all stakeholders. We have seen a number of these explanations in the results chapters of this book and a very detailed one in Chap. 12. There we see that the lack of effectivity of the track and tracing apps is explained by the fact that many people get infected at home and there the app does not really add much. Having these explanations available also adds accountability to the results of the simulations. Through the explanations we can account where the results came from and we can also investigate alternative scenarios in case some of the assumptions might not appear realistic in the current crisis.

The big challenge for having real world impact thus seems to be to continue producing good quality simulations that give good insights and can be explained as well. This will increasingly give more trust in the added value of social simulations. This can be increased by having platforms that allow for quick adaptation of scenarios, more interactivity and good support for explaining results.

15.6 Conclusions

In this chapter we have discussed a number of challenges for social simulations for crisis situations. These challenges range from theoretical conceptual model issues to challenges for getting impact in the world outside academia. Some challenges can be addressed by individuals and research groups working on specific topics. However, in order to move to a next level of impact we have to persist as community to develop better tools, methodologies, models and results. This has to be pursued systematically and over a longer period in order to gain acceptance. It does not mean that the next step has to be perfect. Better than what we have now is good enough. But building upon each others work and improving consistently shows that the social simulation community as a whole has something to contribute to the world. For this we need institutional action and many contributors.

One way to create a more sustainable development of social simulations for crises would be to use the agile social simulation platforms for supporting the resilience of society. This is a continuous challenge of society and its decision makers. It involves investigating which measures government can and should take to provide a resilient society. Being able to simulate many different scenarios (representing all kinds of possible disasters) and how different measures can protect or cope with these disasters would be very important. The type of agile social simulation platforms that are also needed for simulations for crisis situations would be a very good support for this endeavour.

Acknowledgements We would like to acknowledge the members of the ASSOCC team for their valuable contributions to this chapter. This research was partially supported by the Wallenberg AI, Autonomous Systems and Software Program (WASP) and WASP-Humanities and Society (WASP-HS) funded by the Knut and Alice Wallenberg Foundation.

References

1. S. Heidari, M. Jensen, F. Dignum, Simulations with values, in *Advances in Social Simulation* (Springer, 2020), pp. 201–215
2. A. Mertens et al., Are we done yet? or When is our model perfect (Enough)? in *Advances in Social Simulation*, ed. by P. Ahrweiler, M. Neumann (Springer, 2021)
3. D. Webber, A.W. Kruglanski, Psychological factors in radicalization: a “3 N” approach, in *The Handbook of the Criminology of Terrorism* (2016), pp. 33–46
4. L. Hamill, G.N. Gilbert, *Agent-Based Modelling in Economics* (Wiley Online Library, 2016)
5. R.W. Goldsmith, The national balance sheet: another tool in the economist’s kit. Anal. J. **12**(1), 9–14 (1956)
6. V. Grimm et al., The ODD protocol: a review and first update. Ecol. Model. **221**(23), 2760–2768 (2010)

Chapter 16
Conclusions

Frank Dignum, Loïs Vanhée, Maarten Jensen, Christian Kammler, René Mellema, Fabian Lorig, Cezara Păstrăv, Mijke van den Hurk, Alexander Melchior, Amineh Ghorbani, Bart de Bruin, Kurt Kreulen, Harko Verhagen, and Paul Davidsson

Abstract We finish the book with a chapter in which we describe what we decided not to do or include in the ASSOCC framework. Where did we stop? And why did we stop? The temptation is to keep adding more and more aspects in order to simulate more scenarios. However, at the same time the simulation would become more complex and unmanageable. We also summarise what we have learned from all the different parts of our experience. And most importantly we give a roadmap of research that would be needed to make social simulations a standard tool to be used for crisis situations.

F. Dignum (✉) · L. Vanhée · M. Jensen · C. Kammler · R. Mellema · C. Păstrăv
Department of Computing Science, Umeå University, 901 87 Umeå, Sweden
e-mail: dignum@cs.umu.se

L. Vanhée
e-mail: lois.vanhee@umu.se

M. Jensen
e-mail: maartenj@cs.umu.se

C. Kammler
e-mail: ckammler@cs.umu.se

R. Mellema
e-mail: renem@cs.umu.se

A. Ghorbani · B. de Bruin · K. Kreulen
Faculty of Technology, Policy & Management, TU Delft, Jaffalaan 5, 2628 Delft, The Netherlands
e-mail: a.ghorbani@tudelft.nl

B. de Bruin
e-mail: bdb785@gmail.com

K. Kreulen
e-mail: kurtkreulen@gmail.com

F. Dignum (ed.), *Social Simulation for a Crisis*, Computational Social Sciences,
https://doi.org/10.1007/978-3-030-76397-8_16

16.1 When to Stop

In the (Northern hemisphere) summer of 2020 the number of infections seem to be going down enough to allow some limited travelling such that people could go in holidays. The question was whether this would increase the number of infected people again. There are a few risks with people going in holidays. First of all the travelling by air-plane. Having many people packed in a small space will increase the risk of one person spreading the virus to many others.

A second risk is the behaviour of the vacationers on their destination. Once they are at their destination they might want to do the "normal" holiday activities. This will include in many cases going to bars, restaurants and nightclubs. Each of these places can be a possible cause for infections to spread. Exactly the fact that people are going in holidays after having been confined for some time, means that they tend to want to be free of all restrictions at least during their holidays. No matter whether this is smart or not, it is a natural reaction to being allowed "out" after having been locked down.

Would the local authorities not take care that people obey the local restrictions, that were often as severe as the ones in their home countries. In principle that is true. However, the traditional holiday destinations are heavily dependent on tourism and therefore wanted to attract the few tourists that still dared to travel. Therefore restrictions were circumvented whenever possible in favour of making some money during the holiday season.

P. Davidsson
Department of Computer Science and Media Technology, Malmö University,
205 06 Malmö, Sweden
e-mail: paul.davidsson@mau.se

M. van Hurk · A. Melchior
Department of Information and Computing Sciences, Utrecht University, Princetonplein 5,
3584 Utrecht, The Netherlands
e-mail: m.vandenhurk@uu.nl

A. Melchior
e-mail: a.t.melchior@uu.nl

F. Lorig
Internet of Things and People Research Center, Malmö University, 205 06 Malmö, Sweden
e-mail: fabian.lorig@mau.se

A. Melchior
Ministry of Economic Affairs and Climate Policy and Ministry of Agriculture,
Nature and Food Quality, Bezuidenhoutseweg 73, 2594 Den Haag, The Netherlands

L. Vanhée
GREYC, Université de Caen de Caen, 14000 Caen, France

H. Verhagen
Department of Computer and Systems Sciences, Stockholm University,
PO Box 7003, 16407 Kista, Sweden
e-mail: verhagen@dsv.su.se

The third risk was the cross contamination. The local people working in tourism were getting in contact with many different people from different places. That means that they had a higher risk to get infected and also, when infected, transmit the virus to many people from different places. Thus there was a higher risk for the virus to be spread both in the holiday destinations and, through returning vacationers, in the home countries.

At a certain moment we were asked whether we could simulate a scenario for this situation. Although highly relevant for the crisis we decided not to do this. Why not? Could we not do it? Actually, the scenario could be simulated in the ASSOCC framework. However, it would require quite a number of new aspects. Most importantly, we would have to model a touristic destination with a number of specific aspects. Would the destination have a beach? Would the beach be open? How were the bars, restaurants, night clubs and hotels located? Would they be all in the same neighbourhood or spread over town? All of these aspects determine how much the local population and tourists would meet and have contacts.

A second important aspect that would need to be modelled is the daily behaviour of the vacationers. How would their typical day (night) look like? And how would this match with local people working in the tourist sector? Finally, we would also need to model the transportation and get parameters of chances of infecting people in air-planes (while wearing masks, etc.).

Altogether we saw that we would have to estimate a lot of these aspects in a way that the result of a simulation would not add a lot of value over a good estimation of the risks based on some statistics. The only way that a simulation would have added value would be by doing a thorough investigation into all aspects and model them properly on some scale that would give reasonable realistic insights. However, this would take too much time (we estimate around 2–3 months to get everything done properly) to be of value anymore as the summer would be over by that time.

The above is a good example of a scenario that did not make it into ASSOCC. One could argue that we have included only those types of scenarios that happen to fit well in our framework. Thus, the claim that the ASSOCC framework is a good example of a more general sandbox that can be used to quickly generate scenarios cannot be substantiated by just the examples in this book. We actually would agree to some extend to this criticism. As we have indicated in several places, the lack of a spatial model in ASSOCC made it both more efficient, but also made it very difficult to model all kinds of restrictions that are closely tied to spatial parameters. It is one of the reasons for Chap. 14 in which we sketch a path towards a more scalable and robust platform that would include spatial parameters as well.

However, the scenario with which we have started this chapter could have been modelled in the ASSOCC framework, albeit with some difficulties. And we have shown that we can run a quite diverse set of scenarios within the ASSOCC framework as well. Thus we believe ASSOCC is a viable stepping stone for a more scalable and robust platform. But creating a more generic framework always raises the question when one admits that the platform is no longer suited for a certain scenario. We have put that boundary for ASSOCC at those cases where spatial aspects would be more important.

Besides the conceptual question on which scenarios could be modelled in the ASSOCC framework, we also had a more practical issue on when to stop. There were many more scenarios we could have run and we did not run (yet). Many would just need a few easy adjustments or additions of some aspects that could be done rather easy. The boundary that we put here was the complexity of the Netlogo model. With every new scenario there would be a few more parameters added to the huge amount of parameters we already had. It also made the Netlogo code more and more difficult to manage and calibration of the simulations also more difficult. So, in July 2020 we decided that we exhausted the resources that could safely be used within Netlogo to keep an ever growing framework running reasonably smooth. In order to create more scenarios with more aspects we would need to port the framework to a new platform that would be capable to support the modularisation, scalability and contextualisation in a proper way.

The last difficulty we have about when to stop is a very pragmatic one. When a project is not funded and has no explicit project plan with a begin and end date there is no natural date on which one stops the project. However, it is clear that after a year of working on ASSOCC besides normal jobs it is no longer feasible to keep going at the same pace as in the beginning. Thus creating this book as a lasting result of the project that can function as a stepping stone for further work is a good way to create an end point of the project. However, it is not the end of the research, but rather the beginning. But in the future this will have to be done through the more regular channels and without relying on the commitment of so many people to do something good for the world in crisis.

In the next sections we will first recapitulate the lessons we have learned from all the different aspects of this project. We close the chapter and the book with giving a kind of roadmap for research on social simulations for crisis situations.

16.2 Lessons Learned

16.2.1 Part I: Foundations

One of the main pillars of the ASSOCC framework is the abstract agent decision making model that is based on the foundations described in Chap. 2. We started with a very fundamental figure where we see that the actions of an agent are influenced by three aspects: values, motives and affordances. These three influences are mediated through social rules as pictured in Fig. 16.1.

Several members of the ASSOCC team have been working on modelling parts of these principles in social simulations. So, we all agreed that these principles were a good start of the agent decision making model. However, it was also clear that this model would be too complex for any larger scale simulation. If in every decision the agent has to consider values, motives and affordances as separate influences that would have to be prioritised and moreover check which social rules might govern

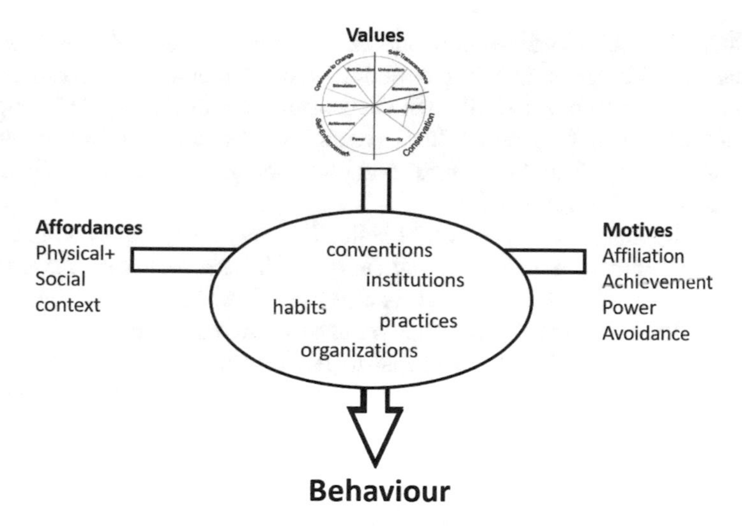

Fig. 16.1 Principles for agent deliberation

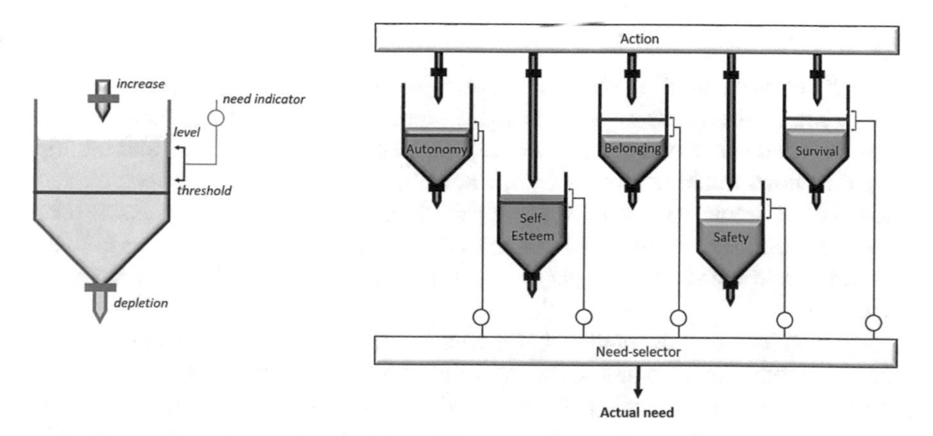

Fig. 16.2 Needs model

the current situation it is clear that this would not be efficient enough for any sizeable simulation.

We therefore decided to package these principles into a homeostatic needs model (Fig. 16.2). The needs model is still an abstract model to determine the agent behaviour over time. However, we now only have one type of influence. Using a homeostatic model has the advantage that the needs can adjust to changing circumstances and also gives a simple, indirect planning mechanism. Using this needs model as the core of the ASSOCC framework has made the complex model possible. However, being the central decision making component that connects all aspects of the environment, it also became the bottleneck of the framework.

In Chap. 3 we described the implementation of the needs model in details and it is clear how this seemingly simple model already becomes very complex when implemented. So, we think that with ASSOCC we are right on the limits of complexity of abstract decision making models that have a single abstract component connecting all aspects of life. In Chap. 14 we argue that more complex models can only be built by contextualising the deliberation.

The last important lesson from part I is that it is very important to build a proper user interface for the simulations. Using Unity for the user interface allows for nice graphics and interactivity. However, it is also clear that Netlogo and Unity are not meant to work together in this way. Thus we had to solve many practical issues along the way. Moreover, the Unity model had to replicate in some way aspects of the Netlogo simulation and thus had to be kept up-to-date all the time, which put a heavy burden on the development team. For a future social simulation platform a tighter integration of the feature of Unity and Netlogo would be desirable.

16.2.2 Part II: Results

We will not repeat all the results of Chaps. 5–10 here. What is noteworthy is that in most scenarios the results were at first sight surprising and sometimes seemed counter intuitive. Given the fact that we could explain all the results in a robust and intuitive way this shows the importance of using social simulations of the type we made in the ASSOCC framework. Apparently our intuitions on the effects of measures taken by the government are often not correct. Due to the complex interactions between the people and influences that play a role the behaviour of people is not always as expected!

The next lesson we learned is that if your simulations show results that are deviating from other models and moreover might not seem intuitive at first sight it is very hard to convince people that your results are interesting. Being the new comer on the block and using a model based on human behaviour rather than epidemiological concepts puts one (understandably) at a disadvantage. It therefore becomes very important to be able to explain where the differences come from and why your results are acceptable. It was clear that we were not well prepared for this. It took us about two months before we could show the analysis of our results in a way that could convince other people. It should be noted that the explanations were not simple or straightforward. They came from combinations of factors. For the analysis and explanation of the results of the effectiveness of the track and tracing apps we collected around 80 pages of graphs! Since that time we have been much better prepared as we had all software in place to generate graphs of all kinds of aspects of a scenario. Thus when we later on ran a scenario on the effects of curfews on the crisis, we could produce results and an analysis and explanation within two days!

Another general lesson we learned from the scenarios is that the inter-dependencies between the different aspects on life on the effects of restrictions are indeed very important. Acknowledgement of these influences when announcing or deciding on new restrictions is of utmost importance for the effectiveness of the restriction! I.e. the way the message is given can determine the success of the measure. The need for belonging and autonomy are fundamental needs of people that cannot be just put aside for the greater good of health of the society at large. They will lead to violations of lockdowns, isolation and curfews. Giving room for these social needs might lead to much better overall result of a measure! E.g. opening terraces down-town in a restricted way when the weather in spring becomes very nice can alleviate the pressure on parks and beaches. Terraces are easier to control because everyone is seated and the distance between people therefore better regulated. This is more difficult in parks where people tend to sit, walk, bike, play, etc.

In general most restrictions from governments have been very crude. Due to uncertainty and fear of new waves of the pandemic restrictions often were very general and overreached. Using social simulations would allow for experimenting with more focused restrictions. These restrictions can leave people's normal life more intact while still giving maximum protection against the virus. E.g. shops could be kept open if systems are found to guarantee distance keeping between customers and customers and personnel. How much would be safe can be experimented through simulations. In a similar way we already did a simulation to check whether basic schools should be closed or not. It appears that closing basic school is actually not very effective. Relieving parents from the care of young children for a couple of hours a day while working at home and having the children socialise at school prevents a lot of stress and can help people abiding by other restrictions.

16.2.3 Part III: ASSOCC Project

In the third part of the book we discussed a number of issues that we learned from the project and were not directly on the theory or implementation of the simulation. In Chap. 11 we discussed the real world impact of the ASSOCC simulations. We have learned that in order to have real world impact with a social simulation we should not just have a good simulation. That, of course, is quite obvious, for all people that have tried to have real world impact with their research.

We did do some things good in our project. We built up good relations with journalists and provide them with lots of information about our simulations. We also had contacts with a number of politicians in that way. We had an extensive website showing all information about the ASSOCC project, its background, foundations and results plus the code. Giving this amount of transparency helps to get trust in the framework!

Another aspect of achieving real world impact can only come from a long term concerted effort to establish social simulation as an important support tool for crisis management. Having tools that properly support decision makers in a crisis is one important requirement. But being part of the crisis management community is another one. Having a seat at the table of advisory committees means that you can also promote the use of social simulations in a continuous way and thus preparing the decision makers for the use of social simulations. This cannot be done once a crisis has started!

In Chap. 12 we have discussed the validity of the results of the ASSOCC simulations. Validation of simulations in situations when little real world data is available is very difficult. However, comparing simulations might give a good basis for an indirect validation. This type of validation is quite costly and intensive. Thus doing more of these types of validations outside crisis times could already build up some credibility of the foundations of a model that is used.

In Chap. 13 we have provided an extensive reflexive study on the software engineering aspects of the ASSOCC project. The main purpose of this chapter was to investigate whether the limitations that we found in the implementation were caused by the platforms used or more a result of a non optimal way of software development. We have shown that, despite the big time pressure, we have adhered to good software development practices, that we have used every possible trick in Netlogo to squeeze out some efficiency and kept the software also as structured as possible in the circumstances of a crisis. In short, the limitations we found of the simulations seem to be structural and not due to the software engineering. So, we also are convince we cannot scale the simulations without making some fundamental new choices that necessitate the use of different platforms. These requirements have been discussed in Chap. 14 and form the starting point of the challenges and future work that we will discuss in the next section.

16.3 A Road Map

Given that we have made a very nice first step, where do we go from here? There is still a lot of work to be done before social simulations will be accepted as part of the normal support tools for crisis management. We can roughly divide the work into four areas:

1. Development of a theory on computational social reality
2. Development of tools for social simulations for crisis situations
3. Development of a methodology to create social simulation for crisis situations
4. Move from supporting crisis response to supporting a resilient society

We will discuss each of these areas in the next sections.

16.3.1 Theory

As we have discussed before, there is no generally accepted social psychological theory that can be used as the basis for the decision making model of the agents and that could give handles to model all kinds of social rules that play a role in the interactions between the agents. The theories that we have used as the foundations for the model of the ASSOCC framework are certainly not the only ones around. We also did not discuss the theories describing norms, institutions, conventions, habits, social practices, etc. All of these concepts play a role in regulating interactions and form contexts in which interactions take place.

We do not argue that all of these concepts have to be implemented in a social simulation platform in order to create good social simulations. However, we should have good computational models of most of these concepts and also their connections to the decision making processes of the agents. Having these rich computational theories as background we can make informed decisions about which of these concepts are more important in a particular crisis situation and combine these concepts in a simulation.

The steps that we outlined in Chap. 14 on contextual reasoning give a first step for an approach that would be able to combine different theories depending on the situation. It starts from the assumptions that social psychological theories are not right or wrong, but are more or less suited for a certain situation. They emphasise different aspects of human behaviour and thus are particular useful in those situations where those aspects play a big role. This combinatorial approach gives new perspectives on how to compose computational social reality models. We have seen in the ASSOCC framework how the combination of values, motives and affordances into a needs model can lead to good results for the COVID- 19 crisis. We should not take this as proof that these concepts are the right ones for all simulations, but rather that a careful combination of theories into a computational model is helpful to create useful social simulations for crises.

16.3.2 Tools

In several chapters in this book (3, 4, 13, 14, 15) we have indicated the limitations of the existing social simulation tools and how we have worked around them in the ASSOCC project. There are many areas where a platform with the right tools integrated would have made a lot of difference in the speed and accuracy of building the simulations, analysing them and explaining them.

We will not repeat all of the issues, but highlight the more fundamental ones. First, we need a platform that supports large scale simulations with complex cognitive agents. Some work is already being done in creating platforms that can easily make use of HPC's, but nothing yet for social simulations where the agents have abstract cognitive decision making models. It requires a modular approach that is both flexible and is founded on sound theoretical social theories.

Being able to specify scenarios on a high level and creating populations for them based on demographic and other data is the next step. The scenarios in the ASSOCC project contained between 150 and 200 parameters. Setting all of them by hand becomes impossible. Similarly creating agents for the scenarios becomes difficult if one runs simulations with 100.000 agents or more. The profiles of the agents should be taken from demographic data, but also their decision making processes initiated based on the right distribution of parameter values.

Finally, the tools should provide interfaces for inspecting the simulation run, being able to interact with it (in a limited way), analysing it and explaining it. We have managed these tasks by connecting Unity and R with the Netlogo platform. However, this requires still a lot of clever programming and engineering skills that are not always available when simulations have to be run in a crisis. Thus a proper integration of these facilities in the simulation platform is crucial.

All of the above issues require not just some engineering of tools, but also fundamental computer science research into appropriate techniques and algorithms to optimally support the purposes of the simulation platform.

16.3.3 Methodology

Creating very complex simulation platforms where one can use combinations of different social theories sounds very nice, but, of course, will become unusable very quick. Whereas methodologies to build social simulations in general are already pretty scarce there is no methodology that could readily be used to create simulations as we have done in the ASSOCC project and envision in the future for crisis management.

During the ASSOCC project the project leader (Frank Dignum) was frequently questioned why particular choices were made in the way they were. Most often the answer was that this was based on experience and on intuition which aspects are important. Although the choices seem to have been correct the basis is hardly scientific. As an example, we have tried to argue why the particular foundational theories chosen for the ASSOCC project are appropriate. Although the arguments may sound convincing this does not imply they are the best! Having some methodology that supports people that have to build social simulations for crisis situations will help a greater uptake of the approach. Rather than a magical, experience based method to create a simulation it should be based on sound principles.

Given that we can come up with a solid methodology for creating the simulations, it also becomes much easier to provide the right support tools. Each step in the methodology can be supported by its own set of tools and if the methodology is based on sound principles the tools can also check on completeness and consistency of the results of each step! Thus creating such a methodology is not just a nicety, but actually a necessity for the development of social simulations for crisis situations.

16.3.4 From Crisis to Resilience

The last area that needs attention is not just a research issue, but requires a concerted effort of the whole social simulation community. While working in an academic community it is very convenient to develop social simulations that in one or another way serve some scientific purpose. They can explain the workings of a social theory, they can illustrate it for specific situations, etc. Once in a while we can have an applied project where we show the usefulness of social simulation for some practical purpose. E.g. there are simulations about energy transitions [1], land [2] and water use [3], etc. But most of these projects are still build with an academic interest to see whether the results of the social simulations might shed new light on some real world issues.

There are some exceptions such as the work done with ComMOd [4] which takes companion modelling as starting point to create simulations for sustainable development. Similarly the GAMA platform [5] which is particular well suited for urban development. However, what is missing is a development of theory and methodologies that indicate how to create social simulations for specific application areas and which tools would be suitable to support this. In short, we need a more fundamental social simulation theory that gives guidance and methodologies for creating social simulations, but also drives the development of platforms and methodologies for new areas.

Based on such a more broad and fundamental social simulation theory it becomes easier to convince other people of the value of social simulations. It also becomes possible to continuously build on tools and methodologies for a wide range of applications. Such a research environment would facilitate that we do not wait for a next crisis to further develop the social simulation tools to support managing that crisis. But we start developing these tools and methodologies right now. In that way we can move from crisis support to support of a resilient society!

16.4 Conclusions

This section has recaptured some of the experiences of the ASSOCC project. We have discussed some of the lessons we learned. Lessons learned directly from the results of the simulations we ran in a diverse range of scenarios. But also lessons learned from running the ASSOCC project. More importantly we have also sketched a number of areas for research and development in order for social simulations for crisis situations to become a standard tool for crisis management. Combining all these experiences and lessons learned from the ASSOCC project we believe that we can end this book with the following optimistic statement:

Agent-based Agile Social Simulation can make a valuable contribution, not only to science, but also to society in times of crisis!

References

1. D. Clarke et al., A critical time for UK energy policy what must be done now to deliver the UK's future energy system: a report for the Council for Science and Technology (2015)
2. R.B. Matthews et al., Agent-based land-use models: a review of applications. Landsc. Ecol. **22.10**, 1447–1459 (2007)
3. A. Ernst, Social simulation: a method to investigate environmental change from a social science perspective, in *Environmental Sociology* (Springer, 2010), pp. 109–122
4. M. Etienne, *Companion Modelling: A Participatory Approach to Support Sustainable Development* (Springer, 2013)
5. A. Drogoul et al., Gama: multi-level and complex environment for agentbased models and simulations, in *12th International Conference on Autonomous Agents and Multi-agent Systems* (Ifaamas, 2013), 2-p

Appendix A
Culture

A.1 Appendix Cultural Model (Chap. 8): Values

Algorithm 1: INFLUENCE OF AGE ON AGENT VALUE SYSTEMS

```
1  λ ← INFLUENCE-OF-AGE-ON-VALUE-SYSTEM
2  if age = "young" then
3  |  set ω_Young = 3 · λ
4  else
5  |  if age = "student" then
6  |  |  set ω_Student = 2 · λ
7  |  else
8  |  |  if age = "worker" then
9  |  |  |  set ω_Worker = 1 · λ
10 |  |  else
11 |  |  |  set ω_Retired = 0 · λ
12 |  |  end
13 |  end
14 end
```


F. Dignum (ed.), *Social Simulation for a Crisis*, Computational Social Sciences,
https://doi.org/10.1007/978-3-030-76397-8

Equation A.1 **specifies a collection of variables used in the Algorithm** 2.

$$
\begin{aligned}
UAI_{pos} &= \text{UNCERTAINTY- AVOIDANCE}\\
PDI_{pos} &= \text{POWER- DISTANCE}\\
MAS_{pos} &= \text{MASCULINITY- VS- FEMININITY}\\
IDV_{pos} &= \text{INDIVIDUALISM- VS- COLLECTIVISM}\\
IVR_{pos} &= \text{INDULGENCE- VS- RESTRAINT}\\
LTO_{pos} &= \text{LONG- VS- SHORT- TERMISM}\\
UAI_{neg} &= 100 - \text{UNCERTAINTY- AVOIDANCE}\\
PDI_{neg} &= 100 - \text{POWER- DISTANCE}\\
MAS_{neg} &= 100 - \text{MASCULINITY- VS- FEMININITY}\\
IDV_{neg} &= 100 - \text{INDIVIDUALISM- VS- COLLECTIVISM}\\
IVR_{neg} &= 100 - \text{INDULGENCE- VS- RESTRAINT}\\
LTO_{neg} &= 100 - \text{LONG- VS- SHORT- TERMISM}
\end{aligned}
\tag{A.1}
$$

Algorithm 2: MAPPING Hofstede Dimensions to Schwartz BVT Values

Input: Scores on Hofstede Cultural Dimensions.
Output: Means for Schwartz BVTValues.

1 **if** *hofstede-schwartz-mapping-mode = "theoretical"* **then**
2 | **set** $\mu_{HED} = IVR_{pos}$
3 | **set** $\mu_{STM} = UAI_{neg}$
4 | **set** $\mu_{SD} = IDV_{pos}$
5 | **set** $\mu_{UNI} = \frac{PDI_{neg}+MAS_{neg}}{2}$
6 | **set** $\mu_{BEN} = MAS_{neg}$
7 | **set** $\mu_{CT} = \frac{PDI_{pos}+IDV_{neg}+LTO_{neg}+IVR_{neg}}{4}$
8 | **set** $\mu_{SEC} = UAI_{pos}$
9 | **set** $\mu_{POW} = \frac{PDI_{pos}+MAS_{pos}}{2}$
10 | **set** $\mu_{ACH} = MAS_{pos}$
11 **end**
12 **if** *hofstede-schwartz-mapping-mode = "empirical & theoretical"* **then**
13 | **set** $\mu_{HED} = IVR_{pos}$
14 | **set** $\mu_{STM} = \mathbf{X}$
15 | **set** $\mu_{SD} = IDV_{pos}$
16 | **set** $\mu_{UNI} = \mathbf{X}$
17 | **set** $\mu_{BEN} = \mathbf{X}$
18 | **set** $\mu_{CT} = \frac{PDI_{pos}+IDV_{neg}+LTO_{neg}+IVR_{neg}}{4}$
19 | **set** $\mu_{SEC} = UAI_{pos}$
20 | **set** $\mu_{POW} = \frac{PDI_{pos}+MAS_{pos}}{2}$
21 | **set** $\mu_{ACH} = MAS_{pos}$
22 **end**

See (Fig. A.1).

Schwartz Values (columns) × **Hofstede Dimensions** (rows)

	HED	STM	SD	UNI	BEN	CT	SEC	POW	ACH
PDI				NS		S [POS]		S	
IDV	[NEG]		S			S			[NEG]
MAS				NS	NS		[POS]	S	S [POS]
UAI	[POS]	NS					S [POS]		
LTO						S [NEG]			[NEG]
IVR	S [POS]			[POS]		S		[NEG]	

Negative = Red
Positive = Green
Neutral = Grey

S = Supported
NS = Not Supported

[POS] = a positive & statistically significant (p < 0.05) relationship
[NEG] = a negative & statistically significant (p < 0.05) relationship

Fig. A.1 Empirical support for HCD and Schwartz BVT values linkages

	Hedonism	Stimulation	Self-direction	Universalism	Benevolence	Conformity & Tradition	Security	Power	Achievement
(Intercept)	3.13 ***	3.67 ***	2.62 ***	2.34 ***	2.29 ***	2.69 ***	2.34 ***	3.87 ***	3.13 ***
	[3.02, 3.24]	[3.57, 3.77]	[2.54, 2.69]	[2.26, 2.42]	[2.19, 2.39]	[2.63, 2.74]	[2.26, 2.42]	[3.74, 4.00]	[3.04, 3.22]
PDI	0.10	0.00	0.04	-0.07	-0.01	-0.13	-0.06	-0.02	-0.05
	[-0.06, 0.25]	[-0.14, 0.14]	[-0.07, 0.15]	[-0.18, 0.04]	[-0.15, 0.13]	[-0.21, -0.05]	[-0.17, 0.05]	[-0.20, 0.16]	[-0.17, 0.08]
IDV	0.16	-0.08	-0.06	-0.05	-0.06	0.07	0.09	0.00	0.17
	[0.01, 0.30]	[-0.21, 0.05]	[-0.17, 0.04]	[-0.15, 0.06]	[-0.19, 0.07]	[-0.01, 0.14]	[-0.02, 0.19]	[-0.17, 0.17]	[0.05, 0.29]
MAS	0.06	-0.01	-0.01	0.00	-0.02	-0.05	-0.10	-0.08	-0.12
	[-0.06, 0.18]	[-0.12, 0.09]	[-0.09, 0.07]	[-0.08, 0.08]	[-0.13, 0.08]	[-0.11, 0.01]	[-0.19, -0.01]	[-0.22, 0.06]	[-0.21, -0.02]
UAI	-0.12	-0.02	-0.07	-0.08	-0.08	-0.02	-0.09	0.04	-0.04
	[-0.23, -0.00]	[-0.12, 0.08]	[-0.15, 0.01]	[-0.16, 0.00]	[-0.19, 0.02]	[-0.08, 0.04]	[-0.18, -0.01]	[-0.10, 0.18]	[-0.14, 0.05]
LTO	-0.11	0.02	0.08	0.03	0.05	0.14	0.05	0.03	0.11
	[-0.24, 0.02]	[-0.10, 0.13]	[-0.01, 0.17]	[-0.06, 0.12]	[-0.07, 0.16]	[0.08, 0.21]	[-0.04, 0.15]	[-0.12, 0.18]	[0.00, 0.21]
IVR	-0.18	0.05	-0.03	-0.11	-0.10	0.06	-0.02	0.29	0.07
	[-0.32, -0.04]	[-0.07, 0.18]	[-0.13, 0.06]	[-0.21, -0.01]	[-0.22, 0.03]	[-0.01, 0.13]	[-0.12, 0.08]	[0.12, 0.45]	[-0.04, 0.19]
N	58	58	58	58	58	58	58	58	58
R^2	0.26	0.05	0.23	0.21	0.17	0.60	0.35	0.28	0.43

Fig. A.2 Regression results for HCD and Schwartz BVT values linkages

With regards to Fig. A.2, one must be aware of the fact that LOW (HIGH) scores on Schwartz BVT values indicate HIGH (LOW) importance. Therefore, negative coefficients are labelled GREEN, whereas positive coefficients are labelled RED. Moreover, coefficients marked with a thick boundary are statistically significant ($p < 0.05$).

Algorithm 3: Computation Value- Based Social Distancing Profile

1 **set** $extroversion = 0.1 \cdot \frac{Val_{ACH}+Val_{STM}+(0.5 \cdot Val_{HED})}{2.5}$
2 **set** $introversion = 0.1 \cdot \frac{Val_{CT}+Val_{SEC}}{2}$
3 **if** *extroversion* > *introversion* **then**
4     $\mathbf{X} \leftarrow X \sim N(extroversion, \sigma[SDP])$
5   **set** social-distancing-profile = **X**
6 **else**
7     $\mathbf{X} \leftarrow X \sim N(introversion, \sigma[SDP])$
8   **set** social-distancing-profile = **X**
9 **end**

A.2 Appendix Cultural Model (Chap. 8): Needs

Algorithm 4: MAPPING VALUES TO NEEDS

1 *Initiate variables:*
2 $\lambda_1 \leftarrow$ maslow-multiplier
3 $\lambda_2 \leftarrow$ weight-survival-needs
4 $\omega_{Esteem} = 1$
5 $\omega_{Belonging} = \omega_{Esteem} \cdot (1 + \lambda_1)$
6 $\omega_{Safety} = \omega_{Belonging} \cdot (1 + \lambda_1)$
7 $\omega_{Survival} = \omega_{Safety} \cdot (1 + \lambda_1)$
8
9 *Set priority-weights of needs within the* ***survival*** *category:*
10 **set** $\omega_{FoodSafety} = \lambda_2 \cdot \omega_{Survival}$
11 **set** $\omega_{FinancialSurvival} = \lambda_2 \cdot \omega_{Survival}$
12 **set** $\omega_{Health} = \lambda_2 \cdot \omega_{Survival}$
13 **set** $\omega_{Sleep} = \lambda_2 \cdot \omega_{Survival}$
14
15 *Set priority-weights of needs within the* ***safety*** *category:*
16 **set** $\omega_{FinancialStability} = (Imp[POW] \cdot 0.01) \cdot \omega_{Safety}$
17 **set** $\omega_{RiskAvoidance} = (Imp[SEC] \cdot 0.01) \cdot \omega_{Safety}$
18 **set** $\omega_{Compliance} = ((Imp[CT] + Imp[SEC]) \cdot 0.005) \cdot \omega_{Safety}$
19 **set** $\omega_{Conformity} = (Imp[CT] \cdot 0.01)) \cdot \omega_{Safety}$
20
21 *Set priority-weights of needs within the* ***belonging*** *category:*
22 **set** $\omega_{Belonging} = (Imp[BEN] + Imp[CT]) \cdot 0.005)) \cdot \omega_{Belonging}$
23
24 *Set priority-weights of needs within the* ***esteem*** *category:*
25 **set** $\omega_{Luxury} = (Imp[HED] + Imp[POW]) \cdot 0.005)) \cdot \omega_{Esteem}$
26 **set** $\omega_{Leisure} = (Imp[HED] + Imp[STM]) \cdot 0.005)) \cdot \omega_{Esteem}$
27 **set** $\omega_{Autonomy} = (Imp[SD] + Imp[ACH]) \cdot 0.005)) \cdot \omega_{Esteem}$

A.3 Appendix Cultural Model (Chap. 8): Model Parameters

See (Table A.1).

Table A.1 Descriptive overview of relevant global model parameters

Model parameter	Default setting [Range]
CONTAGION-FACTOR	7 [1,20]
SOCIAL-DISTANCING-DENSITY-FACTOR	0.4 [0,1]
POLICY-SCENARIO	"No policy scenario"
R0-BASED-TRIGGER	12
R0-BASED-LIFTER	12
STDEV-SOCIAL-DISTANCING-PROFILE	10
MASLOW-MULTIPLIER	0 [0,2]
SURVIVAL-MULTIPLIER	2.5 [0,5]
CULTURAL-TIGHTNESS-FUNCTION- MODIFIER	0.1 [0,0.2]
VALUE-SYSTEM-CALIBRATION-FACTOR	20 [0,40]
INFLUENCE-OF-AGE-ON-VALUES	5
HOFSTEDE-SCHWARTZ-MAPPING	"Empirical & theoretical"
MAX-RANDOM-VALUE	60
MIN-RANDOM-VALUE	20
COUNTRY-DEMOGRAPHIC-SETTINGS	"custom"

*Note that the ranges presented in in table are denoted as [min, max]

Appendix B
General Parameters

This appendix will provide the general settings shared by every scenario, splited by module. Each of the different modules has been explained in detail in Chap. 3, therefore we will not go into a detailed description in this appendix. The goal is to present the parameters for every module used by every scenario. The list below provides an overview over which model can be found in which table. The different possible interventions that can be explored with our model are omitted in this appendix, since these are explored in the different scenarios. The important fact that needs to be noted here is that all interventions are turned off so they do not interfere the with intervention under investigation.

- Table B.1: Disease Model
- Table B.2: Transport Model
- Table B.3: Social Network Model
- Table B.4: Migration Model
- Table B.5: Proxemics Model
- Table B.6: Awareness of Measures Model
- Table B.7: Economic Model
- Table B.8: Social Distancing
- Table B.9: Household Model
- Table B.10: Work personnel distribution

F. Dignum (ed.), *Social Simulation for a Crisis*, Computational Social Sciences,
https://doi.org/10.1007/978-3-030-76397-8

Table B.1 Disease model, for the transitions between the different states see Sect. 3.5.1.3

Parameter	Value
disease-fsm-model	Oxford
contagion-model	Oxford
with-infected?	On
propagation-risk	0.15
daily-risk-believe-experiencing-fake-symptoms	0.00
#beds-in-hospital	11

Table B.2 Transport model

Parameter	Value
ratio-children-public-transport	0.75
ratio-children-shared-car	0.00
ratio-student-public-transport	0.60
ratio-student-shared-car	0.10
ratio-worker-public-transport	0.40
ratio-worker-shared-car	0.15
ratio-retired-public-transport	0.20
ratio-retired-shared-car	0.50
#bus-per-timeslot	27
#max-people-per-bus	20
density-walking-outside	0.05
density-factor-queuing	0.60
density-factor-public-transports	0.50
density-factor-shared-cars	0.80

Table B.3 Social network model

Parameter	Value
network-generation-method	Value-similarity
peer-group-friend-links	7
percentage-of-agents-with-random-link	0.14

Table B.4 Migration model

Parameter	Value
migration?	Off
probability-infection-when-abroad	0.00
probability-going-abroad	0.00
owning-solo-transportation-probability	1.00
probability-getting-back-when-abroad	0.00

Table B.5 Proxemics model

Parameter	Value
density-factor-homes	1.00
#hospital-gp	4
density-factor-hospitals	0.80
#schools-gp	12
density-factor-schools	1.00
#universities-gp	4
density-factor-universities	0.20
#workplaces-gp	25
density-factor-workplaces	0.20
#public-leisure-gp	20
density-factor-public-leisure	0.10
#private-leisure-gp	60
density-factor-private-leisure	0.30
#essential-shops-gp	10
density-factor-essential-shops	0.30
#non-essential-shops-gp	10
density-factor-non-essential-shops	0.60

Table B.6 Awareness of measures

Parameter	Value
percentage-news-watchers	0.75
Aware-of-working-at-home-at-start-of-simulation?	On
Aware-of-social-distancing-at-start-of-simulation?	On

Table B.7 Economic model

Parameter	Value
OVERRIDE-ECONOMY?	Off
amount-of-rations-I-buy-when-going-to-shops	6
close-services-luxury?	Off
days-of-rations-bought	3
workers-wages	12.5
export-value-decay-factor	0.10
goods-produced-by-work-performed	12
government-initial-reserve-of-capital	100000
government-pays-wages?	Off
government-sector-subsidy-ratio	0.00
interest-rate-by-tick	0.0010
max-stock-of-goods-in-a-shop	500
parent-individual-subsidy-per-child-per-tick	2.5
price-of-rations-in-essential-shops	2.8
price-of-rations-in-non-essential-shops	4.0
productivity-at-home	0.5
ratio-of-wage-paid-by-the-government	0.80
ratio-tax-on-essential-shops	0.76
ratio-tax-on-non-essential-shops	0.85
ratio-tax-on-workers	0.42
ratio-tax-on-workplaces	0.60
retirees-tick-subsidy	3.5
services-luxury-ratio-of-expenditures-when-closed	0.20
services-luxury-ratio-of-income-when-closed	0.00
starting-amount-of-capital-retired	50
starting-amount-of-capital-students	45
starting-amount-of-capital-workers	60
students-tick-subsidy	3
unit-price-of-goods	2.9

Table B.8 Social distancing

Parameter	Value
ratio-omniscious-infected-that-trigger-social-distancing-measure	1
social-distancing-density-factor	0.08

Table B.9 Composition of households for different countries

	Adult (%)	Family (%)	Retired (%)	Multi-generational (%)
Belgium	27.8	37.1	31.5	3.6
Canada	44.0	31.0	23.0	2.0
France	30.2	37.5	30.0	2.3
Germany	29.1	45.7	23.4	1.8
Great Britain	29.2	36.0	31.2	3.6
Italy	30.9	34.4	29.8	4.9
Korea South	35.2	16.3	43.1	5.4
Netherlands	27.2	43.2	27.6	2.0
Norway	25.3	47.3	25.6	1.8
Singapore	58.6	12.8	19.1	9.5
Spain	25.8	34.7	33.6	5.9
Sweden	29.5	41.9	27.0	1.6
USA	40.4	31.5	25.9	2.2

Table B.10 Worker distribution

Parameter	Value
probability-hospital-personel	0.03
probability-school-personel	0.03
probability-university-personel	0.04
probability-shopkeeper	0.04

Appendix C
Full Need and Actions Model

This appendix shows the full need and actions model, i.e. what effect the actions have on the needs (see Table C.1). The effect can either be decreasing the need (−), no effect or increasing (+, ++ or +++). The pluses on their own do not have a strict meaning, but give the relative satisfaction given for all the different needs for a group of individuals. So if working from the workplace is ++, and working from home is +, this does not mean that working from home satisfies half as much as working from the workplace, but just that it satisfies it less. The idea of the table is not to be very accurate, rather to give a global idea of how the needs influence the individuals.

For some actions additional increases or decreases apply when it is a working day (Mo-Fr, indicated in red) For example being home in the afternoon on a working day will decrease *complying to rules* as the individual should be working, but this is not the case during the weekend. If there is not a lockdown going on (NL), complying to rules gets a passive bonus, so while there is not a lockdown, all + should be read as ++, all empty fields are +, etc.

Since the individuals can have varying need satisfaction, the actions have coloured columns per agent type, these are indicated at the bottom of the graph. Workers can go to the workplace or work from home, young go to schools and students to the university (however students do not get penalised when not going to the university). For workers working at the university and other locations, the *be at work* action is chosen and not the *university* action. The retired individuals do not get penalised as they are not supposed to be at a working place or specific location during working days. The private and public leisure actions have the same need satisfaction and are therefore taken together.

Conformity is special as a need, because that is dependant upon what the rest of your social network does. Therefore, we did not include this in the table, since every action can give you a decrease or increase for conformity. There are more exceptions

F. Dignum (ed.), *Social Simulation for a Crisis*, Computational Social Sciences,
https://doi.org/10.1007/978-3-030-76397-8

Table C.1 Complete needs and actions model

Morning and afternoon																									
Need	**Subneed \ Actions & Motives**	**Buy at essential shops**				**Buy at non-essential shops**				**School**	**Univer-sity**	**Be at work**	**Work at home**	**Be at a private or public leisure place**				**Be at home**				**Getting treatment at hospital**			
Safety	Risk avoidance		-	-	-		-	-	-	-	-	-	+	-	-	-	-	+	+	+	+	+(S)	+(S)	+(S)	+(S)
	Complying to rules			-				-		+	+	++(NL)	+(L)	-		-		-		-		-		-	
	Financial stability		-	-	-		-	-	-			+	+												
Belonging	No subneed						+	+	+	+	+		+	+	+	+	+	+	+	+	+				
Self-Esteem	Leisure													++	++	++	++	+	+	+	+				
	Luxury						+	+	+																
	Autonomy	-	-	-		-	-	-		+	+	++	+	-	-	-		-	-	-		-	-	-	
Survival	Food safety		+	+	+																				
	Financial survival		-	-	-		-	-	-			+	+												
	Health	-(S)	-(S)	-(S)	-(S)	-(S)	-(S)	-(S)	-(S)	-(S)	-(S)	-(S)	-(S)	-(S)	-(S)	-(S)	-(S)	+(S)	+(S)	+(S)	+(S)	++(S	++(S	++(S	++(S
	Sleep																	+	+	+	+	+	+	+	+

Evening																									
Need	**Subneed \ Actions & Motives**	**Buy at essential shops**				**Buy at non-essential shops**				**School**	**Univer-sity**	**Be at work**	**Work at home**	**Be at a private or public leisure place**				**Be at home**				**Getting treatment at hospital**			
Safety	Risk avoidance		-	-	-		-	-	-	No action	No action	No action	No action	-	-	-	-	+	+	+	+	+(S)	+(S)	+(S)	+(S)
	Complying to rules																								
	Financial stability		-	-	-		-	-	-																
Belonging	No subneed						+	+	+					+	+	+	+	+	+	+	+				
Self-Esteem	Leisure													++	++	++	++	+	+	+	+				
	Luxury						+	+	+																
	Autonomy																								
Survival	Food safety		+	+	+																				
	Financial survival		-	-	-		-	-	-																
	Health	-(S)	-(S)	-(S)	-(S)	-(S)	-(S)	-(S)	-(S)					-(S)	-(S)	-(S)	-(S)	+(S)	+(S)	+(S)	+(S)	++(S	++(S	++(S	++(S
	Sleep																	+	+	+	+	+	+	+	+

Night																									
Need	**Subneed \ Actions & Motives**	**Buy at essential shops**				**Buy at non-essential shops**				**School**	**Univer-sity**	**Be at work**	**Work at home**	**Be at a private or public leisure place**				**Be at home**				**Getting treatment at hospital**			
Safety	Risk avoidance	No action				No action				No action	No action	No action	No action	-	-	-	-	+	+	+	+	+(S)	+(S)	+(S)	+(S)
	Complying to rules																								
	Financial stability																								
Belonging	No subneed													+	+	+	+	+	+	+	+				
Self-Esteem	Leisure													++	++	++	++	+	+	+	+				
	Luxury																								
	Autonomy																	+	+	+	+				
Survival	Food safety																								
	Financial survival																								
	Health													-(S)	-(S)	-(S)	-(S)	+(S)	+(S)	+(S)	+(S)	++(S	++(S	++(S	++(S
	Sleep																	+++	+++	+++	+++	++	++	++	++
	Agent types:	Young	Student	Worker	Retired						(NL): no lockdown			(L): lockdown				(S): if the agent is sick							

in the needs, for example risk avoidance will be decreased when at a location with 10 or more individuals (not social distancing) and 40 or more individuals (social distancing). When an individual thinks it is contagious the need will decrease even more. Another exception is that an individual will sleep worse when being sick (S). For more information see the *Expected Reward* section (Sect. 3.7.5.5).

Printed in the United States
by Baker & Taylor Publisher Services